Nonlinear $\mathcal{H}_\infty$-Control, Hamiltonian Systems and Hamilton-Jacobi Equations

Nonlinear $\mathcal{H}_\infty$-Control, Hamiltonian Systems and Hamilton-Jacobi Equations

M.D.S. Aliyu

Ecole Polytechnique de Montreal, Montreal, Canada

CRC Press
Taylor & Francis Group
Boca Raton   London   New York

CRC Press is an imprint of the
Taylor & Francis Group, an **informa** business

CRC Press
Taylor & Francis Group
6000 Broken Sound Parkway NW, Suite 300
Boca Raton, FL 33487-2742

First issued in paperback 2017

© 2011 by Taylor and Francis Group, LLC
CRC Press is an imprint of Taylor & Francis Group, an Informa business

No claim to original U.S. Government works

ISBN 13: 978-1-138-07275-6 (pbk)
ISBN 13: 978-1-4398-5483-9 (hbk)

Visit the Taylor & Francis Web site at
http://www.taylorandfrancis.com

and the CRC Press Web site at
http://www.crcpress.com

Preface

This book is about nonlinear $\mathcal{H}_\infty$-control theory, Hamiltonian systems, and Hamilton-Jacobi equations. It is the culmination of one decade of the author's endeavors on the subject, and is addressed to practicing professionals, researchers and graduate students interested in optimal and robust control of nonlinear systems. The prerequisites for understanding the book are graduate-level background courses in linear systems theory and/or classical optimal control, or calculus of variation; linear algebra; and advanced calculus. It can be used for a specialized or seminar course in robust and optimal control of nonlinear systems in typical electrical, mechanical, aerospace, systems/industrial engineering and applied mathematics programs. In extreme cases, students from management and economics can also benefit from some of the material.

The theory of nonlinear $\mathcal{H}_\infty$-control which started around 1990, almost a decade after Zames' formulation of the linear theory, is now complete. Almost all the problems solved in the linear case have been equivalently formulated and solved for the nonlinear case; in most cases, the solutions are direct generalizations of the linear theory, while in some other cases, the solutions involve more sophisticated tools from functional analysis and differential games. However, few challenging problems still linger, prominent among which include the long-standing problem of how to efficiently solve the Hamilton-Jacobi equations, which are the cornerstones of the whole theory. Nevertheless, the enterprise has been successful, and by-and-large, the picture is complete, and thus the publishing of this book is timely.

The authors' interest in nonlinear control systems and the $\mathcal{H}_\infty$-control problem in particular, was inspired by the pace-setting book of Prof. H. K. Khalil, *Nonlinear Systems*, Macmillan Publishing Company, 1992, which he used for a first graduate course in nonlinear systems, and the seminal paper, "$\mathcal{L}_2$-Gain Analysis of Nonlinear Systems and Nonlinear State-Feedback $\mathcal{H}_\infty$-Control," IEEE Transactions on Automatic Control, vol. 37, no. 6, June, 1992, by Prof. A. J. van der Schaft. The clarity of the presentations, qualitative exposition, and the mathematical elegance of the subject have captivated his love, interest, and curiosity on the subject which continues till today.

The book presents the subject from a traditional perspective, and is meant to serve as a self-contained reference manual. Therefore, the authors have endeavored to include all the relevant topics on the subject, as well as make the book accessible to a large audience. The book also presents the theory for both continuous-time and discrete-time systems, and thus it is anticipated that it will be the most comprehensive on the subject.

A number of excellent texts and monographs dealing entirely or partially on the subject have already been published, among which are

1. *$\mathcal{H}_\infty$-Optimal Control and Related Minimax Design Problems*, by T. Basar and P. Bernhard, 2nd Ed., Systems and Control: Foundations and Applications, Birkhauser, 1996;

2. *Feedback Design for Discrete-time Nonlinear Control Systems*, by W. Lin and C. I. Byrnes, Systems and Control: Foundations and Applications, Birkhauser, 1996;

3. *Nonlinear Control Systems*, by A. Isidori, 3rd Ed., Springer Verlag, 1997;

4. *Extending $\mathcal{H}_\infty$-Control to Nonlinear Systems*, by J. W. Helton and M. R. James, SIAM Frontiers in Applied Mathematics, 1999;

5. *$\mathcal{L}_2$-gain and Passivity Techniques in Nonlinear Control*, by A. J. van der Schaft, Springer Verlag, 2nd Ed., 2000.

However, after going through the above texts, one finds that they are mostly written in the form of research monographs addressed to experts. Most of the details and the nitty-gritty, including topics on the basics of differential games, nonlinear $\mathcal{H}_\infty$-filtering, mixed $\mathcal{H}_2/\mathcal{H}_\infty$ nonlinear control and filtering, and singular nonlinear $\mathcal{H}_\infty$-control, are not discussed at all, or only touched briefly. Also, algorithms for solving the ubiquitous Hamilton-Jacobi equations as well as practical examples have not been presented.

It is thus in the light of the above considerations that the author decided to summarize what he has accumulated on the subject, and to complement what others have already done. However, in writing this book, the author does not in any way purport that this book should outshine the rest, but rather covers some of the things that others have left out, which may be trivial and unimportant to some, but not quite to others. In this regard, the book draws strong parallels with the text [129] by Helton and James at least up to Chapter 7, but Chapters 8 through 13 are really complementary to all the other books. The link between the subject and analytical mechanics as well as the theory of partial-differential equations is also elegantly summarized in Chapter 4. Moreover, it is thought that such a book would serve as a reference manual, rich in documented results and a guide to those who wish to apply the techniques and/or delve further into the subject.

The author is solely responsible for all the mistakes and errors of commission or omission that may have been transmitted inadvertently in the book, and he wishes to urge readers to please communicate such discoveries whenever and whereever they found them preferably to the following e-mail address: **dikko1@hotmail.com**. In this regard, the author would like to thank the following individuals: Profs. K. Zhou and L. Smolinsky, and some anonymous referees for their valuable comments and careful reading of the manuscript which have tremendously helped streamline the presentation and reduce the errors to a minimum. The author is also grateful to Profs. A. Astolfi and S. K. Nguang for providing him with some valuable references.

M. D. S. Aliyu
Montréal, Québec

Acknowledgments

The author would like to thank Louisiana State University, King Fahd University of Petroleum and Minerals, and Ecole Polytechnique de Montréal for the excellent facilities that they provided him during the writing of the book. The assistance of the staff of the Mathematics Dept., Louisiana State University especially, and in particular, Mr. Jeff Sheldon, where this project began is also gratefully acknowledged.

The author is also forever indebted to his former mentor, Prof. El-Kebir Boukas, without whose constant encouragement, support and personal contribution, this book would never have been a success. Indeed, his untimely demise has affected the final shape of the book, and it is unfortunate that he couldn't live longer to see it published.

Lastly, the author also wishes to thank his friends and family for the encouragement and moral support that they provided during the writing of the book, and in particular would like to mention Mr. Ahmad Haidar for his untiring assistance and encouragement.

Dedication

To my son **Baqir**

List of Figures

Contents

1

Introduction

During the years 1834 to 1845, Hamilton found a system of ordinary differential equations which is now called the *Hamiltonian canonical system*, equivalent to the Euler-Lagrange equation (1744). He also derived the Hamilton-Jacobi equation (HJE), which was improved/modified by Jacobi in 1838 [114, 130]. Later, in 1952, Bellman developed the discrete-time equivalent of the HJE which is called the *dynamic programming principle* [64], and the name Hamilton-Jacobi-Bellman equation (HJBE) was coined (see [287] for a historical perspective). For a century now, the works of these three great mathematicians have remained the cornerstone of analytical mechanics and modern optimal control theory.

In mechanics, Hamilton-Jacobi-theory (HJT) is an extension of Lagrangian mechanics, and concerns itself with a directed search for a coordinate transformation in which the equations of motion can be easily integrated. The equations of motion of a given mechanical system can often be simplified considerably by a suitable transformation of variables such that all the new position and momemtum coordinates are constants. A special type of transformation is chosen in such a way that the new equations of motion retain the same form as in the former coordinates; such a transformation is called *canonical* or *contact* and can greatly simplify the solution to the equations. Hamilton in 1838 has developed the method for obtaining the desired transformation equations using what is today known as *Hamilton's principle*. It turns out that the required transformation can be obtained by finding a smooth function S called a *generating function* or *Hamilton's principal function*, which satisfies a certain nonlinear first-order partial-differential equation (PDE) also known as the *Hamilton-Jacobi equation* (HJE).

Unfortunately, the HJE, being nonlinear, is very difficult to solve; and thus, it might appear that little practical advantage has been gained in the application of the HJT. Nevertheless, under certain conditions, and when the Hamiltonian (to be defined later) is independent of time, it is possible to separate the variables in the HJE, and the solution can then always be reduced to quadratures. In this event, the HJE becomes a useful computational tool only when such a separation of variables can be achieved.

Subsequently, it was long recognized from the Calculus of variation that the variational approach to the problems of mechanics could be applied equally efficiently to solve the problems of optimal control [47, 130, 164, 177, 229, 231]. Thus, terms like "Lagrangian," "Hamiltonian" and "Canonical equations" found their way and were assimilated into the optimal control literature. Consequently, it is not suprising that the same HJE that governs the behavior of a mechanical system also governs the behavior of an optimally controlled system. Therefore, time-optimal control problems (which deal with switching curves and surfaces, and can be implemented by relay switches) were extensively studied by mathematicians in the United States and the Soviet Union. In the period 1953 to 1957, Bellman, Pontryagin et al. [229, 231] and LaSalle [157] developed the basic theory of *minimum-time* problems and presented results concerning the existence, uniqueness and general properties of time-optimal control.

However, classical variational theory could not readily handle "hard" constraints usually associated with control problems. This led Pontryagin and co-workers to develop the famous *maximum principle*, which was first announced at the *International Congress of Mathemati-*

cians held in 1958 at Edinburgh. Thus, while the maximum principle may be surmised as an outgrowth of the Hamiltonian approach to variational problems, the method of dynamic programming of Bellman, may be viewed as an off-shoot of the Hamilton-Jacobi approach to variational problems.

In recent years, as early as 1990, there has been a renewed interest in the application of the HJBE to the control of nonlinear systems. This has been motivated by the successful development of the $\mathcal{H}_\infty$-control theory for linear systems and the pioneering work of Zames [290]. Under this framework, the HJBE became modified and took on a different form essentially to account for disturbances in the system. This Hamilton-Jacobi equation was derived by Isaacs [57, 136] from a differential game perspective. Hence the name Hamilton-Jacobi-Isaacs equation (HJIE) was coined, and has since been widely recognized as the nonlinear counterpart of the Riccati equations characterizing the solution of the $\mathcal{H}_\infty$-control problem for linear systems.

Invariably however, as in mechanics, the biggest bottle-neck to the practical application of the nonlinear equivalent of the $\mathcal{H}_\infty$-control theory [53, 138]-[145], [191, 263, 264] has been the difficulty in solving the Hamilton-Jacobi-Isaacs partial-differential equations (or inequalities); there is no systematic numerical approach for solving them. Various attempts have however been made in this direction with varying success. Moreover, recent progress in developing computational schemes for solving the HJE [6]-[10, 63] are bringing more light and hope to the application of HJT in both mechanics and control. It is thus in the light of this development that the author is motivated to write this book.

1.1   Historical Perspective on Nonlinear $\mathcal{H}_\infty$-Control

The breakthrough in the derivation of the elegant state-space formulas for the solution of the standard linear $\mathcal{H}_\infty$-control problem in terms of two Riccati equations [92] spurned activity to derive the nonlinear counterpart of this solution. This work, unlike earlier works in the $\mathcal{H}_\infty$-theory [290, 101] that emphasized factorization of transfer functions and Nevanlinna-Pick interpolation in the frequency domain, operated exclusively in the time domain and drew strong parallels with established LQG control theory. Consequently, the nonlinear equivalent of the $\mathcal{H}_\infty$-control problem [92] has been developed by the important contributions of Basar [57], Van der Schaft [264], Ball and Helton [53], Isidori [138] and Lin and Byrnes [182]-[180]. Basar's dynamic differential game approach to the linear $\mathcal{H}_\infty$-control led the way to the derivation of the solution of the nonlinear problem in terms of the HJI equation which was derived by Isaacs and reported in Basar's books [57, 59]. However, the first systematic solution to the state-feedback $\mathcal{H}_\infty$-control problem for affine-nonlinear systems came from Van der Schaft [263, 264] using the theory of dissipative systems which had been laid down by Willems [274] and Hill and Moylan [131, 134] and Byrnes et al. [77, 74]. He showed that, for time-invariant affine nonlinear systems which are smooth, the state-feedback $\mathcal{H}_\infty$-control problem is solvable by smooth feedback if there exists a smooth positive-semidefinite solution to a dissipation inequality or equivalently an infinite-horizon (or stationary) HJB-inequality. Coincidently, this HJB-inequality turned out to be the HJI-inequality reported by Basar [57, 59]. Later, Lu and Doyle [191] also presented a solution for the state-feedback problem using nonlinear matrix inequalities for a certain class of nonlinear systems. They also gave a parameterization of all stabilizing controllers.

The solution of the output-feedback problem with dynamic measurement-feedback for affine nonlinear systems was presented by Ball et al. [53], Isidori and Astolfi [138, 139, 141], Lu and Doyle [191, 190] and Pavel and Fairman [223]. While the solution for a general class of nonlinear systems was presented by Isidori and Kang [145]. At the same time, the solution of the discrete-time state and dynamic output-feedback problems were presented by Lin and Byrnes [77, 74, 182, 183, 184] and Guillard et al. [125, 126]. Another approach to the discrete-time problem using risk-sensitive control and the concept of information-state for output-feedback dynamic games was also presented by James and Baras [151, 150, 149] for a general class of discrete-time nonlinear systems. The solution is expressed in terms of dissipation inequalities; however, the resulting controller is infinite-dimensional. In addition, a control Lyapunov-function approach to the global output-regulation problem via measurement-feedback for a class of nonlinear systems in which the nonlinear terms depend on the output of the system, was also considered by Battilotti [62].

Furthermore, the solution of the problem for the continuous time-varying affine nonlinear systems was presented by Lu [189], while the mixed $\mathcal{H}_2/\mathcal{H}_\infty$ problem for both continuous-time and discrete-time nonlinear systems using state-feedback control was solved by Lin [180]. Moreover, the robust control problem with structured uncertainties in system matrices has been extensively considered by many authors [6, 7, 61, 147, 148, 208, 209, 245, 261, 265, 284] both for the state-feedback and output-feedback problems. An inverse-optimal approach to the robust control problem has also been considered by Freeman and Kokotovic [102].

Finally, the filtering problem for affine nonlinear systems has been considered by Berman and Shaked [66, 244], while the continuous-time robust filtering problem was discussed by Nguang and Fu [210, 211] and by Xie et al. [279] for a class of discrete-time affine nonlinear systems.

A more general case of the problem though is the singular nonlinear $\mathcal{H}_\infty$-control problem which has been considered by Maas and Van der Schaft [194] and Astolfi [42, 43] for continuous-time affine nonlinear systems using both state and output-feedback. Also, a closely related problem is that of $\mathcal{H}_\infty$-control of singularly-perturbed nonlinear systems which has been considered by Fridman [104]-[106] and by Pan and Basar [218] for the robust problem. Furthermore, an adaptive approach to the problem for a class of nonlinear systems in parametric strict-feedback form has been considered again by Pan and Basar [219], while a fault-tolerant approach has also been considered by Yang et al. [285].

A more recent contribution to the literature has considered a factorization approach to the problem, which had been earlier initiated by Ball and Helton [52, 49] but discounted because of the inherent difficulties with the approach. This was also the case for the earlier approaches to the linear problem which emphasized factorization and interpolation in lieu of state-space tools [290, 101]. These approaches are the J-j-inner-outer factorization and spectral-factorization proposed by Ball and Van der Schaft [54], and a chain-scattering matrix approach considered by Baramov and Kimura [55], and Pavel and Fairman [224]. While the former approach tries to generalize the approach in [118, 117] to the nonlinear case (the solution is only given for stable invertible continuous-time systems), the latter approach applies the method of conjugation and chain-scattering matrix developed for linear systems in [163] to derive the solution of the nonlinear problem. However, an important outcome of the above endeavors using the factorization approach, has been the derivation of state-space formulas for coprime-factorization and inner-outer factorization of nonlinear systems [240, 54] which were hitherto unavailable [52, 128, 188]. This has paved the way for employing these state-space factors in balancing, stabilization and design of reduced-order controllers for nonlinear systems [215, 225, 240, 33].

In the next section, we present the general setup and an overview of the nonlinear $\mathcal{H}_\infty$-control problem in its various ramifications.

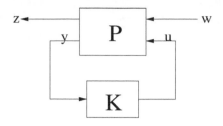

FIGURE 1.1
Feedback Configuration for Nonlinear $\mathcal{H}_\infty$-Control

1.2   General Set-Up for Nonlinear $\mathcal{H}_\infty$-Control Problems

In this section, we present a general framework and an overview of nonlinear $\mathcal{H}_\infty$-control problems using either state-feeback or output measurement-feedback. This setup is shown in Figure 1.1 which represents the general framework for $\mathcal{H}_\infty$-control problems. It shows the feedback interconnection of the plant **P** and the controller **K** with inputs w, u and outputs z, y. The plant can be described by a causal time-invariant finite-dimensional nonlinear state-space system **P**, with superscript "c" for continuous-time (CT) or "d" for discrete-time (DT), defined on a state-space $\mathcal{X} \subseteq \Re^n$ containing the origin $x = \{0\}$:

$$CT: \quad \mathbf{P}^c : \begin{cases} \dot{x} &= f(x, u, w), \quad x(t_0) = x_0 \\ y &= h^y(x, w) \\ z &= h^z(x, u) \end{cases} \tag{1.1}$$

$$DT: \quad \mathbf{P}^d : \begin{cases} x_{k+1} &= f(x_k, u_k, w_k), \quad x(k_0) = x^0 \\ y_k &= h^y(x_k, w_k) \\ z_k &= h^z(x_k, u_k), \quad k \in \mathbf{Z} \end{cases} \tag{1.2}$$

where $x \in \mathcal{X}$ is the state vector, $u : \Re \to \mathcal{U}$, $w : \Re \to \mathcal{W}$ are the control input and the disturbance/reference signal (which is to be rejected/tracked) respectively, which belong to the sets of admissible controls and disturbances $\mathcal{U} \subset \Re^p$, $\mathcal{W} \subset \Re^r$ respectively, $f : \mathcal{X} \times \mathcal{U} \times \mathcal{W} \to \mathcal{X}$ is a smooth C^∞ vector-field (or vector-valued function), while $h^y : \mathcal{X} \times \mathcal{W} \to \mathcal{Y} \subset \Re^m$, $h^z : \mathcal{X} \times \mathcal{U} \to \Re^s$ are smooth functions. The output $y \in \mathcal{Y}$ is the measurement output of the system, while $z \in \Re^s$ is the controlled output or penalty variable which may represent tracking error or a fixed reference position. We assume that the origin $x = 0$ is an equilibrium-point of the system, and for simplicity $f(0, 0, 0) = 0$, $h^z(0, 0) = 0$.

We begin with the following definitions.

Definition 1.2.1 *A solution or trajectory of the system $\mathbf{P}^c$ at any time $t \in \Re$ from an initial state $x(t_0) = x_0$ due to an input $u_{[t_0, t]}$ will be denoted by $x(t, t_0, x_0, u_{[t_0, t]})$ or $\phi(t, t_0, x_0, u_{[t_0, t]})$. Similarly, by $x(k, k_0, x^0, u_{[k_0, k]})$ or $\phi(k, k_0, x^0, u_{[k_0, k]})$ for $\mathbf{P}^d$.*

Definition 1.2.2 *The nonlinear system $\mathbf{P}^c$ is said to have locally $\mathcal{L}_2$-gain from w to z in $N \subset \mathcal{X}$, $0 \in N$, less than or equal to γ, if for any initial state x_0 and fixed feedback $u_{[t_0, T]}$, the response z of the system corresponding to any $w \in \mathcal{L}_2[t_0, T]$ satisfies:*

$$\int_{t_0}^{T} \|z(t)\|^2 dt \leq \gamma^2 \int_{t_0}^{T} \|w(t)\|^2 dt + \kappa(x_0), \quad \forall T > t_0$$

for some bounded function κ such that $\kappa(0) = 0$. The system has $\mathcal{L}_2$-gain $\leq \gamma$ if $N = \mathcal{X}$.

Equivalently, the nonlinear system $\mathbf{P}^d$ has locally ℓ_2-gain less than or equal to γ in N if for any initial state x^0 in N and fixed feedback $u_{[k_0,K]}$, the response of the system due to any $w_{[k_0,K]} \in \ell_2[k_0, K)$ satisfies

$$\sum_{k=k_0}^{K} \|z_k\|^2 \leq \gamma^2 \sum_{k=k_0}^{K} \|w_k\|^2 + \kappa(x^0), \quad \forall K > k_0, K \in \mathbf{Z}.$$

The system has ℓ_2-gain $\leq \gamma$ if $N = \mathcal{X}$.

Remark 1.2.1 *Note that in the above definitions $\|.\|$ means the Euclidean-norm on $\Re^n$.*

Definition 1.2.3 *The nonlinear system $\mathbf{P}^c$ or $[f, h^z]$ is said to be locally zero-state detectable in $\mathcal{O} \subset \mathcal{X}$, $0 \in \mathcal{O}$, if $w(t) \equiv 0$, $u(t) \equiv 0$, $z(t) \equiv 0$ for all $t \geq t_0$, it implies $\lim_{t\to\infty} x(t, t_0, x_0, u) = 0$ for all $x_0 \in \mathcal{O}$. The system is zero-state detectable if $\mathcal{O} = \mathcal{X}$.*

Equivalently, $\mathbf{P}^d$ or $[f, h^y]$ is said to be locally zero-state detectable in $\mathcal{O}$ if $w_k = 0$, $u_k \equiv 0$, $z_k \equiv 0$ for all $k \geq k_0$, it implies $\lim_{k\to\infty} x(k, k_0, x^0, u_k) = 0$ for all $x^0 \in \mathcal{O}$. The system is zero-state detectable if $\mathcal{O} = \mathcal{X}$.

Definition 1.2.4 *The state-space $\mathcal{X}$ of the system $\mathbf{P}^c$ is reachable from the origin $x = 0$ if for any state $x(t_1) \in \mathcal{X}$ at t_1, there exists a time $t_0 \leq t_1$ and an admissible control $u_{[t_0,t_1]} \in \mathcal{U}$ such that, $x(t_1) = \phi(t_1, t_0, \{0\}, u_{[t_0,t_1]})$.*

Equivalently, the state-space of $\mathbf{P}^d$ is reachable from the origin $x = 0$ if for any state $x_{k_1} \in \mathcal{X}$ at k_1, there exists an index $k_0 \leq k_1$ and an admissible control $u_{[k_0,k_1]} \in \mathcal{U}$ such that, $x_{k_1} = \phi(k_1, k_0, \{0\}, u_{[k_0,k_1]})$, $k_0, k_1 \in \mathbf{Z}$.

Now, the state-feedback nonlinear $\mathcal{H}_\infty$-suboptimal control or local disturbance-attenuation problem with internal stability, is to find for a given number $\gamma^* > 0$, a control action $u = \alpha(x)$ where $\alpha \in C^r(\Re^n)$, $r \geq 2$ which renders locally the $\mathcal{L}_2$-gain of the system $\mathbf{P}$ from w to z starting from $x(t_0) = 0$, less or equal to γ^* with internal stability, i.e., all state trajectories are bounded and/or the system is locally asymptotically-stable about the equilibrium point $x = 0$. Notice that, even though $\mathcal{H}_\infty$ is a frequency-domain space, in the time-domain, the $\mathcal{H}_\infty$-norm of the system $\mathbf{P}$ (assumed to be stable) can be interpreted as the $\mathcal{L}_2$-gain of the system from w to z which is the induced-norm from $\mathcal{L}_2$ to $\mathcal{L}_2$:

$$\|\mathbf{P}^c\|_{\mathcal{H}_\infty} = \sup_{0\neq w\in\mathcal{L}_2(0,\infty)} \frac{\|z(t)\|_2}{\|w(t)\|_2}, \quad x(t_0) = 0, \tag{1.3}$$

equivalently the induced-norm from ℓ_2 to ℓ_2:

$$\|\mathbf{P}^d\|_{\mathcal{H}_\infty} = \sup_{0\neq w\in\ell_2(0,\infty)} \frac{\|z\|_2}{\|w\|_2}, \quad x(k_0) = 0, \tag{1.4}$$

where for any $v : [t_0, T] \subset \Re \to \Re^m$ or $\{v\} : [k_0, K] \subset Z \to \Re^m$,

$$\|v\|_{2,[t_0,T]}^2 \triangleq \int_{t_0}^{T} \sum_{i=1}^{m} |v_i(t)|^2 dt, \text{ and } \|\{v_k\}\|_{2,[k_0,K]}^2 \triangleq \sum_{k=k_0}^{K} \sum_{i=1}^{m} |v_{ik}|^2.$$

The $\mathcal{H}_\infty$-norm can be interpreted as the maximum gain of the system for all $\mathcal{L}_2$-bounded (the space of bounded-energy signals) disturbances.

Thus, the problem can be formulated as a finite-time horizon (or infinite-horizon) *minimax* optimization problem with the following cost function (or more precisely functional) [57]:

$$J^c(\mu, w) = \min_{\mu \in \mathcal{U}} \sup_T \sup_{w \in \mathcal{L}_2} \frac{1}{2} \int_{t_0}^{T} [\|z(t)\|^2 - \gamma^2 \|w(t)\|^2] d\tau, \quad T > t_0 \tag{1.5}$$

equivalently

$$J^d(\mu, w) = \min_{\mu \in \mathcal{U}} \sup_k \sup_{w \in \ell_2} \frac{1}{2} \sum_{k=k_0}^{K} [\|z_k\|^2 - \gamma^2 \|w_k\|^2], \quad K > k_0 \in \mathbf{Z} \tag{1.6}$$

subject to the dynamics **P** with internal (or closed-loop) stability of the system. It is seen that, by rendering the above cost function nonpositive, the $\mathcal{L}_2$-gain requirement can be satisfied. Moreover, if in addition, some structural conditions (such as observability or zero-state detectability) are satisfied by the disturbance-free system, then the closed-loop system will be internally stable [131, 268, 274].

The above cost function or performance measure also has a differential game interpretation. It constitutes a two-person zero-sum game in which the minimizing player controls the input u while the maximizing player controls the disturbance w. Such a game has a saddle-point equilibrium solution if the value-function

$$V^c(x, t) = \inf_{\mu \in \mathcal{U}} \sup_{w \in \mathcal{L}_2} \int_t^T [\|z(\tau)\|^2 - \gamma^2 \|w(\tau)\|^2] dt$$

or equivalently

$$V^d(x, k) = \inf_{\mu \in \mathcal{U}} \sup_{w \in \ell_2} \sum_{j=k}^{K} [\|z_j\|^2 - \gamma^2 \|w_j\|^2]$$

is C^1 and satisfies the following dynamic-programming equation (known as Isaacs's equation or HJIE):

$$-V_t^c(x, t) = \inf_u \sup_w \left\{ V_x^c(t, x) f(x, u, w) + [\|z(t)\|^2 - \gamma^2 \|w(t)\|^2] \right\};$$
$$V^c(T, x) = 0, \quad x \in \mathcal{X}, \tag{1.7}$$

or equivalently

$$V^d(x, k) = \inf_{u_k} \sup_{w_k} \left\{ V^d(f(x, u_k, w_k), k+1) + [\|z_k\|^2 - \gamma^2 \|w_k\|^2] \right\};$$
$$V^d(K+1, x) = 0, \quad x \in \mathcal{X}, \tag{1.8}$$

where V_t, V_x are the row vectors of first partial-derivatives with respect to t and x respectively. A pair of strategies $(\mu^\star, \nu^\star)$ provides under feedback-information pattern, a saddle-point equilibrium solution to the above game if

$$J^c(\mu^\star, \nu) \leq J^c(\mu^\star, \nu^\star) \leq J^c(\mu, \nu^\star), \tag{1.9}$$

or equivalently

$$J^d(\mu^\star, \nu) \leq J^d(\mu^\star, \nu^\star) \leq J^d(\mu, \nu^\star). \tag{1.10}$$

For the plant **P**, the above optimization problem (1.5) or (1.6) subject to the dynamics of **P** reduces to that of solving the HJIE (1.7) or (1.8). However, the optimal control $u^\star$ may be difficult to write explicitly at this point because of the nature of the function $f(.,.,.)$.

Therefore, in order to write explicitly the nature of the optimal control and worst-case disturbance in the above minimax optimization problem, we shall for the most part in this book, assume that the plant **P** is affine in nature and is represented by an affine state-space system of the form:

$$\mathbf{P}^{ca} : \begin{cases} \dot{x} &= f(x) + g_1(x)w + g_2(x)u; \ x(0) = x_0 \\ z &= h_1(x) + k_{12}(x)u \\ y &= h_2(x) + k_{21}(x)w \end{cases} \tag{1.11}$$

or

$$\mathbf{P}^{da} : \begin{cases} x_{k+1} &= f(x_k) + g_1(x_k)w_k + g_2(x_k)u_k; \ x_0 = x^0 \\ z_k &= h_1(x_k) + k_{12}(x_k)u_k \\ y_k &= h_2(x_k) + k_{21}(x_k)w_k, \ k \in \mathbf{Z}_+ \end{cases} \tag{1.12}$$

where $f : \mathcal{X} \to \mathcal{X}$, $g_1 : \mathcal{X} \to \mathcal{M}^{n \times r}(\mathcal{X})$, $g_2 : \mathcal{X} \to \mathcal{M}^{n \times p}(\mathcal{X})$, where $\mathcal{M}^{i \times j}(\mathcal{X})$ is the ring of $i \times j$ matrices over $\mathcal{X}$, while $h_1 : \mathcal{X} \to \Re^s)$, $h_2 : \mathcal{X} \to \Re^m$, and $k_{12}, \ k_{21} \in C^r, r \geq 2$ have appropriate dimensions.

Furthermore, since we are more interested in the infinite-time horizon problem, i.e., for a control strategy such that $\lim_{T \to \infty} J^c(u, w)$ (resp. $\lim_{K \to \infty} J^d(u_k, w_k)$) remains bounded and the $\mathcal{L}_2$-gain (resp. ℓ_2-gain) of the system remains finite, we seek a time-independent positive-semidefinite function $V : \mathcal{X} \to \Re$ which vanishes at $x = 0$ and satisfies the following time-invariant (or stationary) HJIE:

$$\min_u \sup_w \left\{ V_x(x)[f(x) + g_1(x)w(t) + g_2(x)u(t)] + \frac{1}{2}(\|z(t)\|^2 - \gamma^2 \|w(t)\|^2) \right\} = 0;$$
$$V(0) = 0, \ x \in \mathcal{X} \tag{1.13}$$

or equivalently the discrete HJIE (DHJIE):

$$V(x) = \min_u \sup_w \left\{ V(f(x) + g_1(x)w + g_2(x)u) + \frac{1}{2}(\|z\|^2 - \gamma^2 \|w\|^2) \right\} = 0;$$
$$V(0) = 0, \ x \in \mathcal{X}. \tag{1.14}$$

The problem of explicitly solving for the optimal control $u^\star$ and the worst-case disturbance $w^\star$ in the HJIE (1.13) (or the DHJI (1.14)) will be the subject of discussion in Chapters 5 and 7 respectively.

However, in the absence of the availability of the state information for feedback, one might be interested in synthesizing a dynamic controller **K** which processes the output measurement $y(t)$, $t \in [t_0, T] \subset [0, \infty)$ (equivalently y_k, $k \in [k_0, K] \subset [0, \infty)$) and generates a control action $u = \alpha([t_0, t])$ (resp. $u = \alpha([k_0, k])$) that renders locally the $\mathcal{L}_2$-gain (equivalently ℓ_2-gain) of the system about $x = 0$ less than or equal to $\gamma > 0$ with internal stability. Such a controller can be represented in the form:

$$\mathbf{K}^c : \quad \dot{\xi} = \eta(\xi, y), \ \xi(0) = \xi_0$$
$$u = \theta(\xi, y)$$
$$\mathbf{K}^d : \quad \xi_{k+1} = \eta(\xi_k, y_k), \ \xi_0 = \xi^0$$
$$u_k = \theta(\xi_k, y_k), \ k \in \mathbf{Z}_+,$$

where $\xi : [0, \infty) \to \mathcal{O} \subseteq \mathcal{X}$, $\eta : \mathcal{O} \times \mathcal{Y} \to \mathcal{X}$, $\theta : \mathcal{O} \times \mathcal{Y} \to \Re^p$. This problem is then known as the *suboptimal local $\mathcal{H}_\infty$-control problem (or local disturbance-attenuation problem) with measurement-feedback* for the system **P**. The purpose of the control action is to achieve local closed-loop stability and to attenuate the effect of the disturbance or reference input w on the controlled output z.

Often this kind of controller will be like a carbon-copy of the plant which is also called *observer-based controller*, and the feedback interconnnection of the controller $\mathbf{K}$ and plant $\mathbf{P}$ results in the following closed-loop system:

$$\mathbf{P}^{ca} \circ K^c \; : \; \begin{cases} \dot{x} & = & f(x) + g_1(x)w(t) + g_2(x)\theta(\xi,y); \;\; x(0) = \hat{x}_0 \\ \dot{\xi} & = & \eta(\xi,y), \;\; \xi(0) = \xi_0 \\ z & = & h_1(x) + k_{12}(x)\theta(\xi,y) \\ y & = & h_2(x) + k_{21}(x)w(t) \end{cases} , \tag{1.15}$$

or equivalently

$$\mathbf{P}^{da} \circ K^d \; : \; \begin{cases} x_{k+1} & = & f(x_k) + g_1(x_k)w_k + g_2(x_k)\theta(\xi_k,y_k); \;\; x_0 = x^0 \\ \xi_{k+1} & = & \eta(\xi_k,y_k), \;\; \xi_0 = \xi^0 \\ z_k & = & h_1(x_k) + k_{12}(x_k)\theta(\xi_k,y_k) \\ y_k & = & h_2(x_k) + k_{21}(x_k)w_k, \;\; k \in \mathbf{Z}_+. \end{cases} \tag{1.16}$$

Then the problem of optimizing the performance (1.5) or (1.6) subject to the dynamics $\mathbf{P}^{ca} \circ \mathbf{K}^c$ (respectively $\mathbf{P}^{da} \circ K^d$) becomes, by the dynamic programming principle, that of solving the following HJIE:

$$\min_u \sup_w \left\{ W_{(x,\xi)}(x,\xi) f^c(x,\xi) + \frac{1}{2}\|z\|^2 - \frac{1}{2}\gamma^2\|w\|^2 \right\} = 0, \;\; W(0,0) = 0, \tag{1.17}$$

or the DHJIE

$$W(x,\xi) = \min_u \sup_w \left\{ W(f^d(x,\xi)) + \frac{1}{2}\|z\|^2 - \frac{1}{2}\gamma^2\|w\|^2 \right\}, \;\; W(0,0) = 0, \tag{1.18}$$

where

$$f^c(x,\xi) = \begin{pmatrix} f(x) + g_1(x)w + g_2(x)\theta(\xi,y) \\ \eta(\xi,y) \end{pmatrix},$$

$$f^d(x_k,\xi_k) = \begin{pmatrix} f(x_k) + g_1(x_k)w_k + g_2(x_k)\theta(\xi_k,y_k) \\ \eta(\xi_k,y_k) \end{pmatrix},$$

in addition to the HJIE (1.13) or (1.14) respectively. The problem of designing such a dynamic-controller and solving the above optimization problem associated with it, will be the subject of discussion in Chapters 6 and 7 respectively.

An alternative approach to the problem is using the theory of dissipative systems [131, 274, 263, 264] which we hereby introduce.

Definition 1.2.5 *The nonlinear system $\mathbf{P}^c$ is said to be locally dissipative in $M \subset \mathcal{X}$ with respect to the supply-rate $s(w(t), z(t)) = \frac{1}{2}(\gamma^2\|w(t)\|^2 - \|z(t)\|^2)$, if there exists a C^0 positive-semidefinite function (or storage-function) $V : M \to \Re$, $V(0) = 0$, such that the inequality*

$$V(x(t_1)) - V(x(t_0)) \le \int_{t_0}^{t_1} s(w(t), z(t)) dt \tag{1.19}$$

is satisfied for all $t_1 > t_0$, for all $x(t_0)$, $x(t_1) \in M$. The system is dissipative if $M = \mathcal{X}$.

Equivalently, $\mathbf{P}^d$ is locally dissipative in M if there exists a C^0 positive semidefinite function $V : M \to \Re$, $V(0) = 0$ such that

$$V(x_{k_1}) - V(x_{k_0}) \le \sum_{k=k_0}^{k_1} s(w_k, z_k) \tag{1.20}$$

is satisfied for all $k_1 > k_0$, for all x_{k_1}, $x_{k_0} \in M$. The system is dissipative if $M = \mathcal{X}$.

Consider now the nonlinear system $\mathbf{P}$ and assume the states of the system are available for feedback. Consider also the problem of rendering locally the $\mathcal{L}_2$-gain of the system less than or equal to $\gamma > 0$ using state-feedback with internal (or asymptotic) stability for the closed-loop system. It is immediately seen from Definition 1.2.2 that, if the system is locally dissipative with respect to the above supply rate, then it also has locally $\mathcal{L}_2$-gain less than or equal to γ. Thus, the problem of local disturbance-attenuation or local $\mathcal{H}_\infty$ suboptimal control for the system $\mathbf{P}$, becomes that of rendering the system locally dissipative with respect to the supply rate $s(w, z) = \frac{1}{2}(\gamma^2\|w\|^2 - \|z\|^2)$ by an appropriate choice of control action $u = \alpha(x)$ (respectively $u_k = \alpha(x_k)$) with the additional requirement of internal stability. Moreover, this additional requirement can be satisfied, on the basis of Lyapunov-LaSalle stability theorems, if the system can be rendered locally dissipative with a positive-definite storage-function or with a positive-semidefinite storage-function and if additionally it is locally observable.

Furthermore, if we assume the storage-function V in Definition 1.2.5 is $C^1(M)$, then we can go from the integral version of the dissipation inequalities (1.19), (1.20) to their differential or infinitesimal versions respectively, by differentiation along the trajectories of the system $\mathbf{P}$ and with t_0 fixed (equivalently k_0 fixed), and t_1 (equivalently k_1) arbitrary, to obtain:

$$V_x(x)[f(x) + g_1(x)w + g_2(x)u] + \frac{1}{2}(\|z\|^2 - \gamma^2\|w\|^2) \le 0,$$

respectively

$$V(f(x) + g_1(x)w + g_2(x)u) + \frac{1}{2}(\|z\|^2 - \gamma^2\|w\|^2) \le V(x).$$

Next, consider the problem of rendering the system dissipative with the minimum control action and in the presence of the worst-case disturbance. This is essentially the $\mathcal{H}_\infty$-control problem and results in the following dissipation inequality:

$$\inf_{u \in \mathcal{U}} \sup_{w \in \mathcal{W}} \left\{ V_x(x)[f(x) + g_1(x)w + g_2(x)u] + \frac{1}{2}(\|z\|^2 - \gamma^2\|w\|^2) \right\} \le 0 \qquad (1.21)$$

or equivalently

$$\inf_{u \in \mathcal{U}} \sup_{w \in \mathcal{W}} \left\{ V(f(x) + g_1(x)w + g_2(x)u) + \frac{1}{2}(\|z\|^2 - \gamma^2\|w\|^2) \right\} \le V(x). \qquad (1.22)$$

The above inequality (1.21) (respectively (1.22)) is exactly the inequality version of the equation (1.13) (respectively (1.14)) and is known as the HJI-inequality. Thus, the existence of a solution to the dissipation inequality (1.21) (respectively (1.22)) implies the existence of a solution to the HJIE (1.13) (respectively (1.14)).

Conversely, if the state-space of the system $\mathbf{P}$ is reachable from $x = 0$, has an asymptotically-stable equilibrium-point at $x = 0$ and an $\mathcal{L}_2$-gain $\le \gamma$, then the functions

$$V_a^c(x) = \sup_T \sup_{w \in L_2[0,T), x(0)=x} -\frac{1}{2}\int_0^T (\gamma^2\|w(t)\|^2 - \|z(t)\|^2)dt,$$

$$V_r^c(x) = \inf_T \inf_{\substack{w \in \mathcal{L}_2(-T,0], \\ x = x_0, x(-T)=0}} \frac{1}{2}\int_{-T}^0 (\gamma^2\|w(\tau)\|^2 - \|z(\tau)\|^2)dt,$$

or equivalently

$$V_a^d(x) = \sup_T \sup_{w \in \ell_2[0,K], x_0 = x^0} -\frac{1}{2} \sum_{k=0}^K (\gamma^2 \|w_k\|^2 - \|z_k\|^2),$$

$$V_r^d(x) = \inf_T \inf_{\substack{w \in \ell_2[-K,0], \\ x = x^0, x_{-K} = 0}} \frac{1}{2} \sum_{k=-K}^0 (\gamma^2 \|w_k\|^2 - \|z_k\|^2),$$

respectively, are well defined for all $x \in M$ and satisfy the dissipation inequality (1.21) (respectively (1.22)) (see Chapter 3 and [264]). Moreover, $V_a(0) = V_r(0) = 0$, $0 \le V_a \le V_r$. Therefore, there exists at least one solution to the dissipation inequality. The functions V_a and V_r are also known as the *available-storage* and the *required-supply* respectively.

It is therefore clear from the foregoing that a solution to the disturbance-attenuation problem can be derived from this perspective. Moreover, the $\mathcal{H}_\infty$-suboptimal control problem for the system **P** has been reduced to the problem of solving the HJIE (1.13) (respectively (1.14)) or the HJI-inequality (1.21) (respectively (1.22)), and hence we have shown that the differential game approach and the dissipative system's approach are equivalent. Similarly, the measurement-feedback problem can also be tackled using the theory of dissipative systems.

The above approaches to the nonlinear $\mathcal{H}_\infty$-control problem for time-invariant affine state-space systems can be extended to more general time-invariant nonlinear state-space systems in the form (1.1) or (1.2) as well as time-varying nonlinear systems. In this case, the finite-time horizon problem becomes relevant; in fact, it is the most relevant. Indeed, we can consider an affine time-varying plant of the form

$$\mathbf{P}_t^{ca} : \begin{cases} \dot{x}(t) &= f(x,t) + g_1(x,t)w(t) + g_2(x,t)u(t); \quad x(0) = \hat{x}_0 \\ z(t) &= h_1(x,t) + k_{12}(x,t)u(t) \\ y(t) &= h_2(x,t) + k_{21}(x,t)w(t) \end{cases} \tag{1.23}$$

or equivalently

$$\mathbf{P}_k^{da} : \begin{cases} x_{k+1} &= f(x_k,k) + g_1(x_k,k)w_k + g_2(x_k,k)u_k, \quad x_0 = x^0 \\ z_k &= h_1(x_k,k) + k_{12}(x_k,k)u_k \\ y_k &= h_2(x_k,k) + k_{21}(x_k,k)w_k, \quad k \in \mathbf{Z} \end{cases} \tag{1.24}$$

with the additional argument "t" (respectively "k") here denoting time-variation; and where all the variables have their usual meanings, while the functions $f : \mathcal{X} \times \Re \to \mathcal{X}$, $g_1 : \mathcal{X} \times \Re \to \mathcal{M}^{n \times r}(\mathcal{X} \times \Re)$, $g_2 : \mathcal{X} \times \Re \to \mathcal{M}^{n \times p}(\mathcal{X} \times \Re)$, $h_1 : \mathcal{X} \times \Re \to \Re^s$, $h_2 : \mathcal{X} \times \Re \to \Re^m$, and k_{12}, k_{21} of appropriate dimensions, are real $C^{\infty,0}(\mathcal{X}, \Re)$ functions, i.e., are smooth with respect to x and continuous with respect to t (respectively smooth with respect to x_k). Furthermore, we may assume without loss of generality that the system has a unique equilibrium-point at $x = 0$, i.e., $f(0,t) = 0$ and $h_i(0,t) = 0$, $i = 1,2$ with $u = 0$, $w = 0$ (or equivalently $f(0,k) = 0$ and $h_i(0,k) = 0$, $i = 1,2$ with $u_k = 0$, $w_k = 0$).

Then, the finite-time horizon state-feedback $\mathcal{H}_\infty$ suboptimal control problem for the above system $\mathbf{P}^a$ can be pursued along similar lines as the time-invariant case with the exception here that, the solution to the problem will be characterized by an evolution equation of the form (1.7) or (1.8) respectively. Briefly, the problem can be formulated analogously as a two-player zero-sum game with the following cost functional:

$$J_t(\mu, w_{[0,\infty)}) = \min_{\mu} \sup_{w_{[0,\infty)}} \int_{t=t_0}^{T} \frac{1}{2}(\|z(t)\|^2 - \gamma^2\|w(t)\|^2)dt \tag{1.25}$$

$$J_k(\mu, w_{[0,\infty)}) = \min_{\mu} \sup_{w_{[0,\infty)}} \sum_{k=k_0}^{K} \frac{1}{2}(\|z_k\|^2 - \gamma^2\|w_k\|^2) \tag{1.26}$$

subject to the dynamics $\mathbf{P}_t^{ca}$ (respectively $\mathbf{P}_k^{da}$) over some finite-time interval $[0, T]$ (respectively $[0, K]$) using state-feedback controls of the form:

$$u = \beta(x, t) \quad \beta(0, t) = 0$$

or

$$u = \beta(x_k, k), \quad \beta(0, k) = 0$$

respectively.

A pair of strategies $(u^\star(x, t), w^\star(x, t))$ under feedback information pattern provides a saddle-point solution to the above problem such that

$$J_t(u^\star(x, t), w(x, t)) \leq J_t(u^\star(x, t), w^\star(x, t)) \leq J_t(u(x, t), w^\star(x, t)) \tag{1.27}$$

or

$$J_k(u^\star(x_k, k), w(x_k, k)) \leq J_k(u^\star(x_k, k), w^\star(x_k, k)) \leq J_k(u(x_k, k), w^\star(x_k, k)) \tag{1.28}$$

respectively, if there exists a positive definite $C^{1\ 1}$ function $V : \mathcal{X} \times [0, T] \to \Re_+$ (respectively $V : \mathcal{X} \times [0, K] \to \Re_+$) satisfying the following HJIE:

$$
\begin{aligned}
-V_t(x, t) &= \inf_u \sup_w \Big\{ V_x(x, t)[f(x, t) + g_1(x, t)w + g_2(x, t)u] + \frac{1}{2}[\|z(t)\|^2 - \\
&\quad \gamma^2\|w(t)\|^2]\Big\}; \quad V(x, T) = 0 \\
&= V_x(x, t)[f(x, t) + g_1(x, t)w^\star(x, t) + g_2(x, t)u^\star(x, t)] + \\
&\quad \frac{1}{2}[\|z^\star(x, t)\|^2 - \gamma^2\|w^\star(x, t)\|^2]; \quad V(x, T) = 0, \ x \in \mathcal{X}
\end{aligned}
\tag{1.29}
$$

or equivalently the recursive equations (DHJIE)

$$
\begin{aligned}
V(x, k) &= \inf_{u_k} \sup_{w_k} \Big\{ V(f(x, k) + g_1(x, k)w_k + g_2(x, k)u_k, k+1) + \frac{1}{2}[\|z_k\|^2 - \\
&\quad \gamma^2\|w_k\|^2]\Big\}; \ k = 1, \ldots, K, \ V(x, K+1) = 0, \ x \in \mathcal{X} \\
&= V(f(x, k) + g_1(x, k)w^\star(x, k), +g_2(x, k)u(x, k), k+1) + \\
&\quad \frac{1}{2}[\|z^\star(x, k)\|^2 - \gamma^2\|w^\star(x, k)\|^2]; \quad V(x, K+1) = 0, \ x \in \mathcal{X}
\end{aligned}
\tag{1.30}
$$

respectively, where $z^\star(x, t)$ (equivalently $z^\star(x, k)$) is the optimal output. Furthermore, a dissipative-system approach to the problem can also be pursued along similar lines as in the time-invariant case, and the output measurement-feedback problem could be tackled similarly.

Notice however here that the HJIEs (1.29) and DHJIE (1.30) are more involved, in the

---

[1] C^1 with respect to both arguments.

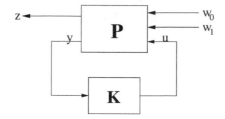

FIGURE 1.2
Feedback Configuration for Nonlinear Mixed $\mathcal{H}_2/\mathcal{H}_\infty$-Control

sense that they are time-varying or evolution PDEs. In the case of (1.29), the solution is a single function of two variables x and t that is also required to satisfy the boundary condition $V(x,T) = 0 \ \forall x \in \mathcal{X}$; while in the case of the DHJIE (1.30), the solution is a set of $K+1$ functions with the last function required to also satisfy the boundary conditions $V(x, K+1) = 0 \ \forall x \in \mathcal{X}$. These equations are notoriously difficult to solve and will be the subject of discussion in the last chapter.

1.2.1   Mixed $\mathcal{H}_2/\mathcal{H}_\infty$-Control Problem

We now consider a different set-up for the $\mathcal{H}_\infty$-control problem; namely, the problem of mixing two cost functions to achieve disturbance-attenuation and at the same time minimizing the output energy of the system. It is well known that we can only solve the suboptimal $\mathcal{H}_\infty$-control problem easily, and therefore $\mathcal{H}_\infty$-controllers are hardly unique. However, $\mathcal{H}_2$-controllers [92, 292] can be designed optimally. It therefore follows that, by mixing the two criterion functions, one can achieve the benefits of both types of controllers, while at the same time, try to recover some form of uniqueness for the controller. This philosophy led to the formulation of the mixed $\mathcal{H}_2/\mathcal{H}_\infty$-control problem.

A typical set-up for this problem is shown in Figure 1.2 with $w = \begin{pmatrix} w_0 \\ w_1 \end{pmatrix}$, where the signal w_0 is a Gaussian white noise signal (or bounded-spectrum signal), while $w_1(t)$ is a bounded-power or energy signal. Thus, the induced norm from the input w_0 to z is the $\mathcal{L}_2$-norm (respectively ℓ_2 -norm) of the plant $\mathbf{P}$, i.e.,

$$\|\mathbf{P}^c\|_{\mathcal{L}_2} \stackrel{\Delta}{=} \sup_{0 \neq w_0 \in \mathcal{S}} \frac{\|z\|_{\mathcal{P}}}{\|w_0\|_{\mathcal{S}}},$$

$$\|\mathbf{P}^d\|_{\ell_2} \stackrel{\Delta}{=} \sup_{0 \neq w_{0,k} \in \mathcal{S}'} \frac{\|z\|_{\mathcal{P}'}}{\|w_0\|_{\mathcal{S}'}},$$

while the induced norm from w_1 to z is the $\mathcal{L}_\infty$-norm (respectively ℓ_∞-norm) of $\mathbf{P}$, i.e.,

$$\|\mathbf{P}^c\|_{\mathcal{L}_\infty} \stackrel{\Delta}{=} \sup_{0 \neq w_1 \in \mathcal{P}} \frac{\|z\|_2}{\|w_1\|_2},$$

$$\|\mathbf{P}^d\|_{\ell_\infty} \stackrel{\Delta}{=} \sup_{0 \neq w_1 \in \mathcal{P}'} \frac{\|z\|_2}{\|w_1\|_2},$$

where

$$
\begin{aligned}
\mathcal{P} &\triangleq \{w(t): \; w \in \mathcal{L}_\infty, \; R_{ww}(\tau), \; S_{ww}(j\omega) \text{ exist for all } \tau \text{ and all } \omega \text{ resp.,} \\
&\quad \|w\|_\mathcal{P}^2 < \infty\}, \\
\mathcal{S} &\triangleq \{w(t): \; w \in \mathcal{L}_\infty, \; R_{ww}(\tau), \; S_{ww}(j\omega) \text{ exist for all } \tau \text{ and all } \omega \text{ resp.,} \\
&\quad \|S_{ww}(j\omega)\|_\infty < \infty\}, \\
\mathcal{P}' &\triangleq \{w: \; w \in \ell_\infty, \; R_{ww}(k), \; S_{ww}(j\omega) \text{ exist for all } k \text{ and all } \omega \text{ resp.,} \\
&\quad \|w\|_{\mathcal{P}'}^2 < \infty\}, \\
\mathcal{S}' &\triangleq \{w: \; w \in \ell_\infty, \; R_{ww}(k), \; S_{ww}(j\omega) \text{ exist for all } k \text{ and all } \omega \text{ resp.,} \\
&\quad \|S_{ww}(j\omega)\|_\infty < \infty\},
\end{aligned}
$$

$$
\|z\|_\mathcal{P}^2 \triangleq \lim_{T\to\infty} \frac{1}{2T} \int_{-T}^{T} \|z(t)\|^2 dt, \quad \|z\|_{\mathcal{P}'}^2 \triangleq \lim_{K\to\infty} \frac{1}{2K} \sum_{k=-K}^{K} \|z_k\|^2,
$$

$$
\|w_0\|_\mathcal{S}^2 = \|S_{w_0 w_0}(j\omega)\|_\infty, \quad \|w_0\|_{\mathcal{S}'}^2 = \|S_{w_0 w_0}(j\omega)\|_\infty,
$$

and $R_{ww}(\tau)$, $S_{ww}(j\omega)$ (equivalently $R_{ww}(k)$, $S_{ww}(j\omega)$) are the autocorrelation and power spectral-density matrices [152] of $w(t)$ (equivalently w_k) respectively. Notice also that $\|(.)\|_\mathcal{P}$ and $\|(.)\|_\mathcal{S}$ are seminorms. In addition, if the plant is stable, we replace the induced $\mathcal{L}$-norms (resp. ℓ-norms) above by their equivalent $\mathcal{H}$-subspace norms.

Since minimizing the $\mathcal{H}_\infty$-induced norm as defined above, over the set of admissible controllers is a difficult problem (i.e., the optimal $\mathcal{H}_\infty$-control problem is difficult to solve exactly [92, 101]), whereas minizing the induced $\mathcal{H}_2$-norm and obtaining optimal $\mathcal{H}_2$-controllers is an easy problem, the objective of the mixed $\mathcal{H}_2/\mathcal{H}_\infty$ design philosophy is then to minimize $\|\mathbf{P}\|_{\mathcal{L}_2}$ (equivalently $\|\mathbf{P}\|_{\ell_2}$) while rendering $\|\mathbf{P}^c\|_{\mathcal{L}_\infty} \le \gamma^\star$ (resp. $\|\mathbf{P}^d\|_{\ell_\infty} \le \gamma^\star$) for some prescribed number $\gamma^\star > 0$. Such a problem can be formulated as a two-player nonzero-sum differential game with two cost functionals:

$$
J_1(u,w) = \int_{t_0}^{T} (\gamma^2 \|w(\tau)\|^2 - \|z(\tau)\|^2) d\tau \tag{1.31}
$$

$$
J_2(u,w) = \int_{t_0}^{T} \|z(\tau)\|^2 d\tau \tag{1.32}
$$

or equivalently

$$
J_{1k}(u,w) = \sum_{k=k_0}^{K} (\gamma^2 \|w_k\|^2 - \|z_k\|^2) \tag{1.33}
$$

$$
J_{2k}(u,w) = \sum_{k=k_0}^{K} \|z_k\|^2 \tag{1.34}
$$

for the finite-horizon problem. Here, the first functional is associated with the $\mathcal{H}_\infty$-constraint criterion, while the second functional is related to the output energy of the system or $\mathcal{H}_2$-criterion. Moreover, here we wish to solve an associated mixed $\mathcal{H}_2/\mathcal{H}_\infty$ problem in which w is comprised of a single disturbance signal $w \in \mathcal{W} \subset \mathcal{L}_2([t_0,\infty), \Re^r)$ (equivalently $w \in \mathcal{W} \subset \ell_2([k_0,\infty), \Re^r)$). It can easily be seen that by making $J_1 \ge 0$ (respectively $J_{1k} \ge 0$) then the $\mathcal{H}_\infty$ constraint $\|\mathbf{P}\|_{\mathcal{L}_\infty} \le \gamma$ is satisfied. Subsequently, minimizing J_2 (respectively J_{2k}) will achieve the $\mathcal{H}_2/\mathcal{H}_\infty$ design objective. Moreover, if we assume also that $\mathcal{U} \subset \mathcal{L}_2([0,\infty), \Re^k)$ (equivalently $\mathcal{U} \subset \ell_2([0,\infty), \Re^k)$) then under closed-loop perfect information, a Nash-equilibrium solution to the above game is said to exist if we can find a pair of strategies $(u^\star, w^\star)$ such that

$$J_1(u^\star, w^\star) \leq J_1(u^\star, w) \quad \forall w \in \mathcal{W}, \tag{1.35}$$

$$J_2(u^\star, w^\star) \leq J_2(u, w^\star) \quad \forall u \in \mathcal{U}. \tag{1.36}$$

Equivalently for $J_{1k}(u^\star, w^\star)$, $J_{2k}(u^\star, w^\star)$.

Furthermore, by minimizing the first objective with respect to w and substituting in the second objective which is then minimized with respect to u, the pair of Nash-equilibrium strategies can be found. A necessary and sufficient condition for optimality in such a differentail game is provided by a pair of cross-coupled HJIEs (resp. DHJIEs). For the state-feedback problem and the case of the plant $\mathbf{P}^c$ given by equation (1.1) or $\mathbf{P}^d$ given by (1.2), the governing equations can be obtained by the dynamic-programming principle or the theory of dissipative systems to be

$$Y_t(x,t) = -\inf_{w \in \mathcal{W}} \left\{ Y_x(x)f(x, u^\star(x), w(x)) - \gamma^2\|w(x)\|^2 + \|z^\star(x)\|^2 \right\},$$

$$Y(x,T) = 0, \quad x \in \mathcal{X},$$

$$x \in \mathcal{X}$$

$$V_t(x,t) = -\inf_{u \in \mathcal{U}} \left\{ V_x(x)f(x, u(x), w^\star(x)) + \|z^\star(x)\|^2 \right\} = 0, \quad V(x,T) = 0,$$

$$x \in \mathcal{X},$$

or

$$Y(x,k) = \min_{w_k \in \mathcal{W}} \left\{ Y(f_k(x, u_k^\star(x), w_k)), k+1) + \gamma^2\|w(x)\|^2 - \|z^\star(x)\|^2 \right\};$$

$$Y(x, K+1) = 0, \quad k = 1, \ldots, K, \quad x \in \mathcal{X}$$

$$V(x,k) = \min_{u_k \in \mathcal{U}} \left\{ V(f_k(x, u_k(x), w_k^\star(x)), k+1) + \|z^\star(x)\|^2 \right\};$$

$$V(x, K+1) = 0, \quad k = 1, \ldots, K, \quad x \in \mathcal{X},$$

respectively, for some negative-(semi)definite function $Y : \mathcal{X} \to \Re$ and positive-(semi)definite function $V : \mathcal{X} \to \Re$. The solution to the above optimization problem will be the subject of discussion in Chapter 11.

1.2.2   Robust $\mathcal{H}_\infty$-Control Problem

A primary motivation for the development of $\mathcal{H}_\infty$-synthesis methods is the design of robust controllers to achieve robust performance in the presence of disturbances and/or model uncertainty due to modeling errors or parameter variations. For all the models and the design methods that we have considered so far, we have concentrated on the problem of disturbance-attenuation using either the pure $\mathcal{H}_\infty$-criterion or the mixed $\mathcal{H}_2/\mathcal{H}_\infty$-criterion. Therefore in this section, we briefly overview the second aspect of the theory.

A typical set-up for studying plant uncertainty and the design of robust controllers is shown in Figure 1.3 below. The third block labelled Δ in the diagram represents the model uncertainty. There are also other topologies for representing the uncertainty depending on the nature of the plant [6, 7, 33, 147, 223, 265, 284]; Figure 1.3 is the simplest of such representations. If the uncertainty is significant in the system, such as unmodelled dynamics or perturbation in the model, then the uncertainty can be represented as a dynamic system with input excited by the plant output and output as a disturbance input to the plant, i.e.,

$$\Delta^c : \begin{cases} \dot{\varphi} = \vartheta(\varphi, u, v), & \vartheta(0,0,0) = 0, \quad \varphi(t_0) = 0 \\ e = \varrho(\varphi, u, v), & \varrho(0,0,0) = 0 \end{cases}$$

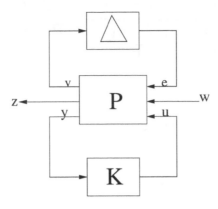

FIGURE 1.3
Feedback Configuration for Robust Nonlinear $\mathcal{H}_\infty$-Control

or

$$\Delta^d \; : \; \begin{cases} \dot{\varphi}_k & = & \vartheta_k(\varphi_k, u_k, v_k), \;\; \vartheta_k(0,0,0) = 0, \;\; \varphi(k_0) = 0 \\ e_k & = & \varrho_k(\varphi, u_k, v_k), \;\; \varrho_k(0,0,0) = 0. \end{cases}$$

The above basic model can be further decomposed (together with the plant) into a coprime factor model as is done in [34, 223, 265].

On the other hand, if the uncertainty is due to parameter variations caused by, for instance, aging or environmental conditions, then it can be represented as a simple norm-bounded uncertainty as in [6, 7, 261, 284]. For the case of the affine system $\mathbf{P}^{ca}$ or $\mathbf{P}^{da}$, such a model can be represented in the most general way in which the uncertainty or perturbation comes from the system input and output matrices as well as the drift vector-field f as

$$\mathbf{P}^{ca}_\Delta \; : \; \begin{cases} \dot{x} & = & [f(x) + \Delta f(x)] + g_1(x)w + [g_2(x) + \Delta g_2(x)]u; \;\; x(0) = x_0 \\ z & = & h_1(x) + k_{12}(x)u \\ y & = & [h_2(x) + \Delta h_2(x)] + k_{21}(x)w \end{cases}$$

or

$$\mathbf{P}^{da}_\Delta \; : \; \begin{cases} x_{k+1} & = & [f(x_k) + \Delta f(x_k)] + g_1(x_k)w_k + [g_2(x_k) + \Delta g_2(x_k)]u_k; \\ & & x_0 = x^0 \\ z_k & = & h_1(x_k) + k_{12}(x_k)u_k \\ y_k & = & [h_2(x_k) + \Delta h_2(x_k)] + k_{21}(x_k)w_k, \;\; k \in \mathbf{Z} \end{cases}$$

respectively, where Δf, Δg_2, Δh_2 belong to some suitable admissible sets.

Whichever type of representation is chosen for the plant, the problem of robustly designing an $\mathcal{H}_\infty$ controller or mixed $\mathcal{H}_2/\mathcal{H}_\infty$-controller for the plant $\mathbf{P}^{ca}_\Delta$ (respectively $\mathbf{P}^{da}_\Delta$) can be pursued along similar lines as in the case of the nominal model $\mathbf{P}^{ca}$ (respectively $\mathbf{P}^{da}$) using either state-feedback or output-feedback with some additional complexity due to the presence of the uncertainty. This problem will be discussed for the continuous-time state-feedback case in Chapter 5 and for the the measurement-feedback case in Chapter 6.

1.2.3 Nonlinear $\mathcal{H}_\infty$-Filtering

Often than not, the states of a dynamic system are not accessible from its output. Therefore, it is necessary to design a scheme for estimating them for the purpose of feedback or other

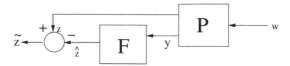

FIGURE 1.4

Configuration for Nonlinear $\mathcal{H}_\infty$-Filtering

use. Such a scheme involves another dynamic system called an "observer" or "filter." It is essentially a carbon-copy of the original system which is error-driven. It takes in the past and present output measurements $y(t)$ of the system for a given time span $[t_0, t] \subset \Re$ and generates an estimate of the desired output z which can be the states or some suitable function of them. It is therefore required to be "causal" so that it is implementable.

A typical set-up for filtering is shown in Figure 1.4. The filter is denoted by $\mathbf{F}$ and is designed to minimize the worst-case gain from w to the error difference between the actual desired output of the system z and the estimated output $\hat{z}$; typically $z = x$ in such applications. We can represent the plant $\mathbf{P}$ by an affine state-space model with the subscript "f" added to denote a filtering model:

$$\mathbf{P}_f^{ca} : \begin{cases} \dot{x} &= f(x) + g_1(x)w; \quad x(t_0) = x_0 \\ z &= h_1(x) \\ y &= h_2(x) + k_{21}(x)w \end{cases} \tag{1.37}$$

or equivalently

$$\mathbf{P}_f^{da} : \begin{cases} x_{k+1} &= f(x_k) + g_1(x_k)w_k; \quad x_{k_0} = x^0 \\ z_k &= h_1(x_k) \\ y_k &= h_2(x_k) + k_{21}(x_k)w_k, \quad k \in \mathbf{Z}, \end{cases} \tag{1.38}$$

where all the variables and functions have their previous meanings and definitions, and $w \in \mathcal{L}_2[t_0, \infty)$ (or equivalently $w \in \ell_2[k_0, \infty)$) is a noise signal that is assumed to corrupt the outputs y and z (respectively y_k and z_k) of the system. Then the objective is to design the filter such that

$$\sup_{w \in \mathcal{L}_2(t_0, \infty)} \frac{\|z - \hat{z}\|_2^2}{\|w\|_2^2} \leq \gamma^\star$$

or equivalently

$$\sup_{w \in \ell_2(t_0, \infty)} \frac{\|z_k - \hat{z}_k\|_2^2}{\|w_k\|_2^2} \leq \gamma^\star$$

for some prescribed number $\gamma^\star > 0$ is achieved. If the filter $\mathbf{F}$ is configured so that it has identical dynamics as the system, then it can be shown that the solution to the filtering problem is characterized by a (D)HJIE that is dual to that of the state-feedback problem.

1.2.4   Organization of the Book

The book contains thirteen chapters and two appendices. It is loosely organized in two parts: Part I, comprising Chapters 1-4 covers mainly introductory and background material, while Part II comprising Chapters 5-13 covers the real subject matter of the book, dealing with all the various types and aspects of nonlinear $\mathcal{H}_\infty$-control problems.

Chapter 1 is an introductory chapter. It covers historical perspectives and gives highlights of the contents of the book. Preliminary definitions and notations are also given to prepare the reader for what is to come. In addition, some introduction on differentiable manifolds, mainly definitions, and Lyapunov-stability theory is also included in the chapter to make the book self-contained.

Chapter 2 gives background material on the basics of differential games. It discusses discrete-time and continuous-time nonzero-sum and zero-sum games from which all of the problems in nonlinear $\mathcal{H}_\infty$-theory are derived and solved. Linear-quadratic (LQ) and two-person games are included as special cases and to serve as examples. No proofs are given for most of the results in this chapter as they are standard, and can be found in wellknown standard references on the subject.

Chapter 3 is devoted to the theory of dissipative systems which is also at the core of the subject of the book. The material presented, however, is well beyond the amount required for elucidating the content matter. The reason being that this theory is not very well known by the control community, yet it pervades almost all the problems of modern control theory, from the LQ-theory and the positive-real lemma, to the $\mathcal{H}_\infty$-theory and the bounded-real lemma. As such, we have given a complete literature review on this subject and endeavored to include all relevant applications of this theory. A wealth of references is also included for further study.

Chapter 4 is also at the core of the book, and traces the origin of Hamilton-Jacobi theory and Hamiltonian systems to Lagrangian mechanics. The equivalence of Hamiltonian and Lagrangian mechanics is stressed, and the motivation behind the Hamiltonian approach is emphasized. The Hamilton-Jacobi equation is derived from variational principles and the duality between it and Hamilton's canonical equations is pointed. In this regard, the method of characteristics for first-order partial-differential equations is also discussed in the chapter, and it is shown that Hamilton's canonical equations are nothing but the characteristic equations for the HJE.

The concept of viscosity and non-smooth solutions of the HJE where smooth solutions and/or Hamiltonians do not exist is also introduced, and lastly the Toda lattice which is a particularly integrable Hamiltonian system, is discussed as an example.

Chapter 5 starts the discussion of nonlinear $\mathcal{H}_\infty$-theory with the state-feedback problem for continuous-time affine nonlinear time-invariant systems. The solution to this problem is derived from the differential game as well as dissipative systems perspective. The parametrization of a class of full-information controllers is given, and the robust control problem in the presence of unmodelled and parametric uncertainty is also discussed. The approach is then extended to time-varying and delay systems, as well as a more general class of nonlinear systems that are not necessarily affine. In addition, the $\mathcal{H}_\infty$ almost disturbance-decoupling problem for affine systems is also discussed.

Chapter 6 continues with the discussion in the previous chapter with the output-feedback problem for affine nonlinear systems. First the output measurement-feedback problem is considered and sufficient conditions for the solvability of this problem are given in terms of two uncoupled HJIEs with an additional side-condition. Moreover, it is shown that the controller that solves this problem is observer-based. A parametrization of all output-feedback controllers is given, and the results are also extended to a more general class of nonlinear systems. The robust measurement-feedback problem in the presence of uncertainties is also considered. Finally, the static output-feedback problem is considered, and sufficient conditions for its solvability are also presented in terms of a HJIE together with some algebraic conditions.

In Chapter 7 the discrete-time nonlinear $\mathcal{H}_\infty$-control problem is discussed. Solution for the full-information, state-feedback and output measurement-feedback problems are given, as well as parametrizations of classes of full-information state and output-feedback controllers. The extension of the solution to more general affine discrete-time systems is also presented. In addition, an approximate and explicit approach to the solution to the problem is also presented.

In Chapter 8 the nonlinear $\mathcal{H}_\infty$-filtering problem is discussed. Solutions for both the continuous-time and the discrete-time problems are given in terms of dual HJIEs and a coupling condition. Some simulation results are also given to show the performance of the nonlinear $\mathcal{H}_\infty$-filter compared to the extended Kalman-filter. The robust filtering problem is also discussed.

Chapter 9 discusses the generalization of the $\mathcal{H}_\infty$-control problems to include singular or ill-posed problems, as well as the $\mathcal{H}_\infty$-control of singularly-perturbed nonlinear systems with small singular parameters. Both the singular state and measurement-feedback problems are discussed, and the case of cascaded systems is also addressed. However, only the continuous-time results are presented. Furhermore, the state-feedback control problem for nonlinear singularly perturbed continuous-time systems is presented, and a class of composite controllers is also discussed. In addition, Chapter 10 continues with the discussion on singularly perturbed systems and presents a solution to the $\mathcal{H}_\infty$ infinite-horizon filtering problem for affine nonlinear singularly perturbed systems. Three types of filters, namely, decomposition, aggregate and reduced-order filters are presented, and again sufficient conditions for the solvability of the problem with each filter are presented.

Chapers 11 and 12 are devoted to the mixed $\mathcal{H}_2/\mathcal{H}_\infty$ nonlinear control and filtering problems respectively. Only the state-feedback control problem is discussed in Chapter 11 and this is complemented by the filtering problem in Chapter 12. The output measurement-feedback problem is not discussed because of its complexity. Moreover, the treatment is exhaustive, and both the continuous and discrete-time results are presented.

Lastly, the book culminates in Chapter 13 with a discussion of some computational approaches for solving Hamilton-Jacobi equations which are the cornerstones of the theory. Iterative as well as exact methods are discussed. But the topic is still evolving, and it is hoped that the chapter will be substantially improved in the future.

1.3   Notations and Preliminaries

In this section, we introduce notations and some important definitions that will be used frequently in the book.

1.3.1   Notation

The notation will be standard most of the times except where stated otherwise. Moreover, **N** will denote the set of natural numbers, while **Z** will denote the set of integers. Similarly, $\Re$, $\Re^n$ will denote respectively, the real line and the n-dimensional real vector space, $t \in \Re$ will denote the time parameter.

$\mathcal{X}$, M, $N,\ldots$ will denote *differentiable-manifolds* of dimension n which are locally Euclidean and $TM = \bigcup_{x \in M} T_x M$, $T^\star M = \bigcup_{x \in M} T_x^\star M$ will denote respectively the *tangent and cotangent bundles* of M with dimensions $2n$. Moreover, π and $\pi^\star$ will denote the natural projections $TM \to M$ and $T^\star M \to M$ respectively.

A $C^r(M)$ *vector-field* is a mapping $f : M \to TM$ such that $\pi \circ f = I_M$ (the identity on M), and f has continuously differentiable partial-derivatives of arbitrary order r. The vector-field is often denoted as $f = \sum_{i=1}^n f_i \frac{\partial}{\partial x_i}$ or simply as $(f_1, \ldots, f_n)^T$ where $f_i, i = 1, \ldots, n$ are the components of f in the coordinate system $x_1, \ldots, x_n$. The vector space of all C^∞ vector-fields over M will be denoted by $V^\infty(M)$.

A vector-field f also defines a differential equation (or a dynamic system) $\dot{x}(t) = f(x)$, $x \in M$, $x(t_0) = x_0$. The *flow (or integral-curve)* of the differential equation $\phi(t, t_0, x_0)$, $t \in \Re$, is the unique solution of the differential equation for any arbitrary initial condition x_0 over an open interval $I \subset \Re$. The flow of a differential-equation will also be referred as the trajectory of the system and will be denoted by $x(t, t_0, x_0)$ or $x(t)$ when the initial condition is immaterial. We shall also assume throughout this book that the vector-fields are complete, and hence the domain of the flow extends over $(-\infty, \infty)$.

The *Lie-bracket* of two vector-fields $f = \sum_{i=1}^n f_i \frac{\partial}{\partial x_i}$, $g = \sum_{i=1}^n g_i \frac{\partial}{\partial x_i}$ is the vector-field $[f, g] : V^\infty(M) \times V^\infty(M) \to V^\infty(M)$ defined by

$$ad_f g \triangleq [f, g] = \sum_{j=1}^n \left(\sum_{i=1}^n \frac{\partial g_j}{\partial x_i} f_i - \frac{\partial f_j}{\partial x_i} g_i \right) \frac{\partial}{\partial x_j}.$$

Furthermore, an *equilibrium-point* of the vector-field f or the differential equation defined by it, is a point $\bar{x}$ such that $f(\bar{x}) = 0$ or $\phi(t, t_0, \bar{x}) = \bar{x} \ \forall t \in \Re$.

An *invariant-set* for the system $\dot{x}(t) = f(x)$, is any set $\mathcal{A}$ such that, for any $x_0 \in \mathcal{A}$, $\Rightarrow \phi(t, t_0, x_0) \in \mathcal{A}$ for all $t \in \Re$. A set $S \subset \Re^n$ is the ω-limit set of a trajectory $\phi(., t_0, x_0)$ if for every $y \in S$, $\exists$ a sequence $t_n \to \infty$ such that $\phi(t_n, t_0, x_0) \to y$.

A *differential k-form* ω_x^k, $k = 1, 2, \ldots$, at a point $x \in M$, is an exterior product from $T_x M$ to $\Re$, i.e., $\omega_x^k : T_x M \times \ldots \times T_x M$ (k copies) $\to \Re$, which is a k-linear skew-symmetric function of k-vectors on $T_x M$. The space of all smooth k-forms on M is denoted by $\Omega^k(M)$.

The set $\mathcal{M}^{p \times q}(M)$ will denote the ring of $p \times q$ matrices over M.

The $\mathcal{F}$-derivative (Fréchet-derivative) of a real-valued function $V : \Re^n \to \Re$ is defined as the function $DV \in \mathcal{L}(\Re^n)$ (the space of linear operators from $\Re^n$ to $\Re^n$) such that $\lim_{v \to 0} \frac{1}{\|v\|} [V(x+v) - V(x) - \langle DV, v \rangle] = 0$, for any $v \in \Re^n$.

For a smooth function $V : \Re^n \to \Re$, $V_x = \frac{\partial V}{\partial x}$ is the row-vector of first partial-derivatives of $V(.)$ with respect to (wrt) x. Moreover, the *Lie-derivative* (or directional-derivative) of the function V with respect to a vector-field X is defined as

$$L_X V(x) = V_x(x) X(x) = X(V)(x) = \sum_{i=1}^n \frac{\partial V}{\partial x_i} (x_1, \ldots, x_n) X_i (x_1, \ldots, x_n).$$

$\|.\| : W \subseteq \Re^n \to \Re$ will denote the Euclidean-norm on W, while $\mathcal{L}_2([t_0, T], \Re^n)$, $\mathcal{L}_2([t_0, \infty), \Re^n)$, $\mathcal{L}_\infty(([t_0, \infty), \Re^n)$, $\mathcal{L}_\infty(([t_0, T], \Re^n)$ will denote the standard Lebesgue-spaces of vector-valued square-integrable and essentially bounded functions over $[t_0, T]$ and $[t_0, \infty)$ respectively, and where for any $v : [t_0, T] \to \Re^n$, $\nu : [t_0, \infty) \to \Re^n$

$$\|v([t_0, T])\|_{\mathcal{L}_2}^2 \triangleq \int_{t_0}^T \sum_{i=1}^n |v_i(t)|^2 dt,$$

$$\|\nu([t_0, \infty))\|_{\mathcal{L}_2}^2 \triangleq \lim_{T \to \infty} \int_{t_0}^T \sum_{i=1}^n |\nu_i(t)|^2 dt.$$

$$\|v([t_0, T])\|_{\mathcal{L}_\infty} \triangleq ess \sup_{[t_0, T]} \{|v_i(t)|, i = 1, \ldots, n\}$$

$$\|\nu([t_0, \infty))\|_{\mathcal{L}_\infty} \triangleq ess \sup_{[t_0, \infty)} \{|\nu_i(t)|, i = 1, \ldots, n\}.$$

Similarly, the spaces $\ell_2([k_0, K], \Re^n)$, $\ell_2([k_0, \infty), \Re^n)$, $\ell_\infty([k_0, K], \Re^n)$, $\ell_\infty([k_0, \infty), \Re^n)$ will denote the corresponding discrete-time spaces, which are the spaces of square-summable and essentially-bounded sequences with the corresponding norms defined for sequences $\{v_k\}$: $[k_0, K] \to \Re^n$, $\{\nu_k\} : [k_0, \infty) \to \Re^n$:

$$\|v([k_0, K])\|_{\ell_2}^2 \triangleq \sum_{k=k_0}^{K} \sum_{i=1}^{n} |v_{ik}|^2,$$

$$\|\nu([k_0, \infty))\|_{\ell_2}^2 \triangleq \lim_{K \to \infty} \sum_{k=k_0}^{K} \sum_{i=1}^{n} |\nu_{ik}|^2,$$

$$\|v([k_0, K])\|_{\ell_\infty} \triangleq ess \sup_{[k_0, K]} \{|v_{ik}|, i = 1, \ldots, n\},$$

$$\|\nu([k_0, \infty))\|_{\ell_\infty} \triangleq ess \sup_{[k_0, \infty)} \{|\nu_{ik}|, i = 1, \ldots, n\}.$$

Most of the times we shall denote the above norms by $\|(.)\|_2$, respectively $\|(.)\|_\infty$ when there is no chance of confusion.

In addition, the spaces $\mathcal{L}_2(j\Re)$ and $\mathcal{L}_\infty(j\Re)$ (equivalently ($\ell_2(j\omega)$ and $\ell_\infty(j\omega)$ are the frequency-domain counterparts of $\mathcal{L}_2([t_0, \infty)$ and $\mathcal{L}_\infty(([t_0, \infty), \Re^n)$, (respectively $\ell_2([k_0, \infty), \Re^n)$, $\ell_\infty([k_0, \infty), \Re^n))$ will be rarely used in the text. However, the subspaces of these spaces which we denote with $\mathcal{H}$, e.g. $\mathcal{H}_2(j\Re)$ and $\mathcal{H}_\infty(j\Re)$ represent those elements of $\mathcal{L}_2(j\Re)$ and $\mathcal{L}_\infty(j\Re)$ respectively that are analytic on the open right-half complex plane, i.e., on $Re(s) > 0$. These will be used to represent symbolically asymptotically-stable input-output maps. Indeed, these spaces, also called *"Hardy-spaces"* (after the name of the mathematician who discovered them), are the origin of the name $\mathcal{H}_\infty$-control. Moreover, the discrete-time spaces are also equivalently represented by $\mathcal{H}_2(j\omega)$ and $\mathcal{H}_\infty(j\omega)$.

For a matrix $A \in \Re^{n \times n}$, $\lambda(A)$ will denote a spectral (or eigen)-value and $\sigma(A) = \lambda^{\frac{1}{2}}(A^T A)$ the singular-value of A. $A \geq B$ ($A > B$) for an $n \times n$ matrix B, implies that $A - B$ is positive-semidefinite (positive-definite respectively).

Lastly, $\mathbf{C}_+$, $\mathbf{C}_-$ and $C^r, r = 0, 1, \ldots, \infty$ will denote respectively the open right-half, left-half complex planes and the set of r-times continuously differentiable functions.

1.3.2   Stability Concepts

In this subsection, we define some of the various notions of stability that we shall be using in the book. The proofs of all the results can be found in [157, 234, 268] from which the material in this subsection is based on. For this purpose, we consider a time-invariant (or autonomous) nonlinear state-space system defined on a manifold $\mathcal{X} \subseteq \Re^n$ in local coordinates $(x_1, \ldots, x_n)$:

$$\dot{x} = f(x); \quad x(t_0) = x_0, \tag{1.39}$$

or

$$x(k+1) = f(x(k)); \quad x(k_0) = x^0, \tag{1.40}$$

where $x \in \mathcal{X} \subset \Re^n$ is the state vector and $f : \mathcal{X} \to V^\infty \mathcal{X}$ is a smooth vector-field (equivalently $f : \mathcal{X} \to \mathcal{X}$ is a smooth map) such that the system is well defined, i.e., satisfies the existence and uniqueness theorem for ordinary differential-equations (ODE) (equivalently difference-equations (DE)) [157]. Further, we assume without any loss of generality (wlog) that the system has a unique equilibrium-point at $x = 0$. Then we have the following definitions.

Definition 1.3.1 *The equilibrium-point $x = 0$ of (1.39) is said to be*

- *stable, if for each $\epsilon > 0$, there exists $\delta = \delta(\epsilon) > 0$ such that*

$$\|x(t_0)\| < \delta \Rightarrow \|x(t)\| < \epsilon \quad \forall t \geq t_0;$$

- *asymptotically-stable, if it is stable and*

$$\|x(t_0)\| < \delta \Rightarrow \lim_{t \to \infty} x(t) = 0;$$

- *exponentially-stable if there exist constants $\kappa > 0$, $\gamma > 0$, such that*

$$\|x(t)\| \leq \kappa e^{-\gamma(t-t_0)} \|x(t_0)\|;$$

- *unstable, if it is not stable.*

Equivalently, the equilibrium-point $x = 0$ of (1.40) is said to be

- *stable, if for each $\epsilon' > 0$, there exists $\delta' = \delta'(\epsilon') > 0$ such that*

$$\|x(k_0)\| < \delta' \Rightarrow \|x(k)\| < \epsilon' \quad \forall k \geq k_0;$$

- *asymptotically-stable, if it is stable and*

$$\|x(k_0)\| < \delta' \Rightarrow \lim_{k \to \infty} x(k) = 0;$$

- *exponentially-stable if there exist constants $\kappa' > 0$, $0 < \gamma' < 1$, such that*

$$\|x(k)\| \leq \kappa' \gamma'^{(k-k_0)} \|x(k_0)\|;$$

- *unstable, if it is not stable.*

Definition 1.3.2 *A continuous function $\alpha : [0, a) \subset \Re_+ \to \Re$ is said to be of class $\mathcal{K}$ if it is strictly increasing and $\alpha(0) = 0$. It is said to be of class $\mathcal{K}_\infty$ if $a = \infty$ and $\alpha(r) \to \infty$ as $r \to \infty$.*

Definition 1.3.3 *A function $V : [0, a) \times D \subset \mathcal{X} \to \Re$ is locally positive-definite if (i) it is continuous, (ii) $V(t, 0) = 0 \, \forall t \geq 0$, and (iii) there exists a constant $\mu > 0$ and a function ψ of class $\mathcal{K}$ such that*

$$\psi(\|x\|) \leq V(t, x), \quad \forall t \geq 0, \forall x \in B_\mu$$

where $B_\mu = \{x \in \Re^n : \|x\| \leq \mu\}$. $V(.,.)$ is positive definite if the above inequality holds for all $x \in \Re^n$. The function V is negative-definite if $-V$ is positive-definite. Further, if V is independent of t, then V is positive-definite (semidefinite) if $V > 0 (\geq 0) \, \forall x \neq 0$ and $V(0) = 0$.

Theorem 1.3.1 *(Lyapunov-stability I). Let $x = 0$ be an equilibrium-point for (1.39). Suppose there exists a C^1-function $V : D \subset \mathcal{X} \to \Re$, $0 \in D$, $V(0) = 0$, such that*

$$V(x) > 0 \ \forall x \neq 0, \tag{1.41}$$
$$\dot{V}(x) \leq 0 \ \forall x \in D. \tag{1.42}$$

Then, the equilibrium-point $x = 0$ is locally stable. Furthermore, if

$$\dot{V}(x) < 0 \ \forall x \in D \setminus \{0\},$$

then $x = 0$ is locally asymptotically-stable.

Equivalently, if V is such that

$$V(x_k) > 0 \ \forall x_k \neq 0, \tag{1.43}$$

$$V(x_{k+1}) - V(x_k) \leq 0 \ \forall x_k \in D. \tag{1.44}$$

Then, the equilibrium-point $x = 0$ of (1.40) is locally stable. Furthermore, if

$$V(x_{k+1}) - V(x_k) < 0 \ \forall x_k \in D \setminus \{0\},$$

then $x = 0$ is locally asymptotically-stable.

Remark 1.3.1 *The function V in Theorem 1.3.1 above is called a Lyapunov-function.*

Theorem 1.3.2 *(Barbashin-Krasovskii). Let $\mathcal{X} = \Re^n$, $x = 0$ be an equilibrium-point for (1.39). Suppose there exists a C^1 function $V : \Re^n \to \Re$, $V(0) = 0$, such that*

$$V(x) > 0 \ \forall x \neq 0, \tag{1.45}$$

$$\|x\| \to \infty \Rightarrow V(x) \to \infty, \tag{1.46}$$

$$\dot{V}(x) < 0 \ \forall x \neq 0. \tag{1.47}$$

Then, $x = 0$ is globally asymptotically-stable.

Equivalently, if V is such that

$$V(x_k) > 0 \ \forall x_k \neq 0, \tag{1.48}$$

$$\|x_k\| \to \infty \Rightarrow V(x_k) \to \infty, \tag{1.49}$$

$$V(x_{k+1}) - V(x_k) < 0 \ \forall x_k \neq 0. \tag{1.50}$$

Then, the equilibrium-point $x = 0$ of (1.40) is globally asymptotically-stable.

Remark 1.3.2 *The function V in Theorem 1.3.2 is called a radially unbounded Lyapunov-function.*

Theorem 1.3.3 *(LaSalle's Invariance-Principle). Let $\Omega \subset \mathcal{X}$ be compact and invariant with respect to the solutions of (1.39). Suppose there exists a C^1-function $V : \Omega \to \Re$, such that $\dot{V}(x) \leq 0 \ \forall x \in \Omega$ and let $\mathcal{O} = \{x \in \Omega \,|\, \dot{V}(x) = 0\}$. Suppose Γ is the largest invariant set in $\mathcal{O}$, then every solution of (1.39) starting in Ω approaches Γ as $t \to \infty$.*

Equivalently, suppose Ω (as defined above) is invariant with respect to the solutions of (1.40) and V (as defined above) is such that $V(x_{k+1}) - V(x_k) \leq 0 \ \forall x_k \in \Omega$. Let $\mathcal{O}' = \{x_k \in \Omega \,|\, V(x_{k+1}) - V(x_k) = 0\}$ and suppose Γ' is the largest invariant set in $\mathcal{O}'$, then every solution of (1.40) starting in Ω approaches Γ' as $k \to \infty$.

The following corollaries are consequences of the above theorem, and are often quoted as the invariance-principle.

Corollary 1.3.1 *Let $x = 0$ be an equilibrium-point of (1.39). Suppose there exists a C^1 function $V : D \subseteq \mathcal{X} \to \Re$, $0 \in D$, such that $\dot{V} \leq 0 \ \forall x \in D$. Let $\mathcal{O} = \{x \in D \,|\, \dot{V}(x) = 0\}$, and suppose that $\mathcal{O}$ contains no nontrivial trajectories of (1.39). Then, $x = 0$ is asymptotically-stable.*

Equivalently, let $x = 0$ be an equilibrium-point of (1.40) and suppose V (as defined above) is such that $V(x_{k+1}) - V(x_k) \leq 0 \ \forall x_k \in D$. Let $\mathcal{O}' = \{x_k \in D \,|\, V(x_{k+1}) - V(x_k) = 0\}$, and suppose that $\mathcal{O}'$ contains no nontrivial trajectories of (1.40). Then, $x = 0$ is asymptotically-stable.

Corollary 1.3.2 *Let $\mathcal{X} = \Re^n$, $x = 0$ be an equilibrium-point of (1.39). Suppose there exists a C^1 radially-unbounded positive-definite function $V : \Re^n \to \Re$, such that $\dot{V} \leq 0 \ \forall x \in \Re^n$. Let $\mathcal{O} = \{x \in D \,|\, \dot{V}(x) = 0\}$, and suppose that $\mathcal{O}$ contains no nontrivial trajectories of (1.39). Then $x = 0$ is globally asymptotically-stable.*

Equivalently, let $\mathcal{X} = \Re^n$, $x = 0$ be an equilibrium-point of (1.40), and suppose V as defined above is such that $V(x_{k+1}) - V(x_k) \leq 0 \ \forall x_k \in \mathcal{X}$. Let $\mathcal{O}' = \{x_k \in \mathcal{X} \,|\, V(x_{k+1}) - V(x_k) = 0\}$, and suppose that $\mathcal{O}'$ contains no nontrivial trajectories of (1.40). Then $x = 0$ is globally asymptotically-stable.

We now consider time-varying (or nonautonomous) systems and summarize the equivalent stability notions that we have discussed above for this class of systems. It is not suprising in fact to note that the above concepts for nonautonomous systems are more involved, intricate and diverse. For instance, the δ in Definition 1.3.1 will in general be dependent on t_0 too in this case, and there is in general no invariance-principle for this class of systems. However, there is something close to it which we shall state in the proceeding.

We consider a nonautonomous system defined on the state-space manifold $\widetilde{\mathcal{X}} \subseteq \Re \times \mathcal{X}$:

$$\dot{x} = f(x, t), \quad x(t_0) = x_0 \tag{1.51}$$

or

$$x(k + 1) = f(x(k), k), \quad x(k_0) = x^0 \tag{1.52}$$

where $x \in \mathcal{X}$, $\widetilde{\mathcal{X}} = \mathcal{X} \times \Re$, $f : \widetilde{\mathcal{X}} \to V^\infty(\widetilde{\mathcal{X}})$ is C^1 with respect to t (equivalently $f : \mathbf{Z} \times \mathcal{X} \to \mathcal{X}$ is C^r with respect to x). Moreover, we shall assume with no loss of generality that $x = 0$ is the unique equilibrium-point of the system such that $f(0, t) = 0 \ \forall t \geq t_0$ (equivalently $f(0, k) = 0 \ \forall k \geq k_0$).

Definition 1.3.4 *A continuous function $\beta : J \subset \Re_+ \times \Re_+ \to \Re_+$ is said to be of class $\mathcal{KL}$ if $\beta(r, .) \in$ class $\mathcal{K}$, $\beta(r, s)$ is decreasing with respect to s, and $\beta(r, s) \to 0$ as $s \to \infty$.*

Definition 1.3.5 *The origin $x = 0$ of (1.51) is*

- *stable, if for each $\epsilon > 0$ and any $t_0 \geq 0$ there exists a $\delta = \delta(\epsilon, t_0) > 0$ such that*

$$\|x(t_0)\| < \delta \Rightarrow \|x(t)\| < \epsilon \quad \forall t \geq t_0;$$

- *uniformly-stable, if there exists a class $\mathcal{K}$ function $\alpha(.)$ and $0 < c \in \Re_+$, such that*

$$\|x(t_0)\| < c \Rightarrow \|x(t)\| \leq \alpha(\|x(t_0)\|) \quad t \geq t_0;$$

- *uniformly asymptotically-stable, if there exists a class $\mathcal{KL}$ function $\beta(., .)$ and $c > 0$ such that*

$$\|x(t_0)\| < c \Rightarrow \|x(t)\| \leq \beta(\|x(t_0)\|, t - t_0) \quad \forall t \geq t_0;$$

- *globally uniformly asymptotically-stable, if it is uniformly asymptotically stable for all $x(t_0)$;*

- *exponentially stable if there exists a $\mathcal{KL}$ function $\beta(r, s) = \kappa r e^{-\gamma s}$, $\kappa > 0$, $\gamma > 0$, such that*

$$\|x(t_0)\| < c \Rightarrow \|x(t)\| \leq \kappa \|x(t_0)\| e^{-\gamma(t - t_0)} \quad \forall t \geq t_0;$$

Equivalently, the origin $x = 0$ of (1.52) is

- *stable, if for each $\epsilon' > 0$ and any $k_0 \geq 0$ there exists a $\delta' = \delta'(\epsilon', k_0) > 0$ such that*

$$\|x(k_0)\| < \delta' \Rightarrow \|x(k)\| < \epsilon' \quad \forall k \geq k_0;$$

- *uniformly-stable, if there exists a class $\mathcal{K}$ function $\alpha'(.)$ and $c' > 0$, such that*

$$\|x(k_0)\| < c' \Rightarrow \|x(k)\| \leq \alpha'(\|x(k_0)\|) \quad \forall k \geq k_0;$$

- *uniformly asymptotically-stable, if there exists a class $\mathcal{KL}$ function $\beta'(.,.)$ and $c' > 0$ such that*

$$\|x(k_0)\| < c' \Rightarrow \|x(k)\| \leq \beta'(\|x(k_0)\|, k - k_0) \quad \forall k \geq k_0;$$

- *globally-uniformly asymptotically-stable, if it is uniformly asymptotically stable for all $x(k_0)$;*

- *exponentially stable if there exists a class $\mathcal{KL}$ function $\beta'(r, s) = \kappa' r \gamma'^s$, $\kappa' > 0, 0 < \gamma' < 1$, and $c' > 0$ such that*

$$\|x(k_0)\| < c' \Rightarrow \|x(k)\| \leq \kappa' \|x(k_0)\| \gamma'^{(k-k_0)} \quad \forall k \geq k_0;$$

Theorem 1.3.4 *(Lyapunov-stability: II). Let $x = 0$ be an equilibrium-point of (1.51), and let $\mathcal{B}(0, r)$ be the open ball with radius r (centered at $x = 0$) on $\mathcal{X}$. Suppose there exists a C^1 function (with respect to both its argument) $V : \mathcal{B}(0, r) \times \Re_+ \to \Re$ such that:*

$$\alpha_1(\|x\|) \leq V(x, t) \leq \alpha_2(\|x\|)$$
$$\frac{\partial V}{\partial t} + \frac{\partial V}{\partial x} f(x, t) \leq -\alpha_3(\|x\|) \quad \forall t \geq t_0, \; \forall x \in \mathcal{B}(0, r),$$

where $\alpha_1, \alpha_2, \alpha_3 \in$ class $\mathcal{K}$ defined on $[0, r)$. Then, the equilibrium-point $x = 0$ is uniformly asymptotically-stable for the system.

If however the above conditions are satisfied for all $x \in \mathcal{X}$ (i.e., as $r \to \infty$) and $\alpha_1, \alpha_2, \alpha_3 \in$ class $\mathcal{K}_\infty$, then $x = 0$ is globally-uniformly asymptotically-stable.

The above theorem can also be stated in terms of exponential-stability.

Theorem 1.3.5 *Let $x = 0$ be an equilibrium-point of (1.51), and let $\mathcal{B}(0, r)$ be the open ball with radius r on $\mathcal{X}$. Suppose there exists a C^1 function (with respect to both its arguments) $V : \mathcal{B}(0, r) \times \Re_+ \to \Re$ and constants $c_1, c_2, c_3, c_4 > 0$ such that:*

$$c_1 \|x\|^2 \leq V(x, t) \leq c_2 \|x\|^2$$
$$\frac{\partial V}{\partial t} + \frac{\partial V}{\partial x} f(x, t) \leq -c_3 \|x\|^2 \quad \forall t \geq t_0, \; \forall x \in \mathcal{B}(0, r),$$
$$\|\frac{\partial V}{\partial x}\| \leq c_4 \|x\|.$$

Then, the equilibrium-point $x = 0$ is locally exponentially-stable for the system.

If however the above conditions are satisfied for all $x \in \mathcal{X}$ (i.e., as $r \to \infty$), then $x = 0$ is globally exponentially-stable.

Remark 1.3.3 *The above theorem is usually stated as a converse theorem. This converse result can be stated as follows: if $x = 0$ is a locally exponentially-stable equilibrium-point for the system (1.51), then the function V with the above properties exists.*

Finally, the following theorem is the time-varying equivalent of the invariance-principle.

Theorem 1.3.6 *Let $\mathcal{B}(0,r)$ be the open ball with radius r on $\mathcal{X}$. Suppose there exists a C^1 function (with respect to both its argument) $V : \mathcal{B}(0,r) \times \Re \to \Re$ such that:*

$$\alpha_1(\|x\|) \le V(x,t) \le \alpha_2(\|x\|)$$
$$\frac{\partial V}{\partial t} + \frac{\partial V}{\partial x} f(x,t) \le -W(x) \le 0 \ \ \forall t \ge t_0, \ \ \forall x \in \mathcal{B}(0,r),$$

where $\alpha_1, \alpha_2 \in$ class $\mathcal{K}$ defined on $[0,r)$, and $W(.) \in C^1(\mathcal{B}(0,r))$. Then all solutions of (1.51) with $\|x(t_0)\| < \alpha_2^{-1}(\alpha_1(r)) \in$ class $\mathcal{K}^2$ are bounded and are such that

$$W(x(t)) \to 0 \ \ as \ t \to \infty.$$

Furthermore, if all of the above assumptions hold for all $x \in \mathcal{X}$ and $\alpha_1(.) \in \mathcal{K}_\infty$, then the above conclusion holds for all $x(t_0) \in \mathcal{X}$, or globally.

The discrete-time equivalents of Theorems 1.3.4-1.3.6 can be stated as follows.

Theorem 1.3.7 *(Lyapunov-stability II). Let $x = 0$ be an equilibrium-point of (1.52), and let $\mathcal{B}(0,r)$ be the open ball on $\mathcal{X}$. Suppose there exists a C^1 function (with respect to both its argument) $V : \mathcal{B}(0,r) \times \mathbf{Z}_+ \to \Re$ such that:*

$$\alpha_1'(\|x\|) \le V(x,k) \le \alpha_2'(\|x\|) \qquad \forall k \in \mathbf{Z}_+$$
$$V(x_{k+1}, k+1) - V(x,k) \le -\alpha_3'(\|x\|) \ \ \forall k \ge k_0, \ \forall x \in \mathcal{B}(0,r),$$

where $\alpha_1', \alpha_2', \alpha_3' \in$ class $\mathcal{K}$ defined on $[0,r)$, then the equilibrium point $x = 0$ is uniformly asymptotically-stable for the system.

If however the above conditions are satisfied for all $x \in \mathcal{X}$ (i.e., as $r \to \infty$) and $\alpha_1', \alpha_2' \in$ class $\mathcal{K}_\infty$, then $x = 0$ is globally-uniformly asymptotically-stable.

Theorem 1.3.8 *Let $x = 0$ be an equilibrium-point of (1.52), and let $\mathcal{B}(0,r)$ be the open ball on $\mathcal{X}$. Suppose there exists a C^1 function (with respect to both its arguments) $V : \mathcal{B}(0,r) \times \mathbf{Z}_+ \to \Re$ and constants $c_1', c_2', c_3', c_4' > 0$ such that:*

$$c_1'\|x\|^2 \le V(x,k) \le c_2'\|x\|^2$$
$$V(x_{k+1}, k+1) - V(x,k) \le -c_3'\|x\|^2 \ \forall k \ge k_0, \ \forall x \in \mathcal{B}(0,r),$$
$$V(x_{k+1}, k) - V(x,k) \le c_4'\|x\| \ \ \forall k \ge k_0$$

then the equilibrium-point $x = 0$ is locally exponentially-stable for the system.

If however the above conditions are satisfied for all $x \in \mathcal{X}$ (i.e., as $r \to \infty$), then $x = 0$ is globally exponentially-stable.

Theorem 1.3.9 *Let $\mathcal{B}(0,r)$ be the open ball on $\mathcal{X}$. Suppose there exists a C^1 function (with respect to both its argument) $V : \mathcal{B}(0,r) \times \mathbf{Z}_+ \to \Re$ such that:*

$$\alpha_1'(\|x\|) \le V(x,k) \le \alpha_2'(\|x\|) \qquad \forall k \in \mathbf{Z}_+$$
$$V(x_{k+1}, k+1) - V(x,k) \le -W(x) \le 0 \ \ \forall k \ge k_0, \ \forall x \in \mathcal{B}(0,r),$$

where $\alpha_1', \alpha_2' \in$ class $\mathcal{K}$ defined on $[0,r)$, and $W(.) \in C^1(\mathcal{B}(0,r))$. Then all solutions of (1.52) with $\|x(k_0)\| < \alpha_2'^{-1}(\alpha_1'(r))$ are bounded and are such that

$$W(x_k) \to 0 \ \ as \ k \to \infty.$$

Furthermore, if all the above assumptions hold for all $x \in \mathcal{X}$ and $\alpha_1'(.) \in \mathcal{K}_\infty$, then the above conclusion holds for all $x(k_0) \in \mathcal{X}$, or globally.

---

[2]If $\alpha_i \in$ class $\mathcal{K}$ defined on $[0,r)$, then α_i^{-1} is defined on $[0, \alpha_i(r))$ and belongs to class $\mathcal{K}$.

1.4   Notes and Bibliography

More background material on differentiable manifolds and differential-forms can be found in Abraham and Marsden [1] and Arnold [38]. The introductory material on stability is based on the well-edited texts by Khalil [157], Sastry [234] and Vidyasagar [268].

2

Basics of Differential Games

The theory of games was developed around 1930 by Von Neumann and Morgenstern [136] and differential games approximately around 1950, the same time that optimal control theory was being developed. The theory immediately found application in warfare and economics; for instance, certain types of battles, airplane dog-fighting, a torpedo pursuing a ship, a missile intercepting an aircraft, a gunner guarding a target against an invader [73], are typical models of differential games. Similarly, financial planning between competing sectors of an economy, or competing products in a manufacturing system, meeting demand with adequate supply in a market system, also fall into the realm of differential games.

Game theory involves multi-person decision-making. It comprises of a task, an objective, and players or decision makers. The task may range from shooting down a plane, to steering of a ship, and to controlling the amount of money in an economy. While the objective which measures the performance of the players may vary from how fast? to how cheaply? or how closely? or a combination of these, the task is accomplished by the player(s).

When the game involves more than one player, the players may play cooperatively or noncooperatively. It is *noncooperative* if each player involved pursues his or her own interest, which are partly conflicting with that of others; and it is *dynamic* if the order in which the decisions are made is important. Whereas when the game involves only one player, the problem becomes that of optimal control which can be solved by the Pontryagin's *"maximum principle"* or Bellman's *"dynamic programming principle"* [73]. Indeed, the theories of differential games, optimal control, the calculus of variation and dynamic programming are all approaches to solving variational problems, and have now become merged into the theory of "modern control."

There are two types of differential games that we shall be concerned with in this chapter; namely, *nonzero-sum* and *zero-sum* noncooperative dynamic games. We shall not discuss static games in this book. Furthermore, we shall only be concerned with games in which the players have perfect knowledge of the current and past states of the system, and their individual decisions are based on this. This is also known as *closed-loop perfect information structure or pattern*. On the other hand, when the decision is only based on the current value of the state of the system, we shall add the adjective *"memoryless."*

We begin with the fundamental *dynamic programming principle* which forms the basis for the solution to almost all unconstrained differential game problems.

2.1 Dynamic Programming Principle

The dynamic programming principle was developed by Bellman [64] as the discrete-time equivalent of Hamilton-Jacobi theory. In fact, the term dynamic programming has become synonymous with Hamilton-Jacobi theory. It also provides a sufficient condition for optimality of an optimal decision process. To derive this condition, we consider a discrete-time dynamic system defined on a manifold $\mathcal{X} \subseteq \Re^n$ which is open and contains the origin $x = 0$, with local coordinates $(x_1, \ldots, x_n)$:

$$x_{k+1} = f_k(x_k, u_k), \quad x_{k_0} = x^0, \quad k \in \{k_0, \ldots\} \subset \mathbf{Z}, \tag{2.1}$$

where $u_k \in \mathcal{U}$ is the control input or decision variable which belongs to the set $\mathcal{U}$ of all admissible controls, $f_k : \mathcal{X} \times \mathcal{U} \to \mathcal{X}$ is a $C^0(\mathcal{X} \times \mathcal{U})$ function of its arguments for each $k \in \mathbf{Z}$. We shall also assume that $f_k(.,.)$ satisfies the following global existence and uniqueness conditions for the solutions of (2.1)

Assumption 2.1.1 *For the function $f_k(x, u)$ in (2.1), for each $k \in \mathbf{Z}$, there exists a constant C_{1k} (depending on k) such that for any fixed u,*

$$\|f_k(x_1, u) - f_k(x_2, u)\| \leq C_{1k} \|x_1 - x_2\| \; \forall x_1, x_2 \in \mathcal{X}, \forall u \in \mathcal{U}, \; k \in \mathbf{Z}.$$

To formulate an optimal decision process (or an optimal control problem), we introduce the following cost functional:

$$J(x^0, k_0; u_{[k_0, K]}) = \sum_{k=k_0}^{K} L_k(x_{k+1}, x_k, u_k) \longrightarrow \min., \tag{2.2}$$

where $L_k : \mathcal{X} \times \mathcal{X} \times \mathcal{U} \to \Re$, $k = k_0, \ldots, K$, are real C^0 functions of their arguments, which is to be minimized over the time span $\{k_0, \ldots, K\}$ as a basis for making the decision. The minimal cost-to-go at any initial state x_k and initial time $k \in \{k_0, \ldots, K\}$ (also known as *value-function*) is then defined as

$$V(x, k) = \inf_{u_{[k, K]} \in \mathcal{U}} \left\{ \sum_{j=k}^{K} L_j(x_{j+1}, x_j, u_j) \right\}, \tag{2.3}$$

and satisfies the boundary condition $V(x, K + 1) = 0$, where $x = x_k$ and $u_{[k, K]} \triangleq \{u_k, \ldots, u_K\}$.

Remark 2.1.1 *For brevity, we shall use the notation $[k_1, k_2]$ henceforth to mean the subset of the integers $\mathbf{Z} \supset \{k_1, \ldots, k_2\}$ and where there will be no confusion.*

We then have the following theorem.

Theorem 2.1.1 *Consider the nonlinear discrete-time system (2.1) and the optimal control problem of minimizing the cost functional (2.2) subject to the dynamics of the system. Then, there exists an optimal control $u_k^\star$, $k \in [k_0, K]$ to the problem if there exist $C^0(\mathcal{X} \times [k_0, K+1])$ functions $V : \mathcal{X} \times [k_0, K]$ which corresponds to the value-function (2.3) for each k, and satisfying the following recursive (dynamic programming) equation subject to the boundary condition:*

$$\begin{aligned} V(x, k) &= \inf_{u_k} \{L_k(f_k(x, u_k), x, u_k) + V(f_k(x, u_k), k + 1)\}; \\ V(x, K + 1) &= 0. \end{aligned} \tag{2.4}$$

Furthermore,

$$J^\star(x^0, k_0; u_{[k_0, K]}^\star) = V(x^0, k_0)$$

is the optimal cost of the decision process.

Proof: The proof of the above theorem is based on the principle of optimality [164] and can be found in [64, 164]. $\square$

Equation (2.4) is the *discrete dynamic programming principle* and governs the solution of most of the discrete-time problems that we shall be dealing with in this book.

Next, we discuss the continuous-time equivalent of the dynamic programming equation which is a first-order nonlinear partial-differential equation (PDE) known as the *Hamilton-Jacobi-Bellman equation* (HJBE). In this regard, consider the continuous-time nonlinear dynamic system defined on an open subset $\mathcal{X} \subseteq \Re^n$ containing the origin $x = 0$, with coordinates $(x_1, \ldots, x_n)$:

$$\dot{x}(t) = f(x(t), u(t), t), \quad x(t_0) = x_0, \tag{2.5}$$

where $f : \mathcal{X} \times \mathcal{U} \times \Re \to \Re^n$ is a C^1 measurable function, and $\mathcal{U} \subset \Re^m$ is the admissible control set which is also measurable, and $x(t_0) = x_0$ is the initial state which is assumed known. Similarly, for the sake of completeness, we have the following assumption for the global existence of solutions of (2.5) [234].

Assumption 2.1.2 *The function $f(., ., .)$ in (2.5) is piece-wise continuous with respect to t and for each $t \in [0, \infty)$ there exist constants C_{1t}, C_{2t} such that*

$$\|f(x_1, u, t) - f(x_2, u, t)\| \leq C_{1t}\|x_1 - x_2\| \quad \forall x_1, x_2 \in \mathcal{X}, \forall u \in \mathcal{U}.$$

For the purpose of optimally controlling the system, we associate to it the following cost functional:

$$J(x_0, t_0; u_{[t_0, T]}) = \int_{t_0}^{T} L(x, u, t)dt \longrightarrow \min., \tag{2.6}$$

for some real C^0 function $L : \mathcal{X} \times \mathcal{U} \times \Re \to \Re$, which is to be minimized over a time-horizon $[t_0, T] \subseteq \Re$. Similarly, define the value-function (or minimum cost-to-go) from any initial state x and initial time t as

$$V(x, t) = \inf_{u_{[t, T]}} \left\{ \int_{t}^{T} L(x(s), u(s), s)ds \right\} \tag{2.7}$$

and satisfying the boundary condition $V(x, T) = 0$. Then we have the following theorem.

Theorem 2.1.2 *Consider the nonlinear system (2.5) and the optimal control problem of minimizing the cost functional (2.6) subject to the dynamics of the system and initial condition. Suppose there exists a $C^1(\mathcal{X} \times [t_0, T])$ function $V : \mathcal{X} \times [t_0, T] \to \Re$, which corresponds to the value-function (2.7), satisfying the Hamilton-Jacobi-Bellman equation (HJBE):*

$$-V_t(x, t) = \min_{u_{[t_0, T]}} \{V_x(x, t)f(x, u, t) + L(x, u, t)\}, \quad V(x, T) = 0. \tag{2.8}$$

Then, there exists an optimal solution $u^\star$ to the problem. Moreover, the optimal cost of the policy is given by

$$J_\star(x_0^\star, t_0; u_{[t_0, T]}^\star) = V(x_0, t_0).$$

Proof: The proof of the theorem can also be found in [47, 175, 164]. $\square$

The above development has considered single player decision processes or optimal control problems. In the next section, we discuss dynamic games, which involve multiple players. Again, we shall begin with the discrete-time case.

2.2   Discrete-Time Nonzero-Sum Dynamic Games

Nonzero-sum dynamic games were first introduced by Isaacs [136, 137] around the years 1954-1956 within the frame-work of two-person zero-sum games. They were later popularized in the works of Starr and Ho [250, 251] and Friedman [107]. Stochastic games were then later developed [59].

We consider a discrete-time deterministic N-person dynamic game of duration $K - k_0$ described by the state equation (which we referred to as the task):

$$x_{k+1} = f_k(x_k, u_k^1, \ldots, u_k^N), \quad x(k_0) = x^0, \quad k \in \{k_0, \ldots, K\}, \tag{2.9}$$

where $x_k \in \mathcal{X} \subset \Re^n$ is the state-vector of the system for $k = k_0, \ldots, K$ which belong to the state-space $\mathcal{X}$ with x^0 the initial state which is known a priori; $u_k^i, i = 1, \ldots, N, N \in \mathbf{N}$ are the decision variables or control inputs of the players $1, \ldots, N$ which belong to the measurable control sets $\mathcal{U}^i \subset \Re^{m_i}, m_i \in \mathbf{N}$; and $f_k : \mathcal{X} \times \mathcal{U}^1 \times \ldots, \mathcal{U}^N \to \Re$ are real C^1 functions of their arguments. To the game is associated the N objectives or pay-offs or cost functionals $J^i : \Gamma^1 \times \ldots \times \Gamma^N \to \Re$, where $\Gamma^i, i = 1 \ldots, N$ are the permissible *strategy spaces* of the players,

$$J^i(u^1, \ldots, u^N) = \sum_{k=k_0}^{K} L_k^i(x_{k+1}, x_k, u_k^1, \ldots, u_k^N), \quad i = 1, \ldots, N, \tag{2.10}$$

and where the $u_k^i \in \Gamma^i, i = 1, \ldots, N$ are functions of x_k, and $L_k^i : \mathcal{X} \times \mathcal{X} \times \mathcal{U}^1 \times \ldots \times \mathcal{U}^N \to \Re$, $i = 1, \ldots, N, k = k_0, \ldots, K$ are real C^1 functions of their arguments. The players play noncooperatively, and the objective is to minimize each of the above pay-off functions. Such a game is called a *nonzero-sum game* or *Nash-game*. The optimal decisions $u_{k,\star}^i, i = 1, \ldots, N$, which minimize each pay-off function at each stage of the game, are called *Nash-equilibrium* solutions.

Definition 2.2.1 *An N-tuple of strategies $\{u_\star^i \in \Gamma^i, i = 1, \ldots, N\}$, where $\Gamma^i, i = 1, \ldots, N$ is the strategy space of each player, is said to constitute a noncooperative Nash-equilibrium solution for the above N-person game if*

$$J_\star^i := J^i(u_\star^1, \ldots, u_\star^N) \le J^i(u_\star^1, \ldots, u^i, u_\star^{i+1}, u_\star^N) \quad i = 1, \ldots, N, \tag{2.11}$$

where $u_\star^i = \{u_{k,\star}^i, k = k_0, \ldots, K\}$ and $u^i = \{u_k^i, k = k_0, \ldots, K\}$.

In this section, we discuss necessary and sufficient conditions for the existence of Nash-equilibrium solutions of such a game under closed-loop (or feedback) no-memory perfect-state-information structure, i.e., when the decision is based on the current state information which is pefectly known. The following theorem based on the dynamic programming principle gives necessary and sufficient conditions for a set of strategies to be a Nash-equilibrium solution [59].

Theorem 2.2.1 *For the N-person discrete-time nonzero-sum game (2.9)-(2.11), an N-tuple of strategies $\{u_\star^i \in \Gamma^i, i = 1, \ldots, N\}$, provides a feedback Nash-equilibrium solution, if and only if, there exist $N \times (K - k_0)$ functions $V^i : \mathcal{X} \times \mathbf{Z} \to \Re$ such that the following* *recursive equations are satisfied:*

$$
\begin{aligned}
V^i(x,k) &= \min_{u_k^i \in \mathcal{U}^i} \left\{ L_k^i(x_{k+1}, x_k, u_{k,\star}^1, \ldots, u_k^i, \ldots, u_{k,\star}^N) + V^i(x_{k+1}, k+1) \right\}, \\
&\quad V^i(x, K+1) = 0, \quad i = 1, \ldots, N, \quad k \in \{k_0, \ldots, K\} \\
&= \min_{u_k^i \in \mathcal{U}^i} \left\{ L_k^i(f_k(x, u_{k,\star}^1, \ldots, u_k^i, \ldots, u_{k,\star}^N), x_k, u_{k,\star}^1, \ldots, u_k^i, \ldots, u_{k,\star}^N) + \right. \\
&\quad \left. V^i(f_k(x, u_{k,\star}^1, \ldots, u_k^i, \ldots, u_{k,\star}^N), k+1) \right\}, \quad V^i(x, K+1) = 0, \\
&\quad i = 1, \ldots, N, \quad k \in \{k_0, \ldots, K\}.
\end{aligned}
\tag{2.12}
$$

Notice that, in the above equations, there are $N \times (K - k_0)$ functions, since at each stage of the decision process, a different function is required for each of the players. However, it is often the case that N functions that satisfy the above recursive equations can be obtained with a single function for each player. This is particularly true in the linear case.

More specifically, let us consider the case of a two-player nonzero-sum game, of which type most of the problems in this book will be, described by the state equation:

$$
x_{k+1} = f_k(x_k, w_k, u_k), \quad x(k_0) = x^0, \quad k \in \{k_0, \ldots, K\}
\tag{2.13}
$$

where $w_k \in \mathcal{W} \subset \Re^{m_w}$ and $u_k \in \mathcal{U} \subset \Re^{m_u}$ represent the decisions of the two players respectively. The pay-offs J^1, J^2 are given by

$$
J^1(w, u) = \sum_{k=k_0}^{K} L_k^1(x_{k+1}, x_k, w_k, u_k),
\tag{2.14}
$$

$$
J^2(w, u) = \sum_{k=k_0}^{K} L_k^2(x_{k+1}, x_k, w_k, u_k),
\tag{2.15}
$$

where $w = w_{[k_0, K]}$, $u = u_{[k_0, K]}$. Then, a pair of strategies $w_\star := \{w_{k,\star}, k = k_0, \ldots, K\}$, $u_\star := \{u_{k,\star}, k = k_0, \ldots, K\}$, will constitute a Nash-equilibrium solution for the game if

$$
J^1(w_\star, u_\star) \leq J^1(w_\star, u) \quad \forall u \in \mathcal{U},
\tag{2.16}
$$

$$
J^2(w_\star, u_\star) \leq J^2(w, u_\star) \quad \forall w \in \mathcal{W}.
\tag{2.17}
$$

Furthermore, the conditions of Theorem 2.2.1 reduce to:

$$
\begin{aligned}
V^1(x,k) &= \min_{u_k \in \mathcal{U}} \left\{ L_k^1(f_k(x, w_{k,\star}, u_k), x, w_{k,\star}, u_k) + V^1(f_k(x, w_{k,\star}, u_k), k+1) \right\}; \\
&\quad V^1(x, K+1) = 0, \quad k = 1, \ldots, K
\end{aligned}
\tag{2.18}
$$

$$
\begin{aligned}
V^2(x,k) &= \min_{w_k \in \mathcal{W}} \left\{ L_k^2(f_k(x, w_k, u_{k,\star}), x, w_k, u_{k,\star}) + V^2(f_k(x, w_k, u_{k,\star}), k+1) \right\}; \\
&\quad V^2(x, K+1) = 0, \quad k = 1, \ldots, K.
\end{aligned}
\tag{2.19}
$$

Remark 2.2.1 *Equations (2.18), (2.19) represent a pair of coupled HJ-equations. They are difficult to solve, and not much work has been done to study their behavior. However, these equations will play an important role in the derivation of the solution to the discrete-time mixed $\mathcal{H}_2/\mathcal{H}_\infty$ control problem in later chapters of the book.*

Another special class of the nonzero-sum games is the *two-person zero-sum* game, in which the objective functionals (2.14), (2.15) are such that

$$
J^1(w, u) = -J^2(w, u) = J(w, u) := \sum_{k=k_0}^{K} L_k(x_{k+1}, x_k, w_k, u_k).
\tag{2.20}
$$

If this is the case, then $J^1(w_\star, u_\star) + J^2(w_\star, u_\star) = 0$ and while u is the minimizer, w is the maximizer. Thus, we might refer to u as the minimizing player and w as the maximizing player. Furthermore, the "Nash-equilibrium" conditions (2.16), (2.17) reduce to the saddle-point conditions:

$$J(w, u_\star) \le J(w_\star, u_\star) \le J(w_\star, u), \quad \forall w \in \mathcal{W}, \, u \in \mathcal{U}, \tag{2.21}$$

where the pair $(w_\star, u_\star)$ is the optimal strategy and is called a "saddle-point." Consequently, the set of necessary and sufficient conditions (2.18), (2.19) reduce to the following condition given in this theorem [57, 59].

Theorem 2.2.2 *For the two-person discrete-time zero-sum game defined by (2.13), (2.20), a pair of strategies $(w_\star, u_\star)$ provides a feedback saddle-point solution if, and only if, there exists a set of $K - k_0$ functions $V : \mathcal{X} \times \mathbf{Z} \to \Re$, such that the following recursive equations are satisfied for each $k \in [k_0, K]$:*

$$
\begin{aligned}
V(x, k) &= \min_{u_k \in \mathcal{U}} \max_{w_k \in \mathcal{W}} \Big\{ L_k(f_k(x, w_k, u_k), x, w_k, u_k) + V(f_k(x, w_k, u_k), k+1) \Big\}, \\
&= \max_{w_k \in \mathcal{W}} \min_{u_k \in \mathcal{U}} \Big\{ L_k(f_k(x, w_k, u_k), x, w_k, u_k) + V(f_k(x, w_k, u_k), k+1) \Big\}, \\
&= L_k(f_k(x, w_{k,\star}, u_{k,\star}), x, w_{k,\star}, u_{k,\star}) + V(f_k(x, w_{k,\star}, u_{k,\star}), k+1), \\
V(x, K+1) &= 0, \tag{2.22}
\end{aligned}
$$

where $x = x_k$. Equation (2.22) is known as *Isaacs equation*, being the discrete-time version of the one developed by Isaacs [136, 59]. Furthermore, the interchangeability of the "max" and "min" operations above is also known as "*Isaacs condition*." The above equation will play a significant role in the derivation of the solution to the $\mathcal{H}_\infty$-control problem for discrete-time nonlinear systems in later chapters of the book.

2.2.1   Linear-Quadratic Discrete-Time Dynamic Games

We now specialize the above results of the N-person nonzero-sum dynamic games to the linear-quadratic (LQ) case in which the optimal decisions can be expressed explicitly. For this purpose, the dynamics of the system is described by a linear state equation:

$$x_{k+1} = f_k(x_k, u_k^1, \ldots, u_k^N) = A_k x_k + \sum_{i=1}^N B_k^i u_k^i, \quad k = [k_0, K] \tag{2.23}$$

where $A_k \in \Re^{n \times n}$, $B_k^i \in \Re^{n \times p_i}$, and all the variables have their previous meanings. The pay-offs $J^i, i = 1, \ldots, N$ are described by

$$J^i(u_{[k_0,K]}^1, \ldots, u_{[k_0,K]}^N) = \sum_{k=1}^N L_k^i(x_{k+1}, u_k^1, \ldots, u_k^N) \tag{2.24}$$

$$L_k^i(x_{k+1}, u_k^1, \ldots, u_k^N) = \frac{1}{2}[x_{k+1}^T Q_{k+1}^i x_{k+1} + \sum_{j=1}^N (u_k^j)^T R_k^{ij} u_k^j] \tag{2.25}$$

where Q_{k+1}^i, R_k^{ij} are matrices of appropriate dimensions with $Q_{k+1}^i \ge 0$ symmetric and $R_k^{ii} > 0$, for all $k = k_0, \ldots, K$, $i, j = 1, \ldots, N$.

The following corollary gives necessary and sufficient conditions for the existence of Nash-equilibrium for the LQ-game [59].

Corollary 2.2.1 *For the N-player LQ game described by the equations (2.23)-(2.25), suppose $Q_{k+1}^i \geq 0$, $i = 1, \ldots, N$ and $R_k^{ij} \geq 0$, $i, j = 1, \ldots, N$, $j \neq i$, $k = k_0, \ldots, K$. Then, there exists a unique feedback Nash-equilibrium solution to the game if, and only if, there exist unique solutions P_k^i, $i = 1, \ldots, N$, $k = k_0, \ldots, K$ to the recursive equations:*

$$[R_k^{ii} + (B_k^i)^T Z_{k+1}^i B_k^i] P_k^i + (B_k^i)^T Z_{k+1}^i \sum_{j=1, j \neq i}^N B_k^j P_k^j = (B_k^i)^T Z_{k+1}^i A_k, \quad i = 1, \ldots, N, \quad (2.26)$$

where

$$Z_k^i = F_k^T Z_{k+1}^i F_k + \sum_{j=1}^N (P_k^j)^T R_k^{ij} P_k^j + Q_k^i, \quad Z_{K+1}^i = Q_{K+1}^i, i = 1, \ldots, N \quad (2.27)$$

$$F_k = A_k - \sum_{i=1}^N B_k^i P_k^i, \quad k = k_0, \ldots, K. \quad (2.28)$$

Furthermore, the equilibrium strategies are given by

$$u_k^i = -P_k^i x_k, \quad i = 1, \ldots, N. \quad (2.29)$$

Remark 2.2.2 *The positive-semidefinite conditions for Q_{k+1}^i, and $R_k^{ij}, j \neq i$ guarantee the convexity of the functionals J^i and therefore also guarantee the existence of a minimizing solution in (2.12) for all $k \in [k_0, K]$. Furthermore, there has not been much work in the literature regarding the existence of solutions to the coupled discrete-Riccati equations (2.26)-(2.28). The only relevant work we could find is given in [84].*

Again, the results of the above corollary can easily be specialized to the case of two players. But more importantly, we shall specialize the results of Theorem 2.2.2 to the case of the two-player LQ zero-sum discrete-time dynamic game described by the state equation:

$$x_{k+1} = A_k x_k + B_k^1 w_k + B_k^2 u_k, \quad x(k_0) = x^0, \ k = [k_0, K] \quad (2.30)$$

and the objective functionals

$$-J^1(u, w) = J^2(u, w) = J(u, w) = \sum_{k=k_0}^K L_k(x_{k+1}, x_k, u, w), \quad (2.31)$$

$$L_k(x_{k+1}, x_k, u, w) = \tfrac{1}{2}[x_{k+1}^T Q_{k+1} x_{k+1} + u_k^T u_k - w_k^T w_k]. \quad (2.32)$$

We then have the following corollary [57, 59].

Corollary 2.2.2 *For the two-player LQ zero-sum dynamic game described by (2.30)-(2.31), suppose $Q_{k+1} \geq 0$, $k = k_0, \ldots, K$. Then there exists a unique feedback saddle-point solution if, and only if,*

$$I + (B_k^2)^T M_{k+1} B_k^2 > 0, \quad \forall k = k_0, \ldots, K, \quad (2.33)$$

$$I - (B_k^1)^T M_{k+1} B_k^1 > 0, \quad \forall k = k_0, \ldots, K, \quad (2.34)$$

where

$$S_k = [I + (B_k^1 B_k^{1^T} - B_k^2 B_k^{2^T}) M_{k+1}], \quad (2.35)$$

$$M_k = Q_k + A_k^T M_{k+1} S_k^{-1} A_k, \quad M_{K+1} = Q_{K+1}. \quad (2.36)$$

Moreover, the unique equilibrium strategies are given by

$$u_{k,\star} = -(B_k^2)^T M_{k+1} S_k^{-1} A_k x_k, \quad (2.37)$$

$$w_{k,\star} = (B_k^1)^T M_{k+1} S_k^{-1} A_k x_k, \quad (2.38)$$

and the corresponding unique state trajectory is given by the recursive equation:

$$x^\star_{k+1} = S_k^{-1} A_k x^\star_k. \tag{2.39}$$

In the next section we consider the continuous-time counterparts of the discrete-time dynamic games.

2.3　Continuous-Time Nonzero-Sum Dynamic Games

In this section, we discuss the continuous-time counterparts of the results that we presented in the previous section for discrete-time dynamic games. Indeed the name "differential game" stems from the continuous-time models of dynamic games.

We consider at the outset N-player nonzero-sum differential game defined by the state equation:

$$\dot{x}(t) = f(x, u^1, \ldots, u^N, t), \quad x(t_0) = x_0, \tag{2.40}$$

where $x(t) \in \mathcal{X} \subseteq \Re^n$ is the state of the system for any $t \in \Re$ which belong to the state-space manifold $\mathcal{X}$ and $x(t_0)$ is the initial state which is assumed to be known a priori by all the players; $u^i \in \mathcal{U}^i, i = 1, \ldots, N$, $N \in \mathbf{N}$ are the decision variables or control inputs of the players $1, \ldots, N$ which belong to the control sets $\mathcal{U}^i \subset \Re^{m_i}$, $m_i \in \mathbf{N}$, which are also measurable; and $f : \mathcal{X} \times \mathcal{U}^1 \times \ldots, \mathcal{U}^N \times \Re \to \Re$ is a real C^1 function of its arguments.

To each player $i = 1, \ldots, N$ is associated an objective functional or pay-off $J^i : \Gamma^1 \times \ldots \times \Gamma^N \to \Re$, where $\Gamma^i, i = 1, \ldots, N$ is the strategy space of the player, which he tries to minimize, and is defined by

$$J^i(u^1, \ldots, u^N) = \phi^i(x(t_f), t_f) + \int_{t_0}^{t_f} L^i(x, u^1, \ldots, u^N, t)dt, \quad i = 1, \ldots, N, \tag{2.41}$$

where $L^i : \mathcal{X} \times \mathcal{U}^1 \times \ldots, \times \mathcal{U}^N \times \Re_+ \to \Re$ and $\phi^i : \mathcal{X} \times \Re_+ \to \Re$, $i = 1, \ldots, N$, are real C^1 functions of their arguments respectively. The final or terminal time t_f may be variable or fixed; here, we shall assume it is fixed for simplicity. The functions $\phi^i(., .), i = 1, \ldots, N$ are also known as the terminal cost functions.

The players play noncooperatively, but each player knows the current value of the state vector as well as all system parameters and cost functions. However, he does not know the strategies of the rival players. Our aim is to derive sufficient conditions for the existence of Nash-equilibrium solutions to the above game under closed-loop memoryless perfect information structure. For this purpose, we redefine Nash-equilibrium for continuous-time dynamic games as follows.

Definition 2.3.1 *An N-tuple of strategies $\{u^i_\star \in \Gamma^i, i = 1, \ldots, N\}$, where $\Gamma^i, i = 1, \ldots, N$ are the set of strategies, constitute a Nash-equilibrium for the game (2.40)-(2.41) if*

$$J^i_\star := J^i(u^1_\star, \ldots, u^N_\star) \le J^i(u^1_\star, \ldots, u^{i-1}_\star, u^i, u^{i+1}_\star, \ldots, u^N_\star) \; \forall i = 1, \ldots, N, \tag{2.42}$$

where $u^i_\star = \{u^i_\star(t), t \in [t_0, t_f]\}$ and $u^i = \{u^i(t), t \in [t_0, t_f]\}$.

To derive the optimality conditions, we consider an N-tuple of piecewise continuous strategies $u := \{u^1, \dots, u^N\}$ and the *value-function* for the i-th player

$$
\begin{aligned}
V^i(x,t) &= \inf_{u \in \mathcal{U}} \left\{ \phi^i(x(t_f), t_f) + \int_t^{t_f} L^i(x, u, \tau) dt \right\} \\
&= \phi^i(x(t_f), t_f) + \int_t^{t_f} L^i(x, u_\star, \tau) d\tau.
\end{aligned}
\tag{2.43}
$$

Then, by applying Definition 2.3.1 and the dynamic-programming principle, it can be shown that the value-functions V^i, $i = 1, \dots, N$ are solutions of the following Hamilton-Jacobi PDEs:

$$
\begin{aligned}
\frac{\partial V^i}{\partial t} &= - \inf_{u^i \in \mathcal{U}^i} H^i(x, t, u_\star^1, \dots, u_\star^{i-1}, u^i, u_\star^{i+1}, \dots, u_\star^N, \frac{\partial V^i}{\partial x}), \\
& \qquad V^i(x(t_f), t_f) = \phi^i(x(t_f), t_f), \tag{2.44} \\
H^i(x, t, u, \lambda_i^T) &= L^i(x, u, t) + \lambda_i^T f(x, u, t), \quad i = 1, \dots, N. \tag{2.45}
\end{aligned}
$$

The strategies $u_\star = (u_\star^1, \dots, u_\star^N)$ which minimize the right-hand-side of the above equations (2.44) are the equilibrium strategies. The equations (2.44) are integrated backwards in time from the terminal manifold $x(t_f)$, and at each (x,t) one must solve a static game for the Hamiltonians $H^i, i = 1, \dots, N$ to find the Nash-equilibrium. This is not always possible except for a class of differential games called *normal* [250]. For this, it is possible to find a unique Nash-equilibrium point $u_\star$ for the vector $H = [H^1, \dots, H^N]$ for all x, λ, t such that when the equations

$$
\frac{\partial V^i}{\partial t}(x,t) = -H^i(x, t, u_\star(x, t, \frac{\partial V^1}{\partial x}, \dots, \frac{\partial V^N}{\partial x}), \frac{\partial V^i}{\partial x})
\tag{2.46}
$$

$$
\dot{x} = f(x, u^1, \dots, u^N, t)
\tag{2.47}
$$

are integrated backward from all the points on the terminal surface, feasible trajectories are obtained. Thus, for a normal game, the following theorem provides sufficient conditions for the existence of Nash-equilibrium solution for the N-player game under closed-loop memoryless perfect-state information pattern [57, 250].

Theorem 2.3.1 *Consider the N-player nonzero-sum continuous-time dynamic game (2.40)-(2.41) of fixed duration $[t_0, t_f]$, and under closed-loop memoryless perfect state information pattern. An N-tuple $u_\star = (u_\star^1, \dots, u_\star^N)$ of strategies provides a Nash-equilibrium solution, if there exist N C^1-functions $V^i : \mathcal{X} \times \Re_+ \to \Re$, $i = 1, \dots, N$ satisfying the HJEs (2.44)-(2.45).*

Remark 2.3.1 *The above result can easily be specialized to the two-person nonzero-sum continuous-time dynamic game. This case will play a significant role in the derivation of the solution to the mixed $\mathcal{H}_2/\mathcal{H}_\infty$ control problem in a later chapter.*

Next, we specialize the above result to the case of the two-person zero-sum continuous-time dynamic game. For this case, we have the dynamic equation and the objective functional described by

$$
\dot{x}(t) = f(x, u, w, t), \quad x(t_0) = x_0,
\tag{2.48}
$$

$$
J^1(u, w) = -J^2(u, w) := J(u, w) = \phi(x(t_f), t_f) + \int_{t_0}^{t_f} L(x, u, w, t) dt,
\tag{2.49}
$$

where $u \in \mathcal{U}$, $w \in \mathcal{W}$ are the strategies of the two players, which belong to the measurable sets $\mathcal{U} \subset \Re^{m_u}$, $\mathcal{W} \subset \Re^{m_w}$ respectively, x_0 is the initial state which is known a priori to both players. Then we have the following theorem which is the continuous-time counterpart of Theorem 2.2.2.

Theorem 2.3.2 *Consider the two-person zero-sum continuous-time differential game (2.48)-(2.49) of fixed duration $[t_0, t_f]$, and under closed-loop memoryless information structure. A pair of strategies $(u_\star, w_\star)$ provides a Nash-feedback saddle-point solution to the game, if there exists a C^1-function $V : \mathcal{X} \times \Re \to \Re$ of both of its arguments, satisfying the PDE:*

$$
\begin{aligned}
-\frac{\partial V(x,t)}{\partial t} &= \inf_{u \in \mathcal{U}} \sup_{w \in \mathcal{W}} \left\{ \frac{\partial V(x,t)}{\partial x} f(x,u,w,t) + L(x,u,w,t) \right\} \\
&= \sup_{w \in \mathcal{W}} \inf_{u \in \mathcal{U}} \left\{ \frac{\partial V(x,t)}{\partial x} f(x,u,w,t) + L(x,u,w,t) \right\} \\
&= \frac{\partial V(x,t)}{\partial x} f(x,u_\star,w_\star,t) + L(x,u_\star,w_\star,t), \\
V(x(t_f),t_f) &= \phi(x(t_f),t_f), \tag{2.50}
\end{aligned}
$$

known as Isaacs equation [136] or Hamilton-Jacobi-Isaacs equation (HJIE).

Remark 2.3.2 *The HJIE (2.50) will play a significant role in the derivation of the solution to the $\mathcal{H}_\infty$-control problem for continuous-time nonlinear systems in later chapters.*

Next, we similarly specialize the above results to linear systems with quadratic objective functions, also known as the "linear-quadratic (LQ)" continuous-time dynamic game.

2.3.1   Linear-Quadratic Continuous-Time Dynamic Games

In this section, we specialize the results of the previous section to the linear-quadratic (LQ) continuous-time dynamic games. For the linear case, the dynamic equations (2.40), (2.41) reduce to the following equations:

$$
\dot{x}(t) = f(x, u^1, \dots, u^N, t) = A(t)x(t) + \sum_{i=1}^{N} B^i(t)u^i(t), \quad x(t_0) = x_0 \tag{2.51}
$$

$$
J^i(u^1, \dots, u^N) = \frac{1}{2}x^T(t_f)Q_f^i x(t_f) + \frac{1}{2}\int_{t_0}^{t_f}[x^T(t)Q^i x(t) + \sum_{j=1}^{N}(u^j)^T(t)R^{ij}(t)u^j(t)]dt, \tag{2.52}
$$

where $A(t) \in \Re^{n \times n}$, $B^i(t) \in \Re^{n \times p_i}$ and Q^i, Q_f^i, $R^{ij}(t)$ are matrices of appropriate dimensions for each $t \in [t_0, t_f]$, $i, j = 1, \dots, N$. Moreover, $Q^i(t)$, Q_f^i are symmetric positive-semidefinite and R^{ii} is positive-definite for $i = 1, \dots, N$. It is then possible to derive a closed-form expression for the optimal strategies as stated in the following corollary to Theorem 2.3.1.

Corollary 2.3.1 *Consider the N-person LQ-continuous-time dynamic game (2.51), (2.52). Suppose $Q^i(t) \geq 0$, $Q_f^i \geq 0$ and symmetric, $R^{ij}(t) > 0$ for all $t \in [t_0, t_f]$, $i, j = 1, \dots, N$, $i \neq j$. Then there exists a linear feedback Nash-equilibrium solution to the differential game under closed-loop memoryless perfect state information structure, if there exists a set of symmetric solutions $P^i(t) \geq 0$ to the N-coupled time-varying matrix Riccati ordinary* *differential-equations (ODEs):*

$$\dot{P}^i(t) + P^i(t)\tilde{A}(t) + \tilde{A}^T(t)P^i(t) + \sum_{j=1}^{N} P^j(t)B^j(t)(R^{jj}(t))^{-1}R^{ij}(t)(R^{jj}(t))^{-1}(B^j)^T P^j(t) +$$

$$Q^i(t) = 0, \quad P^i(t_f) = Q^i_f, \quad i = 1, \ldots, N \tag{2.53}$$

$$\tilde{A}(t) = A(t) - \sum_{i=1}^{N} B^i(t)(R^{ii}(t))^T(B^i)^T P^i(t). \tag{2.54}$$

Furthermore, the optimal strategies and costs for the players are given by

$$u^i_\star(t) = -(R^{ii}(t))^{-1}(B^i(t))^T P^i(t)x(t), \quad i = 1, \ldots, N, \tag{2.55}$$

$$J^i_\star = \frac{1}{2}x_0^T P^i(0)x_0, \quad i = 1, \ldots, N. \tag{2.56}$$

The equations (2.53) represent a system of coupled ODEs which could be integrated backwards using numerical schemes such as the Runge-Kutta method. However, not much is known about the behavior of such a system, although some work has been done for the two-player case [2, 108, 221]. Since these results are not widely known, we shall review them briefly here.

Consider the system equation with two players given by

$$\dot{x}(t) = A(t)x(t) + B^1(t)w(t) + B^2(t)u(t), \quad x(t_0) = x_0, \tag{2.57}$$

and the cost functions:

$$J^1(u, w) = x(t_f)^T Q^1_f x(t_f) + \int_0^{t_f} (x^T Q^1 x + w^T R^{11} w + u^T R^{12} u)dt, \tag{2.58}$$

$$J^2(u, w) = x(t_f)^T Q^2_f x(t_f) + \int_0^{t_f} (x^T Q^2 x + u^T R^{22} u + w^T R^{21} w)dt, \tag{2.59}$$

where $Q^i_f \geq 0$, $Q^i \geq 0$, $R^{ij} \geq 0$, $i \neq j$, $R^{ii} > 0$, $i, j = 1, 2$. Further, assume that the matrices $A(t) = A$, $B^1(t) = B^1$, and $B^2(t) = B^2$ are constant and of appropriate dimensions. Then, the system of coupled ODEs corresponding to the two-player nonzero-sum game is given by

$$\dot{P}^1 = -A^T P^1 - P^1 A + P^1 S^{11} P^1 + P^1 S^{22} P^2 + P^2 S^{22} P^1 - P^2 S^{12} P^2 - Q^1,$$
$$P^1(t_f) = Q^1_f, \tag{2.60}$$

$$\dot{P}^2 = -A^T P^2 - P^2 A + P^2 S^{22} P^2 + P^2 S^{11} P^1 + P^1 S^{11} P^2 - P^1 S^{21} P^1 - Q^2,$$
$$P^2(t_f) = Q^2_f, \tag{2.61}$$

where

$$S^{ij} = B^j(R^{jj})^{-1}R^{ij}(R^{jj})^{-1}(B^j)^T, \quad i, j = 1, 2.$$

In [221] global existence results for the solution of the above system (2.60)-(2.61) is established for the special case $B^1 = B^2$ and $R^{11} = R^{22} = -R^{12} = -R^{21} = I_n$, the identity matrix. It is shown that $P^1 - P^2$ satisfies a linear Lyapunov-type ODE, while $P^1 + P^2$ satisfies a standard Riccati equation. Therefore, $P^1 + P^2$ and hence P^1, P^2 cannot blow up in any finite time, and as $t_f \to \infty$, $P^1(0) + P^2(0)$ goes to a stabilizing, positive-semidefinite solution of the corresponding algebraic Riccati equation (ARE). In the following result, we derive upper and lower bounds for the solutions P^1 and P^2 of (2.60), (2.61).

Lemma 2.3.1 *Suppose P^1, P^2 are solutions of (2.60)-(2.61) on the interval $[t_0, t_f]$. Then, $P^1(t) \geq 0$, $P^2(t) \geq 0$ for all $t \in [t_0, t_f]$.*

Proof: Let $x(t_0) = x_1 \neq 0$ and let x be a solution of the initial-value problem

$$\dot{x}(t) = [A - S^{11} P^1(t) - S^{22} P^2(t)]x(t), \quad x(t_0) = x_1. \tag{2.62}$$

Then for $i = 1, 2$

$$
\begin{aligned}
\frac{d}{dt}[x^T P^i(t)x] &= \dot{x}^T P^i(t)x + x^T \dot{P}^i(t)x + x^T P^i(t)\dot{x} \\
&= x^T \Big\{ (A^T P^i(t) - P^1(t)S^{11}P^i(t) - P^2(t)S^{22}P^i(t)) + \\
&\quad (P^i(t)A - P^i(t)S^{11}P^1(t) - P^i(t)S^{22}P^2(t)) - A^T P^i(t) - \\
&\quad P^i(t)A - Q^i + P^i(t)S^{ii}P^i(t) - \sum_{j \neq i} P^j(t)S^{ij}P^j(t) + \\
&\quad \sum_{j \neq i} (P^i(t)S^{jj}P^j(t) + P^j(t)S^{jj}P^i(t)) \Big\} x \\
&= -x^T \{ Q^i + P^1(t)S^{i1}P^1(t) + P^2(t)S^{i2}P^2(t) \} x
\end{aligned}
$$

where the previous equation follows from (2.60), (2.61) and (2.62). Now integrating from t_0 to t_f yields

$$x_1^T P^i(t_0)x_1 = \int_{t_0}^{t_f} x^T(\tau)\tilde{Q}^i(\tau)x(\tau)d\tau + x^T(t_f)Q_f^i x(t_f) \geq 0 \tag{2.63}$$

where $\tilde{Q}^i := Q^i + P^1 S^{i1} P^1 + P^2 S^{i2} P^2$. Since x_1 is arbitrary and Q^i, S^{i1}, S^{i2}, $P^i(t_f) \geq 0$, the result follows. $\square$

The next theorem gives sufficient conditions for no finite-escape times for the solutions P^1, P^2.

Theorem 2.3.3 *Let $Q \in \Re^{n \times n}$ be symmetric and suppose for all $t \leq t_f$*

$$
\begin{aligned}
F(P^1(t), P^2(t), Q) &:= Q + (P^1(t) + P^2(t))(S^{11} + S^{22})(P^1(t) + P^2(t)) - \\
&\quad P^1(t)S^{21}P^1(t) - P^2(t)S^{12}P^2(t) - (P^1(t)S^{22}P^1(t) + \\
&\quad P^2(t)S^{11}P^2(t)) \geq 0, \tag{2.64}
\end{aligned}
$$

and when $P^1(t)$, $P^2(t)$ exist, then the solutions $P^1(t)$, $P^2(t)$ of (2.60), (2.61) exist for all $t < t_f$ with

$$0 \leq P^1(t) + P^2(t) \leq R_Q(t), \tag{2.65}$$

for some unique positive-semidefinite matrix function $R_Q(t)$ solution to the terminal value problem:

$$\dot{R}_Q(t) = -R_Q(t)A - A^T R_Q(t) - (Q^1 + Q^2 + Q), \quad R_Q(t_f) = Q_f^1 + Q_f^2 \tag{2.66}$$

which exists for all $t < t_f$.

Proof: The first inequality in (2.65) follows from Lemma 2.3.1. Further

$$(\dot{P}^1 + \dot{P}^2) = -A^T(P^1 + P^2) - (P^1 + P^2 A - (Q^1 + Q^2 + Q) + F(P^1(t), P^2(t), Q),$$

and by the monotonicity of solutions to Riccati equations [292], it follows that the second inequality holds while $F(P^1(t), P^2(t), Q) \geq 0$. $\square$

Remark 2.3.3 *For a more extensive discussion of the existence of the solutions P^1, P^2 to the coupled Riccati-ODEs, the reader is referred to reference [108]. However, we shall take up the subject for the infinite-horizon case where $t_f \to \infty$ in a later chapter.*

We now give an example of a pursuit-evasion problem which can be solved as a two-player nonzero-sum game [251].

Example 2.3.1 *We consider a pursuit-evasion problem with the following dynamics*

$$\dot{r} = v, \quad \dot{v} = a_p - a_e$$

where r is the relative position vector of the two adversaries, a_p, a_e are the accelerations of the pursuer and evader respectively, which also serve as the controls to the players.
   The cost functions are taken to be

$$
\begin{aligned}
J_p &= \frac{1}{2}q_p^2 r_f^T r_f + \frac{1}{2}\int_0^T (\frac{1}{c_p}a_p^T a_p + \frac{1}{c_{pe}}a_e^T a_e)dt \\
J_e &= -\frac{1}{2}q_e^2 r_f^T r_f + \frac{1}{2}\int_0^T (\frac{1}{c_e}a_e^T a_e + \frac{1}{c_{ep}}a_p^T a_p)dt
\end{aligned}
$$

where $r_f = r(T)$ and the final time is fixed, while c_p, c_e, c_{pe}, c_{ep} are weighting constants. The Nash-equilibrium controls are obtained by applying the results of Corollary 2.3.1 as

$$a_p = -c_p[0 \ I]P^p \begin{bmatrix} r \\ v \end{bmatrix}, \quad a_e = c_e[0 \ I]P^e \begin{bmatrix} r \\ v \end{bmatrix},$$

where P^p, P^e are solutions to the coupled Riccati-ODES:

$$
\begin{aligned}
\dot{P}^p(t) &= -P^p \begin{bmatrix} 0 & I \\ 0 & 0 \end{bmatrix} - \begin{bmatrix} 0 & 0 \\ I & 0 \end{bmatrix} P^p + c_p P^p \begin{bmatrix} 0 & 0 \\ 0 & I \end{bmatrix} P^p + c_e P^p \begin{bmatrix} 0 & 0 \\ 0 & I \end{bmatrix} P^e + \\
&\quad c_e P^e \begin{bmatrix} 0 & 0 \\ 0 & I \end{bmatrix} P^p - \frac{c_e^2}{c_{pe}} P^e \begin{bmatrix} 0 & 0 \\ 0 & I \end{bmatrix} P^e \\
\dot{P}^e(t) &= -P^e \begin{bmatrix} 0 & I \\ 0 & 0 \end{bmatrix} - \begin{bmatrix} 0 & 0 \\ I & 0 \end{bmatrix} P^e + c_e P^e \begin{bmatrix} 0 & 0 \\ 0 & I \end{bmatrix} P^e + c_p P^e \begin{bmatrix} 0 & 0 \\ 0 & I \end{bmatrix} P^p + \\
&\quad c_p P^p \begin{bmatrix} 0 & 0 \\ 0 & I \end{bmatrix} P^e - \frac{c_p^2}{c_{ep}} P^p \begin{bmatrix} 0 & 0 \\ 0 & I \end{bmatrix} P^p
\end{aligned}
$$

$$P^p(T) = q_p^2 \begin{bmatrix} I & 0 \\ 0 & 0 \end{bmatrix}, \quad P^e(T) = -q_e^2 \begin{bmatrix} I & 0 \\ 0 & 0 \end{bmatrix}.$$

It can be easily verified that the solutions to the above ODEs are given by

$$P^p(t) = \frac{1}{c_p}\tilde{p}(t) \begin{bmatrix} I & \tau I \\ \tau I & \tau^2 I \end{bmatrix}, P^e(t) = \frac{1}{c_e}\tilde{e}(t) \begin{bmatrix} I & \tau I \\ \tau I & \tau^2 I \end{bmatrix}$$

where $\tau = T - t$ is the time-to-go and $\tilde{p}(t)$, $\tilde{e}(t)$ are solutions to the following ODES:

$$
\begin{aligned}
\frac{d\tilde{p}}{d\tau} &= -\tau^2(\tilde{p}^2 + 2\tilde{p}\tilde{e} - a\tilde{e}^2), \quad \tilde{p}(0) = c_p q_p^2, \quad a = \frac{c_p}{c_{pe}} \\
\frac{d\tilde{e}}{d\tau} &= -\tau^2(\tilde{e}^2 + 2\tilde{p}\tilde{e} - b\tilde{p}^2), \quad \tilde{e}(0) = -c_e q_e^2, \quad b = \frac{c_e}{c_{ep}}.
\end{aligned}
$$

Again we can specialize the result of Theorem 2.3.2 to the LQ case. We consider for this case the system equation given by (2.57) and cost functionals represented as

$$-J^1(u,w) \quad = \quad J^2(u,w) := J = \frac{1}{2}x^T(t_f)Q_f x(t_f) + \frac{1}{2}\int_{t_0}^{t_f}[x^T(t)Qx(t) +$$

$$u^T(t)R_u u(t) - w^T(t)R_w w(t)]dt, \tag{2.67}$$

where u is the minimizing player and w is the maximizing player, and without any loss of generality, we can assume $R_w = I$. Then if we assume $V(x(t_f),t_f) = \phi(x(t_f),t_f) = \frac{1}{2}x^T Q_f x$ is quadratic, and also

$$V(x,t) = \frac{1}{2}x^T(t)P(t)x(t), \tag{2.68}$$

where $P(t)$ is symmetric and $P(T) = Q_f$, then substituting these in the HJIE (2.50), we get the following matrix Riccati ODE:

$$\dot{P}(t) + P(t)\tilde{A}(t) + \tilde{A}^T(t)P(t) + P(t)[B^1(t)(B^1(t))^T -$$

$$B^2(t)R_u^{-1}(B^2(t))^T]P(t) + Q = 0, \quad P(t_f) = Q_f. \tag{2.69}$$

Furthermore, we have the following corollary to Theorem 2.3.2.

Corollary 2.3.2 *Consider the two-player zero-sum continuous-time LQ dynamic game described by equations (2.57) and (2.67). Suppose there exist bounded symmetric solutions to the matrix ODE (2.69) for all $t \in [t_0, t_f]$, then there exists a unique feedback saddle-point solution to the dynamic game under closed-loop memoryless information-structure. Further, the unique strategies for the players are given by*

$$u_\star(t) \quad = \quad -R_u^{-1}(B^2(t))^T P(t)x(t), \tag{2.70}$$

$$w_\star(t) \quad = \quad (B^1(t))^T P(t)x(t), \tag{2.71}$$

and the optimal value of the game is given by

$$J(u_\star, w_\star) = \frac{1}{2}x_0^T P(0)x_0. \tag{2.72}$$

Remark 2.3.4 *In Corollary 2.3.1, the requirement that $Q_f \geq 0$ and $Q^i \geq 0$, $i = 1,\ldots,N$ has been included to guarantee the existence of the Nash-equilibrium solution; whereas, in Corollary 2.3.2 the existence of bounded solutions to the matrix Riccati ODE (2.69) is used. It is however possible to remove this assumption if in addition to the requirement that $Q_f \geq 0$ and $Q \geq 0$, we also have*

$$[(B^1(t))^T B^1(t) - (B^2(t))^T R_u^{-1} B^2(t)] \geq 0 \ \forall t \in [t_0, t_f].$$

Then the matrix Riccati equation (2.69) admits a unique positive-semidefinite solution [59].

2.4   Notes and Bibliography

The first part of the chapter on dynamic programmimg principle is based on the books [47, 64, 164, 175], while the material on dynamic games is based in most part on the books by Basar and Bernhard [57] and by Basar and Olsder [59]. The remaining material is based on the papers [250, 251]. On the other hand, the discussion on the existence, definiteness and boundedness of the solutions to the coupled matrix Riccati-ODEs are based on the Reference [108]. Moreover, the results on the stochastic case can also be found in [2].

We have not discussed the *Minimum principle* of Pontryagin [47] because all the problems discussed here pertain to unconstrained problems. However, for such problems, the minimum principle provides a set of necessary conditions for optimality. Details of these necessary conditions can be found in [59]. Furthermore, a treatment of constrained problems using the minimum principle can also be found in [73].

3

Theory of Dissipative Systems

The important concept of dissipativity developed by Willems [272, 273, 274], and later studied by Hill and Moylan [131]-[134], [202] has proven very successful in many feedback design synthesis problems [30]-[32], [233, 264, 272, 273]. This concept which was originally inspired from electrical network considerations, in particular passive circuits [30, 31, 32], generalizes many other important concepts of physical systems such as positive-realness [233], passivity [77] and losslessness [74]. As such, many important mathematical relations of dynamical systems such as the bounded-real lemma, positive-real lemma, the existence of spectral-factorization and finite $\mathcal{L}_p$-gain of linear and nonlinear systems have been shown to be consequences of this important theory. Moreover, there has been a renewed interest lately on this important concept as having been instrumental in the derivation of the solution to the nonlinear $\mathcal{H}_\infty$-control problem [264]. It has been shown that a sufficient condition for the solvability of this problem is the existence of a solution to some dissipation-inequalities.

Our aim in this chapter is to review the theory of dissipative systems as far as is relevant to the subject of this book, and to expand where possible to areas that complement the material. We shall first give an exposition of the theory for continuous-time systems for convenience, and then discuss the equivalent results for discrete-time systems.

In Section 1, we give basic definitions and prove fundamental results about continuous-time dissipative systems for a general class of nonlinear state-space systems and then for affine systems. We discuss the implications of dissipativity on the stability of the system and on feedback interconnection of such systems. We also derive the nonlinear version of the *bounded-real lemma* which provides a necessary and sufficient condition for an affine nonlinear system to have finite $\mathcal{L}_2$-gain.

In Section 2, we discuss passivity and stability of continuous-time passive systems. We also derive the nonlinear generalization of the Kalman-Yacubovitch-Popov (KYP) lemma which also provides a necessary and sufficient condition for a linear system to be passive or positive-real. We also discuss the implications of passivity on stability of the system. In Section 3, we discuss the problem of feedback equivalence to a passive system, i.e., how to render a given continuous-time nonlinear system passive using static state feedback only. Because of the nice stability and stabilizability properties of passive systems, such as (global) asymptotic stabilizability by pure-gain output-feedback [77], it is considerably desirable to render certain nonlinear systems passive. A complete solution to this problem is provided.

In Section 4, we discuss the equivalent properties of dissipativity and passivity for discrete-time nonlinear systems, and finally in Section 5, we discuss the problem of feedback-equivalence to a lossless system. In contrast with the continuous-time case, the problem of feedback-equivalence to a passive discrete-time system seems more difficult and more complicated to achieve than the continuous-time counterpart. Hence, we settle for the easier problem of feedback-equivalence to a lossless-system. A rather complete answer to this problem is also provided under some mild regularity assumptions.

3.1   Dissipativity of Continuous-Time Nonlinear Systems

In this section, we define the concept of dissipativity of a nonlinear system. We consider a nonlinear time-invariant state-space system Σ defined on some manifold $\mathcal{X} \subseteq \Re^n$ containing the origin $x = \{0\}$ in coordinates $x = (x_1, \ldots, x_n)$:

$$\Sigma : \left\{ \begin{array}{rcl} \dot{x} & = & f(x, u), \quad x(t_0) = x_0 \\ y & = & h(x) \end{array} \right. \tag{3.1}$$

where $x \in \mathcal{X}$ is the state vector, $u \in \mathcal{U}$ is the input function belonging to an input space $\mathcal{U} \subset \mathcal{L}_{loc}(\Re^p)$, $y \in \mathcal{Y} \subset \Re^m$ is the output function which belongs to the output space $\mathcal{Y}$. The functions $f : \mathcal{X} \times \mathcal{U} \to \mathcal{X}$ and $h : \mathcal{X} \to \mathcal{Y}$ are real C^r functions of their arguments such that there exists a unique solution $x(t, t_0, x_0, u)$ to the system for any $x_0 \in \mathcal{X}$ and $u \in \mathcal{U}$. We begin with the following definitions.

Definition 3.1.1 *The state-space $\mathcal{X}$ of the system Σ is reachable from the state x_{-1} if for any state $x \in \mathcal{X}$ at t_0, there exists an admissible input $u \in \mathcal{U}$ and a finite time $t_{-1} \leq t_0$ such that $x = \phi(t_0, t_{-1}, x_{-1}, u)$.*

Definition 3.1.2 *A function $s(u(t), y(t)) : \mathcal{U} \times \mathcal{Y} \to \Re$ is a supply-rate to the system Σ, if $s(., .)$ is piecewise continuous and locally integrable, i.e.,*

$$\int_{t_0}^{t_1} |s(u(t), y(t))| dt < \infty \tag{3.2}$$

for any $(t_0, t_1) \in \Re_+^2$.

Remark 3.1.1 *The supply-rate $s(., .)$ is a measure of the instantaneous power into the system. Part of this power is stored as internal energy and part of it is dissipated.*

It follows from the above definition of supply-rate that, to infer about the internal behavior of the system, it is sufficient to evaluate the expected total amount of energy expended by the system over a finite time interval. This leads us to the following definition.

Definition 3.1.3 *The system Σ is dissipative with respect to the supply-rate $s(t) = s(u(t), y(t))$ if for all $u \in \mathcal{U}$, $t_1 \geq t_0$, and $x(t_0) = 0$,*

$$\int_{t_0}^{t_1} s(u(t), y(t)) dt \geq 0 \tag{3.3}$$

when evaluated along any trajectory of the system starting at t_0 with $x(t_0) = 0$.

The above definition 3.1.3 being an inequality, postulates the existence of a *storage-function* and a possible *dissipation-rate* for the system. It follows that, if the system is assumed to have some stored energy which is measured by a function $\Psi : \mathcal{X} \to \Re$ at t_0, then for the system to be dissipative, it is necessary that in the transition from t_0 to t_1, the total amount of energy stored is less than the sum of the amount initially stored and the amount supplied. This suggests the following alternative definition of dissipativity.

Definition 3.1.4 *The system Σ is said to be locally dissipative with respect to the supply-rate $s(u(t), y(t))$ if for all $(t_0, t_1) \in \Re_+^2$, $t_1 \geq t_0$, there exists a positive-semidefinite function (called a storage-function) $\Psi : N \subset \mathcal{X} \to \Re_+$ such that the inequality*

$$\Psi(x_1) - \Psi(x_0) \leq \int_{t_0}^{t_1} s(u(t), y(t)) dt; \quad \forall t_1 \geq t_0 \tag{3.4}$$

is satisfied for all initial states $x_0 \in N$*, where* $x_1 = x(t_1, t_0, x_0, u)$*. The system is said to be dissipative if it is locally dissipative for all* x_0*,* $x_1 \in \mathcal{X}$*.*

Remark 3.1.2 *Clearly Definition 3.1.4 implies Definition 3.1.3, while 3.1.3 implies 3.1.4 if in addition* $\Psi(t_0) = 0$ *or* $\Psi(x_1) \geq \Psi(x_0) = 0$*. The inequality (3.4) is known as the dissipation-inequality. If* $\Psi(.)$ *is viewed as a generalized energy function, then we may assume that there exists a point of minimum storage,* x^e*, at which* $\Psi(x^e) = \inf_{x \in \mathcal{X}} \Psi(x)$*, and* Ψ *can be nomalized so that* $\Psi(x^e) = 0$*.*

Finally, the above inequality (3.4) can be converted to an equality by introducing the dissipation-rate $\delta : \mathcal{X} \times \mathcal{U} \to \Re$ according to the following equation

$$\Psi(x_1) - \Psi(x_0) = \int_{t_0}^{t_1} [s(t) - \delta(t)]dt \ \ \forall t_1 \geq t_0. \tag{3.5}$$

Remark 3.1.3 *The dissipation-rate is nonnegative if the system is dissipative. Moreover, the dissipation-rate uniquely determines the storage-function* $\Psi(.)$ *up to a constant [274].*

We now define the concept of available-storage, the existence of which determines whether the system is dissipative or not.

Definition 3.1.5 *The available-storage* $\Psi_a(x)$ *of the system* Σ *is the quantity:*

$$\Psi_a(x) = \sup_{\substack{x_0 = x, u \in \mathcal{U}, \\ t \geq 0}} \left(-\int_0^t s(u(\tau), y(\tau))d\tau \right) \tag{3.6}$$

where the supremum is taken over all possible inputs $u \in \mathcal{U}$*, and states* x *starting at* $t = 0$*.*

Remark 3.1.4 *If the system is dissipative, then the available-storage is well defined (i.e., it is finite) at each state* x *of the system. Moreover, it determines the maximum amount of energy which may be extracted from the system at any time. This is formally stated in the following theorem.*

Theorem 3.1.1 *For the nonlinear system* Σ*, the available-storage,* $\Psi_a(.)$*, is finite if, and only if, the system is dissipative. Furthermore, any other storage-function is lower bounded by* $\Psi_a(.)$*, i.e.,* $0 \leq \Psi_a(.) \leq \Psi(.)$*.*

Proof: Notice that $\Psi_a(.) \geq 0$ since it is the supremum over a set with the zero element (at $t = 0$). Now assume that $\Psi_a(.) < \infty$. We have to show that the system is dissipative, i.e., for any $(t_0, t_1) \in \Re_+^2$

$$\Psi_a(x_0) + \int_{t_0}^{t_1} s(u(\tau), y(\tau))d\tau \geq \Psi_a(x_1) \ \ \forall x_0, x_1 \in \mathcal{X}. \tag{3.7}$$

In this regard, notice that from (3.6)

$$\Psi_a(x_0) = \sup_{x_1, u_{[t_0,t_1]}} \left(-\int_{t_0}^{t_1} s(u(t), y(t))dt + \Psi_a(x_1) \right).$$

This implies that

$$\Psi_a(x_0) + \int_{t_0}^{t_1} s(u(t), y(t))dt \geq \Psi_a(x_1)$$

and hence $\Psi_a(.)$ satisfies the dissipation-inequality (3.4).

Conversely, assume that Σ is dissipative, then the dissipation-inequality (3.4) implies that for $t_0 = 0$

$$\Psi(x_0) + \int_0^{t_1} s(u(t), y(t))dt \geq \Psi(x_1) \geq 0 \quad \forall x_0, \; x_1 \in \mathcal{X}, \; \forall t_1 \in \Re_+ \tag{3.8}$$

by definition. Therefore,

$$\Psi(x_0) \geq -\int_0^{t_1} s(u(t), y(t))dt$$

which implies that

$$\Psi(x_0) \geq \sup_{\substack{x = x_0, u \in \mathcal{U}, \\ t_1 \geq 0}} -\int_0^{t_1} s(u(t), y(t))dt = \Psi_a(x_0).$$

Hence $\Psi_a(x) < \infty \; \forall \, x \in \mathcal{X}. \; \Box$

Remark 3.1.5 *The importance of the above theorem in checking dissipativeness of the non-linear system Σ cannot be overemphasized. It follows that, if the state-space of the system is reachable from the origin $x = 0$, then by an appropriate choice of an input $u(t)$, equivalently $s(u(t), y(t))$, such that $\Psi_a(.)$ is finite, then it can be rendered dissipative. However, evaluating $\Psi_a(.)$ is a difficult task without the output of the system specified a priori, or solving the state equations. This therefore calls for an alternative approach for determining the dissipativeness of the system.*

We next introduce the complementary notion of *required-supply*. This is the amount of supply (energy) required to bring the system from the point of minimum storage x^e to the state $x(t_0)$, $t_0 = 0$.

Definition 3.1.6 *The required-supply $\Psi_r : \mathcal{X} \to \Re$, of the system Σ with supply-rate $s(u(t), y(t))$ is defined by*

$$\Psi_r(x) = \inf_{x^e \rightsquigarrow x, u \in \mathcal{U}} \int_{t_{-1}}^0 s(u(t), y(t))dt, \quad \forall t_{-1} \leq 0, \tag{3.9}$$

where the infimum is taken over all $u \in \mathcal{U}$ and such that $x = x(t_0, t_{-1}, x^e, u)$.

Remark 3.1.6 *The required-supply is the minimum amount of energy required to bring the system to its present state from the state x^e.*

The following theorem defines dissipativeness of the system in terms of the required-supply.

Theorem 3.1.2 *Assume that the system Σ is reachable from x_{-1}. Then*

(i) Σ *is dissipative with respect to the supply-rate $s(u(t), y(t))$ if, and only if, there exists a constant $\kappa > -\infty$ such that*

$$\inf_{x_{-1} \rightsquigarrow x} \int_{t_{-1}}^0 s(u(t), y(t))dt \geq \kappa \quad \forall x \in \mathcal{X}, \; \forall t_{-1} \leq 0. \tag{3.10}$$

Moreover,

$$\Psi(x) = \Psi_a(x_{-1}) + \inf_{x_{-1} \rightsquigarrow x} \int_{t_{-1}}^0 s(u(t), y(t))dt$$

is a possible storage-function.

(ii) *Let Σ be dissipative, and x^e be an equilibrium-point of the system such that $\Psi(x^e) = 0$. Then $\Psi_r(x^e) = 0$, and $0 \leq \Psi_a \leq \Psi \leq \Psi_r$. Moreover, if Σ is reachable from x^e, then $\Psi_r < \infty$ and is a possible storage-function.*

Proof: (i) By reachability and Theorem 3.1.1, Σ is dissipative if, and only if, $\Psi_a(x_{-1}) < \infty$. Thus, if we take $\kappa \leq -\Psi_a(x_{-1})$, then the above inequality in (i) is satisfied. Next, we show that $\Psi(x)$ given above is a possible storage-function. Note that this function is clearly nonnegative. Further, if we take the system along a path from x_{-1} to x_0 at t_0 in the state-space $\mathcal{X}$ via x_1 at t_1, then

$$
\begin{aligned}
\Psi(x_1) - \Psi(x_0) &= \inf_{x_{-1} \rightsquigarrow x_1} \int_{t_{-1}}^{t_1} s(u(t), y(t))dt - \inf_{x_{-1} \rightsquigarrow x_0} \int_{t_{-1}}^{0} s(u(t), y(t)) \\
&\leq \inf_{x_0 \rightsquigarrow x_1} \int_{t_0}^{t_1} s(u(t), y(t))dt \\
&\leq \int_{t_0}^{t_1} s(u(t), y(t))dt.
\end{aligned}
$$

(ii) That $\Psi_r(x^e) = 0$ follows from Definition 3.1.6. Moreover, for any $u \in \mathcal{U}$ which transfers x^e at t_{-1} to x at $t = 0$, the dissipation-inequality

$$
\Psi(x) \quad \Psi(x^e) \leq \int_{t_{-1}}^{0} s(u(t), y(t))dt
$$

is satisfied. Hence

$$
\begin{aligned}
\Psi(x) &\leq \int_{t_{-1}}^{0} s(u(t), y(t))dt \\
&\leq \inf_{x_{-1} \rightsquigarrow x} \int_{t_{-1}}^{0} s(u(t), y(t))dt = \Psi_r(x).
\end{aligned}
$$

Furthermore, if the state-space $\mathcal{X}$ of Σ is reachable, then clearly $\Psi_r < \infty$. Finally, that Ψ_r is a possible storage-function follows from (i). $\square$

Remark 3.1.7 *The inequality $\Psi_a \leq \Psi \leq \Psi_r$ implies that a dissipative system can supply to the outside only a fraction of what it has stored, and can store only a fraction of what has been supplied.*

The following theorem characterizes the set of all possible storage-functions for a dissipative system.

Theorem 3.1.3 *The set of possible storage-functions of a dissipative dynamical system forms a convex set. Therefore, for a system whose state-space is reachable from x^e, the function $\Psi = \lambda \Psi_a + (1 - \lambda)\Psi_r$, $\lambda \in [0, 1]$ is a possible storage-function for the system.*

Proof: The proof follows from the dissipation-inequality (3.4). $\square$

While in electrical networks resistors are clearly dissipative elements, ideal capacitors and inductors are not. Such elements do not dissipate power and consequently networks made up of such ideal elements are called *lossless*. The following definition characterizes such systems.

Definition 3.1.7 *The dissipative dynamical system Σ is said to be lossless with respect to the supply-rate $s(u(t), y(t))$ if there exists a storage-function $\Psi : \mathcal{X} \to \Re_+$ such that for all $t_1 \geq t_0$, $x_0 \in \mathcal{X}$, $u \in \mathcal{U}$,*

$$\Psi(x_0) + \int_{t_0}^{t_1} s(u(t), y(t))dt = \Psi(x_1) \tag{3.11}$$

where $x_1 = x(t_1, t_0, x_0, u)$.

Several interesting results about dissipative systems can be studied for particular supply-rates and differentiable or C^1 storage-functions; of particular significance are the following supply-rates:

(i) General: $s_1(u, y) = y^T Q y + 2 y^T S u + u^T R u$, where $Q \in \Re^{m \times m}$, $S \in \Re^{m \times k}$, and $R \in \Re^{k \times k}$;

(ii) Finite-gain systems: $s_2(u, y) = \gamma^2 u^T u - y^T y$, where $0 < \gamma < \infty$;

(iii) Passive systems: $s_3(u, y) = y^T u$.

Notice that (ii) and (iii) are special cases of (i). We study the implications of the quadratic supply-rates $s_1(., .)$ and $s_2(., .)$ and defer the study of $s_3(., .)$ to the next section. For this purpose, we consider the affine nonlinear system Σ^a defined on $\mathcal{X} \subset \Re^n$ in coordinates $(x_1, \ldots, x_n)$:

$$\Sigma^a : \begin{cases} \dot{x} &= f(x) + g(x)u, \quad x(t_0) = x_0 \\ y &= h(x) + d(x)u \end{cases} \tag{3.12}$$

where all the variables have their previous meanings, and $f : \mathcal{X} \to V^\infty(\mathcal{X})$, $g : \mathcal{X} \to \mathcal{M}^{n \times k}(\mathcal{X})$, $h : \mathcal{X} \to \Re^m$, $d : \mathcal{X} \to \mathcal{M}^{m \times k}(\mathcal{X})$. Moreover, $f(0) = 0$, $h(0) = 0$. Then we have the following result.

Theorem 3.1.4 *Assume the state-space of the system Σ^a is reachable from the origin $x = 0$ and the available-storage for the sytem if it exists is a C^1-function. Then, a necessary and sufficient condition for the system Σ^a to be dissipative with respect to the supply-rate $s(u, y) = y^T Q y + 2 y^T S u + u^T R u$, where Q, S, have appropriate dimensions, is that there exists a C^1-positive-semidefinite storage-function $\psi : \mathcal{X} \to \Re$, $\psi(0) = 0$, and C^0-functions $l : \mathcal{X} \to \Re^q$, $W : \mathcal{X} \to \Re^{q \times k}$ satisfying*

$$\left. \begin{aligned} \psi_x(x)f(x) &= h^T(x)Qh(x) - l^T(x)l(x) \\ \tfrac{1}{2}\psi_x(x)g(x) &= h^T(x)\hat{S}(x) - l^T(x)W(x) \\ \hat{R}(x) &= W^T(x)W(x) \end{aligned} \right\} \tag{3.13}$$

for all $x \in \mathcal{X}$, where

$$\hat{R} = R + d^T(x)S + S^T d(x) + d^T(x)Qd(x), \quad \hat{S}(x) = Qd(x) + S.$$

Proof: (Sufficiency): Suppose that $\psi(.)$, $l(.)$, and $W(.)$ exist and satisfy (3.13). Right-multiplying the second equation in (3.13) by u on both sides and adding to the first equation, we get

$$\begin{aligned} \psi_x(x)(f(x) + g(x)u) &= h^T(x)Qh(x) + 2h^T(x)\hat{S}(x)u - 2l^T(x)W(x)u - l^T(x)l(x) \\ &= [(h + d(x)u)^T Q(h(x) + d(x)u) + 2(h(x) + d(x)u)^T Su + u^T Ru] - \\ &\quad (l(x) + W(x)u)^T (l(x) + W(x)u). \end{aligned}$$

Then, integrating the above equation for any $u \in \mathcal{U}$, $t_1 \geq t_0$, and any $x(t_0)$, we get

$$
\begin{aligned}
\psi(x(t_1)) - \psi(x(t_0)) &= \int_{t_0}^{t_1} (y^T Q y + 2 y^T S u + u^T R u) dt \\
&\quad - \int_{t_0}^{t_1} (l(x) + W(x)u)^T (l(x) + W(x)u) dt \\
&= \int_{t_0}^{t_1} s(u, y) dt - \int_{t_0}^{t_1} (l(x) + W(x)u)^T (l(x) + W(x)u) dt \\
&\leq \int_{t_0}^{t_1} s(u(t), y(t)) dt.
\end{aligned}
$$

Hence the system is dissipative by Definition 3.1.3.

(Necessity): We assume the system is dissipative and proceed to show that the available-storage

$$
\psi_a(x) = \sup_{u(.), T \geq 0} \left(- \int_0^T s(u, y) dt \right) \tag{3.14}
$$

is finite by Theorem 3.1.1, and satisfies the system of equations (3.13) for some appropriate functions $l(.)$ and $W(.)$.

Note that, by reachability, for any initial state x_0 at $t_0 = 0$, there exists a time $t_{-1} < 0$ and a $u : [-t_{-1}, 0] \to \mathcal{U}$ such that $x(t_{-1}) = 0$ and $x(0) = x_0$. Then by dissipativity,

$$
\int_0^T s(u, y) dt \geq - \int_{t_{-1}}^0 s(u, y) dt.
$$

The right-hand-side (RHS) of the above equation depends on x_0 and since u can be chosen arbitrarily, then there exists a function $\kappa : \mathcal{X} \to \Re$ such that

$$
\int_0^T s(u, y) dt \geq \kappa(x_0) > -\infty
$$

whenever $x(0) = x_0$. By taking the "sup" on the left-hand-side (LHS) of the above equation, we get $\psi_a < \infty$ for all $x \in \mathcal{X}$. Moreover, $\psi_a(0) = 0$.

It remains to show that ψ_a satisfies the system (3.13). However, by dissipativity of the system, ψ_a satisfies the dissipation-inequality

$$
\psi_a(x_1) - \psi_a(x_0) \leq \int_{t_0}^{t_1} s(u, y) dt \tag{3.15}
$$

for any $t_1 \geq t_0$, $x_1 = x(t_1)$, and x_0. Differentiating the above inequality along the trajectories of the system and introducing a dissipation-rate $\delta : \mathcal{X} \times \mathcal{U} \to \Re$, we get

$$
\psi_{a,x}(x)(f(x) + g(x)u) - s(u, y) = -\delta(x, u). \tag{3.16}
$$

Since $s(u, y)$ is quadratic in u, then $\delta(.,.)$ is also quadratic, and may be factorized as

$$
\delta(x, u) = [l(x) + W(x)u]^T [l(x) + W(x)u]
$$

for some suitable functions $l : \mathcal{X} \to \Re^q$, $W : \mathcal{X} \to \Re^{q \times m}$ and some appropriate integer $q \leq k + 1$. Subsituting now the expression for $\delta(.,.)$ in (3.16) gives

$$
-\psi_{a,x}(x)f(x) - \psi_{a,x}(x)g(x)u + h^T(x)Qh(x) + 2h^T(x)\hat{S}(x)u + u^T \hat{R} u = l^T(x)l(x) + 2l^T(x)W(x)u + u^T W^T(x)W(x)u
$$

for all $x \in \mathcal{X}$, $u \in \mathcal{U}$. Finally, equating coefficients of corresponding terms in u gives (3.13) with $\psi = \psi_a$. $\square$

Remark 3.1.8 *The system of equations (3.13) is also referred to as the differential or infinitesimal version of the dissipation-inequality (3.4). A more general case of this differential version for affine nonlinear systems of the form Σ^a with any supply rate $s(u, y)$ and a C^1 storage-function is:*

$$\Psi_x(f(x) + g(x)u) \leq s(u, y). \tag{3.17}$$

Furthermore, a special case of the above result of Theorem 3.1.4 is for finite-gain systems (see [157, 234, 268] for a definition). For this class of nonlinear systems we set $Q = -I$, $S = 0$ and $R = \gamma^2 I$, where the scalar $\gamma > 0$ is the gain of the system. Then (3.13) becomes

$$\left. \begin{array}{rcl} \psi_x(x)f(x) & = & -h^T(x)h(x) - l^T(x)l(x) \\ \frac{1}{2}\psi_x(x)g(x) & = & -h^T(x)d(x) - l^T(x)W(x) \\ \gamma^2 I - d^T(x)d(x) & = & W^T(x)W(x) \end{array} \right\} \tag{3.18}$$

which is the generalization of the Bounded-real lemma for linear systems [30]. This condition gives necessary and sufficient condition for the nonlinear system Σ^a to have finite $\mathcal{L}_2$-gain $\leq \gamma$. We shall discuss this issue in Chapter 5, and the reader is also referred to [135] for additional discussion.

The following corollary is immediate from the theorem.

Corollary 3.1.1 *If the system Σ^a is dissipative with respect to the quadratic supply-rate defined in Theorem 3.1.4 and under the same hypotheses as in the theorem, then there exists a real positive-semidefinite function $\psi(.)$, $\psi(0) = 0$ such that*

$$\frac{d\psi(x)}{dt} = -[l(x) + W(x)u]^T[l(x) + W(x)u] + s(u, y). \tag{3.19}$$

Example 3.1.1 *[274]. Consider the model of an elastic system (or a capacitive network) defined by the state-space equations*

$$\dot{x} = u, \quad y = g(x).$$

where $x \in \Re^n$, $u \in \Re^p$, $g : \Re^n \to \Re^m$ is Lipschitz and all the variables have their previous meanings. The problem is to derive conditions under which the above system is dissipative with respect to the supply-rate $s(u, y) = u^T y$.

Assuming the storage-function for the system is C^2 (by the Lipschitz assumption on g), then applying the results of Theorem 3.1.4 and Remark 3.1.8, we have the dissipation-inequality

$$\psi_x(x).u \leq u^T g(x) \quad \forall u \in \Re^\mu, \quad x \in \Re^n.$$

Then, clearly the above inequality will only hold if and only if $g(x)$ is the gradient of a scalar function. This implies that

$$\frac{\partial g_i(x)}{\partial x_j} = \frac{\partial g_j(x)}{\partial x_i} \quad \forall i, j = 1, \ldots, m \tag{3.20}$$

and the storage-function can be expressed explicitly and uniquely as

$$\psi(x) = \int_{x_0}^{x} g^T(\alpha)d\alpha.$$

Example 3.1.2 *[274]. Consider the previous example with a damping term introduced in the output function which can be described by the equations:*

$$\dot{x} = u, \quad y = g_1(x) + g_2(u); \quad g_2(0) = 0,$$

where $x \in \Re^n$, $u \in \Re^n$, $g_1 : \Re^n \to \Re^m$, $g_2 : \Re^n \to \Re^m$ is Lipschitz. Then if we again assume $\psi \in C^2$, we have the dissipation-inequality

$$\psi_x(x).u \le u^T(g_1(x) + g_2(u)) \quad \forall u \in \Re^p, x \in \Re^n.$$

Thus, the system is dissipative as in the previous example if and only if the condition (3.20) holds for $g = g_1$ and in addition, $u^T g_2(u) \ge 0$. The dissipation-rate $\delta(x, u)$ is also unique and is given by $\delta(x, u) = u^T g_2(u)$.

3.1.1 Stability of Continuous-Time Dissipative Systems

In the previous two sections we have defined the concept of dissipativity of general nonlinear systems Σ and specialized it to affine systems of the form Σ^a. We have also derived necessary and sufficient conditions for the system to be dissipative with respect to some supply-rate. In this section, and motivated by the results in [273], we discuss the implications of dissipativity on stability of the system, i.e., if the system is dissipative with respect to a supply-rate $s(u, y)$ and some storage-function ψ, under what condition can ψ be a Lyapunov-function for the system, and hence guarantee its stability?

Consequently, we would like to first investigate under what conditions is ψ positive-definite, i.e., $\psi(x) > 0$ for all $x \ne 0$? We begin with the following definition.

Definition 3.1.8 *The system (3.12) is said to be locally zero-state observable in $N \subset \mathcal{X}$ containing $x = 0$, if for any trajectory of the system starting at $x_0 \in N$ and such that $u(t) \equiv 0$, $y(t) \equiv 0$ for all $t \ge t_0$, implies $x(t) \equiv 0$.*

We now show that, if the system Σ^a is dissipative with respect to the quadratic supply-rate $s_1(., .)$ and zero-state observable, then the following lemma guarantees that $\psi(.) > 0$ for all $x \in \mathcal{X}$, $x \ne 0$.

Lemma 3.1.1 *If the system Σ^a is dissipative with a C^1 storage-function with respect to the quadratic supply-rate, $s(u, y) = y^T Q y + 2y^T S u + u^T R u$, and zero-state observable, then all solutions of (3.13) are such that $\psi(.) > 0$ for all $x \ne 0$, i.e., Ψ is positive-definite.*

Proof: $\psi(x) \ge \psi_a(x) \ge 0 \ \forall x \in \mathcal{X}$ has already been established in Theorem 3.1.1. In addition, $\psi_a(x) \equiv 0$ implies from (3.14) that $u \equiv 0$, $y \equiv 0$. This by zero-state observability implies $x \equiv 0$. $\square$

We now have the following main stability theorem.

Theorem 3.1.5 *Suppose Σ^a is locally dissipative with a C^1 storage-function with respect to the quadratic supply-rate given in Lemma 3.1.1, and is locally zero-state observable. Then the free system Σ^a with $u \equiv 0$, i.e., $\dot{x} = f(x)$ is locally stable in the sense of Lyapunov if $Q \le 0$, and locally asymptotically-stable if $Q < 0$.*

Proof: By Lemma 3.1.1 and Corollary 3.1.1, there exists a locally positive-definite function $\psi(.)$ such that

$$\frac{d\psi(x)}{dt} = -l^T(x)l(x) + h^T(x)Qh(x) \le 0$$

along the trajectories of $\dot{x} = f(x)$ for all $x \in N \subset \mathcal{X}$. Thus $\dot{x} = f(x)$ is stable for $Q \le 0$

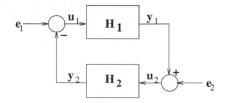

FIGURE 3.1

Feedback-Interconnection of Dissipative Systems

and locally asymptotically-stable if $Q < 0$ by Lyapunov's theorem, Theorem 1.3.1 (see also [157, 268]). □

Remark 3.1.9 *The above theorem implies that a finite-gain system (see [157, 234, 268] for a definition) which is dissipative with a C^1 storage-function with respect to the supply-rate $s_2(u,y) = \gamma^2 u^T u - y^T y$ (for which we can choose $Q = -I$, $S = 0$, $R = \gamma^2 I$ in $s_1(.,.)$) if zero-state observable, is locally asymptotically-stable. Moreover, local stability in the theorem can be replaced by global stability by imposing global observability.*

3.1.2   Stability of Continuous-Time Dissipative Feedback-Systems

In this subsection we continue with the discussion in the previous section on the implications of dissipativity on the stability of feedback systems. We consider the feedback connection shown in Figure 3.1, where $H_1 : u_1 \mapsto y_1$, $H_2 : u_2 \mapsto y_2$ are the input-output maps of the subsystems whose state-space realizations are given by

$$\Sigma_i : \begin{cases} \dot{x}_i &= f_i(x_i) + g_i(x_i)u_i, \quad f_i(0) = 0 \\ y_i &= h_i(x_i) + d_i(x_i)u_i, \quad h_i(0) = 0 \end{cases} \tag{3.21}$$

$i = 1, 2$, where e_1, e_2 are external commands or disturbances, and all the other variables have their usual meanings. Moreover, for the feedback-system to be well defined, it is necessary that $(I + d_2(x_2)d_1(x_1))$ be nonsingular for all x_1, x_2.

Now assuming the two subsystems H_1, H_2 are dissipative, then the following theorem gives a condition under which the feedback-system is stable.

Theorem 3.1.6 *Suppose the subsystems H_1, H_2 are zero-state observable and dissipative with respect to the quadratic supply-rates $s_i(u_i, y_i) = y_i^T Q_i y_i + 2 y_i^T S_i u_i + u_i^T R_i u_i$, $i = 1, 2$. Then the feedback-system (Figure 3.1) is stable (asymptotically-stable) if the matrix*

$$\hat{Q} = \begin{bmatrix} Q_1 + \lambda R_2 & -S_1 + \lambda S_2^T \\ -S_1^T + \lambda S_2 & R_1 + \lambda Q_2 \end{bmatrix} \tag{3.22}$$

is negative-semidefinite (negative-definite) for some $\lambda > 0$.

Proof: Consider as a Lyapunov-function candidate the linear combination of the storage-functions for H_1 and H_2:

$$V(x_1, x_2) = \psi_1(x_1) + \lambda \psi_2(x_2).$$

Since H_1, H_2 are dissipative and zero-state observable, by Corollary 3.1.1, there exist real functions $\psi_i(.) > 0$, $\psi_i(0) = 0$, $l_i(.)$, $W_i(.)$, $i = 1, 2$ such that

$$\frac{d\psi_i(x)}{dt} = -[l_i(x) + W_i(x)u]^T[l_i(x) + W_i(x)u] + s_i(u_i, y_i), \quad i = 1, 2.$$

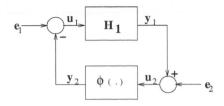

FIGURE 3.2
Feedback-Interconnection of Dissipative Systems

Setting $e_1 = e_2 = 0$, $u_1 = -y_2$, $u_2 = y_1$, and differentiating $V(.,.)$ along a trajectory of the feedback-system we get

$$\frac{dV(x_1, x_2)}{dt} \leq s_1(u_1, y_1) + \lambda s_2(u_2, y_2) = [y_1^T \ y_2^T]\hat{Q}\left[\begin{array}{c} y_1 \\ y_2 \end{array}\right].$$

The result now follows from standard Lyapunov argument. □

Corollary 3.1.2 *Take the same assumptions as in Theorem 3.1.6, and suppose $\hat{Q} \leq 0$, $S_1 = \lambda S_2^T$. Then the feedback-system (Figure 3.1) is asymptotically-stable if either of the following conditions hold:*

(i) *the matrix $(Q_1 + \lambda R_2)$ is nonsingular and the composite system $H_1(-H_2)$ is zero-state observable;*

(ii) *the matrix $(R_1 + \lambda Q_2)$ is nonsingular and the composite system $H_2 H_1$ is zero-state observable.*

Proof: (i) If $\hat{Q} \leq 0$, then $\dot{V}(x_1, x_2) \leq 0$ with equality only if $y_1 = 0$. Thus, if $(Q_1 + \lambda R_2)$ is nonsingular and $H_1(-H_2)$ is zero-state observable, then $y_1(t) \equiv 0$ implies that $x_1(t) \equiv x_2(t) \equiv 0$ and asymptotic stability follows from LaSalle's invariance-principle. Case (ii) also follows using similar arguments. □

The results of the previous theorem can also be extended to the case where H_2 is a memoryless nonlinearity, known as the *Luré* problem in the literature as shown in Figure 3.2. Assume in this case that the subsystem H_2 is defined by the input-output relation

$$H_2 \ : \ y_2 = \phi(u_2)$$

where $\phi : \mathcal{U} \rightarrow \mathcal{Y}$ is an unknown memoryless nonlinearity, but is such that the feedback-interconnection is well defined, while H_1 is assumed to have still the representation (3.21). Assume in addition that H_1 is dissipative with respect to the quadratic supply-rate

$$s_1(u_1, y_1) = y_1^T Q_1 y_1 + 2y_1^T S_1 u_1 + u_1^T R_1 u_1 \tag{3.23}$$

and H_2 is dissipative in the sense that

$$s_2(u_2, y_2) = y_2^T Q_2 y_2 + 2y_2^T S_2 u_2 + u_2^T R_2 u_2 \geq 0 \ \forall u_2, \tag{3.24}$$

then the feedback-system in Figure 3.2 will be stable (asymptotically-stable) as in Theorem 3.1.6 and Corollary 3.1.2 with some additional constraints on the nonlinearity $\phi(.)$. The following result is a generalized version of the *Luré* problem.

Theorem 3.1.7 *Suppose $d_1(x_1) = 0$ for the subsystem H_1 in Figure 3.2, the subsystem is* *zero-state observable and dissipative with a C^1 storage-function with respect to the quadratic supply-rate $s_1(.,.)$ (3.23). Let $\phi(.)$ satisfy*

(i) *the condition (3.24); and*

(ii) *$\frac{\partial}{\partial u}\{\phi(u)\} = \frac{\partial^T}{\partial u}\{\phi(u)\}$.*

Then, the feedback-system (Figure 3.2) is stable (asymptotically-stable) if $\hat{Q} \le 0$ ($\hat{Q} < 0$), where $\hat{Q}$ is as given in equation (3.22).

Proof: Consider the Lyapunov-function candidate

$$V(x_1, x_2) = \psi_1(x_1) + \lambda\psi_2(x_2), \quad \lambda > 0$$

where x_2 is a fictitious state and $\psi_1(.)$, $\psi_2(.)$ are C^1-functions. Taking the time-derivative of $V(.,.)$ along the trajectories of the feedback-system and by Corollary 3.1.1,

$$\dot{V}(x_1, x_2) \le s_1(u_1, y_1) + \lambda s_2(u_2, y_2), \quad \lambda > 0.$$

Finally, setting $e_1 = e_2 = 0$, $u_2 = y_1$ and $u_1 = -\phi(y_1)$, we get

$$\dot{V}(x_1, x_2) \le [y_1^T \ y_2^T]\hat{Q} \begin{bmatrix} y_1 \\ y_2 \end{bmatrix}.$$

The result now follows as in Theorem 3.1.6 and using standard Lyapunov arguments. $\square$

Remark 3.1.10 *Condition (ii) in the theorem above ensures that ϕ is the gradient of a scalar function.*

Notice also that the conditions on $\hat{Q}$ are the same as in Theorem 3.1.6. Similarly, the following corollary is the counterpart of corollary 3.1.2 for the memoryless nonlinearity.

Corollary 3.1.3 *Take the same assumptions as in Theorem 3.1.7 above, suppose $\hat{Q} \le 0$, $S_1 = \lambda S_2^T$. Then the feedback (Figure 3.2) is asymptotically-stable if either one of the following conditions holds:*

(i) *the matrix $Q_1 + \lambda R_2$ is nonsingular and $\phi(0) = 0$; or*

(ii) *the matrix $R_1 + \lambda Q_2$ is nonsingular and $\phi(\sigma) = 0$ implies $\sigma = 0$.*

Proof: The proof follows using the same arguments as in Corollary 3.1.2. $\square$

Remark 3.1.11 *Other interesting corollaries of Theorem 3.1.7 could be drawn for different configurations of the subsystems H_1 and H_2 including the case of a time-varying nonlinearity $\phi(t, x)$. In this case, the corresponding assumption of uniform zero-state observability will then be required.*

3.2 $\mathcal{L}_2$-Gain Analysis for Continuous-Time Dissipative Systems

In this section, we discuss the relationship between the dissipativeness of a system and its finite $\mathcal{L}_2$-gain stability. Again we consider the affine nonlinear system Σ^a defined by the state equations (3.12). We then have the following definition.

Definition 3.2.1 *The nonlinear system Σ^a is said to have locally $\mathcal{L}_2$-gain $\leq \gamma > 0$, in $N \subset \mathcal{X}$, if*

$$\int_{t_0}^{T} \|y(t)\|^2 dt \leq \gamma^2 \int_{t_0}^{T} \|u(t)\|^2 dt + \beta(x_0) \tag{3.25}$$

for all $T \geq t_0$, $x_0 \in N$, $u \in \mathcal{L}_2[t_0, T]$, $y(t) = h(x(t, t_0, x_0, u)) + d(x(t, t_0, x_0, u))u$, $t \in [t_0, T]$, and for some function $\beta : N \to \Re$, $\beta(0) = 0$ [268]. Moreover, the system is said to have $\mathcal{L}_2$-gain $\leq \gamma > 0$ if $N = \mathcal{X}$.

Without any loss of generality, we can take $t_0 = 0$ in the above definition henceforth. We now have the following important theorem which is also known as the *Bounded-real lemma* for continuous-time nonlinear systems.

Theorem 3.2.1 *Consider the system Σ^a and suppose $d(x) = 0$. Let $\gamma > 0$ be given, then we have the following list of implications: (a) $\to$ (b) $\leftrightarrow$ (c) $\to$ (d), where:*

(a) *There exists a smooth solution $V : \mathcal{X} \to \Re_+$ of the Hamilton-Jacobi equation (HJE):*

$$V_x(x)f(x) + \frac{1}{2\gamma^2}V_x(x)g(x)g^T(x)V_x^T(x) + \frac{1}{2}h^T(x)h(x) = 0, \quad V(0) = 0. \tag{3.26}$$

(b) *There exists a smooth solution $V \geq 0$ of the Hamilton-Jacobi inequality (HJI):*

$$V_x(x)f(x) + \frac{1}{2\gamma^2}V_x(x)g(x)g^T(x)V_x^T(x) + \frac{1}{2}h^T(x)h(x) \leq 0, \quad V(0) = 0. \tag{3.27}$$

(c) *There exists a smooth solution $V \geq 0$ of the dissipation-inequality*

$$V_x(x)f(x) + V_x(x)g(x)u \leq \frac{1}{2}\gamma^2\|u\|^2 - \frac{1}{2}\|y\|^2, \quad V(0) = 0 \tag{3.28}$$

for all $u \in \mathcal{U}$, with $y = h(x)$.

(d) *The system has $\mathcal{L}^2 - gain \leq \gamma$.*

Conversely, suppose (d) holds, and the system is reachable from $\{0\}$, then the available-storage V_a and required-supply V_r defined by

$$V_a(x) = - \inf_{\substack{x_0 = x, u \in \mathcal{L}_2(0,T), \\ T \geq 0}} \int_0^T (\gamma^2\|u(t)\|^2 - \|y(t)\|^2)dt \tag{3.29}$$

$$V_r(x) = \inf_{\substack{x_0 = x, u \in \mathcal{L}_2(t_{-1},0), \\ t_{-1} \geq 0, x(t_{-1}) = 0}} \int_{t_{-1}}^0 (\gamma^2\|u(t)\|^2 - \|y(t)\|^2)dt \tag{3.30}$$

respectively, are well defined functions for all $x \in \mathcal{X}$. Moreover they satisfy

$$0 \leq V_a \leq V \leq V_r, \quad V_a(0) = V(0) = V_r(0) = 0, \tag{3.31}$$

for any solution V of (3.26), (3.27), (3.28). Further, the functions V_a, V_r are also storage-functions and satisfy the integral dissipation-inequality

$$V(x(t_1)) - V(x(t_0)) \leq \frac{1}{2}\int_{t_0}^{t_1} (\gamma^2\|u(t)\|^2 - \|y(t)\|^2)dt \tag{3.32}$$

for all $t_1 \geq t_0$ and all $u \in \mathcal{L}_2(t_0, t_1)$, where $x(t_1) = x(t_1, t_0, x_0, u)$. In addition, if V_a and V_r are smooth, then they satisfy the HJE (3.26).

Proof: (Sketch) $(a) \to (b)$ is easy. For $(b) \leftrightarrow (c)$: Let V satisfy (3.27), then by completing the squares we can rewrite (3.27) [264] as

$$V_x(x)f(x) + V_x(x)g(x)u \leq \frac{1}{2}\gamma^2\|u\|^2 - \frac{1}{2}\|y\|^2 - \frac{1}{2}\gamma^2\|u - \frac{1}{\gamma^2}g^T V_x^T(x)\|^2$$

from which (3.28) follows. Conversely, let V satisfy (3.28). Then again by completing the squares, we can write (3.28) as

$$V_x(x)f(x) + \|y\|^2 \leq \frac{1}{2}\gamma^2\|u - \frac{1}{\gamma^2}g^T(x)V_x^T(x)\|^2 - \frac{1}{2\gamma^2}V_x(x)g(x)g^T(x)V_x^T(x)$$

for all $u \in \mathcal{U} = \Re^p$. In particular, for $u = \frac{1}{\gamma^2}g^T(x)V_x^T(x)$, we get (3.27).

$\quad$ $(c) \to (d)$: By integrating (3.28) from t_0 to some $T > t_0$ we get the dissipation-inequality (3.32). Since $V \geq 0$ the latter implies $\mathcal{L}^2$-gain $\leq \gamma$. Finally, the fact that V_a and V_r satisfy the dissipation-inequality (3.28) and are well defined has already been proved in Theorems 3.1.1 and 3.1.2 respectively. Notice also that $\sup -(.) = -\inf(.)$. If in addition, V_a, V_r are smooth, then by the dynamic-programming principle, they satisfy the HJE (3.26). $\square$

$\quad$ The implications of dissipativeness and finite $\mathcal{L}_2$-gain on asymptotic-stability are summarized in the following theorem.

Theorem 3.2.2 *Assume the system Σ^a is zero-state observable, $d(x) = 0$, and there is a smooth solution $V \geq 0$ of (3.26), or (3.27), or (3.28). Then, $V > 0$ for all $x \neq \{0\}$ and the free-system $\dot{x} = f(x)$ is locally asymptotically-stable. If however V is proper, i.e., for each $c > 0$ the level set $\{x \in \mathcal{X} \mid 0 \leq V \leq c\}$ is compact, then the free-system $\dot{x} = f(x)$ is globally asymptotically-stable.*

Proof: If Σ^a is zero-state observable then any $V \geq 0$ satisfying (3.28) is such that $V \geq V_a(x) > 0 \, \forall x \neq 0$ by Lemma 3.1.1 and Theorem 3.1.1. Then for $u(t) \equiv 0$ (3.28) implies $V_x f \leq -\frac{1}{2}\|h\|^2$, and local (global) asymptotic-stability follows by LaSalle's invariance-principle (Theorem 1.3.3). $\square$

Remark 3.2.1 *More general results about the relationship between finite $\mathcal{L}^p$-gain and dissipativity on the one hand and asymptotic-stability on the other hand, can be found in reference [132].*

3.3   Continuous-Time Passive Systems

Passive systems are a class of dissipative systems which are dissipative with respect to the supply-rate $s(u, y) = \langle u, y \rangle$. Passivity plays an important role in the analysis and synthesis of networks and systems. The concept of passivity also arises naturally in many areas of science and engineering. Positive-realness which corresponds to passivity for linear time-invariant dynamical systems, proved to be very essential in the development of major control systems results, particularly in adaptive control; positive-realness guarantees the stability of adaptive control systems. The practical significance of positive-real systems is that they represent energy dissipative systems, such as large space structures with colocated rate sensors and actuators. They are both input-output stable, and stable in the sense of

Lyapunov. Furthermore, a negative feedback interconnection of two positive-real systems is always internally stable.

In this section, we study passive systems and their properties. We present the results for the general nonlinear case first, and then specialize them to linear systems. We begin with the following formal definition.

Definition 3.3.1 *The nonlinear system Σ^a is said to be passive if it is dissipative with respect to the supply-rate $s(u,y) = y^T u$ and there exists a storage-function $\Psi : \mathcal{X} \to \Re$ such that $\Psi(0) = 0$. Hence, for a passive system, the storage-function satisfies*

$$\Psi(x_1) - \Psi(x_0) \le \int_{t_0}^{t_1} y^T u \, dt, \quad \Psi(0) = 0 \tag{3.33}$$

for any $t_1 \ge t_0$ and $x_1 = x(t_1) \in \mathcal{X}$.

Remark 3.3.1 *If we set $u = 0$ in equation (3.33), we see that $\Psi(.)$ is decreasing along any trajectory of the free-system $\Sigma^a(u = 0) : \dot{x} = f(x)$. Hence, if Σ^a is passive with a positive-definite storage-function $\Psi(.)$, then it is stable in the sense of Lyapunov. In fact, if $\Psi(.)$ is C^1 and positive-definite, then differentiating it along any trajectory of the system will yield the same conclusion. However, $\Psi(.)$ may not necessarily be C^1. Similarly, setting $y = 0$ in (3.33) will also imply that Ψ is decreasing along any trajectory of the "clamped system" $\Sigma^a(y = 0) : \dot{x} = f(x) + g(x)u$. The dynamics of this resulting system is the internal dynamics of the system Σ^a and is known as its zero-dynamics. Consequently, it implies that a passive system with positive-definite storage-function has a Lyapunov-stable zero-dynamics or is weakly minimum-phase [234, 268].*

Remark 3.3.2 *A passive system is also said to be strictly-passive if the strict inequality holds in (3.33), or equivalently, there exists a positive-definite function $\Xi : \mathcal{X} \to \Re_+$ such that for all $u \in \mathcal{U}$, $x_0 \in \mathcal{X}$, $t_1 \ge t_0$,*

$$\Psi(x_1) - \Psi(x_0) = \int_{t_0}^{t_1} y^T u \, dt - \int_{t_0}^{t_1} \Xi(x(t)) dt$$

where $x_1 = x(t_1)$.

The following theorem is the equivalent of Theorem 3.1.4 for passive systems.

Theorem 3.3.1 *A necessary and sufficient condition for the system Σ^a to be passive is that there exists a C^1-positive-semidefinite storage-function $\psi : \mathcal{X} \to \Re$, $\psi(0) = 0$, and C^0-functions $l : \mathcal{X} \to \Re^q$, $W : \mathcal{X} \to \Re^{q \times m}$ satisfying*

$$\left. \begin{array}{rcl} \psi_x(x)f(x) & = & -l^T(x)l(x) \\ \frac{1}{2}\psi_x(x)g(x) & = & h^T(x) - l^T(x)W(x) \\ d(x) + d^T(x) & = & W^T(x)W(x) \end{array} \right\} \tag{3.34}$$

for all $x \in \mathcal{X}$. Moreover, if $d(.)$ is constant, then W may be constant too.

Proof: The proof follows along similar lines as Theorem 3.1.4. $\square$

Example 3.3.1 *[131]. Consider the second-order system*

$$\ddot{x} + \alpha(x)\dot{x} + \beta(x) = u$$

where $x \in \Re$, $\alpha, \beta : \Re \to \Re$ are locally Lipschitz, $u \in \Re$ is the input to the system. This model represents many second-order systems including the Van-der-Pol equation. Letting $x_1 = x$, $x_2 = \dot{x}$, $F(x) := \int_0^x \alpha(\sigma)d\sigma$, the system can be represented in state-space as

$$\dot{x}_1 = -F(x_1) + x_2 \tag{3.35}$$

$$\dot{x}_2 = -\beta(x_1) + u. \tag{3.36}$$

Suppose also the output of the system is defined by

$$y = ax_2 - bF(x_1), \ 0 \le b \le a \in \Re.$$

For convenience, define also $G(x) := \int_0^x \beta(\sigma)d\sigma$. Then it can be checked using the conditions (3.34), that for $a = \frac{1}{2}$ and $b = 0$, the system is passive with storage-function

$$\Psi_1(x) = \frac{1}{2}(\dot{x}_1 + F(x_1))^2 + G(x_1)$$

if

$$G(x_1) \ge 0, \ \text{and} \ \beta(x_1)F(x_1) \ge 0.$$

And for $a = b = \frac{1}{2}$, if

$$G(x_1) \ge 0, \ \alpha(x_1) \ge 0$$

with storage-function

$$\Psi_2(x) = \frac{1}{2}\dot{x}_1^2 + G(x_1).$$

In the case of the linear time-invariant (LTI) system:

$$\Sigma^l : \begin{cases} \dot{x} = Ax + Bu \\ y = Cx + Du \end{cases} \tag{3.37}$$

where $A \in \Re^{n \times n}$, $B \in \Re^{n \times m}$, $C \in \Re^{m \times n}$, and $D \in \Re^{m \times m}$, the system is passive with storage-function $\psi(x) = \frac{1}{2}x^T Px$, for some matrix $P = P^T \ge 0$. The conditions (3.34) reduce to the following matrix equations [30, 31] also known as the *positive-real lemma*

$$\left. \begin{array}{rcl} A^T P + PA &=& -L^T L \\ PB &=& C^T - L^T W \\ D + D^T &=& W^T W \end{array} \right\} \tag{3.38}$$

for some constant matrices L, W of appropriate dimensions. Furthermore, the available-storage and required-supply may be evaluated by setting up the infinite-horizon variational problems:

$$\Psi_a(x_0) = -\lim_{t_1 \to \infty} \inf_{u \in \mathcal{U}} \int_0^{t_1} \langle u(t), y(t) \rangle dt, \ \text{subject to } \Sigma^l, \ x(0) = x_0,$$

$$\Psi_r(x_0) = \lim_{t_{-1} \to -\infty} \inf_{u \in \mathcal{U}} \int_{t_{-1}}^0 \langle u(t), y(t) \rangle dt, \ \text{subject to } \Sigma^l, \ x(t_{-1}) = 0,$$

$$x(0) = x_0.$$

Such infinite-horizon least-squares optimal control problems have been studied in Willems [272] including the fact that because of the nature of the cost function, the optimization problem may be singular. This is especially so whenever $D + D^T$ is singular. However, if we assume that $D + D^T$ is nonsingular, then the problem reduces to a standard optimal control problem which may be solved by finding a suitable solution to the algebraic-Riccati equation (ARE):

$$PA + A^T P + (PB - C^T)(D + D^T)^{-1}(B^T P - C) = 0. \tag{3.39}$$

The following lemma, which is an intermediate result, gives a necessary and sufficient condition for Σ^l to be passive in terms of the solution to the ARE (3.39).

Lemma 3.3.1 *Assume (A, B, C, D) is a minimal realization for Σ^l and $D + D^T$ is nonsingular. Then the ARE (3.39) has a real symmetric positive-semidefinite solution if and only if Σ^l is passive. In this case, there exists a unique real symmetric solution P^- to the ARE such that $A^- = A + B(D + D^T)^{-1}(B^T P^- - C)$ has $Re\,\lambda_i(A^-) \leq 0$, $i = 1, \ldots, n$ and similarly a unique real symmetric solution P^+ to the ARE such that $A^+ = A + B(D + D^T)^{-1}(B^T P^+ - C)$ has $Re\,\lambda_i(A^+) \geq 0$, $i = 1, \ldots, n$. Moreover, $P^- \leq P \leq P^+$.*

Proof: The proof of the lemma is lengthy but can be found in [272] (see Lemma 2). $\square$

Remark 3.3.3 *If we remove the assumption $D + D^T$ is nonsingular in the above lemma, then the problem becomes singular and the proof of the lemma becomes more complicated; nevertheless it can be tackled using a limiting process (see reference [272]).*

The following theorem then gives estimates of the available-storage and required-supply.

Theorem 3.3.2 *Suppose (A, B, C, D) is a minimal-realization of Σ^l which is passive. Then the solutions P^+ and P^- of the ARE (3.39) given in Lemma 3.3.1 and Remark 3.3.3 are well defined, and the available-storage and the required-supply are given by $\Psi_a(x) = \frac{1}{2}x^T P^- x$ and $\Psi_r(x) = \frac{1}{2}x^T P^+ x$ respectively. Moreover, there exist $\epsilon > 0$ and a number $c > 0$ such that $\epsilon\|x\|^2 \leq \Psi_a(x) \leq \Psi(x) \leq \Psi_r(x) \leq c\|x\|^2$.*

Proof: Again the proof of the above theorem is lengthy but can be found in [274]. $\square$

The conditions (3.38) of Theorem 3.3.1 for the linear system Σ^l can also be restated in terms of matrix inequalities and related to the solutions P^- and P^+ of the ARE (3.39) in the following theorem.

Theorem 3.3.3 *Suppose (A, B, C, D) is a minimal-realization of Σ^l. Then the matrix inequalities*

$$\begin{bmatrix} A^T P + PA & PB - C^T \\ B^T P - C & -D - D^T \end{bmatrix} \leq 0, \quad P = P^T \geq 0 \tag{3.40}$$

have a solution if, and only if, Σ^l is passive. Moreover, $\Psi(x) = \frac{1}{2}x^T P x$ is a quadratic storage-function for the system, and the solutions P^- and P^+ of (3.39) also satisfy (3.40) such that $0 < P^- \leq P \leq P^+$.

Proof: Differentiating $\Psi(x) = \frac{1}{2}x^T P x$ along the solutions of $\dot{x} = Ax + Bu$, gives

$$\dot{\Psi}(x) = \frac{1}{2}x^T(A^T P + PA)x + u^T B^T P x.$$

Substituting this in the dissipation-inequality (3.33) for the system Σ^l implies that $\Psi(x) = \frac{1}{2}x^T P$ must satisfy the inequality

$$\frac{1}{2}x^T(A^T P + PA)x + u^T B^T P x - x^T C^T u - u^T D u \leq 0,$$

which is equivalent to the inequality (3.40). Finally, if (A, B, C, D) is minimal and passive, then by Theorem 3.3.2, P^- and P^+ exist such that

$$\frac{1}{2}x^T P^- x = \Psi_a(x) \le \Psi(x) = \frac{1}{2}x^T P x \le \frac{1}{2}x^T P^+ x = \Psi_r(x). \ \square$$

Remark 3.3.4 *A frequency-domain criteria for passivity (positive-realness) in terms of the transfer-function $G(s) = D + C(sI - A)^{-1}B$ also exists [30, 31, 274].*

Definition 3.3.2 *A passive system Σ^a is lossless if there exists a storage-function $\Psi : \mathcal{X} \to \Re$, $\Psi(0) = 0$ such that for all $u \in \mathcal{U}$, $x_0 \in \mathcal{X}$, $t_1 \ge t_0$,*

$$\Psi(x_1) - \Psi(x_0) = \int_{t_0}^{t_1} y^T u \, dt, \tag{3.41}$$

where $x_1 = x(t_1, t_0, x_0, u)$.

In the case of the linear system Σ^l, we can state the following theorem for losslessness.

Theorem 3.3.4 *Assume (A, B, C, D) is a minimal-realization of Σ^l. Then Σ^l is lossless if and only if there exists a unique solution $P = P^T \ge 0$ of the system*

$$\left. \begin{array}{rcl} A^T P + PA & = & 0 \\ PB & = & C^T \\ D + D^T & = & 0. \end{array} \right\} \tag{3.42}$$

Moreover, if P exists, then $P = P^+ = P^-$ and defines a unique storage-function $\Psi(x) = \frac{1}{2}x^T P x$.

Proof: Proof follows from Theorem 3.3.1 and the system (3.38) by setting $L = 0$, $W = 0$. $\square$

A class of dissipative systems that are closely related and often referred to as passive systems is called *positive-real*.

Definition 3.3.3 *The system Σ^a is said to be positive-real if for all $u \in \mathcal{U}$, $t_1 \ge t_0$,*

$$\int_{t_0}^{t_1} y^T u \, dt \ge 0$$

whenever $x_0 = 0$.

Remark 3.3.5 *A dissipative system Σ^a with respect to the supply-rate $s(u, y) = u^T y$ is positive-real if and only if its available-storage satisfies $\Psi_a(0) = 0$. Consequently, a passive system is positive-real, and conversely, a positive-real system with a C^0 available-storage in which any state is reachable from the origin $x = 0$, is passive.*

Linear positive-real systems satisfy a celebrated property called the Kalman-Yacubovitch-Popov (KYP) lemma [268]. A nonlinear version of this lemma can also be stated.

Definition 3.3.4 *A passive system Σ^a with $d(x) = 0$ is said to have the KYP property if there exists a C^1 storage-function $\Psi : \mathcal{X} \to \Re$, $\Psi(0) = 0$ such that*

$$\left. \begin{array}{rcl} \Psi_x(x)f(x) & \le & 0 \\ \Psi_x(x)g(x) & = & h^T(x). \end{array} \right\} \tag{3.43}$$

Remark 3.3.6 *The above relations (3.43) are nothing more than a restatement of the conditions (3.34) which are also the differential or infinitesimal version of the dissipation-inequality (3.33). Similarly, for a lossless system with C^1 storage-function we have $\Psi_x f(x) = 0$ and for a strictly-passive system with C^1 storage-function, $\Psi_x f(x) = -\Xi(x) < 0$. Thus, a strictly passive system with a positive-definite storage-function has an asymptotically-stable equilibrium-point at the origin $x = 0$.*

Remark 3.3.7 *It follows from the above and as can be easily shown, that a system Σ^a which has the KYP property is passive with storage-function $\Psi(.)$, and conversely, a passive system with a C^1 storage-function has the KYP property.*

Remark 3.3.8 *For the case of the linear system*

$$\Sigma^l \; : \begin{cases} \dot{x} & = & Ax + Bu \\ y & = & Cx, \end{cases} \tag{3.44}$$

which is passive with the C^2 storage-function $\Psi(x) = \frac{1}{2}x^T P x$, $P = P \geq 0$, the KYP equations (3.43) yield

$$A^T P + PA = -Q$$
$$B^T P = C^T$$

for some matrix $Q \geq 0$.

The following corollary is the main stability result for passive systems. Additional results on passive systems could also be obtained by specializing most of the results in the previous section for the quadratic supply-rate to passive systems, in which $Q = R = 0$ and $S = I$.

Corollary 3.3.1 *A passive system of the form Σ^a with $u = 0$ is stable.*

Proof: Follows from Theorem 3.1.5 by taking $Q = R = 0$, $S = I$ in $s_1(.,.)$ to get

$$\frac{d\phi(x)}{dx} = -l^T(x)l(x).$$

The rest follows from a Lyapunov argument and the fact that $l(.)$ is arbitrary. $\square$

We consider the following example to illustrate the ideas.

Example 3.3.2 *[202]. Consider the second-order system described by the equation*

$$\ddot{x} + \alpha(x) + \beta(\dot{x}) = u, \quad y = \dot{x}$$

where $x \in \Re$, $\alpha, \beta : \Re \to \Re$ are locally Lipschitz, $u \in \Re$ is the input to the system, and $y \in \Re$ is the output of the system. In addition

$$\alpha(0) = 0, \; z\alpha(z) > 0, \; \beta(0) = 0, \; z\beta(z) > 0 \; \forall z \neq 0, \; z \in (-a, a) \in \Re.$$

Letting $x_1 = x$, $x_2 = \dot{x}$, we can represent the model above in state-space as

$$\begin{aligned} \dot{x}_1 & = & x_2 \\ \dot{x}_2 & = & -\alpha(x_1) - \beta(x_2) + u \\ y & = & x_2. \end{aligned}$$

Then it can be checked that the system is passive and satisfies the KYP conditions with the storage-function:

$$V(x) = \int_0^{x_1} \alpha(s)ds + \frac{1}{2}x_2^2.$$

Next, we discuss the problem of rendering the nonlinear system Σ^a, which may not be passive, to a passive system using feedback.

3.4   Feedback-Equivalence to a Passive Continuous-Time Nonlinear System

In this section, we discuss the problem of feedback-equivalence to a passive nonlinear system, i.e., how a nonlinear system can be rendered passive using static state-feedback control. Because passive systems represent a class of systems for which feedback analysis and design is comparatively simpler, more intuitive and better understood, the problem of feedback-equivalence to a passive nonlinear system is a less stringent yet very appealing version of the problem of feedback linearization. Graciously, this objective is achievable under some mild regularity assumption. We shall be considering the following smooth affine nonlinear system defined on the state-space manifold $\mathcal{X} \subseteq \Re^n$ in coordinates $x = (x_1, \ldots, x_n)$ with no feedthrough term:

$$\Sigma^a : \begin{cases} \dot{x} &= f(x) + g(x)u \\ y &= h(x), \end{cases} \tag{3.45}$$

where $x \in \mathcal{X}$, $g : \mathcal{X} \to \mathcal{M}^{n \times m}(\mathcal{X})$, $h : \mathcal{X} \to \Re^m$, i.e., Σ^a is square, and all the variables have their usual meanings. We begin with the following definitions (see also the references [234, 268] for further details).

Definition 3.4.1 *The nonlinear system Σ^a is said to have vector relative-degree $\{r_1, \ldots, r_m\}$ in a neighborhood $\mathcal{O}$ of $x = 0$ if*

$$L_{g_i} L_f^k h_i(x) = 0, \quad \text{for all } 0 \le k \le r_i - 2, \ i = 1, \ldots, m \tag{3.46}$$

and the matrix

$$A^r(x) = \begin{bmatrix} L_{g_1} L_f^{r_1-1} h_1 & \cdots & L_{g_m} L_f^{r_1-1} h_1 \\ \vdots & \ddots & \vdots \\ L_{g_1} L_f^{r_m-1} h_m & \cdots & L_{g_m} L_f^{r_m-1} h_m \end{bmatrix} \tag{3.47}$$

is nonsingular for all $x \in \mathcal{O}$, where,

$$L_\xi \phi = \frac{\partial \phi}{\partial x} \xi = (\nabla_x \phi(x))^T \xi, \quad \text{for any vector field } \xi : \mathcal{X} \to V^\infty(\mathcal{X}), \ \phi : \mathcal{X} \to \Re,$$

is the Lie-derivative of ϕ in the direction of ξ.

Remark 3.4.1 *Thus, the system Σ^a has vector relative-degree $\{1, \ldots, 1\}$ in $\mathcal{O}$ if the matrix $A^1(x) = L_g h(x)$ is nonsingular for all x in $\mathcal{O}$.*

Definition 3.4.2 *If the system Σ^a has relative-degree $r = r_1 + \ldots + r_m < n$, then it can be represented in the normal-form by choosing new coordinates as*

$$\zeta_1^j = h_j, \ j = 1, \ldots, m \tag{3.48}$$
$$\zeta_i^j = L_f^{i-1} h_i, \ i = 2, \ldots, r_i, \ j = 2, \ldots, m. \tag{3.49}$$

Then the r-differentials

$$dL_f^j h_i, \ j = 0, \ldots, r_i - 1, \ i = 1, \ldots, m,$$

are linearly independent. They can be completed into a basis, if the distribution $\{g_1, \ldots, g_m\}$

is involutive, by choosing an additional $n - r$ linearly independent real-valued functions $\eta_1, \ldots, \eta_{n-r}$ locally defined in a neighborhood of $x = 0$ and vanishing at $x = 0$ such that, in this new coordinate the system is represented in the following normal-form

$$\left. \begin{array}{rcl} \dot{\eta} & = & c(\eta, \zeta) \\ \dot{\zeta} & = & b(\eta, \zeta) + a(\eta, \zeta)u \\ y & = & [\zeta_1^1, \ldots, \zeta_1^m]^T \end{array} \right\} \tag{3.50}$$

for some functions $c(.,.)$, $b(.,.)$ and $a(.,.)$ which are related to f, g, h by the diffeomorphism $\Phi : x \mapsto (\zeta, \eta)$ and $a(\zeta, \eta)$ is nonsingular in the neighborhood of $(\eta, \zeta) = (0,0)$.

Definition 3.4.3 *The zero-dynamics of the system Σ^a is the dynamics of the system described by the condition $y = \zeta = 0$ in (3.50), and is given by*

$$\Sigma_0^a * \; : \; \dot{\eta} = c^*(\eta, 0). \tag{3.51}$$

Moreover, it evolves on the $(n - r)$-dimensional submanifold

$$\mathcal{Z}^* = \{x \in \mathcal{O} \; : \; y \equiv 0\}.$$

Remark 3.4.2 *If the system Σ^a has vector relative-degree $\{1, \ldots, 1\}$, in a neighborhood $\mathcal{O}$ of $x = 0$, then $r = m$ and the zero-dynamics manifold has dimension $n - m$.*

In the x coordinates, the zero-dynamics can be described by the system

$$\Sigma^a * \; : \; \dot{x} = f^*(x), \quad x \in \mathcal{Z}^* \tag{3.52}$$

where

$$f^*(x) = (f(x) + g(x)u^*)|_{\mathcal{Z}^*}$$

with

$$u^* = -[L_g h(x)]^{-1} L_f h(x).$$

We have already defined in Remark 3.3.1 a *minimum-phase* system as having an asymptotically-stable zero-dynamics. We further expand on this definition to include the following.

Definition 3.4.4 *The system Σ^a is said to be locally weakly minimum-phase if its zero-dynamics is stable about $\eta = 0$, i.e., there exists a $C^{\hat{r}}$, $\hat{r} \geq 2$ positive-definite function $W(\eta)$ locally defined around $\eta = 0$, with $W(0) = 0$ such that $L_{c^*}W \leq 0$ for all η around $\eta = 0$.*

The passivity of the system Σ^a also has implications on its vector relative-degree. The following theorem goes to show that if the system is passive with a C^2 storage-function, then it will necessarily have a nonzero vector relative-degree at any *regular point* (a point where $rank(L_g h(x))$ is constant).

In the sequel we shall assume without any loss of generality that $rank\{g(0)\} = rank\{dh(0)\} = m$.

Theorem 3.4.1 *Suppose Σ^a is passive with a C^2 positive-definite storage-function, and let the origin $x = 0$ be a regular point for the system. Then $L_g h(0)$ is nonsingular and the system has vector relative-degree $\{1, \ldots, 1\}$ locally at $x = 0$.*

Proof: The reader is referred to [77] for the proof. $\square$

Remark 3.4.3 *A number of interesting corollaries of the theorem can also be drawn. In particular, it can be shown that if either the storage-function is nondegenerate at $x = 0$*

(i.e., the Hessian matrix of V is nonsingular at $x = 0$) or $x = 0$ is a regular point of the system Σ^a, then the system necessarily has a vector relative-degree $\{1, \ldots, 1\}$ about $x = 0$ and is locally weakly minimum-phase.

The main result of this section is how to render a nonpassive system Σ^a into a passive one using a regular static state-feedback of the form

$$u = \alpha(x) + \beta(x)v, \tag{3.53}$$

where $\alpha(x)$ and $\beta(x)$ are smooth functions defined locally around $x = 0$, with $\beta(x)$ nonsingular for all x in this neighborhood. The necessary conditions to achieve this, are that the system is locally weakly minimum-phase and has vector relative-degree $\{1, \ldots, 1\}$. Moreover, these two properties are also invariant under static-feedback of the form (3.53), and it turns out that these two conditions are indeed also sufficient for "local feedback-equivalence to a passive system." We summarize the result in the following theorem.

Theorem 3.4.2 *Suppose $x = 0$ is a regular point for the system Σ^a. Then Σ^a is locally feedback-equivalent to a passive system with a C^2 storage-function if, and only if, it has vector relative-degree $\{1, \ldots, 1\}$ at $x = 0$ and is weakly minimum-phase.*

Proof: Choose as new state variables the outputs of the system $y = h(x)$ and complete them with the $n - m$ functions $\eta = \phi(x)$ to form a basis, so that in the new coordinates, the system Σ^a is represented as

$$
\begin{aligned}
\dot{\eta} &= c(\eta, y) + d(\eta, y)u \\
\dot{y} &= b(\eta, y) + a(\eta, y)u
\end{aligned}
$$

for some smooth functions $a(.,), b(.,.), c(.,.)$ and $d(.,.)$ of appropriate dimensions, with $a(.,.)$ nonsingular for all (η, y) in the neighborhood of $(0, 0)$. Now apply the feedback-control

$$u = a^{-1}(\eta, y)[-b(\eta, y) + v],$$

where v is an auxiliary input. Then the closed-loop system becomes

$$
\begin{aligned}
\dot{\eta} &= \theta(\eta, y) + \vartheta(\eta, y)v \\
\dot{y} &= v.
\end{aligned}
$$

Notice that, in the above transformation, the zero-dynamics of the system is now contained in the dynamics of the first equation $\dot{\eta}$. Finally, apply the following change of variables

$$z = \eta - \vartheta(\eta, 0)y$$

to get the resulting system

$$\dot{z} = f^*(z) + P(z, y)y + \left(\sum_{i=1}^{m} q_i(z, y)y_i \right) v \tag{3.54}$$

$$\dot{y} = v \tag{3.55}$$

where $f^\star(z)$ is the zero-dynamics vector-field expressed in the z coordinate, and $P(z, y)$, $q_i(z, y), i = 1, \ldots, m$ are matrices of appropriate dimensions. Now, if Σ^a is locally weakly minimum-phase at $x = 0$, then there exists a positive-definite C^2 Lyapunov-function $W^\star(z)$,

$W^*(0) = 0$, such that $L_{f^*} W^*(z) \leq 0$ for all z in the neighborhood of $z = 0$. Consequently, define the matrix

$$Q(z, y) = \begin{bmatrix} L_{q_1(z,y)} W^*(z) \\ \vdots \\ L_{q_m(z,y)} W^*(z) \end{bmatrix}. \tag{3.56}$$

Then, $Q(0, y) = 0$ and

$$v = [I + Q(z, y)]^{-1}[-(L_{P(z,y)} W^*(z))^T + w]$$

is well defined in a neighborhood of $(0, 0)$. The resulting closed-loop system (3.54),(3.55) with this auxiliary control is of the form

$$\begin{bmatrix} \dot{z} \\ \dot{y} \end{bmatrix} = \tilde{f}(z, y) + \tilde{g}(z, y) w \tag{3.57}$$

where $\tilde{f}(z, y)$ and $\tilde{g}(z, y)$ are suitable matrices of appropriate dimension. Finally, consider the positive-definite C^2 storage-function candidate for the closed-loop system

$$V(z, y) = W^\star(z) + \frac{1}{2} y^T y.$$

Taking the time-derivative of $V(., .)$ along the trajectories of (3.57) yields

$$\begin{aligned}
\dot{V}(z, y) &= L_{\tilde{f}} V(z, y) + L_{\tilde{g}} V(z, y) w = L_{f^*} W^*(z) + y^T [(L_{P(z,y)} W^*(z))^T + \\
&\quad Q(z, y) v] + y^T v \\
&= L_{f^*} W^*(z) + y^T w \leq y^T w.
\end{aligned}$$

Thus, the closed-loop system is passive with the C^2 storage-function $V(., .)$ as desired. $\square$

Remark 3.4.4 *For the linear system*

$$\Sigma^l : \begin{cases} \dot{x} &= Ax + Bu \\ y &= Cx, \end{cases} \tag{3.58}$$

if $rank(B) = m$, then $x = 0$ is always a regular point and a normal-form can be defined for the system. It follows then from the result of the theorem that Σ^l is feedback-equivalent to a passive linear system with a positive-definite storage-function $V(x) = \frac{1}{2} x^T P x$ if, and only if, CB is nonsingular and Σ^l is weakly minimum-phase.

Remark 3.4.5 *A global version of the above theorem also exists. It is easily seen that, if all the local assumptions on the system are replaced by their global versions, i.e., if there exists a global normal-form for the system and it is globally weakly minimum-phase (or globally minimum-phase[77]), then it will be globally feedback-equivalent to a passive system with a C^2 storage-function which is positive-definite. Furthermore, a synthesis procedure for the linear case is given in [233, 254].*

3.5 Dissipativity and Passive Properties of Discrete-Time Nonlinear Systems

In this section, we discuss the discrete-time counterparts of the previous sections. For this purpose, we again consider a discrete-time nonlinear state-space system defined on $\mathcal{X} \subseteq \Re^n$

containing the origin $x = \{0\}$ in coordinates $(x = x_1, \ldots, x_n)$:

$$\Sigma^d \; : \left\{ \begin{array}{rcl} x_{k+1} & = & f(x_k, u_k), \quad x(k_0) = x^0 \\ y_k & = & h(x_k, u_k) \end{array} \right. \tag{3.59}$$

where $x_k \in \mathcal{X}$ is the state vector, $u_k \in \mathcal{U} \subseteq \Re^m$ is the input function belonging to an input space $\mathcal{U}$, $y_k \in \mathcal{Y} \subseteq \Re^m$ is the output function which belongs to the output space $\mathcal{Y}$ (i.e., Σ^d is square). The functions $f : \mathcal{X} \times \mathcal{U} \to \mathcal{X}$ and $h : \mathcal{X} \times \mathcal{U} \to \mathcal{Y}$ are real C^r functions of their arguments such that there exists a unique solution $x(k, k_0, x_0, u_k)$ to the system for any $x_0 \in \mathcal{X}$ and $u_k \in \mathcal{U}$.

Definition 3.5.1 *A function $s(k) = s(u_k, y_k)$ is a supply-rate to the system Σ^d if it is locally absolutely summable, i.e., $\sum_{k=k_0}^{k_1} |s(k)| < \infty$ for all $(k_0, k_1) \in \mathbf{Z} \times \mathbf{Z}$.*

Then we have the following definition of dissipativity for the system Σ^d.

Definition 3.5.2 *The nonlinear system Σ^d is locally dissipative with respect to the supply-rate $s(u_k, y_k)$ if there exists a C^0 positive-semidefinite storage-function $\Psi : N \subset \mathcal{X} \to \Re$ such that*

$$\Psi(x_{k_1}) - \Psi(x_{k_0}) \le \sum_{k=k_0}^{k_1} s(u_k, y_k) \tag{3.60}$$

for all $k_1 \ge k_0$, $u_k \in \mathcal{U}$, and $x(k_0), x(k_1) \in N$. The system is said to be dissipative if it is locally dissipative for all $x_{k_0}, x_{k_1} \in \mathcal{X}$.

Again, we can move from the integral (or summation) version of the dissipation-inequality (3.60) to its differential (or infinitesimal) version by differencing to get

$$\Psi(x_{k+1}) - \Psi(x_k) \le s(u_k, y_k), \tag{3.61}$$

which is the discrete-time equivalent of (3.17). Furthermore, we can equally define the dissipation-inequality as in (3.1.3) by

$$\sum_{k=k_0}^{k_1} s(u_k, y_k) \ge 0, \quad x^0 = 0 \;\; \forall k_1 \ge k_0. \tag{3.62}$$

The available-storage and required-supply of the system Σ_a^d can also be defined as follows:

$$\Psi_a(x) \;\; = \;\; \sup_{\substack{x^0 = x, u_k \in \mathcal{U}, \\ K \ge 0}} \; -\sum_{k=0}^{K} s(u_k, y_k) \tag{3.63}$$

$$\Psi_r(x) \;\; = \;\; \inf_{\substack{x^e \to x, \\ u_k \in \mathcal{U}}} \; \sum_{k=k_{-1}}^{k=0} s(u_k, y_k), \quad \forall k_{-1} \le 0, \tag{3.64}$$

where $x^e = \arg\inf_{x \in \mathcal{X}} \Psi(x)$. Consequently, the equivalents of Theorems 3.1.1, 3.1.2 and 3.1.3 can be derived for the system Σ^d. However, of particular interest to us among discrete-time dissipative systems, are those that are passive because of their nice properties which are analogous to the continuous-time case. Hence, for the remainder of this section, we shall concentrate on this class and the affine nonlinear discrete-time system:

$$\Sigma^{da} \; : \left\{ \begin{array}{rcl} x_{k+1} & = & f(x_k) + g(x_k)u_k, \quad x(k_0) = x_0 \\ y_k & = & h(x_k) + d(x_k)u_k \end{array} \right. \tag{3.65}$$

where $g : \mathcal{X} \to \mathcal{M}^{n \times m}(\mathcal{X})$, $d : \mathcal{X} \to \Re^m$, and all the other variables have their previous meanings. We shall also assume for simplicity that $f(0) = 0$ and $h(0) = 0$. We proceed with the following definition.

Definition 3.5.3 *The system Σ^{da} is passive if it is dissipative with supply-rate $s(u_k, y_k) = y_k^T u_k$ and the storage-function $\Psi : \mathcal{X} \to \Re$ satisfies $\Psi(0) = 0$. Equivalently, Σ^{da} is passive if, and only if, there exists a positive-semidefinite storage-function Ψ satisfying*

$$\Psi(x_{k_1}) - \Psi(x_{k_0}) \leq \sum_{k=k_0}^{k_1} y_k^T u_k, \quad \Psi(0) = 0 \tag{3.66}$$

for all $k_1 \geq k_0 \in \mathbf{Z}$, $u_k \in \mathcal{U}$, or the infinitesimal version:

$$\Psi(x_{k+1}) - \Psi(x_k) \leq y_k^T u_k, \quad \Psi(0) = 0 \tag{3.67}$$

for all $k_1 \geq k_0 \in \mathbf{Z}$, $u_k \in \mathcal{U}$.

For convenience of the presentation, we shall concentrate on the infinitesimal version of the passivity-inequality (3.67).

Remark 3.5.1 *Similarly, the discrete-time system Σ^{da} is strictly-passive if the strict inequality in (3.66) or (3.67) is satisfied, or there exists a positive-definite function $\Xi : \mathcal{X} \to \Re_+$ such that*

$$\Psi(x_{k+1}) - \Psi(x_k) \leq y_k^T u_k - \Xi(x_k) \quad \forall u_k \in \mathcal{U}, \forall k \in \mathbf{Z}. \tag{3.68}$$

We also have the following definition of losslessness.

Definition 3.5.4 *A passive system Σ^{da} with storage-function Ψ, is lossless if*

$$\Psi(x_{k+1}) - \Psi(x_k) = y_k^T u_k \quad \forall u_k \in \mathcal{U}, k \in \mathbf{Z}. \tag{3.69}$$

The following lemma is the discrete-time version of the nonlinear KYP lemma given in Theorem 3.3.1 and Definition 3.3.4 for lossless systems.

Lemma 3.5.1 *The nonlinear system Σ^{da} with a C^2 storage-function $\Psi(.)$ is lossless if, and only if,*

(i)

$$\Psi(f(x)) = \Psi(x) \tag{3.70}$$

$$\frac{\partial \Psi(\alpha)}{\partial \alpha}\Big|_{\alpha=f(x)} g(x) = h^T(x) \tag{3.71}$$

$$g^T(x)\frac{\partial^2 \Psi(\alpha)}{\partial \alpha^2}\Big|_{\alpha=f(x)} g(x) = d^T(x) + d(x) \tag{3.72}$$

(ii) $\Psi(f(x) + g(x)u)$ *is quadratic in u.*

Proof: (Necessity): If Σ^{da} is lossless, then there exists a positive-semidefinite storage-function $\Psi(.)$ such that

$$\Psi(f(x) + g(x)u) - \Psi(x) = (h(x) + d(x)u)^T u. \tag{3.73}$$

Setting $u = 0$, we immediately get (3.70). Differentiating both sides of the above equation (3.73) with respect to u once and twice, and setting $u = 0$ we also arrive at the equations

(3.71) and (3.72), respectively. Moreover, (ii) is also obvious from (3.72).
(Sufficiency:) If (ii) holds, then there exist functions $p, q, r : \mathcal{X} \to \Re$ such that

$$\Psi(f(x) + g(x)u) = p(x) + q(x)u + u^T r(x)u \quad \forall u \in \mathcal{U} \tag{3.74}$$

where the functions $p(.)$, $q(.)$, $r(.)$ correspond to the Taylor-series expansion of $\Psi(f(x) + g(x)u)$ about u:

$$
\begin{aligned}
p(x) &= \Psi(f(x)) &&(3.75)\\
q(x) &= \left.\frac{\partial \Psi(f(x) + g(x)u)}{\partial u}\right|_{u=0} = \left.\frac{\partial \Psi(\alpha)}{\partial \alpha}\right|_{\alpha=f(x)} g(x) &&(3.76)\\
r(x) &= \left.\frac{1}{2}\frac{\partial^2 \Psi(f(x) + g(x)u)}{\partial u^2}\right|_{u=0} = g(x)\left.\frac{\partial^2 \Psi(\alpha)}{\partial \alpha^2}\right|_{\alpha=f(x)} g(x). &&(3.77)
\end{aligned}
$$

Therefore equation (3.74) implies by (3.70)-(3.72),

$$
\begin{aligned}
\Psi(f(x) + g(x)u) - \Psi(x) &= (q(x) + u^T r(x))u \quad \forall u \in \mathcal{U}\\
&= \{h^T(x) + \frac{1}{2}u^T(d^T(x) + d(x))\}u\\
&= y^T u \quad \forall u \in \mathcal{U}.
\end{aligned}
$$

Hence Σ^{da} is lossless. $\square$

Remark 3.5.2 *In the case of the linear discrete-time system*

$$\Sigma^{dl} : \begin{cases} x_{k+1} &= Ax_k + Bu_k\\ y_k &= Cx_k + Du_k \end{cases}$$

where A, B, C and D are matrices of appropriate dimensions. The system is lossless with C^2 storage-function $\Psi(x_k) = \frac{1}{2}x_k^T P x_k$, $P = P^T$ and the KYP equations (3.70)-(3.72) yield:

$$\left.\begin{aligned} A^T PA &= P\\ B^T PA &= C\\ B^T PB &= D^T + D. \end{aligned}\right\} \tag{3.78}$$

Remark 3.5.3 *Notice from the above equations (3.78) that if the system matrix $D = 0$, then since $B \neq 0$, it implies that $C = 0$. Thus Σ^{dl} can only be lossless if $y_k = 0$ when $D = 0$. In addition, if $rank(B) = m$, then Σ^{dl} is lossless only if $P \geq 0$.*

Furthermore, $D^T + D \geq 0$ and $D + D^T > 0$ if and only if $rank(B) = m$ [74]. Thus, $D > 0$ and hence nonsingular. This assumption is also necessary for the nonlinear system Σ^{da}, i.e., $d(x) > 0$ and will consequently be adopted in the sequel.

The more general result of Lemma 3.5.1 for passive systems can also be stated.

Theorem 3.5.1 *The nonlinear system Σ^{da} is passive with storage-function $\Psi(.)$ which is positive-definite if, and only if, there exist real functions $l : \mathcal{X} \to \Re^i$, $W : \mathcal{X} \to \mathcal{M}^{i \times j}(\mathcal{X})$ of appropriate dimensions such that*

$$\Psi(f(x)) - \Psi(x) = -\frac{1}{2}l^T(x)l(x) \tag{3.79}$$

$$\left.\frac{\partial \Psi(\alpha)}{\partial \alpha}\right|_{\alpha=f(x)} g(x) + l^T(x)W(x) = h^T(x) \tag{3.80}$$

$$d^T(x) + d(x) - g^T(x)\left.\frac{\partial^2 \Psi(\alpha)}{\partial \alpha^2}\right|_{\alpha=f(x)} g(x) = W^T(x)W(x). \tag{3.81}$$

Proof: The proof follows from Lemma 3.5.1. $\square$

Remark 3.5.4 *In the case of the linear system Σ^{dl} the equations (3.79)-(3.81) reduce to the following wellknown conditions for Σ^{dl} to be positive-real:*

$$
\begin{aligned}
A^T P A - P &= -L^T L \\
A^T P B + L^T W &= C^T \\
D^T + D - B^T P B &= W^T W.
\end{aligned}
$$

The KYP lemma above can also be specialized to discrete-time bilinear systems. Furthermore, unlike for the general nonlinear case above, the assumption that $\Psi(f(x) + g(x)u)$ is quadratic in u can be removed. For this purpose, let us consider the following bilinear state-space system defined on $\mathcal{X} \subseteq \Re^n$:

$$
\Sigma^{dbl} : \begin{cases} x_{k+1} &= Ax_k + [B(x_k) + E]u_k \\ y_k &= h(x_k) + d(x_k)u_k, \ h(0) = 0, \end{cases} \tag{3.82}
$$

where all the previously used variables and functions have their meanings, with $A \in \Re^{n \times n}$, $E \in \Re^{n \times m}$, $C \in \Re^{m \times n}$ constant matrices, while $B : \mathcal{X} \to \mathcal{M}^{n \times m}(\mathcal{X})$ is a smooth map defined by

$$
B(x) = [B_1 x, \dots, B_m x],
$$

where $B_i \in \Re^{n \times n}, i = 1, \dots, m$. We then have the following theorem.

Theorem 3.5.2 *(KYP-lemma) The bilinear system Σ^{dbl} is passive with a positive-definite storage-function, $\Psi(x) = \frac{1}{2}x^T P x$ if, and only if, there exists a real constant matrix $L \in \Re^{n \times q}$ and a real function $W : \mathcal{X} \to \mathcal{M}^{i \times j}(\mathcal{X})$ of appropriate dimensions, satisfying*

$$
\begin{aligned}
A^T P A - P &= -L^T L \\
x^T[A^T P(B(x) + E) + L^T W(x)] &= h^T(x) \\
d^T(x) + d(x) - (B(x) + E)^T P(B(x) + E) &= W^T(x)W(x)
\end{aligned}
$$

for some symmetric positive-definite matrix P.

3.6 ℓ_2-Gain Analysis for Discrete-Time Dissipative Systems

In this section, we again summarize the relationship between the ℓ_2-gain of a discrete-time system and its dissipativeness, as well as the implications on its stability analogously to the continuous-time case discussed in Section 3.2. We consider the nonlinear discrete-time system Σ^d given by the state equations (3.59). We then have the following definition.

Definition 3.6.1 *The nonlinear system Σ^d is said to have locally in $N \subset \mathcal{X}$, finite ℓ_2-gain less than or equal to $\gamma > 0$ if for all $u_k \in \ell_2[k_0, K]$*

$$
\sum_{k=k_0}^{K} \|y_k\|^2 \le \gamma^2 \sum_{k=k_0}^{K} \|u_k\|^2 + \beta'(x_0) \tag{3.83}
$$

for all $K \ge k_0$, and all $x_0 \in N$, $y_k = h(x_k, u_k)$, $k \in [k_0, K]$, and for some function $\beta' : \mathcal{X} \to \Re_+$. Moreover, the system is said to have ℓ_2-gain $\le \gamma$ if $N = \mathcal{X}$.

Without any loss of generality, we can take $k_0 = 0$ in the above definition. We now have the following important theorem which is the discrete-time equivalent of Theorems 3.2.1, 3.2.2 or the Bounded-real lemma for discrete-time nonlinear systems.

Theorem 3.6.1 *Consider the discrete-time system Σ^d. Then Σ^d has finite ℓ_2-gain $\leq \gamma$ if Σ^d is finite-gain dissipative with supply-rate $s(u_k, y_k) = (\gamma^2 \|u_k\|^2 - \|y_k\|^2)$, i.e., it is dissipative with respect to $s(.,.)$ and $\gamma < \infty$. Conversely, Σ^d is finite-gain dissipative with supply-rate $s(u_k, y_k)$ if Σ^d has ℓ_2-gain $\leq \gamma$ and is reachable from $x = 0$.*

Proof: By dissipativity (Definition 3.5.2)

$$0 \leq V(x_{K+1}) \leq \sum_{k=0}^{K}(\gamma^2 \|u_k\|^2 - \|y_k\|^2) + V(x_0). \tag{3.84}$$

$$\implies \sum_{k=0}^{K} \|y_k\|^2 \leq \gamma^2 \sum_{k=0}^{K} \|u_k\|^2 + V(x_0). \tag{3.85}$$

Thus, Σ^d has locally ℓ_2-gain $\leq \gamma$ by Definition 3.6.1. Conversely, if Σ^d has ℓ_2-gain $\leq \gamma$, then

$$\sum_{k=0}^{K}(\|y_k\|^2) - \gamma^2 \|u_k\|^2) \leq \beta'(x_0)$$

$$\implies V_a(x) = - \inf_{\substack{x(0) = x, u_k \in \mathcal{U}, \\ K \geq 0}} \sum_{k=0}^{K}(\|u_k\|^2 - \|y_k\|^2) \leq \beta'(x_0) < \infty,$$

the available-storage, is well defined for all x in the reachable set of Σ^d from $x(0)$ with $u_k \in \ell_2$, and for some function β'. Consequently, Σ^d is finite-gain dissipative with some storage-function $V \geq V^a$ and supply-rate $s(u_k, y_k)$. $\square$

The following theorem is the discrete-time counterpart of Theorems 3.2.1, 3.2.2 relating the dissipativity of the discrete-time system (3.65), its ℓ_2-gain and its stability, which is another version of the Bounded-real lemma.

Theorem 3.6.2 *Consider the affine discrete-time system Σ^{da}. Suppose there exists a C^2 positive-definite function $V : U \to \Re_+$ defined locally in a neigborhood U of $x = 0$ satisfying*

(H1)

$$g^T(0)\frac{\partial^2 V}{\partial x^2}(0)g(0) + d^T(0)d(0) - \gamma^2 I < 0;$$

(H2)

$$0 = V(f(x) + g(x)\mu(x)) - V(x) + \frac{1}{2}(\|h(x) + d(x)\mu(x)\|^2 - \gamma^2 \|\mu(x)\|^2) \tag{3.86}$$

where $u = \mu(x)$, $\mu(0) = 0$ is the unique solution of

$$\left.\frac{\partial V}{\partial \lambda}\right|_{\lambda = f(x) + g(x)u} g(x) + u^T(d^T(x)d(x) - \gamma^2 I) = -h^T(x)d(x);$$

(H3) *the affine system (3.65) is zero-state detectable, i.e., $y_k|_{u_k=0} = h(x_k) = 0 \Rightarrow \lim_{k\to\infty} x_k = 0$,* *then Σ^{da} is locally asymptotically-stable and has ℓ_2-gain $\leq \gamma$.*

Proof: Consider the Hamiltonian function for (3.65) under the ℓ_2 cost criterion $J = \sum_{k=0}^{\infty}[\|y_k\|^2 - \gamma^2\|w_k\|^2]$:

$$H(x, u) = V(f(x) + g(x)u) - V(x) + \frac{1}{2}(\|y\|^2 - \gamma^2\|u\|^2).$$

Then

$$\frac{\partial H}{\partial u}(x, u) = \left.\frac{\partial V}{\partial \lambda}\right|_{\lambda=f(x)+g(x)u} g(x) + u^T(d^T(x)d(x) - \gamma^2 I) + h^T(x)d(x) \qquad (3.87)$$

and

$$\frac{\partial^2 H}{\partial u^2}(x, u) = g^T(x)\left.\frac{\partial^2 V}{\partial \lambda^2}\right|_{\lambda=f(x)+g(x)u} g(x) + d^T(x)d(x) - \gamma^2 I.$$

It is easy to check that the point $(x^0, u^0) = (0, 0)$ is a critical point of $H(x, u)$ and the Hessian matrix of H is negative-definite at this point by hypothesis $(H1)$. Hence, $\frac{\partial^2 H}{\partial u^2}(0, 0)$ is nonsingular at $(x^0, u^0) = (0, 0)$. Thus, by the Implicit-function Theorem, there exists an open neighborhood $N \subset \mathcal{X}$ of $x^0 = 0$ and an open neighborhood $U \subset \mathcal{U}$ of $u^0 = 0$ such that there exists a C^1 solution $u = \mu(x)$, $\mu : N \to U$, to (3.87).

Expanding $H(., .)$ about $(x^0, u^0) = (0, 0)$ using Taylor-series formula yields

$$H(x, u) = H(x, \mu(x)) + \frac{1}{2}(u - \mu(x))^T\left(\frac{\partial^2 H}{\partial u^2}(x, \mu(x)) + O(\|u - \mu(x)\|)\right)(u - \mu(x)).$$

Again hypothesis $H1$ implies that there exists a neighborhood of $x^0 = 0$ such that

$$\frac{\partial^2 H}{\partial u^2}(x, \mu(x)) = g^T(x)\left.\frac{\partial^2 V}{\partial \lambda^2}\right|_{\lambda=f(x)+g(x)\mu(x)} g(x) + d^T(x)d(x) - \gamma^2 I < 0.$$

This observation together with the hypothesis $(H2)$ implies that $u = \mu(x)$ is a local maximum of $H(x, u)$, and that there exists open neighborhoods N_0 of x^0 and U_0 of u^0 such that

$$H(x, u) \leq H(x, \mu(x)) = 0 \quad \forall x \in N_0 \text{ and } \forall u \in U_0,$$

or equivalently

$$V(f(x) + g(x)u(x)) - V(x) \leq \frac{1}{2}(\gamma^2\|u\|^2 - \|h(x) + d(x)u\|^2). \qquad (3.88)$$

Thus, Σ^{da} is finite-gain dissipative with storage-function V with respect to the supply-rate $s(u_k, y_k)$ and hence has ℓ_2-gain $\leq \gamma$. To show local asymptotic stability, we substitute $x = 0$, $u = 0$ in (3.88) to see that the equilibrium point $x = 0$ is stable and V is a Lyapunov-function for the system. Moreover, by hypothesis $(H3)$ and LaSalle's invariance-principle, we conclude that $x = 0$ is locally asymptotically-stable with $u = 0$. $\square$

Remark 3.6.1 *In the case of the linear discrete-time system Σ^{dl}, the conditions $(H1)$, $(H2)$ in the above theorem reduce to the algebraic-equation and DARE respectively:*

$$B^T P B + D^T D - \gamma^2 I \leq 0,$$
$$A^T P A - P - (B^T P A + D^T C)^T (B^T P B + D^T D - \gamma^2)^{-1}(B^T P A + D^T C) + C^T C = 0.$$

3.7 Feedback-Equivalence to a Discrete-Time Lossless Nonlinear System

In this section, we derive the discrete-time analogs of the results of Subsection 3.4; namely, when can a discrete-time system of the form Σ^{da} be rendered lossless (or passive) via smooth state-feedback? We present necessary and sufficient conditions for feedback-equivalence to a lossless discrete-time system, and the results could also be modified to achieve feedback-equivalence to a passive system. The results are remarkably analogous to the continuous-time case, and the necessary conditions involve some mild regularity conditions and the requirement of lossless zero-dynamics. The only apparent anomaly is the restriction that $d(x)$ be nonsingular for x in $\mathcal{O} \ni \{0\}$.

We begin by extending the concepts of relative-degree and zero-dynamics to the discrete-time case.

Definition 3.7.1 *The nonlinear system Σ^{da} is said to have vector relative-degree $\{0, \ldots, 0\}$ at $x = 0$ if $d(0)$ is nonsingular. It is said to have uniform vector relative-degree $\{0, \ldots, 0\}$ if $d(x)$ is nonsingular for all $x \in \mathcal{X}$.*

Definition 3.7.2 *If the system Σ^{da} has vector relative-degree $\{0, \ldots, 0\}$ at $x = \{0\}$, then there exists an open neighborhood $\mathcal{O}$ of $x = \{0\}$ such that $d(x)$ is nonsingular for all $x \in \mathcal{O}$ and hence the control*

$$u_k^* = -d^{-1}(x_k)h(x_k), \quad \forall x \in \mathcal{O}$$

renders $y_k \equiv 0$ and the resulting dynamics of the system

$$\Sigma_0^{da} * \ : \ x_{k+1} = f^*(x_k) = f(x_k) + g(x_k)u_k^*, \quad \forall x_k \in \mathcal{O} \subseteq \mathcal{X} \tag{3.89}$$

is known as the zero-dynamics of the system. Furthermore, the n-dimensional submanifold

$$\mathcal{Z}^* = \{x \in \mathcal{O} \ : \ y_k = 0, \ k \in \mathbf{Z}\} \equiv \mathcal{O} \subseteq \mathcal{X}$$

is known as the zero-dynamics submanifold.

Remark 3.7.1 *If Σ^{da} has uniform vector relative-degree $\{0, \ldots, 0\}$, then*

$$\mathcal{Z}^* = \{x \in \mathcal{O} \ : \ y_k = 0, \ k \in \mathbf{Z}\} \equiv \mathcal{O} \equiv \mathcal{X}.$$

Notice that unlike in the continuous-time case with vector relative degree $\{1, \ldots, 1\}$ in which the zero-dynamics evolves on an $n - m$-dimensional submanifold, for the discrete-time system, the zero-dynamics evolves on an n-dimensional submanifold.

The following notion of lossless zero-dynamics replaces that of minimum-phase for feedback-equivalence to a lossless system.

Definition 3.7.3 *Suppose $d(0)$ is nonsingular. Then the system Σ^{da} is said to have locally lossless zero-dynamics if there exists a C^2 positive-definite Lyapunov-function V locally defined in the neighborhood $\mathcal{O}$ of $x = \{0\}$ such that*

(i) $V(f^*(x)) = V(x), \quad \forall x \in \mathcal{O}.$

(ii) $V(f^*(x) + g(x)u)$ *is quadratic in u.*

The system is said to have globally lossless zero-dynamics if $d(x)$ is nonsingular for all $x \in$

$\Re^n$ and there exists a C^2 positive-definite Lyapunov-function V that satisfies the conditions (i), (ii) above for all $x \in \Re^n$.

The following lemma will be required in the sequel.

Lemma 3.7.1 *(Morse-lemma [1]): Let p be a nondegenerate critical point for a real-valued function v. Then there exists a local coordinate system $(y^1, \ldots, y^n)$ in a neighborhood N about p such that $y_i(p) = 0$ for $i = 1, \ldots, n$ and v is quadratic in N:*

$$v(y) = v(p) - y_1^2 - \ldots - y_l^2 + y_{l+1}^2 + \ldots + y_n^2$$

for some integer $0 \leq l \leq n$.

Lemma 3.7.2 *Consider the zero-dynamics $\Sigma_0^{da}*$ of the system Σ^{da}. Suppose there is a Lyapunov-function V which is nondegenerate at $x = 0$ and satisfies $V(f^*(x)) = V(x)$ for all $x \in \mathcal{O}$ neighborhood of $x = 0$. Then there exists a change in coordinates $\tilde{x} = \varphi(x)$ such that $\Sigma^{da}*$ is described by*

$$\tilde{x}_{k+1} = \tilde{f}^*(\tilde{x}_k),$$

where $\tilde{f} := \varphi \circ f^* \circ \varphi^{-1}$, and has a positive-definite Lyapunov-function $\tilde{V}$ which is quadratic in $\tilde{x}$, i.e., $\tilde{V}(\tilde{x}) = \tilde{x}^T P \tilde{x}$ for some $P > 0$, and satisfies $\tilde{V}(\tilde{f}^*(\tilde{x})) = \tilde{V}(\tilde{x})$.

Proof: Applying the Morse-lemma, there exists a local change of coordinates $\tilde{x} = \varphi(x)$ such that

$$\tilde{V}(\tilde{x}) = V(\varphi^{-1}(\tilde{x})) = \tilde{x}^T P \tilde{x}$$

for some $P > 0$. Thus,

$$\tilde{V}(\tilde{f}^*(\tilde{x})) = V(f^* \circ \varphi^{-1}(\tilde{x})) = V(\varphi^{-1}(\tilde{x})) = \tilde{V}(\tilde{x}). \quad \square$$

Since the relative-degree of the system Σ^{da} is obviously so important for the task at hand, and as we have seen in the continuous-time case, the passivity (respectively losslessness) of the system has implications on its vector relative-degree, therefore, we shall proceed in the next few results to analyze the relative-degree of the system. We shall present sufficient conditions for the system to be lossless with a $\{0, \ldots, 0\}$ relative-degree at $x = 0$. We begin with the following preliminary result.

Proposition 3.7.1 *Suppose Σ^{da} is lossless with a C^2 positive-definite storage-function Ψ which is nondegenerate at $x = 0$. Then*

(i) $rank\{g(0)\} = m$ *if and only if $d(0) + d^T(0) > 0$.*

(ii) *Moreover, as a consequence of (i) above, $d(0)$ is nonsingular, and Σ^{da} has vector relative-degree $\{0, \ldots, 0\}$ at $x = 0$.*

Proof: (i)(only if): By Lemma 3.5.1, substituting $x = 0$ in equation (3.72), we have

$$g^T(0) \frac{\partial^2 \Psi(\lambda)}{\partial \lambda^2}\bigg|_{\lambda=0} g(0) = d^T(0) + d(0).$$

Since the Hessian of V evaluated at $\lambda = 0$ is positive-definite, then

$$x^T(d^T(0) + d(0))x = (g(0)x)^T \frac{\partial^2 \Psi(\lambda)}{\partial \lambda^2}\bigg|_{\lambda=0} (g(0)x).$$

Hence, $rank\{g(0)\} = m$ implies that $d(0) + d^T(0) > 0$.
(if): Conversely if $d(0) + d^T(0) > 0$, then

$$(g(0)x)^T \frac{\partial^2 \Psi(\lambda)}{\partial \lambda^2}\bigg|_{\lambda=0} (g(0)x) > 0 \Longrightarrow g(0)x \neq 0 \; \forall x \in \Re^m \setminus \{0\}.$$

Therefore $g(0)x = 0$ has a unique solution $x = 0$ when $rank\{g(0)\} = m$.

Finally, (ii) follows from (i) by positive-definiteness of $d(0) + d^T(0)$. $\square$

Remark 3.7.2 *Notice that for the linear discrete-time system Σ^{dl} with a positive-definite storage-function $\Psi(x) = \frac{1}{2}x^T P x$, $P > 0$, Ψ is always nondegenerate at $x = 0$ since the Hessian of V is $P > 0$.*

The following theorem is the counterpart of Theorem 3.4.1 for the discrete-time system Σ^{da}.

Theorem 3.7.1 *Suppose that $rank\{g(0)\} = m$ and the system Σ^{da} is lossless with a C^2 positive-definite storage-function $\Psi(.)$ which is nondegenerate at $x = 0$. Then Σ^{da} has vector relative-degree $\{0, \ldots, 0\}$ and a lossless zero-dynamics at $x = 0$.*

Proof: By Proposition 3.7.1 Σ^{da} has relative-degree $\{0, \ldots, 0\}$ at $x = 0$ and its zero-dynamics exists locally in the neighborhood $\mathcal{O}$ of $x = 0$. Furthermore, since Σ^{da} is lossless

$$\Psi(f(x) + g(x)u) - \Psi(x) = y^T u, \quad \forall u \in \mathcal{U}. \tag{3.90}$$

Setting $u = u^* = -d^{-1}(x)h(x)$ gives

$$\Psi(f^*(x)) = \Psi(x) \; \forall x \in \mathcal{Z}^*.$$

Finally, if we substitute $u = u^* + \tilde{u}$ in (3.74), it follows that $\Psi(f^*(x) + g(x)\tilde{u})$ is quadratic in $\tilde{u}$. Thus, the zero-dynamics is lossless. $\square$

Remark 3.7.3 *A number of interesting corollaries which relate the losslessness of the system Σ^{da} with respect to a positive-definite storage-function which is nondegenerate at $x = \{0\}$, versus the $rank\{g(0)\}$ and its relative-degree, could be drawn. It is sufficient here however to observe that, under mild regularity conditions, the discrete-time system Σ^{da} with $d(x) \equiv 0$ cannot be lossless; that Σ^{da} can only be lossless if it has vector relative-degree $\{0, \ldots, 0\}$ and $d(x)$ is nonsingular. Conversely, under some suitable assumptions, if Σ^{da} is lossless with a positive-definite storage-function, then it necessarily has vector relative-degree $\{0, \ldots, 0\}$ at $x = 0$.*

We now present the main result of the section; namely, a necessary and sufficient condition for feedback-equivalence to a lossless system using regular static state-feedback control of the form:

$$u_k = \alpha(x_k) + \beta(x_k)v_k, \quad \alpha(0) = 0 \tag{3.91}$$

for some smooth C^∞ functions $\alpha : N \subseteq \mathcal{X} \to \Re^m$, $\beta : N \to \mathcal{M}^{m \times m}(\mathcal{X})$, $0 \in N$, with β invertible over N.

Theorem 3.7.2 *Suppose Ψ is nondegenerate at $x = \{0\}$ and $rank\{g(0)\} = m$. Then Σ^{da} is locally feedback-equivalent to a lossless system with a C^2 storage-function which is positive-definite, if and only if, Σ^{da} has vector relative-degree $\{0, \ldots, 0\}$ at $x = \{0\}$ and lossless zero-dynamics.*

Proof: (Necessity): Suppose there exists a feedback control of the form (3.91) such that Σ^{da} is feedback-equivalent to the lossless system

$$\widetilde{\Sigma}^{da} \; : \; \begin{aligned} x_{k+1} &= \tilde{f}(x_k) + \tilde{g}(x_k)w_k \\ y_k &= \tilde{h}(x_k) + \tilde{d}(x_k)w_k \end{aligned}$$

with a C^2 storage-function Ψ, where w_k is an auxiliary input and

$$\tilde{f}(x) = f(x) + g(x)\alpha(x), \quad \tilde{g}(x) = g(x)\beta(x) \tag{3.92}$$
$$\tilde{h}(x) = h(x) + d(x)\alpha(x), \quad \tilde{d}(x) = d(x)\beta(x). \tag{3.93}$$

Since $\beta(0)$ is nonsingular, then $rank\{\tilde{g}(0)\} = m$. Similarly, by Proposition 3.7.1 $\tilde{d}(0)$ is nonsingular, and therefore $d(0)$. Hence Σ^{da} has relative-degree $\{0,\dots,0\}$ at $x = 0$. Next we show that Σ^{da} has locally lossless zero-dynamics.

The zero-dynamics of the equivalent system $\widetilde{\Sigma}^{da}$ are governed by

$$\widetilde{\Sigma}^{da}_0 * \; : \; x_{k+1} = \tilde{f}^*(x_k), \;\; x \in \mathcal{Z}^*$$

where $\tilde{f}^*(x) := \tilde{f} - \tilde{g}(x)\tilde{d}^{-1}(x)\tilde{h}(x)$. These are identical to the zero-dynamics of the original system Σ^{da} with $f^*(x) = f(x) - g(x)d^{-1}(x)h(x)$ since they are invariant under static state-feedback. But $\widetilde{\Sigma}^{da}$ lossless implies by Theorem 3.7.1 that

(i) $\Psi(\tilde{f}^*(x)) = \Psi(x)$

(ii) $\Psi(\tilde{f}^*(x) + \tilde{g}(x)w)$ is quadratic in w.

Therefore, by the invariance of the zero-dynamics, we have $\Psi(f^*(x)) = \Psi(x)$ and $\Psi(f^*(x) + g(x)u)$ is also quadratic in u. Consequently, Σ^{da} has locally lossless zero-dynamics.
(Sufficiency): If Σ^{da} has relative-degree $\{0,\dots,0\}$ at $x = 0$, then there exists a neighborhood $\mathcal{O}$ of $x = \{0\}$ in which $d^{-1}(x)$ is well defined. Applying the feedback

$$u_k = u_k^* + d^{-1}(x_k)v_k = -d^{-1}(x_k)h(x_k) + d^{-1}(x_k)v_k$$

changes Σ^{da} into the system

$$\begin{aligned} x_{k+1} &= f^*(x_k) + g^*(x_k)v_k \\ y_k &= v_k \end{aligned}$$

where $g^*(x) := g(x)d^{-1}(x)$. Now let the output of this resulting system be identical to that of the lossless system $\widetilde{\Sigma}^{da}$ and have the equivalent system

$$\widetilde{\Sigma}^{da}_1 \; : \; \begin{aligned} x_{k+1} &= f^*(x_k) + g^*(x_k)\tilde{h}(x_k) + g^*(x_k)\tilde{d}(x_k)w_k \\ y_k &= \tilde{h}(x_k) + \tilde{d}(x_k)u_k. \end{aligned}$$

By assumption, Ψ is nondegenerate at $x = 0$ and $rank\{g(0)\} = m$, therefore there exists a neighborhood N of $x = 0$ such that $(g^*(x))^T \frac{\partial^2 \Psi}{\partial \lambda^2}\big|_{\lambda=f^*(x)} g^*(x)$ is positive-definite for all $x \in N$. Then

$$\tilde{d}(x) := \left(\frac{1}{2}(g^*(x))^T \frac{\partial^2 \Psi}{\partial \lambda^2}\Big|_{\lambda=f^*(x)} g^*(x) \right)^{-1} \tag{3.94}$$

$$\tilde{h}(x) := \tilde{d}(x) \left(\frac{\partial \Psi}{\partial \lambda}\Big|_{\lambda=f^*(x)} g^*(x) \right)^T \tag{3.95}$$

are well defined in N. It can now be shown that, with the above construction of $\tilde{d}(x)$ and $\tilde{h}(x)$, the system $\widetilde{\Sigma}_1^{da}$ is lossless with a C^2 storage-function Ψ.

By assumption, $\Psi(f^*(x) + g^*(x)u)$ is quadratic in u, therefore by Taylor-series expansion about $f^*(x)$, we have

$$\Psi(f^*(x) + g^*(x)\tilde{h}(x)) = \Psi(f^*(x)) + \left.\frac{\partial \Psi}{\partial \lambda}\right|_{\lambda = f^*(x)} g^*(x)\tilde{h}(x) +$$
$$\frac{1}{2}\tilde{h}^T(x)(g^*(x))^T \left.\frac{\partial^2 \Psi}{\partial \lambda^2}\right|_{\lambda = f^*(x)} g^*(x)\tilde{h}(x). \tag{3.96}$$

It can then be deduced from the above equation (using again first-order Taylor's expansion) that

$$\left.\frac{\partial \Psi}{\partial \lambda}\right|_{\lambda = f^*(x) + g^*(x)\tilde{h}(x)} g^*(x) = \left.\frac{\partial \Psi}{\partial \alpha}\right|_{\lambda = f^*(x)} g^*(x) + \tilde{h}^T(x)(g^*)^T(x)\left.\frac{\partial^2 \Psi}{\partial \lambda^2}\right|_{\lambda = f^*(x)} g^*(x) \tag{3.97}$$

and

$$(g^*(x))^T \left.\frac{\partial^2 \Psi}{\partial \lambda^2}\right|_{\lambda = f^*(x) + g^*(x)\tilde{h}(x)} g^*(x) = (g^*(x))^T \left.\frac{\partial^2 \Psi}{\partial \lambda^2}\right|_{\lambda = f^*(x)} g^*(x). \tag{3.98}$$

Using (3.94) and (3.95) in (3.97), (3.98), we get

$$\left.\frac{\partial \Psi}{\partial \lambda}\right|_{\lambda = f^*(x) + g^*(x)\tilde{h}(x)} g^*(x)\tilde{d}(x) = \tilde{h}^T(x) \tag{3.99}$$

and

$$(g^*(x)\tilde{d}(x))^T \left.\frac{\partial^2 \Psi}{\partial \lambda^2}\right|_{\lambda = f^*(x) + g^*(x)\tilde{h}(x)} (g^*(x)\tilde{d}(x)) = \tilde{d}^T(x) + \tilde{d}(x). \tag{3.100}$$

Similarly, substituting (3.94) and (3.95) in (3.96) gives

$$\begin{aligned}\Psi(f^*(x) + g^*(x)\tilde{h}(x)) &= \Psi(f^*(x)) - \tilde{h}^T(x)\tilde{d}^{-1}(x)\tilde{h}(x) + \tilde{h}^T(x)\tilde{d}^{-1}(x)\tilde{h}(x) \\ &= \Psi(f^*(x)). \end{aligned} \tag{3.101}$$

Recall now that $(f^*(x) + g^*(x)\tilde{h}(x))$ is the zero-dynamics of the system $\widetilde{\Sigma}_{a1}^d$. Together with the fact that $\Psi(f^*(x) + g^*(x)\tilde{h}(x) + g^*(x)\tilde{d}(x)w)$ is quadratic in w and (3.99)-(3.101) hold, then by Lemma 3.5.1 and Definition 3.7.3, we conclude that the system $\widetilde{\Sigma}_1^{da}$ is lossless with a C^2 storage-function Ψ. $\square$

Remark 3.7.4 *In the case of the linear discrete-time system Σ^{dl}, with $rank(B) = m$. The quadratic storage-function $\Psi(x) = \frac{1}{2}x^T P x$, $P > 0$ is always nondegenerate at $x = 0$, and $\Psi(Ax + Bu)$ is always quadratic in u. It therefore follows from the above theorem that any linear discrete-time system of the form Σ^{dl} is feedback-equivalent to a lossless linear system with a positive-definite storage-function $\Psi(x) = \frac{1}{2}x^T P x$ if, and only if, there exists a positive-definite matrix P such that*

$$(A - BD^{-1}C)^T P(A - BD^{-1}C) = P.$$

Remark 3.7.5 *A global version of the above theorem also exists. It is easily seen that, if the local properties of the system are replaced by their global versions, i.e., if the system has globally lossless zero-dynamics and uniform vector relative-degree $\{0, \ldots, 0\}$, then it would be globally feedback-equivalent to a lossless system with a C^2 positive-definite storage-function.*

Remark 3.7.6 *What is however missing, and what we have not presented in this section, is the analogous synthesis procedure for feedback-equivalence of the discrete-time system Σ^{da} to a passive one similar to the continuous-time case. This is because, for a passive system of the form Σ^{da}, the analysis becomes more difficult and complicated. Furthermore, the discrete-time equivalent of the KYP lemma is not available except for the restricted case when $\Psi(f(x) + g(x)u)$ is quadratic in u (Theorem 3.5.1). But $\Psi(f(x) + g(x)u)$ is in general not quadratic in u, and therefore the fundamental question: "When is the system Σ^{da} passive, or can be rendered passive using smooth state-feedback?" cannot be answered in general.*

3.8 Notes and Bibliography

The basic definitions and fundamental results of Section 3.1 of the chapter on continuous-time systems are based on the papers by Willems [274] and Hill and Moylan [131]-[134]. The results on stability in particular are taken from [131, 133], while the continuous-time Bounded-real lemma is from [131, 202]. The stability results for feedback interconnection of dissipative systems are based on [133], and the connection between dissipativity and finite-gain are discussed in References [132, 264] but have not been presented in the chapter. This will be discussed however in Chapter 5.

The results of Sections 3.3, 3.4 on passivity and local feedback-equivalence to a passive continuous-time system are from the paper by Byrnes et al. [77]. Global results can also be found in the same reference, and a synthesis procedure for the linear case is given in [233, 254].

All the results on the discrete-time systems are based on the papers by Byrnes and Lin [74]-[76], while the KYP lemma for bilinear discrete-time systems is taken from Lin and Byrnes [185]. Finally, results on stochastic systems with Markovian-jump disturbances can be found in [8, 9], while in [100] the results for controlled Markov diffusion processes are discussed. More recent developments in the theory of dissipative systems in the behavioral context can be found in the papers by Willems and Trentelman [275, 228, 259], while applications to stabilization of electrical and mechanical systems can be found in the book by Lozano et al. [187] and the references [204, 228].

4

Hamiltonian Mechanics and Hamilton-Jacobi Theory

Hamiltonian mechanics is a transformation theory that is an off-shoot of Lagrangian mechanics. It concerns itself with a systematic search for coordinate transformations which exhibit specific advantages for certain types of problems, notably in celestial and quantum mechanics. As such, the Hamiltonian approach to the analysis of a dynamical system, as it stands, does not represent an overwhelming development over the Lagrangian method. One ends up with practically the same number of equations as the Lagrangian approach. However, the real advantage of this approach lies in the fact that the transformed equations of motion in terms of a new set of position and momentum variables are easily integrated for specific problems, and also the deeper insight it provides into the formal structure of mechanics. The equal status accorded to coordinates and momenta as independent variables provides a new representation and greater freedom in selecting more relevant coordinate systems for different types of problems.

In this chapter, we study Lagrangian systems from the Hamiltonian standpoint. We shall consider natural mechanical systems for which the kinetic energy is a positive-definite quadratic form of the generalized velocities, and the Lagrangian function is the difference between the kinetic energy and the potential energy. Furthermore, as will be reviewed shortly, it will be shown that the Hamiltonian transformation of the equations of motion of a mechanical system always leads to the Hamilton-Jacobi equation (HJE) which is a first-order nonlinear PDE that must be solved in order to obtain the required transformation generating-function. It is therefore our aim in this chapter to give an overview of HJT with emphasis to the HJE.

4.1 The Hamiltonian Formulation of Mechanics

To review the approach, we begin with the following definition.

Definition 4.1.1 *A differentiable manifold M with a fixed positive-definite quadratic form $\langle \xi, \xi \rangle$ on every tangent space TM_x, $x \in M$, is called a Riemannian manifold. The quadratic form is called a Riemannian metric.*

Now, let the configuration space of the system be defined by a smooth n-dimensional Riemannian manifold M. If (φ, U) is a coordinate chart, we write $\varphi = q = (q_1, \ldots, q_n)$ for the local coordinates and $\dot{q}_i = \frac{\partial}{\partial q_i}$ in the tangent bundle $TM|_U = TU$. We shall be considering *natural mechanical systems* which are defined as follows.

Definition 4.1.2 *A Lagrangian mechanical system on a Riemannian manifold is called natural if the Lagrangian function $L : TM \times \Re \to \Re$ is equal to the difference between the kinetic energy and the potential energy of the system defined as*

$$L(q, \dot{q}, t) = T(q, \dot{q}, t) - V(q, t), \tag{4.1}$$

where $T : TM \times \Re \to \Re$ is the kinetic energy which is given by the symmetric Riemannian quadratic form

$$T = \frac{1}{2}\langle v, v\rangle, \quad v \in T_q M$$

and $V : M \times \Re \to \Re$ is the potential energy of the system (which may be independent of time).

More specifically, for natural mechanical systems, the kinetic energy is a positive-definite symmetric quadratic form of the generalized velocities,

$$T(q, \dot{q}, t) = \frac{1}{2}\dot{q}^T \Psi(q, t)\dot{q}. \tag{4.2}$$

Further, it is well known from Lagrangian mechanics and as can be derived using Hamilton's *principle of least action* [37, 115, 122] (see also Theorem 4.2.1), that the equations of motion of a *holonomic conservative*[1] mechanical system satisfy Langrange's equation of motion given by

$$\frac{d}{dt}\left(\frac{\partial L}{\partial \dot{q}_i}\right) - \frac{\partial L}{\partial q_i} = 0, \quad i = 1, \ldots, n. \tag{4.3}$$

Therefore, the above equation (4.3) may always be written in the form

$$\ddot{q} = g(q, \dot{q}, t), \tag{4.4}$$

for some function $g : TU \times \Re \to \Re^n$.

On the other hand, in the Hamiltonian formulation, we choose to replace all the $\dot{q}_i$ by independent coordinates, p_i, in such a way that

$$p_i := \frac{\partial L}{\partial \dot{q}_i}, \quad i = 1, \ldots, n. \tag{4.5}$$

If we let

$$p_i = h(q, \dot{q}), \quad i = 1, \ldots, n, \tag{4.6}$$

then the Jacobian of h with respect to $\dot{q}$, using (4.1), (4.2) and (4.5), is given by $\Psi(q)$ which is positive-definite, and hence equation (4.5) can be inverted to yield

$$\dot{q}_i = f_i(q_1, \ldots, q_n, p_1, \ldots, p_n, t), \quad i = 1, \ldots, n, \tag{4.7}$$

for some continuous functions $f_i, i = 1, \ldots, n$. In this framework, the coordinates $q = (q_1, q_2, \ldots, q_n)^T$ are referred to as the *generalized-coordinates* and $p = (p_1, p_2, \ldots, p_n)^T$ are the *generalized-momenta*. Together, these variables form a new system of coordinates for the system known as the *phase-space* of the system. If (U, φ) where $\varphi = (q_1, q_2, \ldots, q_n)$ is a chart on M, then since $p_i : TU \to \Re$, i=1,...,n, they are elements of $T^\star U$, and together with the q_i's form a system of $2n$ local coordinates $(q_1, \ldots, q_n, p_1, \ldots, p_n)$ for the phase-space $T^\star U$ of the system in U.

Now define the Hamiltonian function of the system $H : T^\star M \times \Re \to \Re$ as the *Legendre transform*[2] of the Lagrangian function with respect to $\dot{q}$ by

$$H(q, p, t) = p^T \dot{q} - L(q, \dot{q}, t), \tag{4.8}$$

---

[1]Holonomic if the constraints on the system are expressible as equality constraints. Conservative if there exists a time-dependent potential.

[2]To be defined later, see also [37].

and consider the differential of H with respect to q, p and t as

$$dH = \left(\frac{\partial H}{\partial p}\right)^T dp + \left(\frac{\partial H}{\partial q}\right)^T dq + \frac{\partial H}{\partial t} dt. \tag{4.9}$$

The above expression must be equal to the total differential of H given by (4.8) for $p = \frac{\partial L}{\partial \dot{q}}$:

$$dH = \dot{q}^T dp - \left(\frac{\partial L}{\partial q}\right)^T dq - \left(\frac{\partial L}{\partial t}\right)^T dt. \tag{4.10}$$

Thus, in view of the independent nature of the coordinates, we obtain a set of three relationships:

$$\dot{q} = \frac{\partial H}{\partial p}, \quad \frac{\partial L}{\partial q} = -\frac{\partial H}{\partial q}, \quad \text{and} \quad \frac{\partial L}{\partial t} = -\frac{\partial H}{\partial t}.$$

Finally, applying Lagrange's equation (4.3) together with (4.5) and the preceeding results, one obtains the expression for $\dot{p}$. Since we used Lagrange's equation, $\dot{q} = \frac{dq}{dt}$ and $\dot{p} = \frac{dp}{dt}$, and the resulting Hamiltonian canonical equations of motion are then given by

$$\frac{dq}{dt} = \frac{\partial H}{\partial p}(q, p, t), \tag{4.11}$$

$$\frac{dp}{dt} = -\frac{\partial H}{\partial q}(q, p, t). \tag{4.12}$$

Therefore, we have proven the following theorem.

Theorem 4.1.1 *The system of Lagrange's equations (4.3) is equivalent to the system of $2n$ first-order Hamilton's equations (4.11), (4.12).*

In addition, for time-independent conservative systems, $H(q, p)$ has a simple physical interpretation. From (4.8) and using (4.5), we have

$$
\begin{aligned}
H(q, p) &= p^T \dot{q} - L(q, \dot{q}) = \dot{q}^T \frac{\partial L}{\partial \dot{q}} - (T(q, \dot{q}) - V(q)) \\
&= \dot{q}^T \frac{\partial T}{\partial \dot{q}} - T(q, \dot{q}) + V(q) \\
&= 2T(q, \dot{q}) - T(q, \dot{q}) + V(q) \\
&= T(q, \dot{q}) + V(q).
\end{aligned}
$$

That is, $H(q, p, t)$ is the total energy of the system. This completes the Hamiltonian formulation of the equations of motion, and can be seen as an off-shoot of the Lagrangian formulation. It can also be seen that, while the Lagrangian formulation involves n second-order equations, the Hamiltonian description sets up a system of $2n$ first-order equations in terms of the $2n$ variables p and q. This remarkably new system of coordinates gives new insight and physical meaning to the equations. However, the system of Lagrange's equations and Hamilton's equations are completely equivalent and dual to one another.

Furthermore, because of the symmetry of Hamilton's equations (4.11), (4.12) and the even dimension of the system, a new structure emerges on the phase-space T^*M of the system. This structure is defined by a nondegenerate closed differential 2-form which in the above local coordinates is defined as:

$$\omega^2 = dp \wedge dq = \sum_{i=1}^{n} dp_i \wedge dq_i. \tag{4.13}$$

Thus, the pair $(T^\star M, \omega^2)$ form a *symplectic-manifold*, and together with the C^r Hamiltonian function $H : T^\star M \to \Re$, define a *Hamiltonian mechanical system*. With this notation, we have the following representation of a Hamiltonian system.

Definition 4.1.3 *Let $(T^\star M, \omega^2)$ be a symplectic-manifold and let $H : T^\star M \to \Re$ be a Hamiltonian function. Then, the vector-field X_H determined by the condition*

$$\omega^2(X_H, Y) = dH(Y) \tag{4.14}$$

for all vector-fields Y, is called the Hamiltonian vector-field *with energy function H. We call the tuple $(T^\star M, \omega^2, X_H)$ a Hamiltonian system.*

Remark 4.1.1 *It is important to note that the nondegeneracy[3] of ω^2 guarantees that X_H exists, and is a C^{r-1} vector-field. Moreover, on a connected symplectic-manifold, any two Hamiltonians for the same vector-field X_H have the same differential (4.14), so differ by a constant only.*

We also have the following proposition [1].

Proposition 4.1.1 *Let $(q_1, \ldots, q_n, p_1, \ldots, p_n)$ be canonical coordinates so that ω^2 is given by (4.13). Then, in these coordinates*

$$X_H = \left(\frac{\partial H}{\partial p_1}, \ldots, \frac{\partial H}{\partial p_n}, -\frac{\partial H}{\partial q_1}, \ldots, -\frac{\partial H}{\partial q_n} \right) = J \cdot \nabla H$$

where $J = \begin{pmatrix} 0 & I \\ -I & 0 \end{pmatrix}$. Thus, $(q(t), p(t))$ is an integral curve of X_H if and only if Hamilton's equations (4.11), (4.12) hold.

4.2   Canonical Transformation

Now suppose that a transformation of coordinates is introduced $q_i \to Q_i$, $p_i \to P_i$, $i = 1, \ldots, n$ such that every Hamiltonian function transforms as $H(q_1, \ldots, q_n, p_1, \ldots, p_n, t) \to K(Q_1, \ldots, Q_n, P_1, \ldots, P_n, t)$ in such a way that the new equations of motion retain the same form as in the former coordinates, i.e.,

$$\frac{dQ}{dt} = \frac{\partial K}{\partial p}(Q, P, t) \tag{4.15}$$

$$\frac{dP}{dt} = -\frac{\partial K}{\partial q}(Q, P, t). \tag{4.16}$$

Such a transformation is called *canonical* and can greatly simplify the solution to the equations of motion, especially if Q, P are selected such that $K(., ., .)$ is a constant independent of Q and P. When this happens, then Q and P will also be constants and the solution to the equations of motion are immediately available (given the transformation). We simply transform back to the original coordinates under the assumption that the transformation is univalent and invertible. It therefore follows from this that:

---

[3]ω^2 is nondegenerate if $\omega^2(X_1, X_2) = 0 \Rightarrow X_1 = 0$ or $X_2 = 0$ for all vector-fields X_1, X_2 which are smooth sections of $TT^\star M$.

1. The identity transformation is canonical;

2. The inverse of a canonical transformation is a canonical transformation;

3. The product of two or more canonical transformations is also a canonical transformation;

4. A canonical transformation must preserve the differential-form $\omega^2 = dp \wedge dq$ or preserve the canonical nature of the equations of motion (4.15), (4.16).

The use of canonical invariants such as *Poisson brackets* [38, 115] can often be used to check whether a given a transformation $(q, p) \to (Q, P)$ is canonical or not. For any two given C^1-functions $u(q, p)$, $v(q, p)$, their Poisson bracket is defined as

$$[u, \; v]_{q,p} = \sum_{i=1}^{n} \left(\frac{\partial u}{\partial q_i} \frac{\partial v}{\partial p_i} - \frac{\partial u}{\partial p_i} \frac{\partial v}{\partial q_i} \right). \tag{4.17}$$

It can then be shown that a transformation $(q, p) \mapsto (Q, P)$ is canonical if and only if:

$$[Q_i, \; Q_k]_{q,p} = 0, \quad [P_i, \; P_k]_{q,p} = 0, \quad [P_i, \; Q_k]_{q,p} = \delta_{ik}, \quad i, k = 1, 2, \ldots, n \tag{4.18}$$

are satisfied, where δ_{ik} is the Kronecker delta.

Hamilton (1838) has developed a method for obtaining the desired transformation equations using what is today known as *Hamilton's principle* which we introduce hereafter.

Definition 4.2.1 *Let $\gamma = \{(t, q) \; : \; q = q(t), \; t_0 \leq t \leq t_1\}$ be a curve in the (t, q) plane. Define the functional $\Phi(\gamma)$ (which we assume to be differentiable) by*

$$\Phi(\gamma) = \int_{t_0}^{t_1} L(q(\tau), \dot{q}(\tau)) d\tau.$$

Then, the curve γ is an extremal of the functional $\Phi(.)$ if $\delta\Phi(\gamma) = 0$ or $d\Phi(\gamma) = 0 \; \forall t \in [t_0, t_1]$, where δ is the variational operator.

Theorem 4.2.1 *(Hamilton's Principle of Least Action) [37, 115, 122, 127]. The motion of a mechanical system with Lagrangian function $L(.,.,.)$, coincides with the extremals of the functional $\Phi(\gamma)$.*

Accordingly, define the Lagrangian function of the system $L : TM \times \Re \to \Re$ as the Legendre transform [37] of the Hamiltonian function by

$$L(q, \dot{q}, t) = p^T \dot{q} - H(q, p, t). \tag{4.19}$$

Then, in the new coordinates, the new Lagrangian function is

$$\bar{L}(Q, \dot{Q}, t) = P^T \dot{Q} - K(Q, P, t). \tag{4.20}$$

Since both $L(.,.,.)$ and $\bar{L}(.,.,.)$ are conserved, each must separately satisfy Hamilton's principle. However, $L(.,.,.)$ and $\bar{L}(.,.,.)$ need not be equal in order to satisfy the above requirement. Indeed, we can write

$$L(q, \dot{q}, t) = \bar{L}(Q, \dot{Q}, t) + \frac{dS}{dt}(q, p, Q, P, t) \tag{4.21}$$

for some arbitrary function $S : \mathcal{X} \times \bar{\mathcal{X}} \times \Re \to \Re$, where $\mathcal{X}, \bar{\mathcal{X}} \subset T^\star M$ (see also [122], page 286). Since dS is an exact differential (i.e., it is the derivative of a scalar function),

$$\delta \left[\int_{t_0}^{t_1} \frac{dS}{dt}(q, p, Q, P, t) dt \right] = S(q, p, Q, P, t)|_{t_0}^{t_1} = 0. \tag{4.22}$$

Now applying Hamilton's principle to the time integral of both sides of equation (4.21), we get

$$\delta\left[\int_{t_0}^{t_1} L(q,\dot{q},t)dt\right] = \delta\left[\int_{t_0}^{t_1} \bar{L}(Q,\dot{Q},t)dt\right] + \delta\left[\int_{t_0}^{t_1} \frac{dS}{dt}(q,p,Q,P,t)dt\right] = 0; \qquad (4.23)$$

and therefore by (4.22),

$$\delta\left[\int_{t_0}^{t_1} \bar{L}(Q,\dot{Q},t)dt\right] = 0. \qquad (4.24)$$

Thus, to guarantee that a given change of coordinates, say,

$$q_i = \phi_i(Q,P,t) \qquad (4.25)$$
$$p_i = \psi_i(Q,P,t) \qquad (4.26)$$

is canonical, from (4.19), (4.20) and (4.21), it is enough that

$$p^T\dot{q} - H = P^T\dot{Q} - K + \frac{dS}{dt}. \qquad (4.27)$$

This condition is also required [122]. Consequently, the above equation is equivalent to

$$p^T dq - P^T dQ = (H-K)(q,p,Q,P,t)dt + dS(q,p,Q,P,t), \qquad (4.28)$$

which requires on the expression on the left side to be also an exact differential. Further, it can be verified that the presence of $S(.)$ in (4.21) does not alter the canonical structure of the Hamiltonian equations. Applying Hamilton's principle to the right-hand-side of (4.21), we have from (4.24), the Euler-Lagrange equation (4.3), and the argument following it

$$\frac{dQ}{dt} = \frac{\partial K}{\partial p}(Q,P,t) \qquad (4.29)$$

$$\frac{dP}{dt} = -\frac{\partial K}{\partial q}(Q,P,t). \qquad (4.30)$$

Hence the canonical nature of the equations is preserved.

4.2.1   The Transformation Generating Function

As proposed in the previous section, the equations of motion of a given Hamiltonian system can often be simplified significantly by a suitable transformation of variables such that all the new position and momentum coordinates (Q_i, P_i) are constants. In this subsection, we discuss Hamilton's approach for finding such a transformation.

We have already seen that an arbitrary generating function S does not alter the canonical nature of the equations of motion. The next step is to show that, first, if such a function is known, then the transformation we so anxiously seek follows directly. Secondly, that the function can be obtained by solving a certain partial-differential equation (PDE).

The generating function S relates the old to the new coordinates via the equation

$$S = \int (L - \bar{L})dt = f(q,p,Q,P,t). \qquad (4.31)$$

Therefore, S is a function of $4n + 1$ variables of which only $2n$ are independent. Hence, no more than four independent sets of relationships among the dependent coordinates can exist. Two such relationships expressing the old sets of coordinates in terms of the new set are given by equations (4.25), (4.26). Consequently, only two independent sets of relationships among the coordinates remain for defining S and no more than two of the four sets of coordinates may be involved. Therefore, there are four possibilities:

$$S_1 = f_1(q, Q, t); \quad S_2 = f_2(q, P, t); \tag{4.32}$$
$$S_3 = f_3(p, Q, t); \quad S_4 = f_4(p, P, t). \tag{4.33}$$

Any one of the above four types of generating functions may be selected, and a transformation obtained from it. For example, if we consider the generating function S_1, taking its differential, we have

$$dS_1 = \sum_{i=1}^{n} \frac{\partial S_1}{\partial q_i} dq_i + \sum_{i=1}^{n} \frac{\partial S_1}{\partial Q_i} dQ_i + \frac{\partial S_1}{\partial t} dt. \tag{4.34}$$

Again, taking the differential as defined by (4.28), we have

$$dS_1 = \sum_{i=1}^{n} p_i dq_i - \sum_{i=1}^{n} P_i dQ_i + (K - H)dt. \tag{4.35}$$

Finally, using the independence of coordinates, we equate coefficients, and obtain the desired transformation equations

$$\left. \begin{array}{rcl} p_i & = & \frac{\partial S_1}{\partial q_i}(q, Q, t) \\ P_i & = & -\frac{\partial S_1}{\partial Q_i}(q, Q, t) \\ K - H & = & \frac{\partial S_1}{\partial t}(q, Q, t) \end{array} \right\}, \; i = 1, \ldots, n. \tag{4.36}$$

Similar derivation can be applied to the remaining three types of generating functions, and in addition, we can also apply Legendre transformation. Thus, for the generating functions $S_2(., ., .)$, $S_3(., ., .)$ and $S_4(., ., .)$, we have

$$\left. \begin{array}{rcl} p_i & = & \frac{\partial S_2}{\partial q_i}(q, P, t) \\ Q_i & = & \frac{\partial S_2}{\partial P_i}(q, P, t) \\ K - H & = & \frac{\partial S_2}{\partial t}(q, P, t) \end{array} \right\}, i = 1, \ldots, n, \tag{4.37}$$

$$\left. \begin{array}{rcl} q_i & = & -\frac{\partial S_3}{\partial p_i}(p, Q, t) \\ P_i & = & -\frac{\partial S_3}{\partial Q_i}(p, Q, t) \\ K - H & = & \frac{\partial S_3}{\partial t}(p, Q, t) \end{array} \right\}, i = 1, \ldots, n, \tag{4.38}$$

$$\left. \begin{array}{rcl} q_i & = & -\frac{\partial S_4}{\partial p_i}(p, P, t) \\ Q_i & = & \frac{\partial S_4}{\partial P_i}(p, P, t) \\ K - H & = & \frac{\partial S_4}{\partial t}(p, P, t) \end{array} \right\}, i = 1, \ldots, n, \tag{4.39}$$

respectively. It should however be remarked that most of the canonical transformations that are expressed using arbitrary generating functions often have the consequence that the distinct meaning of the generalized coordinates and momenta is blurred. For example, consider the generating function $S = S_1(q, Q) = q^T Q$. Then, it follows from the foregoing that

$$\left. \begin{array}{rcl} p_i & = & \frac{\partial S_1}{\partial q_i} = Q_i \\ P_i & = & -\frac{\partial S_1}{\partial Q_i} = -q_i \\ K & = & \frac{\partial S_1}{\partial t} = H(-P, Q, t) \end{array} \right\}, i = 1, \ldots, n, \tag{4.40}$$

which implies that P_i and q_i have the same units except for the sign change.

One canonical transformation that allows only the tranformation of corresponding coordinates is called a *point transformation*. In this case, $Q(q,t)$ does not depend on p but only on q and possibly t and the meaning of the coordinates is preserved. This ability of a point transformation can also be demonstrated using the genarating function S_2. Consider for instance the transformation $Q = \psi(q,t)$ of the coordinates among each other such that

$$S = S_2(q,P,t) = \psi^T(q,t)P. \tag{4.41}$$

Then, the resulting canonical equations are given by

$$\left. \begin{aligned} p &= \frac{\partial S_2}{\partial q} = \left(\frac{\partial \psi}{\partial q}\right)^T P, \\ Q &= \frac{\partial S_2}{\partial P} = \psi(q,t), \\ K &= H + \frac{\partial \psi^T(q,t)}{\partial t}P \end{aligned} \right\}, \tag{4.42}$$

and it is clear that the meaning of the coordinates is preserved in this case.

4.2.2   The Hamilton-Jacobi Equation (HJE)

In this subsection, we turn our attention to the last missing link in the Hamiltonian transformation theory, i.e., an approach for determining the transformation generating function, S. There is only one equation available

$$H(q,p,t) + \frac{\partial S}{\partial t} = K(P,Q,t). \tag{4.43}$$

However, there are two unknown functions in this equation, namely, S and K. Therefore, the best we can do is to assume a solution for one and then solve for the other. A convenient and intuitive strategy is to arbitrarily set K to be *identically zero*! Under this condition, $\dot{Q}$ and $\dot{P}$ vanish, resulting in $Q = \alpha$, and $P = \beta$, as constants. The inverse transformation then yields the motion $q(\alpha,\beta,t)$, $p(\alpha,\beta,t)$ in terms of these constants of integration.

Consider now generating functions of the first type. Having forced a solution $K \equiv 0$, we must now solve the PDE:

$$\mathbf{H}(\mathbf{q}, \frac{\partial \mathbf{S}}{\partial \mathbf{q}}, \mathbf{t}) + \frac{\partial \mathbf{S}}{\partial \mathbf{t}} = \mathbf{0} \tag{4.44}$$

for S, where $\frac{\partial S}{\partial q} = (\frac{\partial S}{\partial q_1}, \ldots, \frac{\partial S}{\partial q_n})^T$. This equation is known as the *Hamilton-Jacobi equation (HJE)*, and was improved and modified by Jacobi in 1838. For a given function $H(q,p,t)$, this is a first-order PDE in $n+1$ variables for the unknown function $S(q,\alpha,t)$ which is traditionally called *Hamilton's principal function*. We need a solution for this equation which depends on n arbitrary independent constants of integration $\alpha_1, \alpha_2, \ldots, \alpha_n$. Such a solution $S(q,\alpha,t)$ is called a *"complete solution"* of the HJE (4.44), and solving the HJE is equivalent to finding the solution to the equations of motion (4.11), (4.12). On the other hand, the solution of (4.44) is simply the solution of the equations (4.11), (4.12) using the method of characteristics [95]. However, it is generally not simpler to solve (4.44) instead of (4.11), (4.12).

If a complete solution $S(q,\alpha,t)$ of (4.44) can be found and if the generating function $S = S_1(q,Q,t)$ is used, then one obtains

$$\frac{\partial S_1}{\partial q_i} = p_i, \ i = 1, \ldots, n, \tag{4.45}$$

$$\frac{\partial S_1}{\partial \alpha_i} = -\beta_i, \ i = 1, \ldots, n. \tag{4.46}$$

Moreover, since the constants α_i are independent, the Jacobian matrix $\frac{\partial^2 S_1}{\partial q \partial \alpha}$ is nonsingular and therefore by the Implicit-function Theorem, the above two equations can be solved to recover the original variables $q(\alpha, \beta, t)$ and $p(\alpha, \beta, t)$.

4.2.3 Time-Independent Hamilton-Jacobi Equation and Separation of Variables

The preceding section has laid down a systematic approach to the solution of the equations of motion via a transformation theory that culminates in the HJE. However, implementation of the above procedure is difficult, because the chances of success are limited by the lack of efficient mathematical techniques for solving nonlinear PDEs. At present, the only general technique is the method of *separation of variables*. If the Hamiltonian is explicitly a function of time, then separation of variables is not readily achieved for the HJE. On the other hand, if the Hamiltonian is not explicitly a function of time or is independent of time, which arises in many dynamical systems of practical interest, then the HJE separates easily. The solution to (4.44) can then be formulated in the form

$$S(q, \alpha, t) = W(q, \alpha) - \alpha_1 t. \tag{4.47}$$

Consequently, the use of (4.47) in (4.44) yields the following restricted time-independent HJE in W:

$$H(q, \frac{\partial W}{\partial q}) = \alpha_1, \tag{4.48}$$

where α_1, one of the constants of integration is equal to the constant value of H or is an energy constant (if the kinetic energy of the system is homogeneous quadratic, the constant equals the total energy, E). Moreover, since W does not involve time, the new and the old Hamiltonians are equal, and it follows that $K = \alpha_1$. The function W, known as *Hamilton's characteristic function*, thus generates a canonical transformation in which all the new coordinates are *cyclic*.[4] Further, if we again consider generating functions of the first kind, i.e., $S = S_1(q, Q, t)$, then from (4.45), (4.46) and (4.47), we have the following system

$$\left. \begin{array}{rcl} \frac{\partial W}{\partial q_i} & = & p_i, \quad i = 1, 2 \ldots, n, \\ \frac{\partial W}{\partial \alpha_1} & = & t + \beta_1, \\ \frac{\partial W}{\partial \alpha_i} & = & \beta_i, \quad i = 2, \ldots, n. \end{array} \right\} \tag{4.49}$$

The above system of equations can then be solved for the q_i in terms of α_i, β_i and time t.

At this point, it might appear that little practical advantage has been gained in solving a first-order nonlinear PDE, which is notoriously difficult to solve, instead of a system of $2n$ ODEs. Nevertheless, under certain conditions, and when the Hamiltonian is independent of time, it is possible to separate the variables in the HJE, and the solution can then be obtained by integration. In this event, the HJE becomes a useful computational tool.

Unfortunately, there is no simple criterion (for orthogonal coordinate systems the so-called Staeckel conditions [115] have proven to be useful) for determining when the HJE is separable. For some problems, e.g., the three-body problem, it is impossible to separate the variables, while for others it is transparently easy. Fortunately, a great majority of systems of current interest in quantum mechanics and atomic physics are of the latter class. Moreover, it should also be emphasized that the question of whether the HJE is separable depends on the system of generalized coordinates employed. Indeed, the one-body central force problem is separable in polar coordinates, but not in cartesian coordinates.

---

[4]A coordinate q_i is cyclic if it does not enter into the Lagrangian, i.e., $\frac{\partial L}{\partial q_i} = 0$.

To illustrate the Hamilton-Jacobi technique for the time-independent case, we consider an example of the harmonic oscillator.

Example 4.2.1 *[115]. Consider the harmonic oscillator with Hamiltonian function*

$$H_h = \frac{p^2}{2m} + \frac{kq^2}{2}.$$

The corresponding HJE (4.48) is given by

$$\frac{1}{2m}\left(\frac{\partial W}{\partial q}\right)^2 + \frac{kq^2}{2} = \alpha$$

which can be immediately integrated to yield

$$W(q,\alpha) = \sqrt{mk}\int \sqrt{\frac{2\alpha}{k} - q^2}\, dq.$$

Thus,

$$S(q,\alpha) = \sqrt{mk}\int \sqrt{\frac{2\alpha}{k} - q^2}\, dq - \alpha t,$$

and

$$\beta = \frac{\partial S}{\partial \alpha} = \sqrt{\frac{m}{k}}\int \frac{dq}{\sqrt{\frac{2\alpha}{k} - q^2}} - t.$$

The above equation can now be integrated to yield

$$t + \beta = -\sqrt{\frac{m}{k}}\cos^{-1}\left(q\sqrt{\frac{k}{2\alpha}}\right).$$

Now, if we let $\omega = \sqrt{\frac{k}{m}}$, then the above equation can be solved for q to get the solution

$$q(t) = \sqrt{\frac{2\alpha}{k}}\cos(\omega t + \beta)$$

with α, β constants of integration.

4.3   The Theory of Nonlinear Lattices

In this section, we discuss the theory of nonlinear lattices as a special class and an example of Hamiltonian systems that are integrable. Later, we shall also show how the HJE arising from the $\mathcal{A}_2$-Toda lattice can be solved.

Historically, the exact treatment of oscillations in nonlinear lattices became serious in the early 1950's when Fermi, Pasta and Ulam (FPU) numerically studied the problem of energy partition. Fermi et al. wanted to verify by numerical experiment if there is energy flow between the modes of linear-lattice systems when nonlinear interactions are introduced. He wanted to verify what is called the *equipartition of energy* in statistical mechanics. However, to their disappointment, only a little energy partition occurred, and the state of the systems was found to return periodically to the initial state.

Later, Ford and co-workers [258] showed that by using pertubation and by numerical calculation, though resonance generally enhances energy sharing, it has no intimate connection to a periodic phenomenon, and that nonlinear lattices have rather stable-motion (periodic, when the energy is not too high) or pulses (also known as solitons), which he called the *nonlinear normal modes*. This fact also indicates that there will be some nonlinear lattice which admits rigorous periodic waves, and certain pulses (lattice solitons) will be stable there. This remarkable property led to the finding of an integrable one-dimensional lattice with exponential interaction also known as the *Toda lattice*.

The Toda lattice as a Hamiltonian system describes the motion of n particles moving in a straight line with "exponential interaction" between them. Mathematically, it is equivalent to a problem in which a single particle moves in $\Re^n$. Let the positions of the particles at time t (in $\Re$) be $q_1(t), \ldots, q_n(t)$, respectively. We assume that each particle has mass 1. The momentum of the i-th particle at time t is therefore $p_i = \dot{q}_i$. The Hamiltonian function for the *finite (or non-periodic)* lattice is defined to be

$$H(q, p) = \frac{1}{2} \sum_{j=1}^{n} p_j^2 + \sum_{j=1}^{n-1} e^{2(q_j - q_{j+1})}. \tag{4.50}$$

Therefore, the canonical equations for the system are given by

$$\left. \begin{array}{rcl} \frac{dq_j}{dt} & = & p_j \quad j = 1, \ldots, n, \\ \frac{dp_1}{dt} & = & -2e^{2(q_1 - q_2)}, \\ \frac{dp_j}{dt} & = & -2e^{2(q_j - q_{j+1})} + 2e^{2(q_{j-1} - q_j)}, \quad j = 2, \ldots, n-1, \\ \frac{dp_n}{dt} & = & 2e^{2(q_{n-1} - q_n)}. \end{array} \right\} \tag{4.51}$$

It may be assumed in addition that $\sum_{j=1}^{n} q_j = \sum_{j=1}^{n} p_j = 0$, and the coordinates $q_1, \ldots, q_n$ can be chosen so that this condition is satisfied. While for the periodic lattice in which the first particle interacts with the last, the Hamiltonian function is defined by

$$\tilde{H}(q, p) = \frac{1}{2} \sum_{j=1}^{n} p_j^2 + \sum_{j=1}^{n-1} e^{2(q_j - q_{j+1})} + e^{2(q_n - q_1)}. \tag{4.52}$$

We may also consider the infinite lattice, in which there are infinitely many particles.

Nonlinear lattices can provide models for nonlinear phenomena such as wave propagation in nerve systems, chemical reactions, certain ecological systems and a host of electrical and mechanical systems. For example, it is easily shown that a linear lattice is equivalent to a ladder network composed of capacitors $\mathbf{C}$ and inductors $\mathbf{L}$, while a one-dimensional nonlinear lattice is equivalent to a ladder circuit with nonlinear $\mathbf{L}$ or $\mathbf{C}$. To show this, let I_n denote the current, Q_n the charge on the capacitor, Φ_n the flux in the inductance, and write the equations for the circuit as

$$\left. \begin{array}{rcl} \frac{dQ_n}{dt} & = & I_n - I_{n-1}, \\ \frac{d\Phi_n}{dt} & = & V_n - V_{n+1}. \end{array} \right\}. \tag{4.53}$$

Now assume that the inductors and capacitors are nonlinear in such a way that

$$\begin{array}{rcl} Q_n & = & Cv_0 \ln(1 + V_n/v_0) \\ \Phi_n & = & Li_0 \ln(1 + I_n/i_0) \end{array}$$

where (C, v_0, L, i_0) are constants. Then equations (4.53) give

$$\begin{array}{rcl} \dfrac{dQ_n}{dt} & = & i_0(e^{\frac{\Phi_{n-1}}{Li_0}} - e^{\frac{\Phi_n}{Li_0}}) \\ \dfrac{d\Phi_n}{dt} & = & v_0(e^{\frac{Q_{n-1}}{Cv_0}} - e^{\frac{Q_n}{Cv_0}}) \end{array}$$

which are in the form of a lattice with exponential interaction (or Toda system).

Stimulated by Ford's numerical work which revealed the likely integrability of the Toda lattice, Henon and Flaschka [258] independently showed the integrability of the Toda lattice analytically, and began an analytical survey of the lattice. At the same time, the inverse scattering method of solving the initial value problem for the Kortoweg-de Vries equation (KdV) had been firmly formulated by Lax [258], and this method was applied to the infinite lattice to derive a solution using matrix formalism which led to a simplification of the equations of motion. To introduce this formalism, define the following matrices

$$
L = \begin{pmatrix}
p_1 & Q_{1,2} & 0 & \cdots & 0 & 0 \\
Q_{1,2} & p_2 & Q_{2,3} & \cdots & 0 & 0 \\
0 & Q_{2,3} & p_3 & \cdots & 0 & 0 \\
\vdots & \vdots & \vdots & & \vdots & \vdots \\
0 & 0 & 0 & \cdots & p_{n-1} & Q_{n-1,n} \\
0 & 0 & 0 & \cdots & Q_{n-1,n} & p_n
\end{pmatrix}
\tag{4.54}
$$

$$
M = \begin{pmatrix}
0 & Q_{1,2} & 0 & \cdots & 0 & 0 \\
-Q_{1,2} & 0 & Q_{2,3} & \cdots & 0 & 0 \\
0 & -Q_{2,3} & 0 & \cdots & 0 & 0 \\
\vdots & \vdots & \vdots & & \vdots & \vdots \\
0 & 0 & 0 & \cdots & 0 & Q_{n-1,n} \\
0 & 0 & 0 & \cdots & -Q_{n-1,n} & 0
\end{pmatrix}
\tag{4.55}
$$

where $Q_{i,j} = e^{(q_i - q_j)}$. We then have the following propoposition [123].

Proposition 4.3.1 *The Hamiltonian system for the non-periodic Toda lattice (4.50)-(4.51) is equivalent to the Lax equation $\dot{L} = [L, M]$, where the function L, M take values in $sl(n, \Re)$*[5] *and $[.,.]$ is the Lie bracket operation in $sl(n, \Re)$.*

Using the above matrix formalism, the solution of the Toda system (4.51) can be derived [123, 258].

Theorem 4.3.1 *The solution of the Hamiltonian system for the Toda lattice is given by*

$$
L(t) = Ad(\exp tV)_1^{-1} V,
$$

where $V = L(0)$, $Ad(g)X = \frac{d}{dt} g \exp(tX)g^{-1}\big|_{t=0} = gXg^{-1}$ for any $X \in SL(n, \Re)$, $g \in sl(n, \Re)$, and the subscript 1 represents the projection $(\exp -tW)_1 = \exp -t\pi_1 W = \exp -tW_1$ onto the first component in the decomposition of $W = W_1 W_2 \in sL(n, \Re)$.

The solution can be explicitly written in the case of $n = 2$. Letting $q_1 = -q$, $q_2 = q$, $p_1 = -p$ and $p_2 = p$, we have

$$
L = \begin{pmatrix} p & Q \\ Q & -p \end{pmatrix}, \quad M = \begin{pmatrix} 0 & Q \\ -Q & 0 \end{pmatrix},
\tag{4.56}
$$

where $Q = c^{-2q}$. Then the solution of $\dot{L} = [L, M]$ with

$$
L(0) = \begin{pmatrix} 0 & v \\ v & 0 \end{pmatrix},
$$

---

[5]The Lie-algebra of $SL(n, \Re)$, the special linear group of matrices on $\Re$ with determinant ± 1 [38].

is

$$L(t) = Ad \left[\exp t \begin{pmatrix} 0 & v \\ v & 0 \end{pmatrix} \right]_I^{-1} \begin{pmatrix} 0 & v \\ v & 0 \end{pmatrix}.$$

Now

$$\exp t \begin{pmatrix} 0 & v \\ v & 0 \end{pmatrix} = \begin{pmatrix} \cosh tv & \sinh tv \\ \sinh tv & \cosh tv \end{pmatrix}.$$

The decomposition $SL_2(2, \Re) = SO_2 \hat{N}'_2$ is given by

$$\begin{pmatrix} a & b \\ c & d \end{pmatrix} = \left[\frac{1}{\sqrt{b^2 + d^2}} \begin{pmatrix} d & b \\ -b & d \end{pmatrix} \right] \left[\frac{1}{\sqrt{b^2 + d^2}} \begin{pmatrix} 1 & 0 \\ ab + cd & b^2 + d^2 \end{pmatrix} \right].$$

Hence,

$$\left[\exp t \begin{pmatrix} 0 & v \\ v & 0 \end{pmatrix} \right]_1 = \frac{1}{\sqrt{\sinh^2 tv + \cosh^2 tv}} \begin{pmatrix} \cosh tv & \sinh tv \\ -\sinh tv & \cosh tv \end{pmatrix}.$$

Therefore,

$$L(t) = \frac{v}{\sinh^2 tv + \cosh^2 tv} \begin{pmatrix} -2\sinh tv \cosh tv & 1 \\ 1 & 2\sinh tv \cosh tv \end{pmatrix}.$$

Which means that

$$p(t) = -v\frac{\sinh 2tv}{\cosh 2tv}, \quad Q(t) = \frac{v}{\cosh 2tv}.$$

Furthermore, if we recall that $Q(t) = e^{-2q(t)}$, it follows that

$$\begin{aligned} q(t) &= -\frac{1}{2} \log \left(\frac{v}{\cosh 2tv} \right) \\ &= -\frac{1}{2} \log v + \frac{1}{2} \log \cosh 2vt. \end{aligned} \tag{4.57}$$

4.3.1 The $\mathcal{G}_2$-Periodic Toda Lattice

In the study of the generalized periodic Toda lattice, Bogoyavlensky [72] showed that various models of the Toda lattice which admit the $[L, M]$-Lax representation correspond to certain simple Lie-algebras which he called the $\mathcal{A}, \mathcal{B}, \mathcal{C}, \mathcal{D}$ and $\mathcal{G}_2$ periodic Toda systems. In particular, the $\mathcal{G}_2$ is a two-particle system and corresponds to the Lie algebra g_2 which is 14-dimensional, and has been studied extensively in the literature [3, 4, 201]. The Hamiltonian for the g_2 system is given by

$$H(q, p) = \frac{1}{2}(p_1^2 + p_2^2) + e^{(1/\sqrt{3})q_1} + e^{-(\sqrt{3}/2)q_1 + (1/2)q_2} + e^{-q_2}, \tag{4.58}$$

and the Lax equation corresponding to this system is given by $dA/dt = [A, B]$, where

$$\begin{aligned} A(t) &= a_1(t)(X_{-\beta_3} + X_{\beta_3}) + a_2(t)(X_{-\gamma_1} + X_{\gamma_1}) + a_3(t)(s^{-1}X_{-\gamma_3} + sX_{\gamma_3}) + \\ &\quad b_1(t)H_1 + b_2H_2 \\ B(t) &= a_1(t)(X_{-\beta_3} - X_{\beta_3}) + a_2(t)(X_{-\gamma_1} - X_{\gamma_1}) + a_3(t)(s^{-1}X_{-\gamma_3} - sX_{\gamma_3}), \end{aligned}$$

s is a parameter, $\beta_i, i = 1, 2, 3$ and the $\gamma_j, j = 1, 2, 3$ are the short and long roots of the g_2 root system respectively, while $X_{(-)}$ are the corresponding Chevalley basis vectors. Using the following change of coordinates [201]:

$$a_1(t) = \frac{1}{2\sqrt{6}} e^{(1/2\sqrt{3})q_1(t)}, \quad a_2(t) = \frac{1}{2\sqrt{2}} e^{-(\sqrt{3}/4)q_1(t)+(1/4)q_2(t)},$$

$$a_3(t) = \frac{1}{2\sqrt{2}} e^{-(1/2)q_2(t)},$$

$$b_1(t) = \frac{-1}{2\sqrt{3}} p_1(t) + \frac{1}{4} p_2(t), \quad b_2(t) = \frac{1}{2\sqrt{3}} p_1(t),$$

we can represent the g_2 lattice as

$$\dot{a}_1 = a_1 b_2, \quad \dot{a}_2 = a_2(b_1 - b_2), \quad \dot{a}_3 = a_3(-2b_1 - b_2),$$
$$\dot{b}_1 = 2(a_1^2 - a_2^2 + a_3^2), \quad \dot{b}_2 = -4a_1^2 + 2a_2^2,$$
$$H = \frac{1}{2}\langle A(t), A(t)\rangle = 8(3a_1^2 + a_2^2 + a_3^2 + a_3^2 + b_1^2 + b_1 b_2 + b_2^2).$$

Here, the coordinate $a_2(t)$ may be regarded as superfluous, and can be eliminated using the fact that $4a_1^3 a_2^2 a_3 = c$ (a constant) of the motion.

4.4 The Method of Characteristics for First-Order Partial-Differential Equations

In this section, we present the wellknown *method of characteristics* for solving first-order PDEs. It is by far the most generally known method for handling first-order nonlinear PDEs in n independent variables. It involves converting the PDE into an appropriate system of first-order ordinary differential-equations (ODE), which are in turn solved together to obtain the solution of the original PDE. It will be seen during the development that the Hamilton's canonical equations are nothing but the characteristic equations of the Hamilton-Jacobi equation; and thus, solving the canonical equations is equivalent to solving the PDE and vice-versa. The presentation will follow closely those from Fritz-Johns [109] and Evans [95].

4.4.1 Characteristics for Quasi-Linear Equations

We begin with a motivational discussion of the method by considering quasi-linear equations, and then we consider the general first-order nonlinear equation.

The general first-order equation for a function $v = v(x, y, \ldots, z)$ in n variables is of the form

$$f(x, y, \ldots, z, v, v_x, v_y, \ldots, v_z) = 0, \tag{4.59}$$

where $x, y, \ldots, z \in \Re$, $v : \Re^n \to \Re$. The HJE and many first-order PDEs in classical and continuum mechanics, calculus of variations and geometric optics are of the above type. A simpler case of the above equation is the *quasi-linear equation* in two variables:

$$a(x, y, v)v_x + b(x, y, v)v_y = c(x, y, v) \tag{4.60}$$

in two independent variables x, y. The function $v(x, y)$ is represented by a surface $z = v(x, y)$ called an *integral surface* which corresponds to a solution of the PDE. The functions $a(x, y, z)$, $b(x, y, z)$ and $c(x, y, z)$ define a field of vectors in the xyz-space, while $(v_x, v_y, -1)$ is the normal to the surface $z = v(x, y)$.

We associate to the field of characteristic directions (a, b, c) a family of *characteristic curves* that are tangent to these directions. Along any characteristic curve $(x(t), y(t), z(t))$, where t is a parameter, the following system of ODEs must be satisfied:

$$\frac{dx}{dt} = a(x, y, z), \quad \frac{dy}{dt} = b(x, y, z), \quad \frac{dz}{dt} = c(x, y, z). \tag{4.61}$$

If a surface $S : z = v(x, y)$ is a union of characteristic curves, then S is an integral surface; for then through any point P of S, there passes a characteristic curve Γ contained in S.

Next, we consider the *Cauchy problem* for the quasi-linear equation (4.60). It is desired to find a definite method for finding solutions of the PDE from a given "*data*" on the problem. A simple way of selecting a particular candidate solution $v(x, y)$ out of an infinite set of solutions, consists in prescribing a curve Γ in xyz-space which is to be contained in the integral surface $z = v(x, y)$. Without any loss of generality, we can represent Γ parametrically by

$$x = f(s), \quad y = g(s), \quad z = h(s), \tag{4.62}$$

and we seek for a solution $v(x, y)$ such that

$$h(s) = v(f(s), g(s)), \quad \forall s. \tag{4.63}$$

The above problem is the Cauchy problem for (4.60). Our first aim is to derive conditions for a local solution to (4.60) in the vicinity of $x_0 = f(s_0)$, $y_0 = g(s_0)$. Accordingly, assume the functions $f(s)$, $g(s)$, $h(s) \in C^1$ in the neighborhood of some point P_0 that is parameterized by s_0, i.e.,

$$P_0 = (x_0, y_0, z_0) = (f(s_0), g(s_0), h(s_0)). \tag{4.64}$$

Assume also the coefficients $a(x, y, z)$, $b(x, y, z)$, $c(x, y, z) \in C^1$ near P_0. Then, we can describe Γ near P_0 by the solution

$$x = X(s, t), \quad y = Y(s, t), \quad z = Z(s, t) \tag{4.65}$$

of the characteristic equations (4.61) which reduces to $f(s)$, $g(s)$, $h(s)$ at $t - 0$. Therefore, the functions X, Y, Z must satisfy

$$X_t = a(X, Y, Z), \quad Y_t = b(X, Y, Z), \quad Z_t = c(X, Y, Z) \tag{4.66}$$

identically in s, t and also satisfy the initial conditions

$$X(s, 0) = f(s), \quad Y(s, 0) = g(s), \quad Z(s, 0) = h(s). \tag{4.67}$$

By the theorem on existence and uniqueness of solutions to systems of ODEs, it follows that there exists unique set of functions $X(s, t), Y(s, t), Z(s, t)$ of class C^1 satisfying (4.66), (4.67) for (s, t) near $(s_0, 0)$. Further, if we can solve equation (4.65) for s, t in terms of x, y, say $s = S(x, y)$ and $t = T(x, y)$, then z can be expressed as

$$z = v(x, y) = Z(S(x, y), T(x, y)), \tag{4.68}$$

which represents an integral surface Σ. By (4.64), (4.67), $x_0 = X(s_0, 0)$, $y_0 = Y(s_0, 0)$, and by the Implicit-function Theorem, there exist solutions $s = S(x, y)$, $t = T(x, y)$ of

$$x = X(S(x, y), T(x, y)), \quad y = Y(S(x, y), T(x, y)) \tag{4.69}$$

of class C^1 in a neighborhood of (x_0, y_0) and satisfying

$$s_0 = S(x_0, y_0), \quad 0 = T(x_0, y_0)$$

provided the Jacobian determinant

$$\begin{vmatrix} X_s(s_0,0) & Y_s(s_0,0) \\ X_t(s_0,0) & Y_t(s_0,0) \end{vmatrix} \neq 0. \tag{4.70}$$

By (4.66), (4.67) the above condition is further equivalent to

$$\begin{vmatrix} f_s(s_0) & g_s(s_0) \\ a(x_0,y_0,z_0) & b(x_0,y_0,z_0) \end{vmatrix} \neq 0. \tag{4.71}$$

The above gives the local existence condition for the solution of the Cauchy problem for the quasi-linear equation. Uniqueness follows from the following theorem [109].

Theorem 4.4.1 *Let $P = (x_0, y_0, z_0)$ lie on the integral surface $z = v(x, y)$, and Γ be the characteristic curve through P. Then Γ lies completely on S.*

Example 4.4.1 *[109] Consider the initial value problem for the quasi-linear equation*

$$v_y + cv_x = 0, \quad c \text{ a constant, and } \quad v(x,0) = h(x).$$

Solution:
Parametrize the initial curve Γ corresponding to the initial condition above by

$$x = s, \quad y = 0, \quad z = h(x).$$

Then the characteristic equations are given by

$$\frac{dx}{dt} = c, \quad \frac{dy}{dt} = 1, \quad \frac{dz}{dt} = 0.$$

Solving these equations gives

$$x = X(s,t) = s + ct, \quad y = Y(s,t) = t, \quad z = Z(s,t) = h(s).$$

Finally, eliminating s and t from the above solutions, we get the general solution of the equation

$$z = v(x,y) = h(x - cy).$$

Next, we develop the method for the general first-order equation (4.59) in n independent variables.

4.4.2 Characteristics for the General First-Order Equation

We now consider the general nonlinear first-order PDE (4.59) written in vectorial notation as

$$F(Dv, v, x) = 0, \quad x \in U \subseteq \Re^n, \quad \text{subject to the boundary condition } v = g \text{ on } \mathcal{O} \tag{4.72}$$

where $Dv = (v_{x_1}, v_{x_2}, \ldots, v_{x_n})$, $\mathcal{O} \subseteq \partial U$, $g : \mathcal{O} \to \Re$, and $F \in C^\infty(\Re^n \times \Re \times U)$, $g \in C^\infty(\Re)$.

Now suppose v solves (4.72), and fix any point $x \in U$. We wish to calculate $v(x)$ by finding some curve lying within U, connecting x with a point $x^0 \in \mathcal{O}$ and along which we can compute v. Since $v(x^0) = g(x^0)$, we hope to be able to find v along the curve connecting x^0 and x.

To find the characteristic curve, let us suppose that it is described parametrically by the function $\mathbf{x}(s) = (x^1(s), x^2(s), \ldots, x^n(s))$, the parameter s lying in some subinterval of $\Re$. Assume v is a C^2 solution of (4.72), and let

$$z(s) = v(\mathbf{x}(s)), \tag{4.73}$$

$$\mathbf{p}(s) = Dv(\mathbf{x}(s)); \tag{4.74}$$

i.e., $\mathbf{p}(s) = (p^1(s), p^2(s), \ldots, p^n(s)) = (v_{x_1}(s), v_{x_2}(s), \ldots, v_{x_n}(s))$. Then,

$$\dot{p}^i(s) = \sum_{j=1}^n v_{x_i x_j}(\mathbf{x}(s))\dot{x}^j(s), \tag{4.75}$$

where the differentiation is with respect to s. On the other hand, differentiating (4.72) with respect to x_i, we get

$$\sum_{j=1}^n \frac{\partial F}{\partial p_j}(Dv, v, x)v_{x_j x_i} + \frac{\partial F}{\partial z}(Dv, v, x)v_{x_i} + \frac{\partial F}{\partial x_i}(Dv, v, x) = 0. \tag{4.76}$$

Now, if we set

$$\frac{dx_j}{ds}(s) = \frac{\partial F}{\partial p_j}(\mathbf{p}(s), z(s), \mathbf{x}(s)), \quad j = 1, \ldots, n, \tag{4.77}$$

and assuming that the above relation holds, then evaluating (4.76) at $x = \mathbf{x}(s)$, we obtain the identity

$$\sum_{j=1}^n \frac{\partial F}{\partial p_j}(\mathbf{p}(s), z(s), \mathbf{x}(s))v_{x_j x_i} + \frac{\partial F}{\partial z}(\mathbf{p}(s), z(s), \mathbf{x}(s))p^i(s) + \frac{\partial F}{\partial x_i}(\mathbf{p}(s), z(s), \mathbf{x}(s)) = 0. \tag{4.78}$$

Next, substituting (4.77) in (4.75) and using the above identity (4.78), we get

$$\dot{p}^i(s) = -\frac{\partial F}{\partial x_i}(\mathbf{p}(s), z(s), \mathbf{x}(s)) - \frac{\partial F}{\partial z}(\mathbf{p}(s), z(s), \mathbf{x}(s))p^i(s), \quad i = 1, \ldots, n. \tag{4.79}$$

Finally, differentiating z we have

$$\dot{z}(s) = \sum_{j=1}^n \frac{\partial v}{\partial x_j}(\mathbf{x}(s))\dot{x}^j(s) = \sum_{j=1}^n p^j(s)\frac{\partial F}{\partial p_j}(\mathbf{p}(s), z(s), \mathbf{x}(s)). \tag{4.80}$$

Thus, we finally have the following system of ODEs:

$$\left. \begin{array}{rcl} \dot{\mathbf{p}}(s) &=& -D_x F(\mathbf{p}(s), z(s), \mathbf{x}(s)) - D_z F(\mathbf{p}(s), z(s), \mathbf{x}(s))\mathbf{p}(s) \\ \dot{z}(s) &=& D_p F(\mathbf{p}(s), z(s), \mathbf{x}(s)).\mathbf{p}(s) \\ \dot{\mathbf{x}}(s) &=& D_p F(\mathbf{p}(s), z(s), \mathbf{x}(s)), \end{array} \right\} \tag{4.81}$$

where D_x, D_p, D_z are the derivatives with respect to x, p, z respectively. The above system of $2n + 1$ first-order ODEs comprises the *characteristic equations* of the nonlinear PDE (4.72). The functions $\mathbf{p}(s)$, $z(s)$, $\mathbf{x}(s)$ together are called the *characteristics* while $\mathbf{x}(s)$ is called the *projected characteristic* onto the physical region $U \subseteq \Re^n$. Furthermore, if $v \in C^2$ solves the nonlinear PDE (4.72) in U and assume $\mathbf{x}$ solves the last equation in (4.81), then $\mathbf{p}(s)$ solves the first equation and $z(s)$ solves the second for those s such that $\mathbf{x}(s) \in U$.

Example 4.4.2 *[95] Consider the fully nonlinear equation*

$$v_{x_1}v_{x_2} = v, \quad x \in U = \{x_1 > 0\}$$
$$v = x_2^2 \quad on\ \Gamma = \{x_1 = 0\} = \partial U.$$

Solution

Thus, $F(Dv, v, x) = F(p, z, x) = p_1 p_2 - z$, and the characteristic equations (4.81) become

$$
\begin{aligned}
\dot{p}^1 &= p^1 \\
\dot{p}^2 &= p^2 \\
\dot{x}^1 &= p^2 \\
\dot{x}^2 &= p^1 \\
\dot{z} &= 2p^1 p^2.
\end{aligned}
$$

Integrating the above system, we get

$$
\begin{aligned}
x^1(s) &= p_2^0(e^s - 1), \\
x^2(s) &= x^0 + p_1^0(e^s - 1), \quad x^0 \in \Re \\
p^1(s) &= p_1^0 e^s \\
p^2(s) &= p_2^0 e^s \\
z(s) &= z^0 + p_1^0 p_2^0(e^{2s} - 1), \quad z^0 = (x^0)^2.
\end{aligned}
$$

We must now determine the initial parametrization: $p^0 = (p_1^0, p_2^0)$. Since $v = x_2^2$ on Γ, then $p_2^0 = v_{x_2}(0, x^0) = 2x^0$. Then from the PDE, we get $v_{x_1} = v/v_{x_2} \Rightarrow p_1^0 = (x^0)^2/2x^0 = x^0/2$. Upon substitution now in the above equations, we get

$$
\begin{aligned}
x^1(s) &= 2x^0(e^s - 1) \\
x^2(s) &= \frac{x^0}{2}(e^s + 1) \\
p^1(s) &= \frac{x^0}{2}e^s \\
p^2(s) &= 2x^0 e^s \\
z(s) &= (x^0)^2 e^{2s}.
\end{aligned}
$$

Finally, we must eliminate s and x^0 in the above system to obtain the general solution. In this regard, fix $(x_1, x_2) \in U$ and select x^0 such that $(x_1, x_2) = (x^1(s), x^2(s)) = (2x^0(e^s - 1), \frac{x^0}{2}(e^s + 1))$. Consequently, we get

$$
x^0 = \frac{4x_2 - x_1}{4}, \quad e^s = \frac{x_1 + 4x_2}{4x_2 - x_1},
$$

and

$$
v(x_1, x_2) = z(s) = (x^0)^2 e^{2s} = \frac{(x_1 + 4x_2)^2}{16}.
$$

4.4.3   Characteristics for the Hamilton-Jacobi Equation

Let us now consider the characteristic equations for our Hamilton-Jacobi equation discussed in the beginning of the chapter, which is a typical nonlinear first-order PDE:

$$
G(Dv, , v_t, v, x, t) = v_t + H(Dv, x) = 0, \tag{4.82}
$$

where $Dv = D_x v$ and the remaining variables have their usual meaning. For convenience, let $q = (Dv, v_t) = (p, p_{n+1})$, $y = (x, t)$. Therefore,

$$
G(q, z, y) = p_{n+1} + H(p, x); \tag{4.83}
$$

and

$$D_q G = (D_p H(p, x), 1), \quad D_y G = (D_x H(p, x), 0), \quad D_z G = 0.$$

Thus, the characteristic equations (4.81) become

$$\left. \begin{aligned}
\dot{x}^i(s) &= \frac{\partial H}{\partial p_i}(\mathbf{p}(s), \mathbf{x}(s)), \quad (i = 1, 2, \ldots, n), \\
\dot{x}^{n+1}(s) &= 1, \\
\dot{p}^i(s) &= -\frac{\partial H}{\partial x_i}(\mathbf{p}(s), \mathbf{x}(s)), \quad (i = 1, 2, \ldots, n), \\
\dot{p}^{n+1}(s) &= 0 \\
\dot{z}(s) &= D_p H(\mathbf{p}(s), \mathbf{x}(s)) . \mathbf{p}(s) + p^{n+1},
\end{aligned} \right\} \quad (4.84)$$

which can be rewritten in vectorial form as

$$\left. \begin{aligned}
\dot{\mathbf{p}}(s) &= -D_x H(\mathbf{p}(s), \mathbf{x}(s)) \\
\dot{\mathbf{x}}(s) &= D_p H(\mathbf{p}(s), \mathbf{x}(s)) \\
\dot{z}(s) &= D_p H(\mathbf{p}(s), \mathbf{x}(s)) . \mathbf{p}(s) - H(\mathbf{p}(s), \mathbf{x}(s)).
\end{aligned} \right\} \quad (4.85)$$

The first two of the above equations are clearly Hamilton's canonical equations, while the third equation defines the characteristic surface. The variable z is also termed as the *action-variable* which represents the cost-functional for the variational problem

$$z(t) = \min_{x(.)} \int_0^t L(\mathbf{x}(s), \dot{\mathbf{x}}(s)) ds$$

corresponding to the Hamilton-Jacobi equation, where $L(\mathbf{x}, \dot{\mathbf{x}})$ is the Lagrangian function

$$L(\mathbf{x}, \dot{\mathbf{x}}) = \mathbf{p}\dot{\mathbf{x}} - H(\mathbf{x}, \mathbf{p}).$$

Thus, we have made a connection between Hamilton's canonical equations and the Hamilton-Jacobi equation, and it is clear that a solution for one implies a solution for the other. Nevertherless, neither is easy to solve in general, although, for some systems, the PDE does sometimes offer some leeway, and in fact, this is the motivation behind Hamilton-Jacobi theory.

4.5 Legendre Transform and Hopf-Lax Formula

Though the method of characteristics provides a remarkable way of integrating the HJE, in general the characteristic equations and in particular the Hamilton's canonical equations (4.85) are very difficult to integrate. Thus, other approaches for integrating the HJE had to be sought. One method due to Hopf and Lax [95] which applies to Hamiltonians that are independent of q deserves mention. For simplicity we shall assume that M is an open subset of $\Re^n$, and consider the initial-value problem for the HJE

$$\left. \begin{aligned}
v_t + H(Dv) &= 0 \quad \text{in } \Re^n \times (0, \infty) \\
v &= g \quad \text{on } \Re^n \times \{t = 0\}
\end{aligned} \right\} \quad (4.86)$$

where $g : \Re^n \to \Re$ and $H : \Re^n \to \Re$ is the Hamiltonian function which is independent of q. Let the Lagrangian function $L : TM \to \Re$ satisfy the following assumptions.

Assumption 4.5.1 *The Lagrangian function is convex and* $\lim_{q \to \infty} \frac{L(q)}{|q|} = +\infty$.

Note that the convexity in the above assumption also implies continuity. Furthermore, for simplicity, we have dropped the $\dot{q}$ dependence of L. We then have the following definition.

Definition 4.5.1 *The Legendre transform of L is the Hamiltonian function H defined by*

$$
\begin{aligned}
H(p) &= \sup_{q \in \Re^n} \{p.q - L(q)\}, \quad p \in T_q^\star \Re^n = \Re^n \\
&= p.q^\star - L(q^\star) \\
&= p.\mathbf{q}(p) - L(\mathbf{q}(p)),
\end{aligned}
$$

for some $q^\star = \mathbf{q}(p)$.

We note that the "sup" in the above definition is really a "max," i.e., there exists some $q^\star \in \Re^n$ for which the mapping $q \mapsto p.q - L(q)$ has a maximum at $q = q^\star$. Further, if L is differentiable, then the equation $p = DL(q^\star)$ is solvable for q in terms of p, i.e., $q^\star = \mathbf{q}(p)$, and hence the last expression above.

An important property of the Legendre transform [37] is that it is *involutive*, i.e., if $\mathcal{L}_g$ is the Legendre transform, then $\mathcal{L}_g^2(L) = L$ and $\mathcal{L}_g(H) = L$. A stronger result is the following.

Theorem 4.5.1 *(Convex duality of Hamiltonians and Lagrangians). Assume L satisfies Assumption 4.5.1, and define H as the Legendre transform of L. Then, H also satisfies the following:*

(i) *the mapping $p \mapsto H(p)$ is convex,*

(ii) $\lim_{|p| \to \infty} \frac{H(p)}{|p|} = +\infty.$

We now use the variational principle to obtain the solution of the initial-value problem (4.86), namely, the Hopf-Lax formula. Accordingly, consider the following variational problem of minimizing the action function:

$$
\int_0^t L(\dot{w}(s))ds + g(w(0)) \tag{4.87}
$$

over functions $w : [0, t] \to \Re^n$ with $w(t) = x$. The value-function (or cost-to-go) for this minimization problem is given by

$$
v(x, t) := \inf \left\{ \int_0^t L(\dot{w}(s))ds + g(q) \, | \, w(0) = y, \; w(t) = x \right\}, \tag{4.88}
$$

with the infimum taken over all C^1 functions $w(.)$ with $w(t) = x$. Then we have the following result.

Theorem 4.5.2 *(Hopf-Lax Formula). Assume g is Lipschitz continuous, and if $x \in \Re^n$, $t > 0$, then the solution $v = v(x, t)$ to the variational problem (4.87) is*

$$
v(x, t) = \min_{y \in \Re^n} \left\{ tL\left(\frac{x - y}{t}\right) + g(y) \right\}. \tag{4.89}
$$

The proof of the above theorem can be found in [95].

The next theorem asserts that the Hopf-Lax formula indeed solves the initial-value problem of the HJE (4.86) whenever v in (4.89) is differentiable.

Theorem 4.5.3 *Assume H is convex, $\lim_{|p| \to \infty} \frac{H(p)}{|p|} = \infty$. Further, suppose $x \in \Re^n$, $t > 0$,* *and v in (4.89) is differentiable at a point $(x,t) \in \Re^n \times (0,\infty)$. Then (4.89) satisfies the HJE (4.86) with the initial value $v(x,0) = g(x)$.*

Again the proof of the above theorem can be found in [95]. The Hopf-Lax formula provides a reasonably weak solution (a Lipschitz-continuous function which satisfies the PDE almost everywhere) to the initial-value problem for the HJE. The Hopf-Lax formula is useful for variational problems and mechanical systems for which the Hamiltonians are independent of configuration coordinates, but is very limited for the case of more general problems.

4.5.1 Viscosity Solutions of the HJE

It was long recognized that the HJE being nonlinear, may not admit classical (or smooth solutions) even for simple situations [56, 98, 287]. To overcome this difficulty, Crandall and Lions [186] introduced the concept of *viscosity (or generalized) solutions* in the early 1980s [56, 83, 98, 186, 287] which have had wider application. Under the assumption of differentiability of v, any solution v of the HJE will be referred to as a *classical solution* if it satisfies it for all $x \in \mathcal{X}$. In most cases however, the Hamiltonian function H fails to be differentiable at some point $x \in \mathcal{X}$, and hence may not satisfy the HJE everywhere in $\mathcal{X}$. In such cases, we would like to consider solutions that are closest to being differentiable in an extended sense. The closest such idea is that of Lipschitz continuous solutions. This leads to the concept of generalized solutions which we now define [83, 98].

Definition 4.5.2 *Consider the more general form of the Hamiltonian function $H : T^*M \to \Re$ and the Cauchy problem*

$$H(x, D_x v(x)) = 0, \quad v(x,0) = g(x) \tag{4.90}$$

where $D_x v(x)$ denotes some derivative of v at x, which is not necessarily a classical derivative. Now suppose v is locally Lipschitz on N, i.e., for every compact set $O \subset N$ and $x_1, x_2 \in O$ there exists a constant $k_O > 0$ such that

$$|v(x_1) - v(x_2)| \le k_O \|x_1 - x_2\|$$

(it is Lipschitz if $K_O = k$, independent of O), then v is a generalized solution of (4.90) if it satisfies it for almost all $x \in \mathcal{X}$.

Moreover, since every locally Lipschitz function is differentiable at almost all points $x \in N$, the idea of generalized solutions indeed makes sense. However, the concept also implies the lack of uniqueness of generalized solutions. Thus, there can be infinitely many generalized solutions. In this section, we shall restrict ourselves to the class of generalized solutions referred to as *viscosity solutions*, which are unique. Other types of generalized solutions such as "minimax" and "proximal" are also available in the literature [83]. We define viscosity solutions next.

Assume v is continuous in N, and define the following sets which are respectively the *superdifferential* and *subdifferential* of v at $x \in N$

$$D^+ v(x) = \left\{ p \in \Re^n \ : \ \lim_{\substack{x' \to x \\ x' \in N}} \sup \frac{v(x') - v(x) - p.(x' - x)}{\|x' - x\|} \le 0 \right\}, \tag{4.91}$$

$$D^- v(x) = \left\{ q \in \Re^n \ : \ \lim_{\substack{x' \to x \\ x' \in N}} \inf \frac{v(x') - v(x) - q.(x' - x)}{\|x' - x\|} \ge 0 \right\}. \tag{4.92}$$

Remark 4.5.1 *If both $D^+ v(x)$ and $D^- v(x)$ are nonempty at some x, then $D^+ v(x) =$*

$D^- v(x)$ and v is differentiable at x. We now have the following definitions of viscosity solutions.

Definition 4.5.3 *A continuous function v is a viscosity solution of HJE (4.90) if it is both a viscosity subsolution and supersolution, i.e., it satisfies respectively the following conditions:*

$$H(x, p) \leq 0; \quad \forall x \in N, \; \forall p \in D^+ v(x) \tag{4.93}$$
$$H(x, q) \geq 0; \quad \forall x \in N, \; \forall q \in D^- v(x) \tag{4.94}$$

respectively.

An alternative definition of viscosity subsolutions and supersolutions is given in terms of test functions as follows.

Definition 4.5.4 *A continuous function v is a viscosity subsolution of HJE (4.90) if for any $\varphi \in C^1$,*

$$H(x, D\varphi(x)) \leq 0$$

at any local maximum point x of $v - \varphi$. Similarly, v is a viscosity supersolution if for any $\varphi \in C^1$,

$$H(x, D\varphi(x)) \geq 0$$

at any local minimum point x of $v - \varphi$.

Finally, for the theory of viscosity solutions to be meaningful, it should be consistent with the notion of classical solutions. Thus, we have the following relationship between viscosity solutions and classical solutions [56].

Proposition 4.5.1 *If $v \in C(N)$ is a classical solution of HJE (4.90), then $v(x)$ is a viscosity solution, and conversely if $v \in C^1(N)$ is a viscosity solution of (4.90), then v is a classical solution.*

Which states in essense that, every classical solution is a viscosity solution. Furthermore, the following proposition gives a connection with Lipschitz-continuous solutions [56].

Proposition 4.5.2 *(a) If $v \in C(N)$ is a viscosity solution of (4.90), then*

$$H(x, D_x v) = 0$$

at any point $x \in N$ where v is differentiable; (b) if v is locally Lipschitz-continuous and it is a viscosity solution of (4.90), then

$$H(x, D_x v) = 0$$

almost everywhere in N.

Lastly, the following proposition guarantees uniqueness of viscosity solutions [95, 98].

Proposition 4.5.3 *Suppose $H(x, p)$ satisfies the following Lipschitz conditions:*

$$|H(x, p) - H(x, q)| \leq k\|p - q\|$$
$$|H(x, p) - H(x', p)| \leq k\|x - x'\|(1 + \|p\|)$$

for some $k \geq 0$, $x, x', p, q \in \Re^n$, then there exists at most one viscosity solution to the HJE (4.90).

The theory of viscosity solutions is however not limited to the HJE. Indeed, the theory applies to any first-order equation of the types that we have discussed in the beginning of the chapter and also second-order equations of parabolic type.

4.6   Notes and Bibliography

The material in Sections 1-4 on Hamiltonian mechanics is collected from the References [1, 37, 115, 122, 200] and [127], though we have relied more heavily on [122, 127] and [115]. On the other hand, the geometry of the Hamilton-Jacobi equation and a deeper discussion of its associated Lagrangian-submanifolds can be found in [1, 37, 200]. In fact, these are the standard references on the subject. A more classical treatment of the HJE can also be found in Whittaker [271] and a host of hundreds of excellent books in many libraries.

Section 4.3, dealing with an introduction to nonlinear lattices is mainly from [258], and more advanced discussions on the subject can be found in the References [3]-[5, 72, 123, 201].

Finally, Section 4.4 on first-order PDEs is mainly from [95, 109]. More exhaustive discussion on viscosity and generalized solutions of HJEs can be found in [56, 83] from the deterministic point of view, and in [98, 287] from the stochastic point of view.

5

State-Feedback Nonlinear $\mathcal{H}_\infty$-Control for Continuous-Time Systems

In this chapter, we discuss the nonlinear $\mathcal{H}_\infty$ sub-optimal control problem for continuous-time affine nonlinear systems using state-feedback. This problem arises when the states of the system are available, or can be measured directly and used for feedback. We derive sufficient conditions for the solvability of the problem, and we discuss the results for both time-invariant (or autonomous) systems and time-varying (or nonautonomous) systems, as well as systems with a delay in the state. We also give a parametrization of all full-information stabilizing controllers for each system. Moreover, understanding the state-feedback problem will facilitate the understanding of the dynamic measurement-feedback problem which is discussed in the subsequent chapter.

The problem of robust control in the presence of modelling errors and/or parameter variations is also discussed. Sufficient conditions for the solvability of this problem are given, and a class of controllers is presented.

5.1 State-Feedback $\mathcal{H}_\infty$-Control for Affine Nonlinear Systems

The set-up for this configuration is shown in Figure 5.1, where the plant is represented by an affine causal state-space system defined on a smooth n-dimensional manifold $\mathcal{X} \subseteq \Re^n$ in local coordinates $x = (x_1, \ldots, x_n)$:

$$\Sigma^a \; : \; \begin{cases} \dot{x} & = & f(x) + g_1(x)w + g_2(x)u; \;\; x(t_0) = x_0 \\ y & = & x \\ z & = & h_1(x) + k_{12}(x)u \end{cases} \tag{5.1}$$

where $x \in \mathcal{X}$ is the state vector, $u \in \mathcal{U} \subseteq \Re^p$ is the p-dimensional control input, which belongs to the set of admissible controls $\mathcal{U}$, $w \in \mathcal{W}$ is the disturbance signal, which belongs to the set $\mathcal{W} \subset \mathcal{L}_2([t_0, \infty), \Re^r)$ of admissible disturbances, the output $y \in \Re^n$ is the states-vector of the system which is measured directly, and $z \in \Re^s$ is the output to be controlled. The functions $f : \mathcal{X} \to V^\infty(\mathcal{X})$, $g_1 : \mathcal{X} \to \mathcal{M}^{n \times r}(\mathcal{X})$, $g_2 : \mathcal{X} \to \mathcal{M}^{n \times p}(\mathcal{X})$, $h_1 : \mathcal{X} \to \Re^s$, and $k_{12} : \mathcal{X} \to \mathcal{M}^{p \times m}(\mathcal{X})$ are assumed to be real C^∞-functions of x. Furthermore, we assume without loss of generality that $x = 0$ is a unique equilibrium point of the system with $u = 0$, $w = 0$, and is such that $f(0) = 0$, $h_1(0) = 0$. We also assume that the system is well defined, i.e., for any initial state $x(t_0) \in \mathcal{X}$ and any admissible input $u(t) \in \mathcal{U}$, there exists a unique solution $x(t, t_0, x_0, u)$ to (5.1) on $[t_0, \infty)$ which continuously depends on the initial conditions, or the system satisfies the local existence and uniqueness theorem for ordinary differential-equations [157].

Again Figure 5.1 also shows that for this configuration, the states of the plant are accessible and can be directly measured for the purpose of feedback control. We begin with

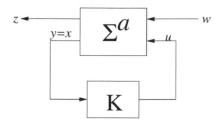

FIGURE 5.1
Feedback Configuration for State-Feedback Nonlinear $\mathcal{H}_\infty$-Control

the definition of smooth-stabilizability and also recall the definition of $\mathcal{L}_2$-gain of the system Σ^a.

Definition 5.1.1 *(Smooth-stabilizability). The nonlinear system Σ^a (or simply $[f, g_2]$) is locally smoothly-stabilizable if there exists a C^0-function $F : U \subset \mathcal{X} \to \Re^p$, $F(0) = 0$, such that $\dot{x} = f(x) + g_2(x)F(x)$ is locally asymptotically stable. The system is smoothly-stabilizable if $U = \mathcal{X}$.*

Definition 5.1.2 *The nonlinear system Σ^a is said to have locally $\mathcal{L}_2$-gain from w to z in $U \subset \mathcal{X}$, less than or equal to γ, if for any $x_0 \in U$ and fixed u, the response z of the system corresponding to any $w \in \mathcal{W}$ satisfies:*

$$\int_{t_0}^{T} \|z(t)\|^2 dt \le \gamma^2 \int_{t_0}^{T} \|w(t)\|^2 dt + \beta(x_0), \quad \forall T > t_0,$$

for some bounded C^0 function $\beta : U \to \Re$ such that $\beta(0) = 0$. The system has $\mathcal{L}_2$-gain $\le \gamma$ if the above inequality is satisfied for all $x \in \mathcal{X}$, or $U = \mathcal{X}$.

Since we are interested in designing smooth feedback laws for the system to make it asymptotically or internally stable, the requirement of smooth-stabilizability for the system will obviously be necessary for the solvability of the problem before anything else. The suboptimal state-feedback nonlinear $\mathcal{H}_\infty$-control or local disturbance-attenuation problem with internal stability, can then be formally defined as follows.

Definition 5.1.3 *(State-Feedback Nonlinear $\mathcal{H}_\infty$ (Suboptimal)-Control Problem (SFBNL-HICP)). The state-feedback $\mathcal{H}_\infty$ suboptimal control or local disturbance-attenuation problem with internal stability for the system Σ^a, is to find a static state-feedback control function of the form*

$$u = \alpha(x, t), \quad \alpha : \Re_+ \times N \to \Re^p, \quad N \subset \mathcal{X} \tag{5.2}$$

for some smooth function α depending on x and possibly t only, such that the closed-loop system:

$$\Sigma_{clp}^a : \begin{cases} \dot{x} & = & f(x) + g_1(x)w + g_2(x)\alpha(x, t); \quad x(t_0) = x_0 \\ z & = & h_1(x) + k_{12}(x)\alpha(x, t) \end{cases} \tag{5.3}$$

has, for all initial conditions $x(t_0) \in N$, locally $\mathcal{L}_2$-gain from the disturbance signal w to the output z less than or equal to some prescribed number $\gamma^\star > 0$ with internal stability, or equivalently, the closed-loop system achieves local disturbance-attenuation less than or equal to $\gamma^\star$ with internal stability.

Internal stability of the system in the above definition means that all internal signals in the system, or trajectories, are bounded, which is also equivalent to local asymptotic-stability of the closed-loop system with $w = 0$ in this case.

Remark 5.1.1 *The optimal problem in the above definition is to find the minimum $\gamma^\star > 0$ for which the $\mathcal{L}_2$-gain is minimized. This problem is however more difficult to solve.*

One way to measure the $\mathcal{L}_2$-gain (or with an abuse of the terminology, $\mathcal{H}_\infty$-norm) of the system (5.1), is to excite it with a periodic input $w_T \in W$, where $W \subset \mathcal{W}$ is the subspace of periodic continuous-time functions (e.g., a sinusoidal signal), and to measure the steady-state output response $z_{ss}(.)$ corresponding to the steady-state state response $x_{ss}(.)$. Then the $\mathcal{L}_2$-gain can be calculated as

$$\|\Sigma^a\|_{\mathcal{H}_\infty} = \sup_{w \in W} \frac{\|z_{ss}\|_T}{\|w\|_T}$$

where

$$\|w\|_T = \frac{1}{T}\left(\int_{t_0}^{t_0+T} \|w(s)\|^2 ds\right)^{\frac{1}{2}}, \quad \|z_{ss}\|_T = \frac{1}{T}\left(\int_{t_0}^{t_0+T} \|w(s)\|^2 ds\right)^{\frac{1}{2}}.$$

Returning now to the *SFBNLHICP*, to derive sufficient conditions for the solvability of this problem, we apply the theory of differential games developed in Chapter 2. It is fairly clear that the problem of choosing a control function $u^\star(.)$ such that the $\mathcal{L}_2$-gain of the closed-loop system from w to z is less than or equal to $\gamma > 0$, can be formulated as a two-player zero-sum differential game with u the minimizing player's decision, w the maximizing player's decision, and the objective functional:

$$\min_{u \in \mathcal{U}} \max_{w \in \mathcal{W}} J(u, w) = \frac{1}{2}\int_{t_0}^{T}[\|z(t)\|^2 - \gamma^2\|w(t)\|^2]dt, \qquad (5.4)$$

subject to the dynamical equations (5.1) over a finite time-horizon $T > t_0$.

At this point, we separate the problem into two subproblems; namely, (i) achieving local disturbance-attenuation and (ii) achieving local asymptotic-stability. To solve the first problem, we allow w to vary over all possible disturbances including the worst-case disturbance, and search for a feedback control function $u : \mathcal{X} \times \Re \to \mathcal{U}$ depending on the current state information, that minimizes the objective functional $J(.,.)$ and renders it nonpositive for all w starting from $x_0 = 0$. By so doing, we have the following result.

Proposition 5.1.1 *Suppose for $\gamma = \gamma^\star$ there exists a locally defined feedback-control function $u^\star : N \times \Re \to \Re^p$, $0 \in N \subset \mathcal{X}$, which is possibly time-varying, and renders $J(.,.)$ nonpositive for the worst possible disturbance $w^\star \in \mathcal{W}$ (and hence for all $w \in \mathcal{W}$) for all $T > 0$. Then the closed-loop system has locally $\mathcal{L}_2$-gain $\leq \gamma^\star$.*

Proof:

$$J(u^\star, w^\star) \leq 0 \Rightarrow \|z\|_{\mathcal{L}_2[t_0, T]} \leq \gamma^\star \|w\|_{\mathcal{L}_2[t_0, T]} \quad \forall T > 0. \quad \square$$

To derive the sufficient conditions for the solvability of the first sub-problem, we define the value-function for the game $V : \mathcal{X} \times [0, T] \to \Re$ as

$$V(t, x) = \inf_u \sup_w \frac{1}{2}\int_t^T[\|z(\tau)\|^2 - \gamma\|w(\tau)\|^2]d\tau$$

and apply Theorem 2.4.2 from Chapter 2. Consequently, we have the following theorem.

Theorem 5.1.1 *Consider the SFBNLHICP problem as a two-player zero-sum differential game with the cost functional (5.4). A pair of strategies $(u^\star(x,t),\ w^\star(x,t))$ provides, under feedback information structure, a saddle-point solution to the game such that*

$$J(u^\star, w) \leq J(u^\star, w^\star) \leq J(u, w^\star),$$

if the value-function V is C^1 and satisfies the HJI-PDE (HJIE):

$$
\begin{aligned}
-V_t(x,t) \quad &= \quad \min_u \sup_w \left\{ V_x(x,t)[f(x) + g_1(x)w + g_2(x)u] + \frac{1}{2}(\|z\|^2 - \gamma^2\|w\|^2) \right\} \\
&= \quad \sup_w \min_u \left\{ V_x(x,t)[f(x) + g_1(x)w + g_2(x)u] + \frac{1}{2}\|z\|^2 - \frac{1}{2}\gamma^2\|w\|^2 \right\} \\
&= \quad V_x(x,t)[f(x) + g_1(x)w^\star(x,t) + g_2(x)u^\star(x,t)] + \frac{1}{2}\|h_1(x) + k_{12}(x)u^\star(x,t)\|^2 - \\
&\qquad \frac{1}{2}\gamma^2\|w^\star(x,t)\|^2; \quad V(x,T) = 0.
\end{aligned}
\tag{5.5}
$$

Next, to find the pair of feedback strategy $(u^\star, w^\star)$ that satisfies Isaac's equation (5.5), we form the Hamiltonian function $H : T^\star \mathcal{X} \times \mathcal{U} \times \mathcal{W} \to \Re$ for the problem:

$$H(x,p,u,w) = p^T(f(x) + g_1(x)w + g_2(x)u) + \frac{1}{2}\|h_1(x) + k_{12}(x)u\|^2 - \frac{1}{2}\gamma^2\|w\|^2, \tag{5.6}$$

and search for a unique saddle-point $(u^\star, w^\star)$ such that

$$H(x,p,u^\star,w) \leq H(x,p,u^\star,w^\star) \leq H(x,p,u,w^\star) \tag{5.7}$$

for each (u,w) and each (x,p), where p is the adjoint variable.

Since the function $H(.,.,.,.)$ is C^2 in both u and w, the above problem can be solved by applying the necessary conditions for an unconstrained optimization problem. However, the only problem that might arise is if the coefficient matrix of u is singular. This more general problem will be discussed in Chapter 9. But in the meantime to overcome this problem, we need the following assumption.

Assumption 5.1.1 *The matrix*

$$R(x) = k_{12}^T(x)k_{12}(x)$$

is nonsingular for all $x \in \mathcal{X}$.

Under the above assumption, the necessary conditions for optimality for u and w provided by the minimum (maximum) principle [175] are

$$\frac{\partial H}{\partial u}(u^\star, w) = 0, \quad \frac{\partial H}{\partial w}(u, w^\star) = 0$$

for all (u,w). Application of these conditions gives

$$
\begin{aligned}
u^\star(x,p) \quad &= \quad -R^{-1}(x)(g_2^T(x)p + k_{12}^T(x)h_1(x)), &(5.8) \\
w^\star(x,p) \quad &= \quad \frac{1}{\gamma^2}g_1^T(x)p. &(5.9)
\end{aligned}
$$

Moreover, since by assumption $R(.)$ is nonsingular and therefore positive-definite, and $\gamma > 0$, the above equilibrium-point is clearly an optimizer of $J(u, w)$. Further, we can write

$$H(x, p, u, w) = H^\star(x, p) + \frac{1}{2}\|u - u^\star\|_{R(x)}^2 - \frac{1}{2}\gamma^2\|w - w^\star\|^2, \qquad (5.10)$$

where

$$H^\star(x, p) = H(x, p, u^\star(x, p), w^\star(x, p))$$

and the notation $\|a\|_Q$ stands for $a^T Q a$ for any $a \in \Re^n$, $Q \in \Re^{n \times n}$. Substituting $u^\star$ and $w^\star$ in turns in (5.10) show that the saddle-point conditions (5.7) are satisfied.

Now assume that there exists a C^1 positive-semidefinite solution $V : \mathcal{X} \to \Re$ to Isaac's equation (5.5) which is defined in a neighborhood N of the origin, that vanishes at $x = 0$ and is time-invariant (this assumption is plausible since $H(., ., ., .)$ is time-invariant). Then the feedbacks $(u^\star, w^\star)$ necessarily exist, and choosing

$$p - V_x^T(x)$$

in (5.10) yields the identity:

$$
\begin{aligned}
H(x, V_x^T(x), w, u) &= V_x(x)(f(x) + g_1(x)w + g_2(x)u) + \frac{1}{2}\|h_1(x) + k_{12}(x)u\|^2 - \frac{1}{2}\gamma^2\|w\|^2 \\
&= H^\star(x, V_x^T(x)) + \frac{1}{2}\|u - u^\star\|_{R(x)}^2 + \frac{1}{2}\gamma^2\|w - w^\star\|^2.
\end{aligned}
$$

Finally, notice that for $u = u^\star$ and $w = w^\star$, the above identity yields

$$H(x, V_x^T(x), w^\star, u^\star) = H^\star(x, V_x^T(x))$$

which is exactly the right-hand-side of (5.5), and for this equation to be satisfied, $V(.)$ must be such that

$$H^\star(x, V_x^T(x)) = 0. \qquad (5.11)$$

The above condition (5.11) is the time-invariant HJIE for the disturbance-attenuation problem. Integration of (5.11) along the trajectories of the closed-loop system with $\alpha(x) = u^\star(x, V_x^T(x))$ (independent of t!) starting from $t = t_0$ and $x(t_0) = x_0$, to $t = T > t_0$ and $x(T)$ yields

$$V(x(T)) - V(x_0) \le \frac{1}{2}\int_{t_0}^T (\gamma^2\|w\|^2 - \|z\|^2)dt \ge 0 \quad \forall w \in \mathcal{W}.$$

This means $J(u^\star, w)$ is nonpositive for all $w \in \mathcal{W}$, and consequently implies the $\mathcal{L}_2$-gain of the system is less than or equal to γ. This also solves part (i) of the state-feedback suboptimal $\mathcal{H}_\infty$-control problem. Before we consider part (ii) of the problem, we make the following simplifying assumption.

Assumption 5.1.2 *The output vector $h_1(.)$ and weighting matrix $k_{12}(.)$ are such that*

$$k_{12}^T(x)k_{12}(x) = I$$

and

$$h_1^T(x)k_{12}(x) = 0$$

for all $x \in \mathcal{X}$. Equivalently, we shall henceforth write $z = \begin{bmatrix} h_1(x) \\ u \end{bmatrix}$ under this assumption.

Remark 5.1.2 *The above assumption implies that there are no cross-product terms in the performance or cost-functional (5.4), and the weighting on the control is unity.*

Under the above assumption 5.1.2, the HJIE (5.11) becomes

$$V_x(x)f(x) + \frac{1}{2}V_x(x)[\frac{1}{\gamma^2}g_1(x)g_1^T(x) - g_2(x)g_2^T(x)]V_x^T(x) + \frac{1}{2}h_1^T(x)h_1(x) = 0, \quad V(0) = 0,$$
(5.12)

and the feedbacks (5.8), (5.9) become

$$u^\star(x) \;=\; -g_2^T(x)V_x^T(x) \tag{5.13}$$

$$w^\star(x) \;=\; \frac{1}{\gamma^2}g_1^T(x)V_x^T(x). \tag{5.14}$$

Thus, the above condition (5.12) together with the associated feedbacks (5.13), (5.14) provide a sufficient condition for the solvability of the state-feedback suboptimal $\mathcal{H}_\infty$ problem on the infinite-time horizon when $T \to \infty$.

On the other hand, let us consider the *finite-horizon* problem as defined by the cost functional (5.4) with $T < \infty$. Assuming there exists a time-varying positive-semidefinite C^1 solution $V : \mathcal{X} \times \Re \to \Re$ to the HJIE (5.5) such that

$$p = V_x^T(x, t),$$

then substituting in (5.8), (5.9) and the HJIE (5.5) under the Assumption 5.1.2, we have

$$u^\star(x, t) \;=\; -g_2^T(x)V_x^T(x, t) \tag{5.15}$$

$$w^\star(x, t) \;=\; \frac{1}{\gamma^2}g_1^T(x)V_x^T(x, t), \tag{5.16}$$

where V satisfies the HJIE

$$V_t(x, t) + V_x(x, t)f(x) + \frac{1}{2}V_x(x, t)[\frac{1}{\gamma^2}g_1(x)g_1^T(x) - g_2(x)g_2^T(x)]V_x^T(x, t) +$$

$$\frac{1}{2}h^T(x)h_1(x) = 0, \quad V(x, T) = 0. \tag{5.17}$$

Therefore, the above HJIE (5.17) gives a sufficient condition for the solvability of the finite-horizon suboptimal $\mathcal{H}_\infty$ control problem and the associated feedbacks.

Let us consider an example at this point.

Example 5.1.1 *Consider the nonlinear system with the associated penalty function*

$$\dot{x}_1 \;=\; x_2$$

$$\dot{x}_2 \;=\; -x_1 - \frac{1}{2}x_2^3 + x_2 w + u$$

$$z \;=\; \begin{bmatrix} x_2 \\ u \end{bmatrix}.$$

The HJIE (5.12) corresponding to this system and penalty function is given by

$$x_2 V_{x_1} - x_1 V_{x_2} - \frac{1}{2}x_2^3 V_{x_2} + \frac{1}{2}\frac{x_2^2(x_2^2 - \gamma^2)}{\gamma^2} + \frac{1}{2}x_2^2 = 0.$$

Let $\gamma = 1$ and choose

$$V_{x_1} = x_1, \quad V_{x_2} = x_2.$$

Then we see that the HJIE is solved with $V(x) = \frac{1}{2}(x_1^2 + x_2^2)$ which is positive-definite. The associated feedbacks are given by

$$u^\star = -x_2, \quad w^\star = x_2^2.$$

It is also interesting to notice that the above solution V to the HJIE (5.12) is also a Lyapunov-function candidate for the free system: $\dot{x}_1 = x_2$, $\dot{x}_2 = -x_1 - \frac{1}{2}x_2^3$.

Next, we consider the problem of asymptotic-stability for the closed-loop system (5.3), which is part (ii) of the problem. For this, let

$$\alpha(x) = u^\star(x) = -g_2^T(x)V_x^T(x),$$

where $V(.)$ is a smooth positive-semidefinite solution of the HJIE (5.12). Then differentiating V along the trajectories of the closed-loop system with $w = 0$ and using (5.12), we get

$$\begin{aligned}
\dot{V}(x) &= V_x(x)(f(x) - g_2(x)g_2^T(x)V_x(x)) \\
&= -\frac{1}{2}\|u^\star\|^2 - \frac{1}{2}\gamma^2\|w^\star\|^2 - \frac{1}{2}h^T(x)h_1(x) \le 0,
\end{aligned}$$

where use has been made of the HJIE (5.12). Therefore, $\dot{V}$ is nonincreasing along trajectories of the closed-loop system, and hence the system is stable in the sense of Lyapunov. To prove local asymptotic-stability however, an additional assumption on the system will be necessary.

Definition 5.1.4 *The nonlinear system Σ^a is said to be locally zero-state detectable if there exists a neighborhood $U \subset \mathcal{X}$ of $x = 0$ such that, for all $x(t_0) \in U$, if $z(t) \equiv 0$, $u(t) \equiv 0$ for all $t \ge t_0$, it implies that $\lim_{t \to \infty} x(t, t_0, x_0, u) = 0$. It is zero-state detectable if $U = \mathcal{X}$.*

Thus, if we assume the system Σ^a to be locally zero-state detectable, then it is seen that for any trajectory of the system $x(t) \in U$ such that $\dot{V}(x(t)) \equiv 0$ for all $t \ge t_s$ for some $t_s \ge t_0$, it is necessary that $u(t) \equiv 0$ and $z(t) \equiv 0$ for all $t \ge t_s$. This by zero-state detectability implies that $\lim_{t \to \infty} x(t) = 0$. Finally, since $x = 0$ is the only equilibrium-point of the system in U, by LaSalle's invariance-principle, we can conclude local asymptotic-stability.

The above result is summarized as the solution to the state-feedback $\mathcal{H}_\infty$ sub-optimal control problem (SFBNLHICP) in the next theorem after the following definition.

Definition 5.1.5 *A nonnegative function $V : \mathcal{X} \to \Re$ is proper if the level set $V^{-1}([0, a]) = \{x \in \mathcal{X} | 0 \le V(x) \le a\}$ is compact for each $a > 0$.*

Theorem 5.1.2 *Consider the nonlinear system Σ^a and the SFBNLHICP for the system. Assume the system is smoothly-stabilizable and locally zero-state detectable in $N \subset \mathcal{X}$. Suppose also there exists a smooth positive-semidefinite solution to the HJIE (5.12) in N. Then the control law*

$$u^\star = \alpha(x) = -g_2^T(x)V_x^T(x), \quad x \in N \tag{5.18}$$

solves the SFBNLHICP locally in N. If in addition Σ^a is globally zero-state detectable and V is proper, then $u^\star$ solves the problem globally.

Proof: The first part of the theorem has already been proven in the above developments. For the second part regarding global asymptotic-stability, note that, if V is proper, then V is a global solution of the HJIE (5.11), and the result follows by application of LaSalle's invariance-principle from Chapter 1 (see also the References [157, 268]). $\square$

The existence of a C^2 solution to the HJIE (5.12) is related to the existence of an invariant-manifold for the corresponding Hamiltonian system:

$$X_{H_\gamma^\star} : \begin{cases} \frac{dx}{dt} = \frac{\partial H_\gamma^\star(x,p)}{\partial p} \\ \frac{dp}{dt} = -\frac{\partial H_\gamma^\star(x,p)}{\partial x}, \end{cases} \tag{5.19}$$

where

$$H_\gamma^\star(x,p) = p^T f(x) + \frac{1}{2}p^T[\frac{1}{\gamma^2}g_1(x)g_1^T(x) - g_2(x)g_2^T(x)]p + \frac{1}{2}h_1^T(x)h_1(x).$$

It can be seen then that, if V is a C^2 solution of the Isaacs equation, then differentiating $H^\star(x,p)$ in (5.11) with respect to x we get

$$\left(\frac{\partial H_\gamma^\star}{\partial x}\right)_{p=V_x^T} + \left(\frac{\partial H_\gamma^\star}{\partial p}\right)_{p=V_x^T} \frac{\partial V_x^T}{\partial x} = 0,$$

and since the Hessian matrix

$$\frac{\partial V_x^T}{\partial x}$$

is symmetric, it implies that the submanifold

$$\mathbf{M} = \{(x,p) : p = V_x^T(x)\} \tag{5.20}$$

is invariant under the flow of the Hamiltonian vector-field $X_{H_\gamma^\star}$, i.e.,

$$\left(\frac{\partial H_\gamma^\star}{\partial x}\right)_{p=V_x^T} = -\left(\frac{\partial H_\gamma^\star}{\partial p}\right)_{p=V_x^T} \frac{\partial V_x^T}{\partial x}.$$

The above developments have considered the $SFBNLHICP$ from a differential games perspective. In the next section, we consider the same problem from a dissipative point of view.

Remark 5.1.3 *With $p = V_x^T(x)$, for some smooth solution $V \geq 0$ of the HJIE (5.12), the disturbance $w^\star = \frac{1}{\gamma^2}g_1(x)V_x^T(x)$, $x \in \mathcal{X}$ is referred to as the worst-case disturbance affecting the system. Hence the title "worst-case" design for $\mathcal{H}_\infty$-control design.*

Let us now specialize the results of Theorem 5.1.2 to the linear system

$$\Sigma^l : \begin{cases} \dot{x} = Fx + G_1 w + G_2 u; \quad x(0) = x_0 \\ z = \begin{bmatrix} H_1 x \\ u \end{bmatrix}, \end{cases} \tag{5.21}$$

where $F \in \Re^{n \times n}$, $G_1 \in \Re^{n \times r}$, $G_2 \in \Re^{n \times p}$, and $H_1 \in \Re^{m \times n}$ are constant matrices. Also, let the transfer function $w \mapsto z$ be T_{zw}, and assume $x(0) = 0$. Then the $\mathcal{H}_\infty$-norm of the system from w to z is defined by

$$\|T_{zw}\|_\infty \overset{\Delta}{=} \sup_{0 \neq w \in \mathcal{L}_2[0,\infty)} \frac{\|z\|_2}{\|w\|_2}.$$

We then have the following corollary to the theorem.

Corollary 5.1.1 *Consider the linear system (5.21) and the SFBNLHICP for it. Assume (F, G_2) is stabilizable and (H_1, F) is detectable. Further, suppose for some $\gamma > 0$, there* *exists a symmetric positive-semidefinite solution $P \geq 0$ to the algebraic-Riccati equation (ARE):*

$$F^T P + PF + P[\frac{1}{\gamma^2}G_1 G_1^T - G_2 G_2^T]P + H_1^T H_1 = 0. \tag{5.22}$$

Then the control law

$$u = -G_2^T P x$$

solves the SFBNLHICP for the system Σ^l, i.e., renders its $\mathcal{H}_\infty$-norm less than or equal to a prescribed number $\gamma > 0$ and $(F - G_2 G_2^T P)$ is asymptotically-stable or Hurwitz.

Remark 5.1.4 *Note that the assumptions (F, G_2) stabilizable and (H_1, F) detectable in the above corollary actually guarantee the existence of a symmetric solution $P \geq 0$ to the Riccati equation (5.22) [292]. Moreover, any solution $P = P^T \geq 0$ of (5.22) is stabilizing.*

Remark 5.1.5 *Again, the assumption (H_1, F) detectable in the corollary can be replaced by the linear equivalent of the zero-state detectability assumption for the nonlinear case, which is*

$$rank \begin{pmatrix} A - j\omega I & G_2 \\ H & I \end{pmatrix} = n + m \ \forall \omega \in \Re.$$

This condition also means that the system does not have a stable unobservable mode on the $j\omega$-axis.

The converse of Corollary 5.1.1 also holds, and is stated in the following theorem which is also known as the Bounded-real lemma [160].

Theorem 5.1.3 *Assume (H_1, F) is detectable and let $\gamma > 0$. Then there exists a linear feedback-control*

$$u = Kx$$

such that the closed-loop system (5.21) with this feedback is asymptotically-stable and has $\mathcal{L}_2$-gain $\leq \gamma$ if, and only if, there exists a solution $P \geq 0$ to (5.22). In addition, if $P = P^T \geq 0$ is such that

$$\sigma \left(F - G_2 G_2^T P + \frac{1}{\gamma^2}G_1 G_1^T P \right) \subset \mathbf{C}^-,$$

where $\sigma(.)$ denotes the spectrum of $(.)$, then $\|T_{zw}\|_\infty < \gamma$.

5.1.1 Dissipative Analysis

In this section, we reconsider the *SFBNLHICP* for the affine nonlinear system (5.1) from a dissipative system's perspective developed in Chapter 3 (see also [131, 223]). In this respect, the first part of the problem (subproblem (i)) can be regarded as that of finding a static state-feedback control function $u = \alpha(x)$ such that the closed-loop system (5.3) is rendered dissipative with respect to the supply-rate

$$s(w(t), z(t)) = \frac{1}{2}(\gamma^2 \|w(t)\|^2 - \|z(t)\|^2)$$

and a suitable storage-function. For this purpose, we first recall the following definition from Chapter 3.

Definition 5.1.6 *The nonlinear system (5.1) is locally dissipative with respect to the* *supply-rate $s(w,z) = \frac{1}{2}(\gamma^2\|w\|^2 - \|z\|^2)$, if there exists a storage-function $V : N \subset \mathcal{X} \to \Re_+$ such that for any initial state $x(t_0) = x_0 \in N$, the inequality*

$$V(x_1) - V(x_0) \leq \int_{t_0}^{t_1} \frac{1}{2}(\gamma^2\|w(t)\|^2 - \|z(t)\|^2)dt \tag{5.23}$$

is satisfied for all $w \in \mathcal{L}_2[t_0, \infty)$, where $x_1 = x(t_1, t_0, x_0, u)$.

Remark 5.1.6 *Rewriting the above dissipation-inequality (5.23) as (since $V \geq 0$)*

$$\frac{1}{2}\int_{t_0}^{t_1} \|z(t)\|^2 dt \leq \frac{1}{2}\int_{t_0}^{t_1} \gamma^2\|w(t)\|^2 dt + V(x_0)$$

and allowing $t_1 \to \infty$, it immediately follows that dissipativity of the system with respect to the supply-rate $s(w,z)$ implies finite $\mathcal{L}_2$-gain $\leq \gamma$ for the system.

We can now state the following proposition.

Proposition 5.1.2 *Consider the nonlinear system (5.3) and the the SFBNLHICP using static state-feedback control. Suppose for some $\gamma > 0$, there exists a smooth solution $V \geq 0$ to the HJIE (5.12) or the HJI-inequality:*

$$V_x(x)f(x) + \frac{1}{2}V_x(x)[\frac{1}{\gamma^2}g_1(x)g_1^T(x) - g_2(x)g_2^T(x)]V_x^T(x) + \frac{1}{2}h_1^T(x)h_1(x) \leq 0, \quad V(0) = 0, \tag{5.24}$$

in $N \subset \mathcal{X}$. Then, the control function (5.18) solves the problem for the system in N.

Proof: The equivalence of the solvability of the HJIE (5.12) and the inequality (5.24) has been shown in Chapter 3. For the local disturbance-attenuation property, rewrite the HJ-inequality as

$$\begin{aligned}\dot{V}(x) &= V_x(x)[f(x) + g_1(x)w - g_2(x)g_2^T(x)V_x^T(x)], \quad x \in N \\ &\leq -\frac{1}{2}\|h\|^2 - \frac{\gamma^2}{2}\|w - \frac{1}{\gamma^2}g_1^T(x)V_x^T(x)\|^2 + \frac{1}{2}\gamma^2\|w\|^2 - \frac{1}{2}\|u^\star\|^2. \end{aligned} \tag{5.25}$$

Integrating now the above inequality from $t = t_0$ to $t = t_1 > t_0$, and starting from $x(t_0)$, we get

$$V(x(t_1)) - V(x(t_0)) \leq \int_{t_0}^{t_1} \frac{1}{2}(\gamma^2\|w\|^2 - \|z^\star\|^2)dt, \quad x(t_0), x(t_1) \in N,$$

where $z^\star = \begin{bmatrix} h_1(x) \\ u^\star \end{bmatrix}$. Hence, the system is locally dissipative with respect to the supply-rate $s(w,z)$, and consequently by Remark 5.1.6 has the local disturbance-attenuation property. $\square$

Remark 5.1.7 *Note that the inequality (5.25) is obtained whether the HJIE is used or the HJI-inequality is used.*

To prove asymptotic-stability for the closed-loop system, part (ii) of the problem, we have the following theorem.

Theorem 5.1.4 *Consider the nonlinear system (5.3) and the SFBNLHICP. Suppose the system is smoothly-stabilizable, zero-state detectable, and the assumptions of Proposition 5.1.1 hold for the system. Then the control law (5.18) renders the closed-loop system (5.3) locally asymptotically-stable in N with $w = 0$ and therefore solves the SFBNLHICP for* *the system locally in N. If in addition the solution $V \geq 0$ of the HJIE (or inequality) is proper, then the system is globally asymptotically-stable with $w = 0$, and the problem is solved globally.*

Proof: Substituting $w = 0$ in the inequality (5.25), it implies that $\dot{V}(t) \leq 0$ and the system is stable. Further, if the system is zero-state detectable, then for any trajectory of the system such that $\dot{V}(x(t)) \equiv 0$, for all $t \geq t_s$ for some $t_s \geq t_0$, it implies that $z(t) \equiv 0$, $u^\star(t) \equiv 0$, for all $t \geq t_s$, which in turn implies that $\lim_{t \to \infty} x(t) = 0$. The result now follows by application of Lasalle's invariance-principle. For the global asymptotic-stability of the system, we note that if V is proper, then V is a global solution of the HJI-inequality (5.24), and the result follows by applying the same arguments as above. $\square$

We consider another example.

Example 5.1.2 *Consider the nonlinear system defined on the half-space $N_{\frac{1}{2}} = \{x | x_1 > \frac{1}{2} x_2\}$*

$$\dot{x}_1 = \frac{-\frac{1}{4}x_1^2 - x_2^2}{2x_1 - x_2} + w$$
$$\dot{x}_2 = x_2 + w + u$$
$$z = [x_1 \; x_2 \; u]^T.$$

The HJI-inequality (5.24) corresponding to this system for $\gamma = \sqrt{2}$ is given by

$$\left(\frac{-\frac{1}{4}x_1^2 - x_2^2}{2x_1 - x_2}\right)V_{x_1} + (x_2)V_{x_2} + \frac{1}{4}V_{x_1}^2 + \frac{1}{2}V_{x_1}V_{x_2} - \frac{1}{4}V_{x_2}^2 + \frac{1}{2}(x_1^2 + x_2^2) \leq 0.$$

Then, it can be checked that the positive-definite function

$$V(x) = \frac{1}{2}x_1^2 + \frac{1}{2}(x_1 - x_2)^2$$

globally solves the above HJI-inequality in N with $\gamma = \sqrt{2}$. Moreover, since the system is zero-state detectable, then the control law

$$u = x_1 - x_2$$

asymptotically stabilizes the system over $N_{\frac{1}{2}}$.

Next, we investigate the relationship between the solvability of the $SFBNLHICP$ for the nonlinear system Σ^a and its linearization about $x = 0$:

$$\bar{\Sigma}^l : \begin{cases} \dot{\bar{x}} = F\bar{x} + G_1\bar{w} + G_2\bar{u}; \quad \bar{x}(0) = \bar{x}_0 \\ \bar{z} = \begin{bmatrix} H_1\bar{x} \\ \bar{u} \end{bmatrix} \end{cases} \quad (5.26)$$

where $F = \frac{\partial f}{\partial x}(0) \in \Re^{n \times n}$, $G_1 = g_1(0) \in \Re^{n \times r}$, $G_2 = g_2(0) \in \Re^{n \times p}$, $H_1 = \frac{\partial h}{\partial x}(0)$, and $\bar{u} \in \Re^p$, $\bar{x} \in \Re^n$, $\bar{w} \in \Re^r$. A number of interesting results relating the $\mathcal{L}_2$-gain of the linearized system $\bar{\Sigma}^l$ and that of the system $\bar{\Sigma}^a$ can be concluded [264]. We summarize here one of these results.

Theorem 5.1.5 *Consider the linearized system $\bar{\Sigma}^l$, and assume the pair (H_1, F) is de-tectable [292]. Suppose there exists a state-feedback $\bar{u} = K\bar{x}$ for some $p \times n$ matrix K, such that the closed-loop system is asymptotically-stable and has $\mathcal{L}_2$-gain from $\bar{w}$ to $\bar{z}$*

less than $\gamma > 0$. Then, there exists a neighborhood $\mathcal{O}$ of $x = 0$ and a smooth positive-semidefinite function $V : \mathcal{O} \to \Re$ that solves the HJIE (5.12). Furthermore, the control law $u^\star = -g_2(x)V_x^T(x)$ renders the $\mathcal{L}_2$-gain of the closed-loop system (5.3) less than or equal to γ in $\mathcal{O}$.

We defer a full study of the solvability and algorithms for solving the HJIE (5.12) which are crucial to the solvability of the $SFBNLHICP$, to a later chapter. However, it is sufficient to observe that, based on the results of Theorems 5.1.3 and 5.1.5, it follows that the existence of a stabilizing solution to the ARE (5.22) guarantees the local existence of a positive-semidefinite solution to the HJIE (5.12). Thus, any necessary condition for the existence of a symmetric solution $P \geq 0$ to the ARE (5.22) becomes also necessary for the local existence of solutions to (5.11). In particular, the stabilizability of (F, G_2) is necessary, and together with the detectability of (H_1, F) are sufficient. Further, it is well known from linear systems theory and the theory of Riccati equations [292, 68] that the existence of a stabilizing solution $P = P^T$ to the ARE (5.22) implies that the two subspaces

$$\mathcal{X}_-(\bar{H}_\gamma^\star) \quad \text{and} \quad Im \begin{bmatrix} 0 \\ I \end{bmatrix}$$

are complementary and the Hamiltonian matrix

$$\bar{H}_\gamma^\star = \begin{bmatrix} F & (\frac{1}{\gamma^2}G_1G_1^T - G_2G_2^T) \\ -H^TH & -F^T \end{bmatrix}$$

does not have imaginary eigenvalues, where $\mathcal{X}_-(\bar{H}_\gamma^\star)$ is the stable eigenspace of $\bar{H}_\gamma^\star$. Translated to the nonlinear case, this requires that the stable invariant-manifold M^- of the Hamiltonian vector-field $X_{H_\gamma^\star}$ through $(x, V_x^T(x)) = (0,0)$ (which is of the form (5.20)) to be n-dimensional and tangent to $\mathcal{X}_-(\bar{H}_\gamma^\star) := span \begin{bmatrix} I \\ P \end{bmatrix}$ at $(x, V_x^T) = (0,0)$, and the matrix $\bar{H}_\gamma^\star$ corresponding to the linearization of $H_\gamma^\star$ does not have purely imaginary eigenvalues. The latter condition is referred to as being *hyperbolic* and this situation will be regarded as the *noncritical* case. Thus, the detectability of (H_1, F) excludes the condition that $\bar{H}_\gamma^\star$ has imaginary eigenvalues, but this is not necessary. Indeed, the HJIE (5.12) can also have smooth solutions in the *critical case* in which the Hamiltonian matrix $\bar{H}_\gamma^\star$ is *nonhyperbolic*. In this case, the manifold $\mathbf{M}$ is not entirely the stable-manifold, but will contain a nontrivial *center-stable manifold*.

Proof: (of Theorem 5.1.5): By Theorem 5.1.3 there exists a solution $P \geq 0$ to (5.22). It follows that the stable invariant manifold M^- is tangent to $\mathcal{X}_-(\bar{H}_\gamma^\star)$ at $(x, p) = (0,0)$. Hence, locally about $x = 0$, there exists a smooth solution V^- to the HJIE (5.12) satisfying $\frac{\partial^2 V^-}{\partial x^2}(0) = P$. In addition, since $F - G_2G_2^TP$ is asymptotically-stable, the vector-field $f - g_2g_2^T\frac{\partial V^-}{\partial x}$ is asymptotically-stable. Rewriting the HJIE (5.12) as

$$V_x^-(x)(f(x) - g_2(x)g_2^T(x)V_x^{-T}(x)) + \frac{1}{2}V_x^-(x)[\frac{1}{\gamma^2}g_1(x)g_1^T(x) + g_2(x)g_2^T(x)]V_x^{-T}(x) +$$

$$\frac{1}{2}h_1^T(x)h_1(x) = 0,$$

it implies by the Bounded-real lemma (Chapter 3) that locally about $x = 0$, $V^- \geq 0$ and the closed-loop system has $\mathcal{L}_2$-gain $\leq \gamma$ for all $w \in \mathcal{W}$ such that $x(t)$ remains in $\mathcal{O}$. $\square$

In the next section, we discuss controller parametrization.

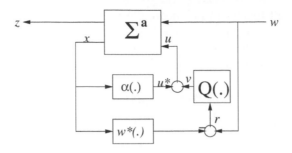

FIGURE 5.2
Controller Parametrization for FI-State-Feedback Nonlinear $\mathcal{H}_\infty$-Control

5.1.2 Controller Parametrization

In this subsection, we discuss the state-feedback $\mathcal{H}_\infty$ controller parametrization problem which deals with the problem of specifying a set (or all the sets) of possible state-feedback controllers that solves the *SFBNLHICP* for the system (5.1) locally.

The basis for the controller parametrization we discuss is the Youla (or Q)-parametrization for all stabilizing controllers for the linear problem [92, 195, 292] which has been extended to the nonlinear case [188, 215, 214]. Although the original Youla-parametrization uses coprime factorization, the modified version presented in [92] does not use coprime-factorization. The structure of the prametrization is shown in Figure 5.2. Its advantage is that it is given in terms of a free parameter which belongs to a linear space, and the closed-loop map is affine in this free parameter. Thus, this gives an additional degree-of-freedom to further optimize the closed-loop maps in order to achieve other design objectives.

Now, assuming Σ^a is smoothly-stabilizable and the disturbance signal $w \in \mathcal{L}_2[0, \infty)$ is fully measurable, also referred to as the *full-information (FI) structure*, then the following proposition gives a parametrization of a family of full-information controllers that solves the *SFBNLHICP* for Σ^a.

Proposition 5.1.3 *Assume the nonlinear system Σ^a is smoothly stabilizable and zero-state detectable. Suppose further, the disturbance signal is measurable and there exists a smooth (local) solution $V \geq 0$ to the HJIE (5.12) or inequality (5.24) such that the SFBNLHICP is (locally) solvable. Let $\mathcal{FG}$ denote the set of finite-gain (in the $\mathcal{L}_2$ sense) asymptotically-stable (with zero input and disturbances) input-affine nonlinear plants, i.e.,*

$$\mathcal{FG} \triangleq \{\Sigma^a | \Sigma^a(u = 0, w = 0) \text{ is asymptotically-stable and has } \mathcal{L}_2\text{-gain} \leq \gamma\}.$$

Then, the set

$$\mathbf{K}_{FI} = \{u|u = u^\star + Q(w - w^\star), \ Q \in \mathcal{FG}, \ Q : inputs \mapsto outputs\} \qquad (5.27)$$

is a paremetrization of all FI-state-feedback controllers that solves (locally) the SFBNLHICP for the system Σ^a.

Proof: Apply $u \in \mathbf{K}_{FI}$ to the system Σ^a resulting in the closed-loop system:

$$\Sigma^a_{u^\star}(Q) : \begin{cases} \dot{x} &= f(x) + g_1(x)w + g_2(x)(u^\star + Q(w - w^\star)); \ x(0) = x_0 \\ z &= \begin{bmatrix} h_1(x) \\ u \end{bmatrix}. \end{cases} \qquad (5.28)$$

If $Q = 0$, then the result follows from Theorem 5.1.2 or 5.1.4. So assume $Q \neq 0$, and since $Q \in \mathcal{FG}$, $r \triangleq Q(w - w^\star) \in \mathcal{L}_2[0, \infty)$. Let $V \geq 0$ be a (local) solution of (5.12) or (5.24) in N for some $\gamma > 0$. Then, differentiating this along a trajectory of the closed-loop system, completing the squares and using (5.12) or (5.24), we have

$$
\begin{aligned}
\frac{d}{dt}V &= V_x[f + g_1 w - g_2 g_2^T V_x^T + g_2 r] \\
&= V_x f + \frac{1}{2}V_x[\frac{1}{\gamma^2}g_1 g_1^T - g_2 g_2^T]V_x^T + \frac{1}{2}\|h_1\|^2 - \frac{1}{2}\|h_1\|^2 - \\
&\quad \frac{\gamma^2}{2}\|w - \frac{1}{\gamma^2}g_1^T V_x^T\|^2 + \frac{1}{2}\gamma^2\|w\|^2 - \frac{1}{2}\|r - g_2^T V_x^T(x)\|^2 + \frac{1}{2}\|r\|^2 \\
&\leq \frac{1}{2}\gamma^2\|w\|^2 - \frac{1}{2}\|h_1\|^2 - \frac{1}{2}\|u\|^2 + \frac{1}{2}\|r\|^2 - \frac{\gamma^2}{2}\|w - \frac{1}{\gamma^2}g_1^T V_x^T\|^2.
\end{aligned}
\tag{5.29}
$$

Now, integrating the above inequality (5.29) from $t = t_0$ to $t = t_1 > t_0$, starting from $x(t_0)$ and using the fact that

$$
\int_{t_0}^{t_1} \|r\|^2 dt \leq \gamma^2 \int_{t_0}^{t_1} \|w - w^\star\|^2 dt \quad \forall t_1 \geq t_0, \forall w \in \mathcal{W},
$$

we get

$$
V(x(t_1)) - V(x(t_0)) \leq \int_{t_0}^{t_1} \frac{1}{2}(\gamma^2\|w\|^2 - \|z\|^2)dt, \quad \forall x(t_0), \ x(t_1) \in N.
\tag{5.30}
$$

This implies that the closed-loop system (5.28) has $\mathcal{L}_2$-gain $\leq \gamma$ from w to z. Finally, the part dealing with local asymptotic-stability can be proven as in Theorems 5.1.2, 5.1.4. $\square$

Remark 5.1.8 *Notice that the set $\mathcal{FG}$ can also be defined as the set of all smooth input-affine plants $Q : r \mapsto v$ with the realization*

$$
\Sigma_Q : \left\{ \begin{aligned} \dot{\xi} &= a(\xi) + b(\xi)r \\ v &= c(\xi) \end{aligned} \right.
\tag{5.31}
$$

where $\xi \in \mathcal{X}$, $a : \mathcal{X} \to V^\infty(\mathcal{X})$, $b : \mathcal{X} \to \mathcal{M}^{n \times p}$, $c : \mathcal{X} \to \Re^m$ are smooth functions, with $a(0) = 0$, $c(0) = 0$, and such that there exists a positive-definite function $\varphi : \mathcal{X} \to \Re_+$ satisfying the bounded-real condition:

$$
\varphi_\xi(\xi)a(\xi) + \frac{1}{2\gamma^2}\varphi_\xi(\xi)b(\xi)b^T(\xi)\varphi_\xi^T(\xi) + \frac{1}{2}c^T(\xi)c(\xi) = 0.
$$

5.2   State-Feedback Nonlinear $\mathcal{H}_\infty$ Tracking Control

In this section, we consider the traditional state-feedback tracking, model-following or servomechanism problem. This involves the tracking of a given reference signal which may be any one of the classes of reference signals usually encountered in control systems, such as steps, ramps, parabolic or sinusoidal signals. The objective is to keep the error between the system output y and the reference signal arbitrarily small. Thus, the problem can be treated in the general framework discussed in the previous section with the penalty variable z representing the tracking error. However, a more elaborate design scheme may be necessary in order to keep the error as desired above.

The system is represented by the model (5.1) with the penalty variable

$$z = \left[\begin{array}{c} h_1(x) \\ u \end{array} \right], \tag{5.32}$$

while the signal to be tracked is generated as the output y_m of a reference model defined by

$$\Sigma_m : \left\{ \begin{array}{rcl} \dot{x}_m & = & f_m(x_m), \quad x_m(t_0) = x_{m0} \\ y_m & = & h_m(x_m) \end{array} \right. \tag{5.33}$$

$x_m \in \Re^l$, $f_m : \Re^l \to V^\infty(\Re^l)$, $h_m : \Re^l \to \Re^m$ and we assume that this system is completely observable [212]. The problem can then be defined as follows.

Definition 5.2.1 *(State-Feedback Nonlinear $\mathcal{H}_\infty$ (Suboptimal) Tracking Control Problem (SFBNLHITCP)). Find if possible, a static state-feedback control function of the form*

$$u = \alpha_{trk}(x, x_m), \quad \alpha_{trk} : N_o \times N_m \to \Re^p \tag{5.34}$$

$N_o \subset \mathcal{X}$, $N_m \subset \Re^l$, for some smooth function α_{trk}, such that the closed-loop system (5.1), (5.34), (5.33) has, for all initial conditions starting in $N_o \times N_m$ neighborhood of $(0,0)$, locally $\mathcal{L}_2$-gain from the disturbance signal w to the output z less than or equal to some prescribed number $\gamma^\star > 0$ and the tracking error satisfies $\lim_{t\to\infty}\{y - y_m\} = 0$.

To solve the above problem, we follow a two-step procedure:

Step 1: Find a feedforward-control law $u_\star = u_\star(x, x_m)$ so that the equilibrium point $x = 0$ of the closed-loop system

$$\dot{x} = f(x) + g_2(x)u_\star(x, 0) \tag{5.35}$$

is exponentially stable, and there exists a neighborhood $U = N_o \times N_m$ of $(0,0)$ such that for all initial conditions $(x_0, x_{m0}) \in U$ the trajectories $(x(t), x_m(t))$ of

$$\left\{ \begin{array}{rcl} \dot{x} & = & f(x) + g_2(x)u_\star(x, x_m) \\ \dot{x}_m & = & f_m(x_m) \end{array} \right. \tag{5.36}$$

satisfy

$$\lim_{t\to\infty} \{h_1(\theta(x_m(t))) - h_m(x_m(t))\} = 0.$$

To solve this step, we seek for an invariant-manifold

$$M_\theta = \{x | x = \theta(x_m)\}$$

and a control law $u_\star = \alpha_f(x, x_m)$ such that the submanifold M_θ is invariant under the closed-loop dynamics (5.36) and $h_1(\theta(x_m(t))) - h_m(x_m(t)) \equiv 0$. Fortunately, there is a wealth of literature on how to solve this problem [143]. Under some suitable assumptions, the following equations give necessary and sufficient conditions for the solvability of this problem:

$$\frac{\partial\theta}{\partial x_m}(x_m)(f_m(x_m) = f(\theta(x_m)) + g_2(\theta(x_m)\bar{u}_\star(x_m) \tag{5.37}$$

$$h_1(\theta(x_m(t)) - h_m(x_m(t)) = 0, \tag{5.38}$$

where $\bar{u}_\star(x_m) = \alpha_f(\theta(x_m), x_m)$.

The next step is to design an auxiliary feedback control v so as to drive the system onto the above submanifold and to achieve disturbance-attenuation as well as asymptotic tracking. To formulate this step, we consider the combined system

$$\begin{cases} \dot{x} & = & f(x) + g_1(x)w + g_2(x)u \\ \dot{x}_m & = & f_m(x_m), \end{cases} \tag{5.39}$$

and introduce the following change of variables

$$\begin{aligned} \xi & = & x - \theta(x_m) \\ v & = & u - \bar{u}_\star(x_m). \end{aligned}$$

Then

$$\begin{aligned} \dot{\xi} & = & F(\xi, x_m) + G_1(\xi, x_m)w + G_2(\xi, x_m)v \\ \dot{x}_m & = & f_m(x_m) \end{aligned}$$

where

$$\begin{aligned} F(\xi, x_m) & = & f(\xi + \theta(x_m)) - \frac{\partial \theta}{\partial x_m}(x_m)f_m(x_m) + g_2(\xi + \theta(x_m))\bar{u}_\star(x_m) \\ G_1(\xi, x_m) & = & g_1(\xi + \theta(x_m)) \\ G_2(\xi, x_m) & = & g_2(\xi + \theta(x_m)). \end{aligned}$$

Similarly, we redefine the tracking error and the new penalty variable as

$$\tilde{z} = \left[\begin{array}{c} h_1(\xi + \theta(x_m)) - h_m(x_m) \\ v \end{array} \right].$$

Step 2: Find an auxiliary feedback control $v_\star = v_\star(\xi, x_m)$ so that along any trajectory $(\xi(t), x_m(t))$ of the closed-loop system (5.40), the $\mathcal{L}_2$-gain condition

$$\int_{t_0}^T \|\tilde{z}(t)\|^2 dt \leq \gamma^2 \int_{t_0}^T \|w(t)\|^2 dt + \kappa(\xi(t_0), x_{m0})$$

$$\Longleftrightarrow \int_{t_0}^T \{\|h_1(\xi + \theta(x_m)) - h_m(x_m)\|^2 + \|v\|^2\}dt \leq \gamma^2 \int_{t_0}^T \|w(t)\|^2 dt + \kappa(\xi(t_0), x_{m0})$$

is satisfied for some function κ, for all $w \in \mathcal{W}$, for all $T < \infty$ and all initial conditions $(\xi(t_0), x_{m0})$ in a neighborhood $\bar{N}_o \times N_m$ of the origin $(0, 0)$. Moreover, if $\xi(t_0) = 0$ and $w(t) \equiv 0$, then we may set $v_\star(t) \equiv 0$ to achieve perfect tracking.

Clearly, the above problem is now a standard state-feedback $\mathcal{H}_\infty$-control problem, and the techniques discussed in the previous sections can be employed to solve it. The following theorem then summarizes the solution to the *SFBNLHITCP*.

Theorem 5.2.1 *Consider the nonlinear system (5.1) and the SFBNLHITCP for this system. Suppose the control law $u_\star = u_\star(x, x_m)$ and invariant-manifold M_θ can be found that solve Step 1 of the solution to the tracking problem. Suppose in addition, there exists a smooth solution $\Psi : \bar{N}_o \times N_m \to \Re$, $\Psi(\xi, x_m) \geq 0$ to the HJI-inequality*

$$\Psi_\xi(\xi, x_m)F(\xi, x_m) + \Psi_{x_m}(\xi, x_m)f_m(x_m) +$$

$$\frac{1}{2}\Psi_\xi(\xi, x_m)\left[\frac{1}{\gamma^2}G_1(\xi, x_m)G_1^T(\xi, x_m) - G_2(\xi, x_m)G_2^T(\xi, x_m)\right]\Psi_\xi^T(\xi, x_m) +$$

$$\frac{1}{2}\|h_1(\xi + x_m) - h_m(x_m)\|^2 \leq 0, \quad x \in \bar{N}_o, \ \xi \in N_m, \quad \Psi(0, 0) = 0. \tag{5.40}$$

Then the SFBNLHITCP is locally solvable with the control laws $u = \bar{u}_\star$ and

$$v_\star = -G_2^T(\xi, x_m)\Psi_\xi^T(\xi, x_m).$$

Moreover, if Ψ is proper with respect to ξ (i.e., if $\Psi(\xi, x_m) \to \infty$ when $\|\xi\| \to \infty$) and the system is zero-state detectable, then $\lim_{t \to \infty} \xi(t) = 0$ also for all initial conditions $(\xi(t_0), x_{m0}) \in \bar{N}_o \times N_m$.

5.3 Robust Nonlinear $\mathcal{H}_\infty$ State-Feedback Control

In this section, we consider the state-feedback $\mathcal{H}_\infty$-control problem for the affine nonlinear system Σ^a in the presence of unmodelled dynamics and/or parameter variations. This problem has been considered in many references [6, 7, 209, 223, 245, 261, 265, 284]. The approach presented here is based on [6, 7] and is known as guaranteed-cost control. It is an extension of quadratic-stabilization, and was first developed by Chang [78] and later popularized by Petersen [226, 227, 235]. For this purpose, the system is represented by the model:

$$\Sigma_\Delta^a : \begin{cases} \dot{x} &= f(x) + \Delta f(x, \theta, t) + g_1(x)w + [g_2(x) + \Delta g_2(x, \theta, t)]u; \\ & x(t_0) = x_0 \\ y &= x \\ z &= \begin{bmatrix} h_1(x) \\ u \end{bmatrix} \end{cases} \tag{5.41}$$

where all the variables and functions have their previous meanings and in addition $\Delta f : \mathcal{X} \to V^\infty(\mathcal{X})$, $\Delta g_2 : \mathcal{X} \to \mathcal{M}^{n \times p}(\mathcal{X})$ are unknown functions which belong to the set Ξ of admissible uncertainties, and $\theta \in \Theta \subset \Re^r$ are the system parameters which may vary over time within the set Θ. Usually, a knowledge of the sets Ξ and Θ is necessary in order to be able to synthesize robust control laws.

Definition 5.3.1 *(Robust State-Feedback Nonlinear $\mathcal{H}_\infty$-Control Problem (RSFBNL-HICP)) Find (if possible!) a static control function of the form*

$$\tilde{u} = \beta(x, t), \quad \beta : N \times \Re \to \Re^p \tag{5.42}$$

for some smooth function β depending on x and t only, such that the closed-loop system:

$$\Sigma_{\Delta, cl}^a : \begin{cases} \dot{x} &= f(x) + \Delta f(x, \theta, t) + g_1(x)w + [g_2(x) + \Delta g_2(x, \theta, t)]\beta(x, t); \\ & x(t_0) = x_0 \\ z &= \begin{bmatrix} h_1(x) \\ \beta(x, t) \end{bmatrix} \end{cases} \tag{5.43}$$

has locally $\mathcal{L}_2$-gain from the disturbance signal w to the output z less than or equal to some prescribed number $\gamma^\star > 0$ with internal-stability, or equivalently, the closed-loop system achieves local disturbance-attenuation less than or equal to $\gamma^\star$ with internal stability, for all perturbations Δf, $\Delta g_2 \in \Xi$ and all parameter variations in Θ.

To solve the above problem, we first characterize the sets of admissible uncertainties of the system Ξ and Θ.

Assumption 5.3.1 *The admissible uncertainties of the system are structured and matched, and they belong to the following sets:*

$$\Xi_\Delta = \{\Delta f, \Delta g_2 \mid \Delta f(x, \theta, t) = H_2(x)F(x, \theta, t)E_1(x), \quad \Delta g_2(x, \theta, t) = g_2(x)F(x, \theta, t)E_2(x),$$
$$\|F(x, \theta, t)\|^2 \le 1 \; \forall x \in \mathcal{X}, \theta \in \Theta, t \in \Re\}$$
$$\Theta = \{\theta | 0 \le \theta \le \theta_u\}$$

where $H_2(.)$, $F(.,.)$, $E_1(.)$, $E_2(.)$ have appropriate dimensions.

Remark 5.3.1 *The conditions of Assumption 5.3.1 are called "matching-conditions" and these types of uncertainties are known as "structured uncertainties."*

Now, define the following cost functional:

$$J_{gc}(u, w) = \frac{1}{2} \int_{t_0}^{T} (\|z(t)\|^2 - \gamma^2 \|w(t)\|^2)dt, \quad T > t_0. \tag{5.44}$$

Then we have the following definition:

Definition 5.3.2 *The function $\beta(.,.)$ is a guaranteed-cost control for the system (5.43) if there exists a positive-(semi)definite C^1 function $V : N \subset \mathcal{X} \to \Re_+$ that satisfies the inequality*

$$\frac{\partial V(x, t)}{\partial t} + \frac{\partial V(x, t)}{\partial x}\left\{f(x) + \Delta f(x, \theta, t) + g_1(x)w + [g_2(x) + \Delta g_2(x, \theta, t)]\beta(x, t)\right\} +$$
$$\frac{1}{2}(\|z\|^2 - \gamma^2\|w\|^2) \le 0 \; \forall x \in N, \;\; \forall w \in \mathcal{L}_2[0, \infty), \;\; \forall \Delta f, \Delta g_2 \in \Xi_\Delta, \theta \in \Theta. \tag{5.45}$$

It can now be observed that, since the cost function J_{gc} is exactly the $\mathcal{H}_\infty$-control cost function (equation (5.4)), then a guaranteed cost control $\beta(.,.)$ which stabilizes the system (5.43) clearly solves the robust $\mathcal{H}_\infty$-control problem. Moreover, integrating the inequality (5.45) from $t = t_0$ to $t = t_1 > t_0$ and starting at $x(t_0)$, we get the dissipation inequality (5.23). Thus, a guaranteed-cost control with cost function (5.44) renders the system (5.43) dissipative with respect to the supply-rate $s(w, z)$. Consequently, the guaranteed-cost framework solves the $RSFBNLHICP$ for the nonlinear uncertain system (5.43) from both perspectives. In addition, we also have the following proposition for the optimal cost of this policy.

Proposition 5.3.1 *If the control law $\beta(.,.)$ satisfies the guaranteed-cost criteria, then the optimal cost $J_{gc}^\star(u^\star, w^\star)$ of the policy is bounded by*

$$V^a(x, t) \le J_{gc}^\star(u^\star, w^\star) \le V(t_0, x_0)$$

where

$$V^a(x, t) = \sup_{x(0)=x, u \in \mathcal{U}, t \ge 0} - \int_0^t s(w(\tau), z(\tau))d\tau$$

is the available-storage of the system defined in Chapter 3.

Proof: Taking the supremum of $-J_{gc}(.,)$ over $\mathcal{U}$ and starting at x_0, we get the lower bound. To get the upper bound, we integrate the inequality (5.45) from $t = t_0$ to $t = t_1$ to get the disssipation inequality (5.23) which can be expressed as

$$V(t_1, x(t_1)) + \int_{t_0}^{t_1} \frac{1}{2}(\|z\|^2 - \gamma^2\|w\|^2)dt \le V(t_0, x(t_0)), \;\; \forall x(t_0), x(t_1) \in N.$$

Since $V(.,.) \geq 0$, the result follows. $\square$

The following lemma is a nonlinear generalization of [226, 227] and will be needed in the sequel.

Lemma 5.3.1 *For any matrix functions $H_2(.)$, $F(.,.,.)$ and $E(.)$ of appropriate dimensions such that $\|F(x, \theta, t)\| \leq 1$ for all $x \in \mathcal{X}$, $\theta \in \Theta$, and $t \in \Re$, then*

$$V_x(x)H_2(x)F(x,\theta,t)E(x) \leq \frac{1}{2}[V_x(x)H_2(x)H_1^T(x)V_x^T(x) + E^T(x)E(x)]$$

$$F(x,\theta,t)E(x) \leq I + \frac{1}{4}E^T(x)E(x)$$

for some C^1 function $V : \mathcal{X} \to \Re$, for all $x \in \mathcal{X}$, $\theta \in \Theta$ and $t \in \Re$.

Proof: For the first inequality, note that

$$0 \leq \|H_2^T(x)V_x^T(x) - F(x,\theta,t)E(x)\|^2 \;=\; V_x(x)H_2(x)H_2^T(x)V_x^T(x) - $$
$$2V_x(x)H_2(x)F(x,\theta,t)E(x) + E^T(x)E(x),$$

from which the result follows. Similarly, to get the second inequality, we have

$$0 \leq \|I - \frac{1}{2}F(x,\theta,t)E(x)\|^2 = I - F(x,\theta,t)E(x) + \frac{1}{4}E^T(x)E(x)$$

and the result follows. $\square$

Next we present one approach to the solution of the *RSFBNLHICP* which is the main result of this section.

Theorem 5.3.1 *Consider the nonlinear uncertain system Σ_Δ^a and the problem of synthesizing a guaranteed-cost control $\beta(.,.)$ that solves the RSFBNLHICP locally. Suppose the system is smoothly-stabilizable, zero-state detectable, and there exists a positive-(semi)definite C^1 function $\widetilde{V} : N \subset \mathcal{X} \to \Re$, $0 \in N$ satisfying the following HJIE(inequality):*

$$\widetilde{V}_x(x)f(x) + \frac{1}{2}\widetilde{V}_x(x)\left[\frac{1}{\gamma^2}g_1(x)g_1^T(x) + H_2(x)H_2^T(x) - g_2(x)g_2^T(x)\right]\widetilde{V}_x^T(x) + $$
$$\frac{1}{2}h_1^T(x)h_1(x) + \frac{1}{2}E_1(x)E_1^T(x) \leq 0, \;\; \widetilde{V}(0) = 0, \;\; x \in N. \tag{5.46}$$

Then the problem is solved by the control law

$$\tilde{u} = \beta(x) = -g_2^T(x)\widetilde{V}_x^T(x). \tag{5.47}$$

Proof: Consider the left-hand-side (LHS) of the inequality (5.45). Using the results of Lemma 5.3.1 and noting that it is sufficient to have the function $\widetilde{V}$ dependent on x only, then

$$LHS \;=\; \widetilde{V}_x(x)f(x) + \frac{1}{2}\widetilde{V}_x(x)H_2(x)H_2^T(x)\widetilde{V}_x^T(x) + \frac{1}{2}E_1^T(x)E_1(x) + $$
$$\widetilde{V}_x(x)g_1(x)w + \widetilde{V}_x(x)g_2(x)(2I + \frac{1}{4}E_2^T(x)E_2(x))\tilde{u} + \frac{1}{2}\|z\|^2 - \gamma^2\|w\|^2$$
$$=\; \widetilde{V}_x(x)f(x) + \frac{1}{2}\widetilde{V}_x(x)[g_1(x)g_1^T(x) + H_2(x)H_2^T(x) - g_2(x)g^T(x)]\widetilde{V}_x^T(x) + $$
$$\frac{1}{2}h_1^T(x)h_1(x) + \frac{1}{2}E_1^T(x)E_1(x) - \frac{1}{2}\|w - \frac{1}{\gamma^2}g_1^T(x)\widetilde{V}_x^T(x)\|^2 - $$
$$\frac{1}{2}\widetilde{V}_x g_2(x)[3I + \frac{1}{2}E_2^T(x)E_2(x)]g_2^T(x)\widetilde{V}_x^T(x), \;\; \forall x \in N, \forall w \in \mathcal{L}_2[0, \infty).$$

Now using the HJIE (5.46), we obtain

$$
\begin{aligned}
LHS \;=\;& -\frac{1}{2}\|w - \frac{1}{\gamma^2}g_1^T(x)\widetilde{V}_x^T(x)\|^2 - \frac{1}{2}\widetilde{V}_x g_2(x)[3I + \frac{1}{2}E_2^T(x)E_2(x)]g_2^T(x)\widetilde{V}_x^T \leq 0, \\
& \forall x \in N, \forall w \in \mathcal{L}_2[0,\infty),
\end{aligned}
$$

which implies that the control law (5.47) is a guaranteed-cost control for the system (5.41) and hence solves the *RSFBNLHICP*. $\square$

Theorem 5.3.1 above gives sufficient conditions for the existence of a guaranteed-cost control law for the uncertain system Σ_Δ^a. With a little more effort, one can obtain a necessary and sufficient condition in the following theorem.

Theorem 5.3.2 *Consider the nonlinear uncertain system Σ_Δ^a and the problem of synthesizing a guaranteed-cost control $\beta(.,.)$ that solves the RSFBNLHICP locally. Assume the system is smoothly-stabilizable and zero-state detectable. Then, a necessary and sufficient condition for the existence of such a control law is that there exists a positive-(semi)definite C^1 function $\widetilde{V} : N \subset \mathcal{X} \to \Re$, $0 \in N$ satisfying the following HJIE (inequality)*

$$
\widetilde{V}_x(x)f(x) + \frac{1}{2}\widetilde{V}_x(x)\Big[\frac{1}{\gamma^2}g_1(x)g_1^T(x) + H_2(x)H_2^T(x) - g_2(x)(2I +
$$

$$
\tfrac{1}{4}E_2^T(x)E_2(x))g_2^T(x)\Big]\widetilde{V}_x^T(x) + \tfrac{1}{2}h_1^T(x)h_1(x) + \tfrac{1}{2}E_1(x)E_1^T(x) \leq 0, \;\; \widetilde{V}(0) = 0. \quad (5.48)
$$

Moreover, the problem is solved by the optimal feedback control law:

$$
\tilde{u}^\star = \beta(x) = -g_2^T(x)\widetilde{V}_x^T(x). \tag{5.49}
$$

Proof: We shall only give the proof for the necessity part of the theorem only. The sufficiency part can be proven similarly to Theorem 5.3.1. Define the Hamiltonian of the system $H : T^\star\mathcal{X} \times \mathcal{U} \times \mathcal{W} \to \Re$ by

$$
\begin{aligned}
H(x, \widetilde{V}_x^T, u, w) \;=\;& \widetilde{V}_x(x)[f(x) + \Delta f(x,\theta,t) + g_1(x)w + \Delta g_2(x)(x,\theta,t)u] + \\
& \frac{1}{2}\|z\|^2 - \frac{1}{2}\gamma^2\|w\|^2.
\end{aligned}
$$

Then from Theorem 5.1.1, a necessary condition for an optimal control is that the inequality

$$
\min_{\mathcal{U}} \sup_{\mathcal{W}, \Xi_\Delta, \Theta} H(x, \widetilde{V}_x, u, w) \leq 0, \tag{5.50}
$$

or equivalently, the saddle-point condition:

$$
J_{gc}(u^\star, w) \leq J_{gc}(u^\star, w^\star) \leq J_{gc}(u, w^\star) \;\; \forall u \in \mathcal{U}, w \in \mathcal{W}, \Delta f, \; \Delta g_2 \in \Xi_\Delta, \; \theta \in \Theta, \tag{5.51}
$$

be satisfied for all admissible uncertainties. Further, by Lemma 5.3.1,

$$
\begin{aligned}
\sup_{\Xi_\Delta, \Theta} H(x, \widetilde{V}_x, u, w) \;\leq\;& \widetilde{V}_x(x)f(x) + \frac{1}{2}\widetilde{V}_x(x)H_2(x)H_2^T(x)\widetilde{V}_x^T(x) + \frac{1}{2}E_1^T(x)E_1(x) + \\
& \widetilde{V}_x(x)g_1(x)w + \widetilde{V}_x(x)g_2(x)(2I + \frac{1}{4}E_2^T(x)E_2(x))u + \\
& \frac{1}{2}(h_1^T(x)h_1(x) + u^T u) - \frac{1}{2}\gamma^2\|w\|^2.
\end{aligned}
$$

Setting the above inequality to an equality and differentiating with respect to u and w respectively, results in the optimal feedbacks:

$$
\begin{aligned}
u^\star \;=\;& -(2I + \frac{1}{4}E_2^T(x)E_2(x))^T g_2^T(x)\widetilde{V}_x^T(x) \\
w^\star \;=\;& \frac{1}{\gamma^2}g_1^T(x)\widetilde{V}_x^T(x).
\end{aligned}
$$

It can also be checked that the above feedbacks satisfy the saddle-point conditions (5.51) and that $u^\star$ minimizes J_{gc} while $w^\star$ maximizes it. Finally, substituting the above optimal feedbacks in (5.50) we obtain the HJIE (5.48). $\square$

The above result, Theorem 5.3.1, can be specialized to linear uncertain systems of the form

$$
\Sigma_\Delta^l : \begin{cases}
\dot{x} = [F + \Delta F(x, \theta, t)]x + G_1 w + [G_2 + \Delta G_2(x, \theta, t)]u; \; x(0) = x_0 \\
z = \begin{bmatrix} H_1 x \\ u \end{bmatrix},
\end{cases}
\tag{5.52}
$$

where the matrices F, G_1, G_2 and H_2 are as defined in (5.21) and $\Delta F(.,.,.)$, $\Delta g_2(.,.,.) \in \Xi_\Delta$ have compatible dimensions. Moreover, Ξ_Δ, Θ in this case are defined as

$$
\begin{aligned}
\Xi_{\Delta,l} &= \{\Delta F, \Delta G_2 | \Delta F(x, \theta, t) = H_2 \tilde{F}(x, \theta, t) E_1, \; \Delta G_2(x, \theta, t) = G_2 \tilde{F}(x, \theta, t) E_2, \\
&\quad \|\tilde{F}(x, \theta, t)\|^2 \leq 1 \; \forall x \in \mathcal{X}, \theta \in \Theta, t \in \Re\}, \\
\Theta_l &= \{\theta | 0 \leq \theta \leq \theta_u\},
\end{aligned}
$$

where H_2, $\tilde{F}(.,.)$, E_1, $E_2(.)$ have appropriate dimensions. Then we have the following corollary to the theorem.

Corollary 5.3.1 *Consider the linear uncertain system (5.52) and the RSFBNLHICP for it. Assume (Γ, G_2) is stabilizable and (H, F) is detectable. Suppose there exists a symmetric positive-semidefinite solution $P \geq 0$ to the ARE:*

$$
F^T P + PF + P[\frac{1}{\gamma^2} G_1 G_1^T + H_2 H_2^T - G_2 G_2^T]P + H_1^T H_1 + E_1^T E_1 = 0.
\tag{5.53}
$$

Then the control law

$$
\tilde{u} = -G_2^T P x
$$

solves the RSFBNLHICP for the system Σ_Δ^l.

The above results for the linear uncertain system can be further extended to a class of nonlinear uncertain systems with nonlinearities in the system matrices ΔF, ΔG_2 and some additional unknown C^0 drift vector-field $f : \mathcal{X} \times \Re \to V^\infty(\Re^q)$, which is Caratheodory,[1] as described by the following model:

$$
\Sigma_{\Delta_\sigma}^l : \begin{cases}
\dot{x} = [F + \Delta F(x, \theta, t)]x + G_1 w + [G_2 + \Delta G_2(x, \theta, t)]u + G_2 f(x, t); \\
\quad x(t_0) = x_0 \\
z = \begin{bmatrix} H_1 x \\ u \end{bmatrix},
\end{cases}
\tag{5.54}
$$

where all the variables and matrices have their previous meanings, and in addition, the set of admissible uncertainties is characterized by

$$
\begin{aligned}
\Xi_{\Delta_f} &= \{\Delta F, \Delta G_2, f | \Delta F(x, \theta, t) = H_2 \tilde{F}(x, \theta, t) E_1, \; \Delta G_2(x, \theta, t) = G_2 R(x, \\
&\quad \theta, t), \; \|\tilde{F}(x, \theta, t)\|^2 \leq \zeta \, \forall t > t_0, \; \|f(x, \theta, t)\| \leq \rho(x, t) \in \text{class } \mathcal{K} \text{ wrt } x, \\
&\quad \text{positive wrt } t \text{ and } \lim_{t \to \infty} \rho(x, t) < \infty, \; \|R(x, \theta, t)\|_\infty \leq \eta, \; 0 \leq \eta < 1 \\
&\quad \forall x \in \mathcal{X}, \theta \in \Theta\}.
\end{aligned}
$$

---

[1]A function $\sigma : \Re^n \times \Re \to \Re^q$ is a Caratheodory function if: (i) $\sigma(x,.)$ is Lebesgue measurable for all $x \in \Re^n$; (ii) $\sigma(.,t)$ is continuous for each $t \in \Re$; and (iii) for each compact set $O \subset \Re^n \times \Re$, there exists a Lebesgue integrable function $m : \Re \to \Re$ such that $\|\sigma(x, t)\| \leq m(t) \, \forall (x, t) \in O$.

We can now state the following theorem.

Theorem 5.3.3 *Consider the nonlinear uncertain system $\Sigma_{\Delta_f}^a$ and the RSFBNLHICP for it. Assume the nominal system (F, G_2) is stabilizable and (H_1, F) is detectable. Suppose further there exists an $\epsilon > 0$ and a symmetric positive-definite matrix $Q \in \Re^{n \times n}$ such that the ARE:*

$$F^T P + PF + P[\epsilon \zeta H_2 H_2^T - 2G_2 G_2^T + \gamma^{-2} G_1 G_1^T]P + \frac{1}{\epsilon} E_1^T E_1 + H_1^T H_1 + Q = 0 \quad (5.55)$$

has a symmetric positive-definite solution P. Then the control law

$$\left. \begin{array}{rcl} u_r & = & -Kx - \frac{1}{(1-\eta)} \phi_r(x, t) \\ \phi_r(x, t) & = & \frac{(\rho(x,t)+\eta\|Kx\|)^2}{Kx(\rho(x,t)+\eta\|Kx\|)+\epsilon^\star\|x\|^2} Kx \\ K & = & G_2^T P \\ \epsilon^\star & < & \frac{\lambda_{min}(Q)}{2} \end{array} \right\} \quad (5.56)$$

solves the RSFBNLHICP for the system.

Proof: Suppose there exist a solution $P > 0$ to the ARE (5.55) for some $\epsilon > 0$, $Q > 0$. Let

$$V(x(t)) = x^T(t) P x(t)$$

be a Lyapunov function candidate for the closed-loop system (5.54), (5.56). We need to show that the closed-loop system is dissipative with this storage-function and the supply-rate $s(w, z) = (\gamma^2 \|w\|^2 - \|z\|^2)$ for all admissible uncertainties and disturbances, i.e.,

$$\dot{V}(x(t)) - (\gamma^2 \|w\|^2 - \|z\|^2) \leq 0, \quad \forall w \in \mathcal{W}, \ \forall \Delta F, \ \Delta G_2, \ f \in \Xi_{\Delta_f}, \ \theta \in \Theta.$$

Differentiating $V(.)$ along a trajectory of this closed-loop system and completing the squares, we get

$$\begin{array}{rcl} \dot{V}(x(t)) & = & x^T[(F+\Delta F)^T P + P(F+\Delta F)]x + 2x^T P(G_2 + \Delta G_2)u + \\ & \triangleq & 2x^T P G_1 w + 2x^T P G_2 f(x,t) A_1(x,t) + A_2(x,t) + A_3(x,t) - \\ & & \|z\|^2 + \gamma^2 \|w\|^2 \end{array}$$

where

$$\begin{array}{rcl} A_1(x,t) & = & x^T[(F+\Delta F)^T P + P(F+\Delta F) + \gamma^2 P G_1 G_1^T P + H_1^T H_1 - \\ & & 2PG_2 G_2^T P]x \\ A_2(x,t) & = & -2x^T P \left[(G_2 + \Delta G_2)\frac{1}{1-\eta}\phi_r(x,t) + \Delta G_2 G_2^T Px - G_2 f(x,t) \right] \\ A_3(x,w) & = & -\gamma^2 [w - \gamma^{-2} G_1^T Px]^T [w - \gamma^{-2} G_1^T Px]. \end{array}$$

Consider the terms $A_1(x,t)$, $A_2(x,t)$, $A_3(x,t)$, and noting that

$$x^T P \Delta F x = x^T P H_2 F(x,t) E_1 x \leq \frac{1}{2} x^T \left\{ \epsilon \zeta P H_2 H_2^T P + \frac{1}{\epsilon} E_1^T E_1 \right\} x \quad \forall x \in \Re^n,$$

$\epsilon > 0$. Then,

$$\begin{array}{rcl} A_1(x,t) & \leq & x^T[(F^T P + PF + \epsilon \zeta P H_2 H_2^T P + \frac{1}{\epsilon} E_1^T E_1 + \gamma^2 P G_1 G_1^T P + H_1^T H_1 - \\ & & 2PG_2 G_2^T P]x \\ & \leq & -x^T Q x, \end{array}$$

where the last inequality follows from the ARE (5.53). Next,

$$
\begin{aligned}
A_2(x,t) &= -2x^T PG_2(I + R(x,t))\frac{1}{1-\eta}\phi_r(x,t) - 2x^T PG_2[R(x,t)Kx - f(x,t)] \\
&\leq -2x^T PG_2\phi_r(x,t) - 2x^T PG_2[R(x,t)Kx - f(x,t)] \\
&\leq 2x^T PG_2[f(x,t) - R(x,t)Kx] - \frac{\|Kx\|^2(\rho(x,t) + \eta\|Kx\|)^2}{\|Kx\|(\rho(x,t) + \eta\|Kx\|) + \epsilon^\star\|x\|^2} \\
&\leq \left\{ \|Kx\|(\rho(x,t) + \eta\|Kx\|) - \frac{\|Kx\|^2(\rho(x,t) + \eta\|Kx\|)^2}{\|Kx\|(\rho(x,t) + \eta\|Kx\|) + \epsilon^\star\|x\|^2} \right\} \\
&\leq 2\left\{ \frac{\|Kx\|(\rho(x,t) + \eta\|Kx\|)\epsilon^\star\|x\|^2}{\|Kx\|(\rho(x,t) + \eta\|Kx\|) + \epsilon^\star\|x\|^2} \right\} \\
&\leq 2\epsilon^\star\|x\|^2.
\end{aligned}
$$

Therefore,

$$
\begin{aligned}
\dot{V}(x(t)) &\leq -x^T Qx + 2\epsilon^\star x^T x - \|z\|^2 + \gamma^2\|w\|^2 \\
&= -\|z\|^2 + \gamma^2\|w\|^2 - x^T(Q - 2\epsilon^\star I)x
\end{aligned}
\tag{5.57}
$$

and

$$
\dot{V}(x(t)) + \|z\|^2 - \gamma^2\|w\|^2 \leq -x^T(Q - 2\epsilon^\star I)x \leq 0.
$$

Thus, the closed-loop system is dissipative and hence has $\mathcal{L}_2$-gain $\leq \gamma$ from w to z for all admissible uncertainties and all disturbances $w \in \mathcal{W}$. Moreover, from the above inequality, with $w = 0$, we have

$$
\dot{V}(x(t)) \leq -x^T(Q - 2\epsilon^\star I)x < 0
$$

since $\epsilon^\star < \frac{\lambda_{min}(Q)}{2}$. Consequently, by Lyapunov's theorem, we have exponential-stability of the closed-loop system. $\square$

Example 5.3.1 *[209]. Consider the system (5.54) with*

$$
F = \begin{bmatrix} 0 & 1 \\ 0 & 0 \end{bmatrix}, \quad G_1 = \begin{bmatrix} 1 \\ 0 \end{bmatrix}, \quad G_2 = \begin{bmatrix} 1 \\ 0 \end{bmatrix}, \quad H_2 = \begin{bmatrix} 0.1 \\ 0.2 \end{bmatrix}
\tag{5.58}
$$

$$
H_1 = [0.5 \; 0], \quad E_1 = [0.1 \; 0.2], \quad \|f(x,t)\| \leq |\tfrac{1}{3}x_2|^3.
$$

Applying the result of Theorem 5.3.3 with $\gamma = 1$, $\epsilon = 1$ and

$$
Q = \begin{bmatrix} 0.5 & 0 \\ 0 & 0.5 \end{bmatrix},
$$

the ARE (5.55) has a symmetric positive-definite solution

$$
P = \begin{bmatrix} 3.7136 & 2.7514 \\ 2.7514 & 2.6671 \end{bmatrix} \implies K = [2.7514 \; 2.6671].
$$

5.4   State-Feedback $\mathcal{H}_\infty$-Control for Time-Varying Affine Nonlinear Systems

In this section, we consider the state-feedback $\mathcal{H}_\infty$-control problem for affine nonlinear time-varying systems ($TVSFBNLHICP$). Such systems are called nonautonomous, as the dynamics of the system is explicitly a function of time and can be represented by the model:

$$
\Sigma_t^a \; : \; \begin{cases} \dot{x}(t) &=& f(x,t) + g_1(x,t)w + g_2(x,t)u; \;\; x(0) = x_0 \\ y(t) &=& x(t) \\ z(t) &=& \left[\begin{array}{c} h(x,t) \\ u \end{array} \right] \end{cases} \tag{5.59}
$$

where all the variables have their usual meanings, while the functions $f : \mathcal{X} \times \Re \to V^\infty$, $g_1 : \mathcal{X} \times \Re \to \mathcal{M}^{n \times r}(\mathcal{X} \times \Re)$, $g_2 : \mathcal{X} \times \Re \to \mathcal{M}^{n \times p}(\mathcal{X} \times \Re)$, $h : \mathcal{X} \times \Re \to \Re^s$, and $h_2 : \mathcal{X} \times \Re \to \Re^m$ are real $C^{\infty,0}(\mathcal{X}, \Re)$ functions, i.e., are smooth with respect to x and continuous with respect to t. We also assume without any loss of generality that $x = 0$ is the only equilibrium point of the system with $u = 0$, $w = 0$, and is such that $f(0,t) = 0$, $h(0,t) = 0$.

Furthermore, since the system is time-varying, we are here interested in finding control laws that solve the *TVSFNLHICP* for any finite time-horizon $[0,T]$. Moreover, since most of the results on the state-feedback $\mathcal{H}_\infty$-control problem for the time-invariant case carry through to the time-varying case with only slight modifications to account for the time variation, we shall only summarize here the main result [189].

Theorem 5.4.1 *Consider the nonlinear system (5.59) and the TVSFBNLHICP for it. Assume the system is uniformly (for all t) smoothly-stabilizable and uniformly zero-state detectable, and for some $\gamma > 0$, there exists a positive-definite function $V : N_1 \subset \mathcal{X} \times [0,T] \to \Re_+$ such that $V(x,T) \geq P_T(x)$ and $V(x,0) \leq P_0(x) \; \forall x \in N_1$, for some nonnegative functions $P_T, P_0 : \mathcal{X} \to \Re_+$, with $P_T(0) = P_0(0) = 0$, which satisfies the time-varying HJIE (inequality):*

$$
V_t(x,t) + V_x(x,t)f(x,t) + \frac{1}{2}V_x(x,t)[\frac{1}{\gamma^2}g_1(x,t)g_1^T(x,t) - g_2(x,t)g_2^T(x,t)]V_x^T(x,t) +
$$

$$
\frac{1}{2}h^T(x,t)h(x,t) \leq 0, \;\; x \in N_1, t \in [0,T]. \tag{5.60}
$$

Then, the problem is solved by the feedback

$$
u^\star = -g_2^T(x,t)V_x^T(x,t),
$$

and

$$
w^\star = \frac{1}{\gamma^2}g_1^T(x,t)V_x^T(x,t)
$$

is the worst-case disturbance affecting the system.

Similarly, a parametrization of a set of full-information state-feedback controls for such systems can be given as

$$
\mathbf{K}_{FI_T} \;\; = \;\; \Big\{ u | u = -g_2^T(x,t)V^T(x,t) + Q(t)(w - g_1^T(x,t)V^T(x,t)), \; Q \in \mathcal{FG}_T,
$$

$$
Q : inputs \mapsto outputs \Big\}
$$

where

$$
\mathcal{FG}_T \;\; \triangleq \;\; \{\Sigma_t^a | \Sigma_t^a(u = 0, w = 0) \text{ is uniformly asymptotically stable and has}
$$

$$
\mathcal{L}_2([0,T])\text{-gain} \leq \gamma \, \forall T > 0\}.
$$

In the next section, we consider the state-feedback problem for affine nonlinear systems with state-delay.

5.5 State-Feedback $\mathcal{H}_\infty$-Control for State-Delayed Affine Nonlinear Systems

In this section, we present the state-feedback $\mathcal{H}_\infty$-control problem and its solution for affine nonlinear state-delayed systems $(SFBNLHICPSDS)$. These are system that have memory and in which the dynamics of the system is affected by past values of the state variable. Therefore their qualitative behavior is significantly richer, and their analysis is more complicated. The dynamics of this class of systems is also intimately related to that of time-varying systems that we have studied in the previous section.

We consider at the outset the following autonomous affine nonlinear state-space system with state-delay defined over an open subset $\mathcal{X}$ of $\Re^n$ with $\mathcal{X}$ containing the origin $x = 0$:

$$\Sigma_{d_0}^a : \begin{cases} \dot{x}(t) &= f(x(t), x(t - d_0)) + g_1(x(t))w(t) + g_2(x(t))u(t); \\ & x(t) = \psi(t), \ t \subset [t_0 - d_0, t_0], \ x(t_0) = \phi(t_0) = x_0, \\ y(t) &= x(t), \\ z(t) &= h_1(x(t)) + k_{12}(x(t))u(t) + k_{13}(x(t))u(t - d_0), \end{cases} \tag{5.61}$$

where $x(.) \in \mathcal{X}$ is the state vector, $u(.) \in \mathcal{U} \subseteq \Re^p$ is the p-dimensional control input, which belongs to the set of admissible controls $\mathcal{U}$, $w(.) \in \mathcal{W}$ is the disturbance signal which has to be tracked or rejected and belongs to the set $\mathcal{W} \subset \Re^r$ of admissible disturbances, the output $y \in \Re^m$ is the measured output of the system, $z(.) \in \Re^s$ is the output to be controlled, $d_0 > 0$ is the state delay which is constant, and $\phi(t) \in C[-d, t_0]$ is the initial function. Further, the functions $f(.,.) : \mathcal{X} \times \mathcal{X} \to V^\infty$, $g_1(.) : \mathcal{X} \to \Re^{n \times r}(\mathcal{X})$, $g_2 : \mathcal{X} \to \Re^{n \times p}(\mathcal{X})$, $h_1 : \mathcal{X} \to \Re^s$, $h_2 : \mathcal{X} \to \Re^m$ and $k_{12}, k_{13} : \mathcal{X} \to \Re^{s \times p}$ are real C^∞ functions of $x(.)$ such that the system (5.61) is well defined. That is, for any initial states $x(t_0 - d)$, $x(t_0) \in \mathcal{X}$ and any admissible input $u(t) \in \mathcal{U}$, there exists a unique solution $x(t, t_0, x_0, x_{t_0 - d}, u)$ to (5.61) on $[t_0, \infty)$ which continuously depends on the initial data, or the system satisfies the local existence and uniqueness theorem for functional differential equations. Without any loss of generality we also assume that the system has an equilibrium at $x = 0$, and is such that $f(0, 0) = 0$, $h_1(0) = 0$.

We introduce the following definition of $\mathcal{L}_2$-gain for the affine delay system (5.61).

Definition 5.5.1 *The system (5.61) is said to have $\mathcal{L}_2$-gain from $u(t)$ to $y(t)$ less than or equal to some positive number $\gamma' > 0$, if for all $(t_0, t_1) \in [-d, \infty)$, initial state vector $x_0 \in \mathcal{X}$, the response of the system $z(t)$ due to any $u(t) \in \mathcal{L}_2[t_0, t_1]$ satisfies*

$$\int_{t_0}^{t_1} \|z(t)\|^2 dt \leq \frac{\gamma'^2}{2} \int_{t_0}^{t_1} (\|w(t)\|^2 + \|w(t - d_0)\|^2) dt + \beta(x_0) \ \forall t_1 \geq t_0 \tag{5.62}$$

and some nonnegative function $\beta : \mathcal{X} \to \Re_+$, $\beta(0) = 0$.

The following definition will also be required in the sequel.

Definition 5.5.2 *The system (5.61) with $u(t) \equiv 0$ is said to be locally zero-state detectable, if for any trajectory of the system with initial condition $x(t_0) \in U \subset \mathcal{X}$ and a neighborhood of the origin, such that $z(t) \equiv 0 \ \forall t \geq t_0 \Rightarrow \lim_{t \to \infty} x(t, t_0, x_0, x_{t_0 - d_0}, 0) = 0$.*

The problem can then be defined as follows.

Definition 5.5.3 *(State-Feedback Suboptimal $\mathcal{H}_\infty$-Control Problem for State-Delayed Systems (SFBHICPSDS)). The problem is to find a smooth state-feedback control law of the form*

$$u(t) = \alpha(t, x(t), x(t - d_0)), \ \alpha \in C^2(\Re \times N \times N), \ \alpha(t, 0, 0) = 0 \ \forall t, \quad N \subset \mathcal{X},$$

which is possibly time-varying, such that the closed-loop system has $\mathcal{L}_2$-gain from $w(.)$ to $z(.)$ less than or equal to some prescribed positive number $\gamma > 0$, and is locally asymptotically-stable with $w(t) = 0$.

For this purpose, we make the following simplifying assumptions on the system.

Assumption 5.5.1 *The matrices $h_1(.)$, $k_{12}(.)$, $k_{13}(.)$ of the system (5.61) are such that $\forall t,\ i \neq j, i,\ j = 2, 3$*

$$h_1(x(t))k_{1j}(x(t)) = 0, \quad k_{1j}^T(x(t))k_{1j}(x(t)) = I, \quad k_{1i}^T(x(t))k_{1j}(x(t)) = 0.$$

Under the above assumption, we can without loss of generality represent $z(.)$ as

$$z(t) = \begin{bmatrix} h_1(x(t)) \\ u(t) \\ u(t - d_0) \end{bmatrix}.$$

Assumption 5.5.2 *The system $\Sigma_{d_0}^a$ or the pair $[f, g_2]$ is locally smoothly-stabilizable, if there exists a feedback control function $\alpha = \alpha(t, x(t), x(t - d_0))$ such that $\dot{x}(t) = f(x(t), x(t - d_0)) + g_2(x(t))\alpha(t, x(t), x(t - d_0))$ is locally asymptotically stable for all initial conditions in some neighborhood U of $x = 0$.*

The following theorem then gives sufficient conditions for the solvability of this problem. Note also that in the sequel we shall use the notation x_t for $x(t)$ and x_{t-d_0} for $x(t - d_0)$ for convenience.

Theorem 5.5.1 *Consider the nonlinear system $\Sigma_{d_0}^a$ and assume it is zero-state detectable. Suppose the Assumptions 5.5.1, 5.5.2 hold, and for some $\gamma > 0$ there exists a smooth positive-(semi)definite solution to the Hamilton-Jacobi-Isaacs inequality (HJII):*

$$V_t(t, x_t, x_{t-d_0}) + V_{x_t}(t, x_t, x_{t-d_0})f(x_t, x_{t-d_0}) + V_{x_{t-d_0}}(t, x_t, x_{t-d_0})f(x_{t-d_0}, x_{t-2d}) +$$

$$\frac{1}{2}\left\{\frac{1}{\gamma^2}[V_{x_t}g_1(x_t)g_1^T(x_t)V_{x_t}^T + V_{x_{t-d_0}}g_1(x_{t-d_0})g_1^T(x_{t-d_0})V_{x_{t-d_0}}^T] - \right.$$

$$\left. [V_{x_t}g_2(x_t)g_2^T(x_t)V_{x_t}^T + V_{x_{t-d_0}}g_2(x_{t-d_0})g_2^T(x_{t-d_0})V_{x_{t-d_0}}^T]\right\} + \frac{1}{2}h_1^T(x_t)h_1(x_t) \leq 0,$$

$$V(t, 0, 0) = 0 \ \forall x_t, x_{t-d_0} \in \mathcal{X}. \tag{5.63}$$

Then, the control law

$$u^\star(t) = -g_2^T(x(t))V_{x_t}^T(t, x(t), x(t - d_0)) \tag{5.64}$$

solves the SFBNLHICPSDS for the system $\Sigma_{d_0}^a$.

Proof: Rewriting the HJII (5.63) as

$$V_t(t, x_t, x_{t-d_0}) + V_{x_t}(t, x_t, x_{t-d_0})[f(x_t, x_{t-d_0}) + g_1(x_t)w(t) + g_2(x_t)u(t)] +$$

$$V_{x_{t-d_0}}[f(x_{t-d_0}, x_{t-2d}) + g_1(x_{t-d_0})w(t - d_0) + g_2(x_{t-d_0})u(t - d_0)] \leq$$

$$\frac{1}{2}\|u(t) - u^\star(t)\|^2 + \frac{1}{2}\|u(t - d_0) - u^\star(t - d_0)\|^2 - \frac{\gamma^2}{2}\|w(t) - w^\star(t)\|^2 -$$

$$\frac{\gamma^2}{2}\|w(t - d_0) - w^\star(t - d_0)\|^2 - \frac{1}{2}\|u(t)\|^2 - \frac{1}{2}\|u(t - d_0)\|^2 + \frac{\gamma^2}{2}\|w(t)\|^2 +$$

$$\frac{\gamma^2}{2}\|w(t - d_0)\|^2 - \frac{1}{2}\|h_1(x_t)\|^2$$

where

$$w^\star(t) = \frac{1}{\gamma^2} g_1^T(x(t)) V_{x_t}^T(x(t), x(t - d_0)),$$

$$u^\star(t - d_0) = -g_2^T(x_{t-d_0}) V_{x_{t-d_0}}^T(t, x(t), x(t - d_0)),$$

and in the above, we have suppressed the dependence of $u^\star(.)$, $w^\star(.)$ on $x(t)$ for convenience. Then, the above inequality further implies

$$V_t(t, x_t, x_{t-d_0}) + V_{x_t}(t, x_t, x_{t-d_0})[f(x_t, x_{t-d_0}) + g_1(x_t)w(t) + g_2(x_t)u(t)] +$$

$$V_{x_{t-d_0}}[f(x_{t-d_0}, x_{t-2d}) + g_1(x_{t-d_0})w(t - d_0) + g_2(x_{t-d_0})u(t - d_0)] \le$$

$$\frac{1}{2}\|u(t) - u^\star(t)\|^2 + \frac{1}{2}\|u(t - d_0) - u^\star(t - d_0)\|^2 - \frac{1}{2}\|u(t)\|^2 - \frac{1}{2}\|u(t - d_0)\|^2 +$$

$$\frac{\gamma^2}{2}\|w(t)\|^2 + \frac{\gamma^2}{2}\|w(t - d_0)\|^2 - \frac{1}{2}\|h_1(x_t)\|^2.$$

Substituting now $u(t) = u^\star(t)$ and integrating from $t = t_0$ to $t = t_1 \ge t_0$, starting from x_0, we get

$$V(t_1, x(t_1), x(t_1 - d)) - V(t_0, x_0, x(t_0 - d_0)) \le \frac{1}{2}\int_{t_0}^{t_1}\Big\{\gamma^2(\|w(t)\|^2 +$$

$$\|w(t - d_0)\|^2) - \|h_1(x_t)\|^2 - \|u^\star(t)\|^2 - \|u^\star(t - d_0)\|^2\Big\}dt,$$

which implies that the closed-loop system is dissipative with respect to the supply rate $s(w(t), z(t)) = \frac{1}{2}[\gamma^2(\|w(t)\|^2 + \|w(t - d_0)\|^2) - \|z(t)\|^2]$, and hence has $\mathcal{L}_2$-gain from $w(t)$ to $z(t)$ less than or equal to γ. Finally, for local asymptotic-stability, differentiating V from above along the trajectories of the closed-loop system with $w(.) = 0$, we get

$$\dot{V} \le -\frac{1}{2}\|z(t)\|^2,$$

which implies that the system is stable. In addition the condition when $\dot{V} \equiv 0 \ \forall t \ge t_s$, corresponds to $z(t) \equiv 0 \forall t \ge t_s$. By zero-state detectability, this implies that $\lim_{t\to\infty} x(t) = 0$, and thus by LaSalle's invariance-principle, we conclude asymptotic-stability. $\square$

A parametrization of all stabilizing state-feedback $\mathcal{H}_\infty$-controllers for the system in the full-information (FI) case (when the disturbance can be measured) can also be given.

Proposition 5.5.1 *Assume the nonlinear system Σ_d^a satisfies Assumptions 5.5.1, 5.5.2 and is zero-state detectable. Suppose the $SFBNLHICPSDS$ is solvable and the disturbance signal $w \in \mathcal{L}_2[0, \infty)$ is fully measurable. Let $\mathcal{FG}$ denote the set of finite-gain (in the $\mathcal{L}_2$ sense) asymptotically-stable (with zero input and disturbances) input-affine nonlinear plants of the form:*

$$\Sigma^a : \begin{cases} \dot{\xi} &= a(\xi) + b(\xi)u' \\ v &= c(\xi). \end{cases} \tag{5.65}$$

Then the set

$$\mathbf{K}_{FI} = \{u(t)|u(t) = u^\star(t) + Q(w(t) - w^\star(t)), \ Q \in \mathcal{FG}, \ Q : inputs \mapsto outputs\} \tag{5.66}$$

is a paremetrization of all FI-state-feedback controllers that solves (locally) the $SFBNLHICPSDS$ for the system Σ_d^a.

Proof: Apply $u(t) \in \mathbf{K}_{FI}$ to the system $\Sigma_{d_0}^a$ resulting in the closed-loop system:

$$\Sigma_d^a(u^\star(Q)) : \begin{cases} \dot{x}(t) &= f(x(t), x(t-d_0)) + g_1(x(t))w(t) + g_2(x(t))(u^\star(t) + \\ & \quad Q(w(t) - w^\star(t))); \quad x(t_0) = x_0 \\ z(t) &= \begin{bmatrix} h_1(x(t)) \\ u(t) \\ u(t-d_0) \end{bmatrix}. \end{cases} \tag{5.67}$$

If $Q = 0$, then the result follows from Theorem 5.5.1. So assume $Q \neq 0$, and since $Q \in \mathcal{FG}$, $r(t) \overset{\Delta}{=} Q(w(t) - w^\star(t)) \in \mathcal{L}_2[0, \infty)$. Then differentiating $V(., ., .)$ along the trajectories of the closed-loop system (5.67),(5.66) and completing the squares, we have

$$
\begin{aligned}
\frac{d}{dt}V &= V_t + V_{x_t}[f(x_t, x_{t-d_0}) + g_1(x_t)w(t) - g_2(x_t)g_2^T(x_t)V_{x_t}^T + g_2(x_t)r(t)] + \\
& \quad V_{x_{t-d_0}}[f(x_{t-d_0}, x_{t-2d_0}) + g_1(x_{t-d_0})w(t-d_0) - g_2(x_{t-d_0})g_2^T(x_{t-d_0}) \times \\
& \quad V_{x_{t-d_0}}^T(t, x_t, x_{t-d_0}) + g_2(x_{t-d_0})r(t-d_0)] \\
&= V_t(t, x_t, x_{t-d_0}) + V_{x_t}(t, x_t, x_{t-d_0})f(x_t, x_{t-d_0}) + \\
& \quad V_{x_{t-d_0}}(t, x_t, x_{t-d_0})f(x_{t-d_0}, x_{t-2d}) - \frac{1}{2}\gamma^2\|w(t) - w^\star(t)\|^2 + \frac{1}{2}\gamma^2\|w(t)\|^2 + \\
& \quad \frac{1}{2}\gamma^2\|w^\star(t)\|^2 - \frac{1}{2}\|r(t) - g_2^T(x_t)V_{x_t}^T(t, x_t, x_{t-d_0})\|^2 + \\
& \quad \frac{1}{2}\|r(t)\|^2 - \frac{1}{2}V_{x_t}g_2(x_t)g_2(x_t)V_{x_t}^T - \frac{1}{2}\gamma^2\|w(t-d_0) - w^\star(t-d_0)\|^2 + \\
& \quad \frac{1}{2}\gamma^2\|w(t-d_0)\|^2 + \frac{1}{2}\gamma^2\|w^\star(t-d_0)\|^2 - V_{x_{t-d_0}}g_2(x_{t-d_0})g_2^T(x_{t-d_0})V_{x_{t-d_0}}^T \\
& \quad -\frac{1}{2}\|r(t-d_0) - g_2^T(x_{t-d_0})V_{x_{t-d_0}}^T\|^2 + \frac{1}{2}\|r(t-d_0)\|^2 + \frac{1}{2}\|u^\star(t-d_0)\|^2.
\end{aligned}
$$

Using now the HJI-inequality (5.63), in the above equation we get

$$
\begin{aligned}
\frac{d}{dt}V &\leq -\frac{1}{2}\gamma^2\|w(t) - w^\star(t)\|^2 + \frac{1}{2}\gamma^2\|w(t)\|^2 - \frac{1}{2}\|u(t)\|^2 + \frac{1}{2}\|r(t)\|^2 - \\
& \quad \frac{1}{2}\gamma^2\|w(t-d_0) - w^\star(t-d_0)\|^2 + \frac{1}{2}\gamma^2\|w(t-d_0)\|^2 - \frac{1}{2}\|u(t-d_0)\|^2 + \\
& \quad \frac{1}{2}\|r(t-d_0)\|^2 - \frac{1}{2}\|h_1(t)\|^2.
\end{aligned}
$$

Integrating now the above inequality from $t = t_0$ to $t = t_1 > t_0$, starting from $x(t_0)$, and using the fact that

$$\int_{t_0}^{t_1} \|r(t)\|^2 dt \leq \gamma^2 \int_{t_0}^{t_1} \|w(t) - w^\star(t)\|^2,$$

we get

$$V(t_1, x(t_1), x(t_1 - d)) - V(t_0, x_0, x(t_0 - d_0)) \leq \frac{1}{2}\int_{t_0}^{t_1}\Big\{\gamma^2(\|w(t)\|^2 +$$

$$\|w(t-d_0)\|^2) - \|y(t)\|^2 - \|u(t)\|^2 - \|u(t-d_0)\|^2\Big\}dt,$$

which implies that the closed-loop system is dissipative with respect to the supply-rate $s(w(t), z(t)) = \frac{1}{2}[\gamma^2(\|w(t)\|^2 + \|w(t-d_0)\|^2) - \|z(t)\|^2]$, and therefore, the system has $\mathcal{L}_2$-gain from $w(t)$ to $z(t)$ less than or equal to γ. Finally, asymptotic-stability of the system can similarly be concluded as in 5.5.1. $\square$

5.6 State-Feedback $\mathcal{H}_\infty$-Control for a General Class of Nonlinear Systems

In this section, we look at the state-feedback problem for a more general class of nonlinear systems which is not necessarily affine. We consider the following class of nonlinear systems defined on a manifold $\mathcal{X} \subseteq \Re^n$ containing the origin in local coordinates $x = (x_1, \ldots, x_n)$:

$$\Sigma^g : \begin{cases} \dot{x} &= F(x, w, u), \quad x(t_0) = x_0 \\ y &= x \\ z &= Z(x, u) \end{cases} \tag{5.68}$$

where all the variables have their previous meanings, while $F : \mathcal{X} \times \mathcal{W} \times \mathcal{U} \to V^\infty$ is the state dynamics function and $Z : \mathcal{X} \times \mathcal{U} \to \Re^s$ is the controlled output function. Moreover, the functions $F(.,.,.)$ and $Z(.,.)$ are smooth C^r, $r \geq 1$ functions of their arguments, and the point $x = 0$ is a unique equilibrium-point for the system Σ^g and is such that $F(0,0,0) = 0$, $Z(0,0) = 0$. The following assumption will also be required in the sequel.

Assumption 5.6.1 *The linearization of the function $Z(x, u)$ is such that*

$$rank(D_{21}) = rank \left(\frac{\partial Z}{\partial u}(0,0) \right) = p.$$

Define now the Hamiltonian function for the above system $\tilde{H} : T^\star \mathcal{X} \times \mathcal{W} \times \mathcal{U} \to \Re$ as

$$\tilde{H}(x, p, w, u) = p^T F(x, w, u) + \frac{1}{2}\|Z(x, u)\|^2 - \frac{1}{2}\gamma^2\|w\|^2. \tag{5.69}$$

Then it can be seen that the above function is locally convex with respect to u and concave with respect to w about $(x, p, w, u) = (0, 0, 0, 0)$, and therefore has a unique local saddle-point (w, u) for each (x, p) in a neighborhood of this point. Thus, by Assumption 5.6.1 and the Implicit-function Theorem, there exist unique smooth functions $w^\star(x, p)$ and $u^\star(x, p)$, defined in a neighborhood of $(0, 0)$ such that $w^\star(0, 0) = 0$, $u^\star(0, 0) = 0$ and satisfying

$$\frac{\partial \tilde{H}}{\partial w}(x, p, w^\star(x, p), u^\star(x, p)) = 0, \tag{5.70}$$

$$\frac{\partial \tilde{H}}{\partial u}(x, p, w^\star(x, p), u^\star(x, p)) = 0. \tag{5.71}$$

Moreover, suppose there exists a nonnegative C^1-function $\tilde{V} : \mathcal{X} \to \Re$, $\tilde{V}(0) = 0$ which satisfies the inequality

$$\tilde{H}^\star(x, \tilde{V}_x^T(x)) = \tilde{H}(x, \tilde{V}_x^T(x), w^\star(x, \tilde{V}_x^T(x)), u^\star(x, \tilde{V}_x^T(x))) \leq 0, \tag{5.72}$$

and define the feedbacks

$$u^\star = \alpha(x) = u^\star(x, \tilde{V}_x^T(x)), \tag{5.73}$$

$$w^\star = w^\star(x, \tilde{V}_x^T(x)). \tag{5.74}$$

Then, substituting $u^\star$ in (5.68) yields a closed-loop system satisfying

$$\tilde{V}_x(x)F(x, w, \alpha(x)) + \frac{1}{2}\|Z(x, \alpha(x))\|^2 - \frac{1}{2}\gamma^2\|w\|^2 \leq 0,$$

which is dissipative with respect to the supply-rate $\tilde{s}(w, z) = \frac{1}{2}(\gamma^2\|w\|^2 - \|z\|^2)$ with storage-function $\tilde{V}$ in the neighborhood of $(x, w) = (0, 0)$.

The local asymptotic-stability of the system with $w = 0$ can also be proven as in the previous sections if the system is assumed to be zero-state detectable, or satisfies the following hypothesis.

Assumption 5.6.2 *Any bounded trajectory $x(t)$ of the system*

$$\dot{x}(t) = F(x(t), 0, u(t))$$

satisfying

$$Z(x(t), u(t)) = 0$$

for all $t \geq t_s$, is such that $\lim_{t \to \infty} x(t) = 0$.

We summarize the above results in the following theorem.

Theorem 5.6.1 *Consider the nonlinear system (5.68) and the SFBNLHICP for it. Assume the system is smoothly-stabilizable and zero-state detectable or satisfies Assumptions 5.6.1 and 5.6.2. Suppose further there exists a C^1 nonnegative function $\tilde{V} : N \subset \mathcal{X} \to \Re_+$, locally defined in a neighborhood N of $x = 0$ with $\tilde{V}(0) = 0$ satisfying the following HJI-inequality:*

$$\tilde{V}_x(x)F(x, w^\star, u^\star) + \frac{1}{2}\|Z(x, u^\star)\|^2 + \frac{1}{2}\gamma^2\|w^\star\|^2 \leq 0, \quad \tilde{V}(0) = 0, \; x \in N. \tag{5.75}$$

Then the feedback control law (5.73) solves the SFBNLHICP for the system.

Proof: It has been shown in the preceding that if $\tilde{V}$ exists and locally solves the HJI-inequality (5.75), then the system is dissipative with $\tilde{V}$ as storage-function and supply-rate $\tilde{s}(w, z)$. Consequently, the system has the local disturbance-attenuation property. Finally, if the system is zero-state detectable or satisfies Assumption 5.6.2, then local asymptotic-stability can be proven along the same lines as in Sections 5.1-5.5. $\square$

Remark 5.6.1 *Similarly, the function $w^\star = w^\star(x, \tilde{V}_x^T(x))$ is also interpreted as the worst-case disturbance affecting the system.*

5.7   Nonlinear $\mathcal{H}_\infty$ Almost-Disturbance-Decoupling

In this section, we discuss the state-feedback nonlinear $\mathcal{H}_\infty$ almost-disturbance-decoupling problem $(SFBNLHIADDP)$ which is very closely related to the $\mathcal{L}_2$-gain disturbance-attenuation problem except that it is formulated from a geometric perspective. This problem is an off-shoot of the geometric (exact) disturbance-decoupling problem which has been discussed extensively in the literature [140, 146, 212, 277] and was first formulated by Willems [276] to characterize those systems for which disturbance-decoupling can be achieved approximately with an arbitrary degree of accuracy.

The $SFBNLHIADDP$ is more recently formulated [198, 199] and is defined as follows. Consider the affine nonlinear system (5.1) with single-input single-output (SISO) represented in the form:

$$\left. \begin{array}{rl} \dot{x} &= f(x) + g_2(x)u + \sum_{i=1}^{r} g_{1i}(x)w_i(t); \quad x(t_0) = x_0 \\ y &= h(x) \end{array} \right\}, \tag{5.76}$$

where all the variables have their previous meanings with $u \in \mathcal{U} \subseteq \Re$ and output function $h : \mathcal{X} \to \Re$.

Definition 5.7.1 *(State-Feedback $\mathcal{L}_2$-Gain Almost-Disturbance-Decoupling Problem (SFB $\mathcal{L}_2$-gainADDP)). Find (if possible!) a parametrized set of smooth state-feedback controls*

$$u = u(x, \lambda), \;\; \lambda \in \Re_+ \; \lambda \; arbitrarily \; large$$

such that for every $t \in [t_0, T]$,

$$\int_{t_0}^{t} y^2(\tau)d\tau \leq \frac{1}{\lambda} \int_{t_0}^{t} \|w(\tau)\|^2 d\tau \tag{5.77}$$

for the closed-loop system with initial condition $x_0 = 0$ and for any disturbance function $w(t)$ defined on an open interval $[t_0, T)$ for which there exists a solution for the system (5.76).

Definition 5.7.2 *(State-Feedback Nonlinear $\mathcal{H}_\infty$ Almost-Disturbance-Decoupling Problem (SFBNLHIADDP)). The $SFBNLHIADDP$ is said to be solvable for the SISO system (5.76) if the $SFBL_2$-gainADDP for the system is solvable with $u = u(x, \lambda)$, $u(0, \lambda) = 0 \, \forall \lambda \in \Re_+$ and the origin is globally asymptotically-stable for the closed-loop system with $w(t) = 0$.*

In the following, we shall give sufficient conditions for the solvability of the above two problems for SISO nonlinear systems that are in the *"strict-feedback"* form. It would be shown that, if the system possesses a structure such that it is strictly feedback-equivalent to a linear system, is globally minimum-phase and has zero-dynamics that are independent of the disturbances, then the $SFBNLHIADDP$ is solvable. We first recall the following definitions [140, 212].

Definition 5.7.3 *The strong control characteristic index of the system (5.43) is defined as the integer ρ such that*

$$L_{g_2} L_f^i h(x) = 0, \;\; 0 \leq i \leq \rho - 2, \, \forall x \in \mathcal{X}$$
$$L_{g_2} L_f^{\rho-1} h(x) \neq 0, \;\; \forall x \in \mathcal{X}.$$

Otherwise, $\rho = \infty$ if $L_{g_2} L_f^i h(x) = 0 \, \forall i, \forall x \in \mathcal{X}$.

Definition 5.7.4 *The disturbance characteristic index of the system (5.43) is defined as the integer ν such that*

$$L_{g_{1j}} L_f^i h(x) = 0, \;\; 1 \leq j \leq r, 0 \leq i \leq \nu - 2, \forall x \in \mathcal{X}$$
$$L_{g_{1j}} L_f^{\nu-1} h(x) \neq 0, \;\; for \; some \; x \in \mathcal{X}, \; and \; some \; j, \, 1 \leq j \leq r.$$

We assume in the sequel that $\nu \leq \rho$.

Theorem 5.7.1 *Suppose for the system (5.76) the following hold:*

(i) ρ *is well defined;*

(ii) $\mathcal{G}_{r-1} = span\{g_2, ad_f g_2, \dots, ad_f^{\rho-1} g_2\}$ *is involutive and of constant rank ρ in $\mathcal{X}$;*

(iii) $ad_{g_{1i}} \mathcal{G}_j \subset \mathcal{G}_j$, $1 \leq i \leq r$, $0 \leq j \leq \rho - 2$, *with $\mathcal{G}_j = span\{g_2, ad_f g_2, \dots, ad_f^j g_2\}$;*

(iv) *the vector-fields*

$$\tilde{f} = f - \frac{1}{L_{g_2} L_f^{\rho-1} h} L_f^\rho h, \quad \tilde{g}_2 = \frac{1}{L_{g_2} L_f^{\rho-1} h} g_2$$

are complete,

then the SFBL$_2$-gainADDP is solvable.

Proof: See Appendix A.

Remark 5.7.1 *Conditions (ii),(iii) of the above theorem require the z_μ-dynamics of the system to be independent of $z_2, \ldots, z_\rho$. Moreover, condition (iii) is referred to as the "strict-feedback" condition in the literature [153].*

From the proof of the theorem, one arrives also at an alternative set of sufficient conditions for the solvability of the SFBL$_2$-gainADDP.

Theorem 5.7.2 *Suppose for the system (5.76) the following hold:*

(i) *ρ is well defined;*

(ii) *$d(L_{g_{1i}} L_f^i) \in span\{dh, d(L_f h), \ldots, d(L_f^i h)\}, \nu - 1 \leq i \leq \rho - 1, 1 \leq j \leq r \forall x \in \mathcal{X};$*

(iii) *the vector-fields*

$$\tilde{f} = f - \frac{1}{L_{g_2} L_f^{\rho-1} h} L_f^\rho h, \quad \tilde{g}_2 = \frac{1}{L_{g_2} L_f^{\rho-1} h} g_2$$

are complete.

Then the SFBL$_2$-gainADDP is solvable.

Proof: Conditions (i)-(iii) guarantee the existence of a global change of coordinates by augmenting the ρ linearly-independent set

$$z_1 = h(x), \; z_2 = L_f h(x), \ldots, z_\rho = L_f^{\rho-1} h(x)$$

with an arbitrary $n - \rho$ linearly-independent set $z_{\rho+1} = \psi_{\rho+1}(x), \ldots, z_n = \psi_n$ with $\psi_i(0) = 0$, $\langle d\psi_i, g_2 \rangle = 0, \rho + 1 \leq i \leq n$. Then the state feedback

$$u = \frac{1}{L_{g_2} L_f^{\rho-1} h(x)} (v - L_f^\rho h(x))$$

globally transforms the system into the form:

$$\begin{aligned}
\dot{z}_i &= z_{i+1} + \Psi_i^T(z)w \quad 1 \leq i \leq \rho - 1, \\
\dot{z}_\rho &= v + \Psi_\rho^T(z)w \\
\dot{z}_\mu &= \psi(z_1, z_\mu) + \Pi^T(z)w,
\end{aligned}$$

where $z_\mu = (z_{\rho+1}, \ldots, z_n)$. The rest of the proof follows along the same lines as Theorem 5.7.1. $\square$

Remark 5.7.2 *Condition (ii) in the above Theorem 5.7.2 requires that the functions $\Psi_1, \ldots, \Psi_\rho$ do not depend on the z_μ dynamics of the system.*

The following theorem now sums-up all the sufficient conditions for the solvability of the *SFBNLHIADDP*.

Theorem 5.7.3 *Assume the conditions (i)-(iv) of Theorem 5.7.1 hold for the system (5.76), and the zero-dynamics*

$$\dot{z}_\mu = \psi(0, z_\mu)$$

are independent of w and globally asymptotically-stable about the origin $z_\mu = 0$ (i.e. globally minimum-phase). Then the SFBNLHIADDP is solvable.

Proof: From the first part of the proof of Theorem 5.7.1 and the fact that the zero-dynamics are independent of w, (5.76) can be transformed into the form:

$$\left. \begin{array}{rcl} \dot{z}_i & = & z_{i+1} + \Psi_i^T(z_1, \ldots, z_i)w \quad 1 \le i \le \rho - 1, \\ \dot{z}_\rho & = & v + \Psi_\rho^T(z_1, \ldots, z_\rho, z_\mu)w \\ \dot{z}_\mu & = & \psi(0, z_\mu) + z_1(\psi_1(z_1, z_\mu) + \Pi_1^T(z_1, z_\mu)w), \end{array} \right\} \tag{5.78}$$

for some suitable functions ψ_1, Π_1. Moreover, since the system is globally minimum-phase, by a converse theorem of Lyapunov [157], there exists a radially-unbounded Lyapunov-function $V_{\mu 0}(z_\mu)$ such that

$$L_{\psi(0, z_\mu)} V_{\mu 0} = \langle dV_{\mu 0}, \psi(0, z_\mu) \rangle < 0.$$

In that case, we can consider the Lyapunov-function candidate

$$V_{\mu 01} = V_{\mu 0}(z_\mu) + \frac{1}{2} z_1^2.$$

Its time-derivative along the trajectories of (5.78) is given by

$$\dot{V}_{\mu 01} = \langle dV_{\mu 0}, \psi(0, z_\mu) \rangle + z_1 \langle dV_{\mu 0}, \psi_1 \rangle + z_1 z_2 + z_1(\Psi_1^T + \langle dV_{\mu 0}, \Pi_1^T(z_1, z_\mu) \rangle) w. \tag{5.79}$$

Now let

$$\bar{\Psi}(z_1, z_\mu) = \Psi_1^T + \langle dV_{\mu 0}, \Pi_1^T(z_1, z_\mu) \rangle,$$

and define

$$z_{02}^*(z_1, z_\mu) = -\langle dV_{\mu 0}, \psi_1 \rangle - z_1 - \frac{1}{4} \lambda z_1(1 + \bar{\Psi}_1^T \bar{\Psi}_1).$$

Subsituting the above in (5.79) yields

$$\dot{V}_{\mu 01} \le -z_1^2 + \frac{1}{\lambda} \|w\|^2 + \langle dV_{\mu 0}, \psi(0, z_\mu) \rangle. \tag{5.80}$$

The last term in the above expression (5.80) is negative, and therefore by standard Lyapunov-theorem, it implies global asymptotic-stability of the origin $z = 0$ with $w = 0$. Moreover, upon integration from $z(t_0) = 0$ to some arbitrary value $z(t)$, we get

$$-\int_{t_0}^t y^2(\tau)d\tau + \frac{1}{\lambda} \int_{t_0}^t \|w(\tau)\|^2 d\tau \ge V_{\mu 01}(z(t)) - V_{\mu 01}(0) \ge 0$$

which implies the $\mathcal{L}_2$-gain condition holds, and hence the $SFB\mathcal{L}_2$-gain$ADDP$ is solvable for the system with $\rho = 1$. Therefore the auxiliary control $v = z_{02}^*(z_1, z_\mu)$ solves the *SFBHIADDP* for the system with $\rho = 1$. Using an inductive argument as in the proof of Theorem 5.7.1 the result can be shown to hold also for $\rho > 1$. $\square$

Remark 5.7.3 *It is instructive to observe the relationship between the SFBNLHIADDP* *and the $SFBNLHICP$ discussed in Section 5.1. It is clear that if we take $z = y = h(x)$ in the ADDP, and seek to find a parametrized control law $u = \alpha_\gamma(x)$, $\alpha(0) = 0$ and the minimum $\gamma = \sqrt{\frac{1}{\lambda}}$ such that the inequality (5.77) is satisfied for all $t \geq 0$ and all $w \in \mathcal{L}_2(0,t)$ with internal-stability for the closed-loop system, then this amounts to a $SFBNLHICP$. However, the problem is "singular" since the function z does not contain u. Moreover, making λ arbitrarily large $\lambda \to \infty$ corresponds to making γ arbitrarily small, $\gamma \to 0$.*

5.8   Notes and Bibliography

The results presented in Sections 5.1 and 5.6 are mainly based on the valuable papers by Isidori et al. [138, 139, 145, 263, 264], while the controller parametrization is based on [188]. Results for the controller parametrization based on coprime factorizations can be found in [215, 214]. More extensive results on the $SFBNLHICP$ for different system configurations along the lines of [92] are given in [223], while results on a special class of nonlinear systems is given in [191]. In addition, a J-dissipative approach is presented in [224]. Similarly, more results on the $RSFBNLHICP$ which uses more or less similar techniques presented in Section 5.3 are given in the following references [192, 147, 148, 245, 261, 223, 284, 285]. Moreover, the results on the tracking problem presented in Section 5.2 are based on the reference [50].

The results for the state-delayed systems are based on [15], and for the general class of nonlinear systems presented in Section 5.6 are based on the paper by Isidori and Kang [145]. While results on stochastic systems, in particular systems with Markovian jump disturbances, can be found in the references [12, 13, 14].

Lastly, Section 5.7 on the $SFBHIADDP$ is based on [198, 199].

6

Output-Feedback Nonlinear $\mathcal{H}_\infty$-Control for Continuous-Time Systems

In this chapter, we discuss the nonlinear $\mathcal{H}_\infty$ sub-optimal control problem for continuous-time nonlinear systems using output-feedback. This problem arises when the states of the system are not available for feedback, and so have to be estimated in some way and then used for feedback, or the output of the system itself is used for feedback. The estimator is basically an observer that satisfies an $\mathcal{H}_\infty$ requirement. In the former case, the observer uses the measured output of the system to estimate the states, and the whole arrangement is referred to as an observer-based controller or more generally a dynamic controller. The set-up is depicted in Figure 6.1 in this case. For the most part, this chapter will be devoted to this problem. On the other hand, the latter problem is referred to as a static output-feedback controller and will be discussed in the last section of the chapter.

We derive sufficient conditions for the solvability of the above problem for time-invariant (or autonomous) affine nonlinear systems, as well as for a general class of systems. The parametrization of stabilizing controllers is also discussed. The problem of robust control in the presence of modelling errors and/or parameter variations is also considered, as well as the reliable control of sytems in the event of sensor and/or actuator failure.

6.1 Output Measurement-Feedback $\mathcal{H}_\infty$-Control for Affine Nonlinear Systems

We begin in this section with the output-measurement feedback problem in which it is desired to synthesize a dynamic observer-based controller using the output measurements. To this effect, we consider first an affine causal state-space system defined on a smooth n-dimensional manifold $\mathcal{X} \subseteq \Re^n$ in local coordinates $x = (x_1, \ldots, x_n)$:

$$\Sigma^a : \begin{cases} \dot{x} &= f(x) + g_1(x)w + g_2(x)u; \quad x(t_0) = x_0 \\ z &= h_1(x) + k_{11}(x)w + k_{12}(x)u \\ y &= h_2(x) + k_{21}(x)w \end{cases} \tag{6.1}$$

where $x \in \mathcal{X}$ is the state vector, $u \in \mathcal{U} \subseteq \Re^p$ is the p-dimensional control input, which belongs to the set of admissible controls $\mathcal{U}$, $w \in \mathcal{W}$ is the disturbance signal, which belongs to the set $\mathcal{W} \subset \mathcal{L}_2([t_0, \infty), \Re^r)$ of admissible disturbances, the output $y \in \mathcal{Y} \subset \Re^m$ is the measured output of the system, and $z \in \Re^s$ is the output to be controlled. The functions $f : \mathcal{X} \to V^\infty(\mathcal{X})$, $g_1 : \mathcal{X} \to \mathcal{M}^{n \times r}(\mathcal{X})$, $g_2 : \mathcal{X} \to \mathcal{M}^{n \times p}(\mathcal{X})$, $h_1 : \mathcal{X} \to \Re^s$, $h_2 : \mathcal{X} \to \Re^m$, and $k_{11} : \mathcal{X} \to \mathcal{M}^{s \times r}$, $k_{12} : \mathcal{X} \to \mathcal{M}^{s \times p}(\mathcal{X})$, $k_{21} : \mathcal{X} \to \mathcal{M}^{m \times r}(\mathcal{X})$ are real C^∞ functions of x. Furthermore, we assume without any loss of generality that the system (6.1) has a unique equilibrium-point at $x = 0$ such that $f(0) = 0$, $h_1(0) = h_2(0) = 0$, and for simplicity we have the following assumptions which are the nonlinear versions of the standing assumptions in [92]:

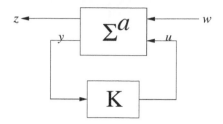

FIGURE 6.1
Configuration for Nonlinear $\mathcal{H}_\infty$-Control with Measurement-Feedback

Assumption 6.1.1 *The system matrices are such that*

$$\left.\begin{array}{rcl}
k_{11}(x) & = & 0 \\
h_1^T(x)k_{12}(x) & = & 0 \\
k_{12}^T(x)k_{12}(x) & = & I \\
k_{21}(x)g_1^T(x) & = & 0 \\
k_{21}(x)k_{21}^T(x) & = & I.
\end{array}\right\} . \tag{6.2}$$

Remark 6.1.1 *The first of these assumptions corresponds to the case in which there is no direct feed-through between w and z; while the second and third ones mean that there are no cross-product terms and the control weighting matrix is identity in the norm function for z respectively. Lastly, the fourth and fifth ones are dual to the second and third ones.*

We begin with the following definition.

Definition 6.1.1 *(Measurement-Feedback (Sub-Optimal) $\mathcal{H}_\infty$-Control Problem (MFBNL-HICP)). Find (if possible!) a dynamic feedback-controller of the form:*

$$\Sigma_{dyn}^c : \left\{ \begin{array}{rcl} \dot{\xi} & = & \eta(\xi, y) \\ u & = & \theta(\xi, y), \end{array} \right. \tag{6.3}$$

where $\xi \in \Xi \subset \mathcal{X}$ a neighborhood of the origin, and $\eta : \Xi \times \Re^m \to V^\infty(\Xi)$, $\eta(0,0) = 0$, $\theta : \Xi \times \Re^m \to \Re^p$, $\theta(0,0) = 0$, are some smooth functions, which processes the measured variable y of the plant (6.1) and generates the appropriate control action u, such that the closed-loop system (6.1), (6.3) has locally $\mathcal{L}_2$-gain from the disturbance signal w to the output z less than or equal to some prescribed number $\gamma^\star > 0$ with internal stability, or equivalently, the closed-loop system achieves local disturbance-attenuation less than or equal to $\gamma^\star$ with internal-stability.

Since the state information is not available for measurement in contrast with the previous chapter, the simplest thing to do is to estimate the states with an observer, and then use this estimate for feedback. From previous experience on the classical theory of state observation, the "estimator" usually comprises of an exact copy of the dynamics of x, corrected by a term proportional to the error between the actual output y of the plant (6.1) and an estimated output $\hat{y}$. Such a system can be described by the equations:

$$\Sigma_e^c : \left\{ \begin{array}{rcl} \dot{\xi} & = & f(\xi) + g_1(\xi)w + g_2(\xi)u + G(\xi)(y - \hat{y}) \\ \hat{y} & = & h_2(\xi) + k_{21}(\xi)w, \end{array} \right. \tag{6.4}$$

where $G(.)$ is the *output-injection gain* matrix which is to be determined. A state-feedback control law can then be imposed on the system based on this estimated state (and the results from Chapter 5) as:

$$u := \alpha_2(\xi) = -g_2^T(\xi)V_\xi^T(\xi), \tag{6.5}$$

where V solves the HJIE (5.12). The combination of (6.4) and (6.5) will then represent our postulated dynamic-controller (6.3). With the above feedback, the estimator (6.4) becomes

$$\dot{\xi} = f(\xi) + g_1(\xi)w + g_2(\xi)\alpha_2(\xi) + G(\xi)(y - \hat{y}). \tag{6.6}$$

Besides the fact that we still have to determine an appropriate value for $G(.)$, we also require an explicit knowledge of the disturbance w. Since the knowledge of w is generally not available, it is reasonable to replace its actual value by the "worst-case" value, which is given (from Chapter 5) by

$$w^\star := \alpha_1(x) = \frac{1}{\gamma^2}g_1^T(x)V_x^T(x)$$

for some $\gamma > 0$. Substituting this expression in (6.6) and replacing x by ξ, results in the following *certainty-equivalence* worst-case estimator

$$\dot{\xi} = f(\xi) + g_1(\xi)\alpha_1(\xi) + g_2(\xi)\alpha_2(\xi) + G(\xi)(y - h_2(\xi) - k_{21}(\xi)\alpha_1(\xi)). \tag{6.7}$$

Next, we return to the problem of selecting an appropriate gain matrix $G(.)$, which is another of our design parameters, to achieve the objectives stated in Definition 6.1.1. It will be shown in the following that, if some appropriate sufficient conditions are fulfilled, then the matrix $G(.)$ can be chosen to achieve this.

We begin with the closed-loop system describing the dynamics of the plant (6.1) and the dynamic-feedback controller (6.7), (6.5) which has the form

$$\Sigma_{clp}^{ac} : \begin{cases} \dot{x} &= f(x) + g_1(x)w + g_2(x)\alpha_2(\xi); \quad x(t_0) = x_0 \\ \dot{\xi} &= \tilde{f}(\xi) + g_2(\xi)\alpha_2(\xi) + G(\xi)(h_2(x) + k_{21}(x)w - \tilde{h}_2(\xi)) \\ z &= h_1(x) + k_{12}(x)\alpha_2(\xi) \end{cases} \tag{6.8}$$

where

$$\begin{aligned} \tilde{f}(\xi) &= f(\xi) + g_1(\xi)\alpha_1(\xi) \\ \tilde{h}_2(\xi) &= h_2(\xi) + k_{21}(\xi)\alpha_1(\xi). \end{aligned}$$

If we can choose $G(.)$ such that the above closed-loop system (6.8) is dissipative with a C^1 storage-function, with respect to the supply-rate $s(w,z) = \frac{1}{2}(\gamma^2\|w\|^2 - \|z\|^2)$ and is locally asymptotically-stable about $(x,\xi) = (0,0)$, then we would have solved the *MFBNLHICP* for (6.1).

To proceed, let us further combine the x and ξ dynamics in the dynamics x^e as follows.

$$\dot{x}^e = f^e(x^e) + g^e(x^e)(w - \alpha_1(x)) \tag{6.9}$$

where

$$x^e = \begin{pmatrix} x \\ \xi \end{pmatrix}, \quad f^e(x^e) = \begin{pmatrix} \tilde{f}(x) + g_2(x)\alpha_2(\xi) \\ \tilde{f}(x) + g_2(x)\alpha_2(\xi) + G(\xi)(\tilde{h}_2(x) - \tilde{h}_2(\xi)) \end{pmatrix},$$

$$g^e(x^e) = \begin{pmatrix} g_1(x) \\ G(\xi)k_{21}(x) \end{pmatrix},$$

and let

$$h^e(x^e) = \alpha_2(x) - \alpha_2(\xi).$$

Apply now the change of input variable $r = w - \alpha_1(x)$ in (6.9) and set the output to v obtaining the following equivalent system:

$$\begin{cases} \dot{x}^e &= f^e(x^e) + g^e(x^e)r \\ v &= h^e(x^e). \end{cases} \tag{6.10}$$

The problem would then be solved if we can choose $G(.)$ such that the above system (6.10) is dissipative with respect to the supply-rate $s(r,v) = \frac{1}{2}(\gamma^2\|r\|^2 - \|v\|^2)$ and internally-stable. A necessary and sufficient condition for this to happen is if the following *bounded-real* condition [139] is satisfied:

$$W_{x^e}(x^e)f^e(x^e) + \frac{1}{2\gamma^2}W_{x^e}(x^e)g^e(x^e)g^{eT}(x^e)W_{x^e}^T(x^e) + \frac{1}{2}h^{eT}(x^e)h^e(x^e) \leq 0 \tag{6.11}$$

for some C^1 nonnegative function $W : N \times N \to \Re$ locally defined in a neighborhood $N \subset \mathcal{X}$ of $x^e = 0$, and which vanishes at $x^e = 0$. The above inequality further implies that

$$W_{x^e}(x^e)(f^e(x^e) + g^e(x^e)r) \leq \frac{1}{2}(\gamma^2\|r\|^2 - \|v\|^2) \quad \forall r \in \mathcal{W}, \tag{6.12}$$

which means, W is a storage-function for the equivalent system with respect to the supply-rate $s(r,v) = \frac{1}{2}(\gamma^2\|r\|^2 - \|v\|^2)$. Now assuming that W exists, then it can be seen that the closed-loop system (6.8) is locally dissipative, with the storage-function

$$\Psi(x^e) = V(x) + W(x^e),$$

with respect to the supply-rate $s(w,z) = \frac{1}{2}(\gamma^2\|w\|^2 - \|z\|^2)$, where $V(.)$ solves the HJIE (5.12). Indeed,

$$\begin{aligned}
\frac{d\Psi(x^e(t))}{dt} + \frac{1}{2}(\|z\|^2 - \gamma^2\|w\|^2) &= \Psi_{x^e}(f^e(x^e) + g^e(x^e)(w - \alpha_1(x))) + \frac{1}{2}(\|z\|^2 - \gamma^2\|w\|^2) \\
&= V_x(x)(f(x) + g_1(x)w + g_2(x)\alpha_2(\xi)) + \\
&\quad W_{x^e}(f^e(x^e) + g^e(x^e)r) + \frac{1}{2}\|h_1(x) + k_{12}(x)\alpha_2(\xi)\|^2 \\
&\quad -\frac{1}{2}\gamma^2\|w\|^2 \\
&\leq \frac{1}{2}\Big\{\|\alpha_2(\xi) - \alpha_2(x)\|^2 - \gamma^2\|w - \alpha_1(x)\|^2 + \gamma^2\|r\|^2 \\
&\quad -\|v\|^2\Big\} \\
&\leq 0,
\end{aligned}$$

which proves dissipativity. The next step is to show that the closed-loop system (6.9) has a locally asymptotically-stable equilibrium-point at $(x,\xi) = (0,0)$. For this, we need additional assumptions on the system (6.1).

Assumption 6.1.2 *Any bounded trajectory $x(t)$ of the system*

$$\dot{x}(t) = f(x(t)) + g_2(x(t))u(t)$$

satisfying

$$h_1(x(t)) + k_{12}(x(t))u(t) = 0$$

for all $t \geq t_s$, is such that $\lim_{t\to\infty} x(t) = 0$.

Assumption 6.1.3 *The equilibrium-point $\xi = 0$ of the system*

$$\dot{\xi} = \tilde{f}(\xi) - G(\xi)\tilde{h}_2(\xi) \tag{6.13}$$

is locally asymptotically-stable.

Then from above, with $w = 0$,

$$
\begin{aligned}
\frac{d\Psi(x^e(t))}{dt} &= \Psi_{x^e}(f^e(x^e) + g^e(x^e)(-\alpha_1(x))) \\
&\leq -\frac{1}{2}\|h_1(x(t)) + k_{12}(x(t))\alpha_2(\xi(t))\|^2
\end{aligned} \tag{6.14}
$$

along any trajectory $(x(.), \xi(.))$ of the closed-loop system. This proves that the equilibrium-point $(x(.), \xi(.)) = (0,0)$ is stable. To prove asymptotic-stability, observe that any trajectory $x^e(t)$ such that

$$\frac{d\Psi(x^o(t))}{dt} \equiv 0, \quad \forall t \geq t_s,$$

is necessarily a trajectory of

$$\dot{x} = f(x) + g_2(x)\alpha_2(\xi)$$

such that $x(t)$ is bounded and

$$h_1(x(t)) + k_{12}(x(t))\alpha_2(\xi(t)) \equiv 0$$

for all $t \geq t_s$. By Assumption 6.1.2, this implies $\lim_{t \to \infty} x(t) = 0$. Moreover, since $k_{12}(x)$ has full column rank, the above also implies that $\lim_{t \to \infty} \alpha_2(\xi(t)) = 0$. Thus, the ω-limit set of such a trajectory is a subset of

$$\Omega_0 = \{(x, \xi) | x = 0, \ \alpha_2(\xi) = 0\}.$$

Any initial condition on this limit-set yields a trajectory

$$x(t) \equiv 0$$

for all $t \geq t_s$, which corresponds to the trajectory $\xi(t)$ of

$$\dot{\xi} = \tilde{f}(\xi) - G(\xi)\tilde{h}_2(\xi).$$

By Assumption 6.1.3, this implies $\lim_{t \to \infty} \xi(t) = 0$, and local asymptotic-stability follows from the invariance-principle.

The above result can now be summarized in the following lemma.

Theorem 6.1.1 *Consider the nonlinear system (6.1) and suppose Assumptions 6.1.1, 6.1.2 hold. Suppose also there exists a C^1 positive-(semi)definite function $V : N \to \Re_+$ locally defined in a neighborhood N of $x = 0$, vanishing at $x = 0$ and satisfying the HJIE (5.12). Further, suppose there exists also a C^1 real-valued function $W : N_1 \times N_1 \to \Re$ locally defined in a neighborhood $N_1 \times N_1$ of $x^e = 0$, vanishing at $x^e = 0$ and is such that $W(x^e) > 0$ for all $x \neq \xi$, together with an $n \times m$ smooth matrix function $G(\xi) : N_2 \to \mathcal{M}^{n \times m}$, locally defined in a neighborhood N_2 of $\xi = 0$, such that*

(i) *the HJIE (6.11) holds;*

(ii) *Assumption 6.1.3 holds;*

(iii) *$N \cap N_1 \cap N_2 \neq \emptyset$.*

Then the MFBNLHICP for the system (6.1) is locally solvable.

A modified result to the above Theorem 6.1.1 as given in [141, 147] can also be proven along the same lines as the above. We first have the following definition.

Definition 6.1.2 *Suppose $f(0) = 0$, $h(0) = 0$, then the pair $\{f, h\}$ is locally zero-state detectable, if there exists a neighborhood O of $x = 0$ such that, if $x(t)$ is a trajectory of the free-system $\dot{x} = f(x)$ with $x(t_0) \in O$, then $h(x(t))$ is defined for all $t \geq 0$ and $h(x(t)) \equiv 0$ for all $t \geq t_0$, implies $\lim_{t\to\infty} x(t) = 0$*

Theorem 6.1.2 *Consider the nonlinear system (6.1) and assume that $k_{21}(x)k_{21}^T(x) = I$ in (6.2). Assume also the following:*

(i) *The pair $\{f, h_1\}$ is locally zero-state detectable.*

(ii) *There exists a smooth positive-(semi)definite function V locally defined in a neighborhood N of the origin $x = 0$ with $V(0) = 0$ and satisfying the HJIE (5.12).*

(iii) *There exists an $n \times m$ smooth matrix function $G(\xi)$ such that the equilibrium-point $\xi = 0$ of the system*

$$\dot{\xi} = f(\xi) + g_1(\xi)\alpha_1(\xi) - G(\xi)h_2(\xi) \tag{6.15}$$

is locally asymptotically-stable.

(iv) *There exists a smooth positive-semidefinite function $W(x, \xi)$ locally defined in a neighborhood $N_1 \times N_1 \subset \mathcal{X} \times \mathcal{X}$ of the origin such that $W(0, \xi) > 0$ for each $\xi \neq 0$, and satisfying the HJIE :*

$$[W_x(x,\xi) \ \ W_\xi(x,\xi)]f_e(x,\xi) +$$
$$\frac{1}{2\gamma^2}[W_x(x,\xi) \ W_\xi(x,\xi)]\begin{bmatrix} g_1(x)g_1^T(x) & 0 \\ 0 & G(\xi)G^T(\xi) \end{bmatrix}\begin{bmatrix} W_x^T(x,\xi) \\ W_\xi^T(x,\xi) \end{bmatrix} +$$
$$\frac{1}{2}h_e^T(x,\xi)h_e(x,\xi) = 0, \quad W(0,0) = 0, \quad (x,\xi) \in N_1 \times N_1. \tag{6.16}$$

Then, the MFBNLHICP for the system is solved by the output-feedback controller:

$$\Sigma_2^c : \begin{cases} \dot{\xi} &= f(\xi) + g_1(\xi)\alpha_1(\xi) + g_2(\xi)\alpha_2(\xi) + G(\xi)(y - h_2(\xi)) \\ u &= \alpha_2(\xi), \end{cases} \tag{6.17}$$

where $f_e(x,\xi)$, $h_e(x,\xi)$, are given by:

$$f_e(x,\xi) = \begin{pmatrix} f(x) + g_1(x)\alpha_1(x) + g_2(x)\alpha_2(\xi) \\ f(\xi) + g_1(\xi)\alpha_1(\xi) + g_2(\xi)\alpha_2(\xi) + G(\xi)(h_2(x) - h_2(\xi)) \end{pmatrix}$$
$$h_e(x,\xi) = \alpha_2(\xi) - \alpha_2(x).$$

Example 6.1.1 *We specialize the result of Theorem 6.1.2 to linear systems and compare it with some standard results obtained in other references, e.g., [92] (see also [101, 292]). Consider the linear time-invariant system (LTI):*

$$\Sigma^l : \begin{cases} \dot{x} &= Fx + G_1 w + G_2 u \\ z &= H_1 x + K_{12} u \\ y &= H_2 x + K_{21} w \end{cases} \tag{6.18}$$

for constant matrices $F \in \Re^{n \times n}$, $G_1 \in \Re^{n \times r}$, $G_2 \in \Re^{n \times p}$, $H_1 \in \Re^{s \times n}$, $H_2 \in \Re^{m \times n}$,

$K_{12} \in \Re^{s \times p}$ and $K_{21} \in \Re^{m \times r}$ *satisfying Assumption 6.1.1. Then, we have the following result.*

Proposition 6.1.1 *Consider the linear system (6.18) and suppose the following hold:*

(a) *the pair* (F, G_1) *is stabilizable;*

(b) *the pair* (F, H_1) *is detectable;*

(c) *there exist positive-definite symmetric solutions* X, Y *of the AREs:*

$$F^T X + XF + X[\frac{1}{\gamma^2} G_1 G_1^T - G_2 G_2^T] + H_1^T H_1 = 0, \tag{6.19}$$

$$Y F^T + FY + Y[\frac{1}{\gamma^2} H_1 H_1^T - H_2 H_2^T] + G_1^T G_1 = 0; \tag{6.20}$$

(d) $\rho(XY) < \gamma^2$.

Then, the hypotheses (i)-(iv) in Theorem 6.1.2 also hold, with

$$G = ZH_2^T$$
$$V(x) = \frac{1}{2} x^T X x$$
$$W(x, \xi) = \frac{1}{2} \gamma^2 (x - \xi) Z^{-1} (x - \xi),$$

where

$$Z = Y \left(I - \frac{1}{\gamma^2} XY \right)^{-1}.$$

Proof: *(i) is identical to (b). To show (ii) holds, we note that, if X is a solution of the ARE (6.19), then the positive-definite function $V(x) = \frac{1}{2} x^T X x$ is a solution of the HJIE (6.16). To show that (iii) holds, observe that, if X is a solution of ARE (6.19) and Y a solution of ARE (6.20), then the matrix*

$$Z = Y(I - \frac{1}{\gamma^2} XY)^{-1}$$

is a solution of the ARE [292]:

$$Z(F + G_1 F_1)^T + (F + G_1 F_1)Z + Z(\frac{1}{\gamma^2} F_2^T F_2 - H_2^T H_2)Z + G_1 G_1^T = 0, \tag{6.21}$$

where $F_1 = \frac{1}{\gamma^2} G_1^T X$ and $F_2 = -G_2^T X$. Moreover, Z is symmetric and by hypothesis (d) Z is positive-definite. Hence we can take $V(\xi) = \xi^T Z \xi$ as a Lyapunov-function for the system

$$\dot{\xi} = (F + G_1 F_1 - GH_2)^T \xi, \tag{6.22}$$

which is the linear equivalent of (6.15). Using the ARE (6.21), it can be shown that

$$\dot{L} = -(\|G^T \xi\|^2 + \|G_1^T \xi\|^2 + \frac{1}{\gamma^2} \|F_2 Z \xi\|^2),$$

which implies that the equilibrium-point $\xi = 0$ for the system (6.22) is stable. Furthermore, the condition $\dot{L}(\xi(t)) = 0$ implies $G_1 \xi(t) = 0$ and $G\xi(t) = 0$, which results in

$$\dot{\xi}(t) = F^T \xi(t).$$

Finally, since (A, G_1) is stabilizable, together with the fact that $G_1\xi(t) = 0$, and the standard invariance-principle, we conclude that $\lim_{t\to\infty} \xi(t) = 0$ or asymptotic-stability of the system (6.22). Thus, (iii) also holds.

Lastly, it remains to show that (iv) also holds. Choosing

$$W(x, \xi) = \frac{1}{2}\gamma^2(x - \xi)^T Z^{-1}(x - \xi)$$

and substituting in the HJIE (6.16), we get four AREs which are all identical to the ARE (6.21), and the result follows. $\square$

Remark 6.1.2 *Based on the result of the above proposition, it follows that the linear dynamic-controller*

$$\begin{aligned}
\dot{\xi} &= (F + G_1 F_1 + G_2 F_2 - GH_2)\xi + Gy \\
u &= F_2\xi
\end{aligned}$$

with $F_1 = \frac{1}{\gamma^2}G_1^T X$, $F_2 = -G_2^T X$ and $G = ZH_2^T$ solves the linear $MFBNLHICP$ for the system (6.18).

The preceding results, Theorems 6.1.1, 6.1.2, establish sufficient conditions for the solvability of the output measurement-feedback problem. However, they are not satisfactory in that, firstly, they do not give any hint on how to select the output-injection gain-matrix $G(.)$ such that the HJIE (6.11) or (6.16) is satisfied and the closed-loop system (6.13) is locally asymptotically-stable. Secondly, the HJIEs (6.11), (6.16) have twice as many independent variables as the HJIE (5.12). Thus, a final step in the above design procedure would be to partially address these concerns.

Accordingly, an alternative set of sufficient conditions can be provided which involve an additional HJI-inequality having the same number of independent variables as the dimension of the plant and not involving the gain matrix $G(.)$. For this, we begin with the following lemma.

Lemma 6.1.1 *Suppose V is a C^3 solution of the HJIE (5.12) and $Q : O \to \Re$ is a C^3 positive-definite function locally defined in a neighborhood of $x = 0$, vanishing at $x = 0$ and satisfying*

$$S(x) < 0$$

where

$$\begin{aligned}
S(x) &= Q_x(x)(\tilde{f}(x) - G(x)\tilde{h}_2(x)) + \frac{1}{2\gamma^2}Q_x(x)(g_1(x) - G(x)k_{21}(x))(g_1(x) - \\
&\quad G(x)k_{21}(x))^T Q_x^T(x) + \frac{1}{2}\alpha_2^T(x)\alpha_2(x),
\end{aligned}$$

for each $x \neq 0$ and such that the Hessian matrix $\frac{\partial^2 S}{\partial x^2}$ is nonsingular at $x = 0$. Then the function

$$W(x^e) = Q(x - \xi)$$

satisfies the conditions (i), (ii) of Theorem 6.1.1.

Proof: Let $W(x^e) = Q(x - \xi)$ and let

$$\Upsilon(x, \xi) = W_{x^e}(x^e)f^e(x^e) + \frac{1}{2\gamma^2}W_{x^e}(x^e)g^e(x^e)g^{eT}(x^e)W_{x^e}^T((x^e) + \frac{1}{2}h^{eT}(x^e)h^e(x^e).$$

Thus, to show that $W(.)$ as defined above satisfies condition (i) of Theorem 6.1.1, is to show that Υ is nonpositive. For this, set

$$e = x - \xi$$

and let

$$\Pi(e,\xi) = [\Upsilon(x,\xi)]_{x=\xi+e}.$$

It can then be shown by simple calculations that

$$\Pi(0,\xi) = 0, \quad \left.\frac{\partial\Pi(e,\xi)}{\partial e}\right|_{e=0} = 0,$$

which means $\Pi(x,\xi)$ can be expressed in the form

$$\Pi(e,\xi) = e^T\Lambda(e,\xi)e$$

for some C^0 matrix function $\Lambda(.,.)$. Moreover, it can also be verified that

$$\Lambda(0,0) = \left.\frac{\partial^2 S(x)}{\partial x^2}\right|_{x=0} < 0$$

by assumption. Thus, $\Pi(e,\xi)$ is nonpositive in the neighborhood of $(x,\xi) = (0,0)$ and (i) follows.

To establish (ii), note that by hypothesis $S(x) < 0$ implies

$$Q_x(x)(\tilde{f}(x) - G(x)\tilde{h}_2(x)) < 0.$$

Therefore, $Q(x) > 0$ is a Lyapunov-function for the system (6.13), and the equilibrium-point $\xi = 0$ of this system is locally asymptotically-stable. Hence the result. $\square$

As a consequence of the above lemma, we can now present alternative sufficient conditions for the solvability of the $MFBNLHICP$. We begin with the following assumption.

Assumption 6.1.4 *The matrix*

$$R_1(x) = k_{21}^T(x)k_{21}(x)$$

is nonsingular (and positive-definite) for each x.

Theorem 6.1.3 *Consider the nonlinear system (6.1) and suppose Assumptions 6.1.1, 6.1.2 and 6.1.4 hold. Suppose also the following hold:*

(i) *there exists a C^3 positive-definite function V, locally defined in a neighborhood N of $x = 0$ and satisfying the HJIE (5.12);*

(ii) *there exists a C^3 positive-definite function $Q : N_1 \subset \mathcal{X} \to \Re_+$ locally defined in a neighborhood N_1 of $x = 0$, vanishing at $x = 0$, and satisfying the HJ-inequality*

$$Q_x(x)f^\sharp(x) + \frac{1}{2\gamma^2}Q_x(x)g_1^\sharp(x)g_1^{\sharp T}(x)Q_x^T(x) + \frac{1}{2}H^\sharp(x) < 0 \tag{6.23}$$

together with the coupling condition

$$Q_x(x)L(x) = \tilde{h}_2^T(x) \tag{6.24}$$

for some $n \times m$ smooth C^2 matrix function $L(x)$, where

$$\begin{aligned}
f^\sharp(x) &= \tilde{f}(x) - g_1(x)k_{21}^T(x)R_1^{-1}(x)\tilde{h}_2(x), \\
g_1^\sharp(x) &= g_1(x)[I - k_{12}^T(x)R_1^{-1}(x)k_{21}(x)], \\
H^\sharp(x) &= (\alpha_2^T(x)\alpha_2(x) - \gamma^2\tilde{h}(x)R_1^{-1}(x)\tilde{h}_2(x)),
\end{aligned}$$

and such that the Hessian matrix of the right-hand-side of (6.23) is nonsingular for all $x \in N_1$.

Then the MFBNLHICP for the system is solvable with the controller (6.7), (6.5) if $G(.)$ is selected as

$$G(x) = (\gamma^2 L(x) + g_1(x)k_{21}^T(x))R_1^{-1}(x). \tag{6.25}$$

Proof: (Sketch, see [139] for details). By standard completion of squares arguments, it can be shown that the function $S(x)$ satisfies the inequality

$$S(x) \geq Q_x(x)(\tilde{f}(x) - g_1(x)k_{21}^T(x)R_1^{-1}(x)\tilde{h}_2(x)) + \frac{1}{2}[\alpha_2^T(x)\alpha_2(x) - \gamma^2\tilde{h}_2^T(x)R_1^{-1}(x)\tilde{h}_2(x)] +$$

$$\frac{1}{2\gamma^2}Q_x(x)g_1(x)[I - k_{21}^T(x)R_1^{-1}(x)k_{21}(x)]g_1^T(x)Q_x^T(x),$$

and equality holds if and only if

$$Q_x(x)G(x) = (\gamma^2\tilde{h}_2(x) + Q_xg_1(x)k_{21}(x))R_1^{-1}(x).$$

Therefore, in order to make $S(x) < 0$, it is sufficient for the new HJ-inequality (6.23) to hold for each x and $G(x)$ to be chosen as in (6.25). In this event, the matrix $G(x)$ exists if and only if $Q(x)$ satisfies (6.24). Finally, application of the results of Theorem 6.1.1, Lemma 6.1.1 yield the result. $\square$

Example 6.1.2 *We consider the case of the linear system (6.18) in which case the simplifying assumptions (6.2) reduce to:*

$$K_{12}^T K_{12} = I, \quad H_1^T K_{12} = 0, \quad K_{21}G_1^T = 0, \quad K_{21}^T K_{21} = R_1 > 0.$$

The existence of a C^3 positive-definite function Q satisfying the strict HJ-inequality (6.23) is equivalent to the existence of a symmetric positive-definite matrix Z satisfying the bounded-real inequality

$$F^{\sharp T}Z + ZF^\sharp + \frac{1}{\gamma^2}ZG_1^\sharp G_1^{\sharp T}Z + H^\sharp < 0 \tag{6.26}$$

where

$$F^\sharp = \tilde{F} - G_1 K_{21}^T R_1^{-1}\tilde{H}_2, \quad G_1^\sharp = G_1(I - K_{21}^T R_1^{-1}K_{21}), \quad H^\sharp = F_2^T F_2 - \gamma^2\tilde{H}_2^T R_1^{-1}\tilde{H}_2$$

and

$$\tilde{F} = F + G_1 F_1, \quad \tilde{H}_2 = H_2 + K_{21}F_1$$

$$F_1 = \frac{1}{\gamma^2}G_1^T P, \quad F_2 = -G_2^T P$$

with P a symmetric positive-definite solution of the ARE (5.22). Moreover, if Z satisfies the bounded-real inequality (6.26), then the output-injection gain G is given by

$$G = (\gamma^2 Z^{-1}\tilde{H}_2^T + G_1 K_{21}^T)R_1^{-1},$$

with $L = \frac{1}{2}Z^{-1}\tilde{H}_2^T$.

Remark 6.1.3 *From the discussion in [264] regarding the necessary conditions for the existence of smooth solutions to the HJE (inequalities) characterizing the solution to the state-feedback nonlinear $\mathcal{H}_\infty$-control problem, it is reasonable to expect that, if the linear*

$MFBNLHICP$ corresponding to the linearization of the system (6.1) at $x = 0$ is solvable, then the nonlinear problem should be locally solvable. As a consequence, the linear approximation of the plant (6.1) at $x = 0$ must satisfy any necessary conditions for the existence of a stabilizing linear controller which solves the linear $MFBNLHICP$.

Remark 6.1.4 *Notice that the controller derived in Theorem 6.1.3 involves the solution of two uncoupled HJ-inequalities and a coupling-condition. This is quite in agreement with the linear results as derived in [92], [101], [292].*

In the next subsection, we discuss the problem of controller parametrization.

6.1.1 Controller Parameterization

In this subsection, we discuss the parametrization of a set of stabilizing controllers that locally solves the $MFBNLHICP$. As seen in the case of the state-feedback problem, this set is a linear set, parametrized by a free parameter system, Q, which varies over the set of all finite-gain asymptotically-stable input-affine systems. However, unlike the linear case, the closed-loop map is not affine in this free parameter.

Following the results in [92] for the linear case, the nonlinear case has also been discussed extensively in references [188, 190, 215]. While the References [188, 190] employ direct state-space tools, the Reference [215] employs coprime factorizations. In this subsection we give a state-space characterization.

The problem at hand is the following. Suppose a controller Σ^c (herein-after referred to as the *"central controller"*) of the form (6.3) solves the $MFBNLHICP$ for the nonlinear system (6.1). Find the set of all controllers (or a subset), that solves the $MFBNLHICP$ for the system.

As in the state-feedback case, this problem can be solved by an affine parametrization of the central controller with a system $Q \in \mathcal{FG}$ having the realization

$$\Sigma_Q : \left\{ \begin{array}{lll} \dot{q} & = & a(q) + b(q)r, \quad a(0) = 0, \\ v & = & c(q), \quad c(0) = 0 \end{array} \right. \tag{6.27}$$

where $q \in \mathcal{X}$ and $a : \mathcal{X} \to V^\infty \mathcal{X}$, $b : \mathcal{X} \to \mathcal{M}^{n \times p}$, $c : \mathcal{X} \to \Re^m$ are smooth functions. Then the family of controllers

$$\Sigma_Q^c : \left\{ \begin{array}{lll} \dot{\xi} & = & \eta(\xi, y) \\ \dot{q} & = & a(q) + b(q)(y - \hat{y}) \\ u & = & \theta(\xi, y) + c(q) \end{array} \right. \tag{6.28}$$

where Σ_Q varies over all $Q \in \mathcal{FG}$ and $\hat{y}$ is the estimated output, also solves the $MFBNLHICP$ for the system (6.1). The structure of this parametrization is also shown on Figure 6.2, and we have the following result.

Proposition 6.1.2 *Assume Σ^a is locally smoothly-stabilizable and locally zero-state detectable. Suppose there exists a controller of the form Σ^c that locally solves the $MFBNLHICP$ for the system, then the family of controllers Σ_Q^c given by (6.28), where $Q \in \mathcal{FG}$ is a parametrization of the set of all input-affine controllers that locally solves the problem for Σ^a.*

Proof: We defer the proof of this proposition to the next theorem, where we give an explicit form for the controller Σ_Q^c.

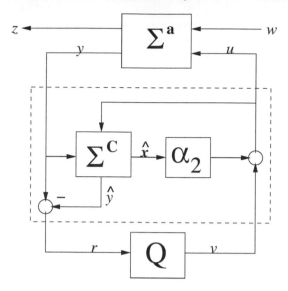

FIGURE 6.2
Parametrization of Nonlinear $\mathcal{H}_\infty$-Controllers with Measurement-Feedback; Adopted from
IEEE Transactions on Automatic Control, vol. 40, no. 9, © 1995, "A state-space approach
to parametrization of stabilizing controllers for nonlinear systems," by Lu W.-M.

Theorem 6.1.4 *Suppose all the hypotheses (i)-(iv) of Theorem 6.1.2 hold, and
$k_{21}(x)k_{21}^T(x) = I$. In addition, assume the following hypothesis also holds:*

(v) *There exists an input-affine system Σ_Q as defined above, whose equilibrium-point $q = 0$
is locally asymptotically-stable, and is such that there exists a smooth positive-definite
function $U : N_4 \to \Re_+$, locally defined in a neighborhood of $q = 0$ in $\mathcal{X}$ satisfying the
HJE:*

$$U_q(q)a(q) + \frac{1}{2\gamma^2}U_q(q)b(q)b^T(q)U_q^T(q) + \frac{1}{2}c^T(q)c(q) = 0. \qquad (6.29)$$

Then the family of controllers

$$\Sigma_Q^c : \begin{cases} \dot{\xi} &=& f^\flat(\xi) + G(y - h_2(\xi)) + \Gamma(\xi)v \\ \dot{q} &=& a(q) + b(q)r \\ v &=& c(q) \\ r &=& y - \hat{y} \\ u &=& \alpha_2(\xi) + v \end{cases} \qquad (6.30)$$

where

$$f^\flat(\xi) = f(\xi) + g_1(\xi)\alpha_1(\xi) + g_2(\xi)\alpha_2(\xi), \quad \hat{y} = \tilde{h}_2(\xi) = h_2(\xi) + k_{21}(\xi)\alpha_1(\xi)$$

solves the $MFBNLHICP$ locally.

Proof: First observe that the controller (6.30) is exactly the controller given in (6.28) with
explicit state-space realization for Σ^c. Thus, proving the result of this theorem also proves
the result of Proposition 6.1.2.

Accordingly, we divide the proof into several steps:

(a) Consider the Hamiltonian function defined in Chapter 5:

$$H(x, w, u) = V_x(x)(f(x) + g_1(x)w + g_2(x)u) + \frac{1}{2}\|h_1^T\|^2 + \frac{1}{2}\|u\|^2 - \frac{1}{2}\gamma^2\|w\|^2$$

which is quadratic in (w, u). Observe that

$$\left(\frac{\partial H}{\partial w}\right)_{w=\alpha_1(x)} = 0, \quad \left(\frac{\partial H}{\partial u}\right)_{u=\alpha_2(x)} = 0.$$

Therefore for every (w, u) we have

$$V_x(x)(f(x) + g_1(x)w + g_2(x)u) = \frac{1}{2}[\|u - \alpha_2(x)\|^2 - \gamma^2\|w - \alpha_1(x)\|^2 - \|u\|^2 + \gamma^2\|w\|^2 - \|h_1\|^2].$$

(b) Consider now the closed-loop system:

$$\left. \begin{array}{rcl} \dot{x} &=& f(x) + g_1(x)w + g_2(x)(\alpha_2(\xi) + v) \\ \dot{\xi} &=& f^b(\xi) + G(\xi)(y - h_2(\xi)) + \Gamma(\xi)v \end{array} \right\} \tag{6.31}$$

and the augmented pseudo-Hamiltonian function

$$\begin{array}{rcl} H^b(x, \xi, v, w, \lambda, \mu) &=& \lambda^T[f(x) + g_1(x)w + g_2(x)(\alpha_2(\xi) + v)] + \\ && \mu^T[f^b(\xi) + G(\xi)(y - h_2(\xi)) + \Gamma(\xi)v] + \\ && \frac{1}{2}\|\alpha_2(\xi) + v - \alpha_2(x)\|^2 - \frac{1}{2}\gamma^2\|w - \alpha_1(x)\|^2. \end{array}$$

By setting $(\tilde{v}, \tilde{w})$ such that

$$\left(\frac{\partial H^b}{\partial v}\right)_{v=\tilde{v}} = 0, \quad \left(\frac{\partial H^b}{\partial w}\right)_{w=\tilde{w}} = 0,$$

then for every (v, w) we have

$$H^b(x, \xi, v, w) = H^b_\star(x, \xi, v, w) + \frac{1}{2}\|v - \tilde{v}\|^2 - \frac{1}{2}\gamma^2\|w - \tilde{w}\|^2,$$

where

$$H^b_\star(x, \xi, v, w) = H^b(x, \xi, \tilde{v}, \tilde{w}, W_x^T, W_\xi^T) \le 0.$$

Furthermore, along any trajectory of the closed-loop system (6.31),

$$\begin{array}{rcl} \dot{W} = W_x\dot{x} + W_\xi\dot{\xi} &=& H^b_\star(x, \xi, v, w) + \frac{1}{2}\|v - \tilde{v}\|^2 - \frac{1}{2}\gamma^2\|w - \tilde{w}\|^2 - \\ && \frac{1}{2}\|\alpha_2(\xi) + v - \alpha_2(x)\|^2 - \frac{1}{2}\gamma^2\|w - \alpha_1(x)\|^2. \end{array}$$

(c) Similarly, consider now the system (6.27). By hypothesis (v), we have

$$U_q(q)(a(q) + b(q)r) = \frac{1}{2}[-\gamma^2\|r - r^\star\|^2 - \|v\|^2 + \gamma^2\|r\|^2]$$

where $r^\star = \frac{1}{\gamma^2}b^T(q)U_q^T(q)$.

(d) Then, consider the Lyapunov-function candidate

$$\Omega(x,\xi,q) = V(x) + W(x,\xi) + U(q),$$

which is locally positive-definite by construction. Differentiating Ω along the trajectories of the closed-loop system (6.31), (6.27), we get

$$
\begin{aligned}
\dot{\Omega} &= V_x(x)\dot{x} + W_x(x,\xi)\dot{x} + W_\xi(x,\xi)\dot{\xi} + U_q(q)\dot{q} \\
&= H_\star^\flat(x,\xi,v,w) + \frac{1}{2}\Big\{[\|u - \alpha_2(x)\|^2 - \gamma^2\|w - \alpha_1(x)\|^2 - \|u\|^2 + \\
&\quad \gamma^2\|w\|^2 - \|h_1(x)\|^2] + [\|v - \tilde{v}\|^2 - \gamma^2\|w - \tilde{w}\|^2 - \|u - \alpha_2(x)\|^2 + \\
&\quad \gamma^2\|w - \alpha_1(x)\|^2] + [-\gamma^2\|r - r^\star\|^2 - \|v\|^2 + \gamma^2\|r\|^2]\Big\}.
\end{aligned}
$$

Notice that

$$\|v - \tilde{v}\|^2 \le \gamma^2\|r - r^\star\|^2$$

since $\|Q\|_{\mathcal{H}_\infty} \le \gamma$. Therefore

$$\dot{\Omega} \le \frac{1}{2}[\gamma^2(\|r\|^2 + \|w\|^2) - \|v\|^2 - \|z\|^2]. \tag{6.32}$$

Integrating now the above inequality from $t = t_0$ to $t = T$, we have

$$\Omega(x(T),T) - \Omega(x(t_0),t_0) \le \frac{1}{2}\int_{t_0}^T [\gamma^2(\|r\|^2 + \|w\|^2) - \|v\|^2 - \|z\|^2]dt.$$

Which implies that the closed-loop system has $\mathcal{L}_2$-gain $\le \gamma$ from $\begin{bmatrix} r \\ w \end{bmatrix}$ to $\begin{bmatrix} v \\ z \end{bmatrix}$ as desired.

(e) To show closed-loop stability, set $w = 0$, $r = 0$, in (6.32) to get

$$\dot{\Omega} \le -\frac{1}{2}(\|z\|^2 + \|v\|^2).$$

This proves that the equilibrium-point $(x,\xi,q) = (0,0,0)$ is stable. Further, the condition that $\dot{\Omega} = 0$ implies that $w = 0$, $r = 0$, $z = 0$, $v = 0 \Rightarrow u = 0$, $h_1(x) = 0$, $\alpha_2(\xi) = 0$. Thus, any trajectory of the system $(x^o(t), \xi^o(t), q^o(t))$ satisfying the above conditions, is necessarily a trajectory of the system

$$
\begin{aligned}
\dot{x} &= f(x) \\
\dot{\xi} &= f(\xi) + g_1(\xi)\alpha_1(\xi) + G(\xi)(h_2(x) - h_2(\xi)) \\
\dot{q} &= a(q).
\end{aligned}
$$

Therefore, by Assumption (i) and the fact that $\dot{q} = a(q)$ is locally asymptotically-stable about $q = 0$, we have $\lim_{t\to\infty}(x(t), q(t)) = 0$. Similarly, by Assumption (iii) and a well known stability property of cascade systems, we also have $\lim_{t\to\infty} \xi(t) = 0$. Finally, by LaSalle's invariance-principle we conclude that the closed-loop system is locally asymptotically-stable. $\square$

6.2 Output Measurement-Feedback Nonlinear $\mathcal{H}_\infty$ Tracking Control

In this section, we discuss the output measurement-feedback tracking problem which was discussed in Section 5.2 for the state-feedback problem. The objective is to design an output-feedback controller for the system (6.1) to track a reference signal which is generated as the output y_m of a reference model defined by

$$\Sigma_m : \begin{cases} \dot{x}_m &= f_m(x_m), \quad x_m(t_0) = x_{m0} \\ y_m &= h_m(x_m) \end{cases} \tag{6.33}$$

$x_m \in \Re^l$, $f_m : \Re^l \to \Re^l$. We assume also for simplicity that this system is completely observable (see References [268, 212]). The problem can then be defined as follows.

Definition 6.2.1 *(Measurement-Feedback Nonlinear $\mathcal{H}_\infty$ (Suboptimal) Tracking Control Problem (MFBNLHITCP)). Find (if possible!) a dynamic output-feedback controller of the form*

$$\begin{cases} \dot{x}_c &= \eta_c(x_c, x_m), \quad x_c(t_0) = x_{c0} \\ u &= \alpha_{dyntrk}(x_c, x_m), \quad \alpha_{dyntrk} : N_1 \times \tilde{N}_m \to \Re^p \end{cases} \tag{6.34}$$

$x_c \subset \Re^n$, $N_1 \subset \mathcal{X}$, $\tilde{N}_m \subset \Re^l$, for some smooth function α_{dyntrk}, such that the closed-loop system (6.1), (6.34), (6.33) is exponentially stable about $(x, x_c, x_m) = (0, 0, 0)$, and has for all initial conditions starting in $N_1 \times \tilde{N}_m$ neighborhood of $(0, 0)$, locally $\mathcal{L}_2$-gain from the disturbance signal w to the output z less than or equal to some prescribed number $\gamma^\star > 0$, as well as the tracking error satisfies $\lim_{t\to\infty}\{y - y_m\} = 0$.

To solve the above problem, we consider the system (6.1) with

$$z = \begin{bmatrix} h_1(x) \\ u \end{bmatrix}$$

and follow as in Section 5.2 a two-step design procedure.

Step 1: We seek a feedforward dynamic-controller of the form

$$\begin{cases} \dot{x}_c &= a(x_c) + b(x_c)y \\ u &= c(x_c, x_m) \end{cases} \tag{6.35}$$

where $x_c \in \Re^n$, so that the equilibrium point $(x, x_c) = (0, 0)$ of the closed-loop system with $w = 0$,

$$\begin{aligned} \dot{x} &= f(x) + g_2(x)c(x_c, x_m) \\ \dot{x}_c &= a(x_c) + b(x_c)h_2(x) \end{aligned}$$

is exponentially stable and there exists a neighborhood $\tilde{U} \subset \mathcal{X} \times \Re^n \times \Re^l$ of $(0, 0, 0)$ such that, for all initial conditions $(x_0, x_{c0}, x_{m0}) \in \tilde{U}$, the solution $(x(t), x_c(t), x_m(t))$ of

$$\begin{cases} \dot{x} &= f(x) + g_2(x)c(x_c, x_m) \\ \dot{x}_c &= a(x_c) + b(x_c)h_2(x) \\ \dot{x}_m &= f_m(x_m) \end{cases}$$

satisfies

$$\lim_{t\to\infty} h_1(\theta(x(t))) - h_m(x_m(t)) = 0.$$

To solve this step, we similarly seek an invariant-manifold

$$\tilde{M}_{\theta,\sigma} = \{x | x = \theta(x_m),\ x_c = \sigma(x_m)\}$$

which is invariant under the closed-loop dynamics (6.36) and is such that the error $e = h_1(x(t)) - h_m(x_m(t))$ vanishes identically on this submanifold. Again, this requires that the following necessary conditions are satisfied by the control law $u = \bar{u}_\star(x_m)$:

$$\frac{\partial \theta}{\partial x_m}(x_m)f_m(x_m) = f(\theta(x_m)) + g_2(\theta(x_m)\bar{u}_\star(x_m)$$

$$\frac{\partial \sigma}{\partial x_m}(x_m)f(x_m) = a(\sigma(x_m)) + b(\sigma(x_m))y_\star(x_m)$$

$$h_1(\theta(x(t))) - h_m(x_m(t)) = 0,$$

where $\bar{u}_\star(x_m) = c(\sigma(x_m), x_m)$, $y_\star(x_m) = h_2(\theta(x_m))$.

Now, assuming the submanifold $\tilde{M}_{\theta,\sigma}$ is found and the control law $\bar{u}_\star(x_m)$ has been designed to maintain the system on this submanifold, the next step is to design a feedback control v so as to drive the system onto the above submanifold and to achieve disturbance-attenuation and asymptotic-tracking. To formulate this step, we reconsider the combined system with the disturbances

$$\begin{cases} \dot{x} & = & f(x) + g_1(x)w + g_2(x)u \\ \dot{x}_c & = & a(x_c) + b(x_c)h_2(x) + b(x_c)k_{21}(x)w \\ \dot{x}_m & = & f_m(x_m) \\ y & = & h_2(x) + k_{21}(x)w, \end{cases} \tag{6.36}$$

and introduce the following change of variables

$$\begin{aligned} \xi_1 & = & x - \theta(x_m) \\ \xi_2 & = & x_c - \sigma(x_m) \\ v & = & u - \bar{u}_\star(x_m). \end{aligned}$$

Then

$$\begin{aligned} \dot{\xi}_1 & = & \tilde{F}_1(\xi_1, \xi_2, x_m) + \tilde{G}_{11}(\xi_1, x_m)w + \tilde{G}_{12}(\xi_1, x_m)v \\ \dot{\xi}_2 & = & \tilde{F}_2(\xi_1, \xi_2, x_m) + \tilde{G}_{21}(\xi_2, x_m)w \\ \dot{x}_m & = & f_m(x_m) \\ y & = & \begin{bmatrix} h_2(\xi_1, x_m) + k_{21}(\xi_1, x_m)w \\ h_m(x_m) \end{bmatrix} \end{aligned}$$

where

$$\tilde{F}_1(\xi_1, x_m) = f(\xi_1 + \theta(x_m)) - \frac{\partial \theta}{\partial x_m}(x_m)f_m(x_m) + g_2(\xi_1 + \theta(x_m))\bar{u}_\star(x_m)$$

$$\tilde{F}_2(\xi_2, x_m) = a(\xi_2 + \sigma(x_m)) + b(\xi_2 + \sigma(x_m))h_2(\xi_1 + \theta(x_m)) - \frac{\partial \sigma}{\partial x_m}(x_m)f_m(x_m)$$

$$\tilde{G}_{11}(\xi_1, x_m) = g_1(\xi_1 + \theta(x_m))$$

$$\tilde{G}_{12}(\xi_1, x_m) = g_2(\xi_1 + \theta(x_m))$$

$$\tilde{G}_{21}(\xi_2, x_m) = b(\xi_2 + \sigma(x_m))k_{21}(\xi_1 + \theta(x_m)).$$

Similarly, we redefine the tracking error and the new penalty variable as

$$\tilde{z} = \begin{bmatrix} h_1(\xi + \theta(x_m)) - h_m(x_m) \\ v \end{bmatrix}.$$

Step 2: Find a dynamic feedback compensator of the form (6.35) with an auxiliary output $v_\star = v_\star(\xi, x_m)$ so that the closed-loop system (6.35), (6.36) is exponentially stable and along any trajectory $(\xi(t), x_m(t))$ of the closed-loop system, the $\mathcal{L}_2$-gain condition

$$\int_0^T \|\tilde{z}(t)\|^2 \leq \gamma^2 \int_0^T \|w(t)\|^2 dt + \tilde{\beta}(\xi(t_0), x_{m0})$$

$$\Leftrightarrow \int_0^T \{\|h_1(\xi + \theta(x_m)) - h_m(x_m)\|^2 + \|v\|^2\} \leq \gamma^2 \int_0^T \|w(t)\|^2 dt + \beta(\xi(t_0), x_{m0})$$

is satisfied for some function $\tilde{\beta}$, for all $w \in \mathcal{W}$, for all $T < \infty$ and all initial conditions $(\xi_1(t_0), \xi_2(t_0), x_{m0})$ in a sufficiently small neighborhood of the origin $(0, 0)$.

The above problem is now a standard measurement-feedback $\mathcal{H}_\infty$ control problem, and can be solved using the techniques developed in the previous section.

6.3 Robust Output Measurement-Feedback Nonlinear $\mathcal{H}_\infty$-Control

In this section, we consider the robust output measurement-feedback nonlinear $\mathcal{H}_\infty$-control problem for the affine nonlinear system (6.1) in the presence of unmodelled dynamics and/or parameter variations. This problem has been considered in many references [34, 208, 223, 265, 284], and the set-up is shown in Figure 6.3. The approach presented here is based on [208]. For this purpose, the system is represented by the model:

$$\Sigma_\Delta^a : \begin{cases} \dot{x} &= f(x) + \Delta f(x, \theta, t) + g_1(x)w + [g_2(x) + \Delta g_2(x, \theta, t)]u; \\ & x(0) = x_0 \\ z &= h_1(x) + k_{12}(x)u \\ y &= [h_2(x) + \Delta h_2(x, \theta, t)] + k_{21}(x)w \end{cases} \tag{6.37}$$

where all the variables and functions have their previous meanings and in addition $\Delta f : \mathcal{X} \to V^\infty(\mathcal{X})$, $\Delta g_2 : \mathcal{X} \to \mathcal{M}^{n \times p}(\mathcal{X})$, and $\Delta h_2 : \mathcal{X} \to \Re^m$ are unknown functions which belong to the set Ξ_1 of admissible uncertainties, and $\theta \in \Theta \subset \Re^r$ are the system parameters which may vary over time within the set Θ.

Definition 6.3.1 *(Robust Output Measurement-Feedback Nonlinear $\mathcal{H}_\infty$-Control Problem (RMFBNLHICP)). Find (if possible!) a dynamic controller of the form (6.3) such that the closed-loop system (6.37), (6.3) has locally $\mathcal{L}_2$-gain from the disturbance signal w to the output z less than or equal to some prescribed number $\gamma^\star > 0$ with internal stability, for all admissible uncertainties $\Delta f, \Delta g_2, \Delta h_2, \in \Xi_1$ and all parameter variations in Θ.*

We begin by first characterizing the sets of admissible uncertainties of the system Ξ_1 and Θ_1 as

Assumption 6.3.1 *The sets of admissible uncertainties of the system are characterized as*

$$\begin{aligned} \Xi_1 &= \{\Delta f, \Delta g_2, \Delta h_2 \,|\, \Delta f(x, \theta, t) = H_1(x)F(x, \theta, t)E_1(x), \\ & \quad \Delta g_2(x, \theta, t) = g_2(x)F(x, \theta, t)E_2(x), \; \Delta h_2 = H_3(x)F(x, \theta, t)E_3(x), \; \|F(x, \theta, t)\|^2 \leq 1, \\ & \quad \forall x \in \mathcal{X}, \; \theta \in \Theta, \; t \in \Re\}, \\ \Theta_1 &= \{\theta | 0 \leq \theta \leq \theta_u, \; \|\theta\| \leq \kappa < \infty\}, \end{aligned}$$

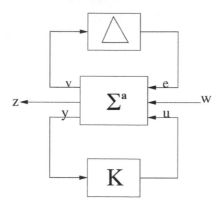

FIGURE 6.3
Configuration for Robust Nonlinear $\mathcal{H}_\infty$-Control with Measurement-Feedback

for some known matrices $H_1(.)$, $F(.,.)$, $E_1(.)$, $E_2(.)$, $H_3(.)$, E_3 of appropriate dimensions.

Then, to solve the $RMFBNLHICP$, we clearly only have to modify the solution to the nominal $MFBNLHICP$ given in Theorems 6.1.1, 6.1.2, 6.1.3, and in particular, modify the HJIEs (inequalities) (6.39), (6.23), (5.12) to account for the uncertainties in Δf, Δg_2 and Δh_2.

The HJI-inequality (5.24) has already been modified in the form of the HJI-inequality (5.46) to obtain the solution to the $RSFBNLHICP$ for all admissible uncertainties Δf, Δg_2 in Ξ. Therefore, it only remains to modify the HJI-inequality (6.23), or equation (6.16) to account for the uncertainties Δf, Δg_2, and Δh_2 in Ξ_1, Θ. If we choose to modify HJIE (6.16) to accomodate the uncertainties, then the result can be stated as a corollary to Theorem 6.1.3 as follows.

Corollary 6.3.1 *Consider the nonlinear system (6.37) and assume that $k_{21}(x)k_{21}^T(x) = I$ in (6.2). Suppose also the following hold:*

(i) *The pair $\{f, h_1\}$ is locally detectable.*

(ii) *There exists a smooth positive-definite function $\widetilde{V}$ locally defined about the origin with $\widetilde{V}(0) = 0$ and satisfying the HJIE (5.46).*

(iii) *There exists an $n \times m$ matrix $G_\Delta : \mathcal{X} \to \mathcal{M}^{n \times m}$ such that the equilibrium $\xi = 0$ of the system*

$$\dot{\xi} = f(\xi) + H_1(\xi)E_1(\xi) + g_1(\xi)\alpha_1(\xi) - G_\Delta(\xi)(h_2(\xi) + H_3(\xi)E_3(\xi)) \tag{6.38}$$

is locally asymptotically-stable.

(iv) *There exists a smooth positive-semidefinite function $\widetilde{W}(x, \xi)$ locally defined in a neighborhood $\tilde{N}_1 \times \tilde{N}_1 \subset \mathcal{X} \times \mathcal{X}$ of the origin $(x, \xi) = (0, 0)$ such that $\widetilde{W}(0, \xi) > 0$ for each*

$\xi \neq 0$, and satisfying the HJIE

$$[\widetilde{W}_x(x,\xi) \; \widetilde{W}_\xi(x,\xi)]f_e(x,\xi) +$$
$$\frac{1}{2\gamma^2}[\widetilde{W}_x(x,\xi) \; \widetilde{W}_\xi(x,\xi)] \begin{bmatrix} g_1(x)g_1^T(x) + H_1^T(x)H_1(x) & 0 \\ 0 & G(\xi)G^T(\xi) + H_1^T(\xi)H_1(\xi) \end{bmatrix} \times$$
$$\begin{bmatrix} \widetilde{W}_x^T(x,\xi) \\ \widetilde{W}_\xi^T(x,\xi) \end{bmatrix} + \frac{1}{2}(h_e^T(x,\xi)h_e(x,\xi) + E_1^T(x)E_1(x) + E_3^T(\xi)E_3(\xi)) = 0,$$
$$\widetilde{W}(0,0) = 0, \quad (x,\xi) \in \tilde{N}_1 \times \tilde{N}_1, \tag{6.39}$$

where $f_e(x,\xi)$, $h_e(x,\xi)$ are as defined previously.

Then, the RMFBNLHICP for the system is solved by the output-feedback controller:

$$\tilde{\Sigma}_2^c : \begin{cases} \dot{\zeta} &= f(\xi) + H_1(\xi)F_1(\xi) + q_1(\xi)\alpha_1(\xi) + g_2(\xi)\alpha_2(\xi) + G_\Delta(\xi)[y - \\ & \quad h_2(\xi) - H_3(\xi)E_3(\xi)] \\ u &= \alpha_2(\xi), \end{cases} \tag{6.40}$$

Proof: The proof can be pursued along the same lines as Theorem 6.1.3. $\square$

In the next section we discuss another aspect of robust control known as reliable control.

6.3.1 Reliable Robust Output-Feedback Nonlinear $\mathcal{H}_\infty$-Control

In this subsection, we discuss reliable control which is another aspect of robust control. The aim however in this case is to maintain control and stability in the event of a set of actuator or sensor failures. Failure detection is also another aspect of reliable control.

For the purpose of elucidating the scheme, let us represent the system on $\mathcal{X} \subset \Re^n$ in the form

$$\Sigma^{ar} : \begin{cases} \dot{x} &= f(x) + g_1(x)w_0 + \sum_{j=1}^p g_{2j}(x)u_j; \quad x(0) = x_0, \\ y_i &= h_{2i}(x) + w_i, \quad i = 1, \dots, m, \\ z &= \begin{bmatrix} h_1(x) \\ u_1 \\ \vdots \\ u_p \end{bmatrix}, \end{cases} \tag{6.41}$$

where all the variables have their usual meanings and dimensions, and in addition $w = [w_0^T \; w_1^T \; \dots \; w_m^T]^T \in \Re^s$ is the overall disturbance input vector,

$$g_2(x) = [g_{21}(x) \; g_{22}(x) \; \dots \; g_{2p}(x)],$$
$$h_2(x) = [h_{21}(x) \; h_{22}(x) \; \dots \; h_{2m}(x)]^T.$$

Without any loss of generality, we can also assume that $f(0) = 0$, $h_1(0) = 0$ and $h_{2i}(0) = 0$, $i = 1, \dots, m$. The problem is the following.

Definition 6.3.2 *(Reliable Measurement-Feedback Nonlinear $\mathcal{H}_\infty$-Control Problem (RLMF-BNLHICP)).* *Suppose $\mathcal{Z}_a \subset \{1, \dots, p\}$ and $\mathcal{Z}_s \subset \{1, \dots, m\}$ are the subsets of actuators and sensors respectively that are prone to failure. Find (if possible!) a dynamic controller of the form:*

$$\Sigma^{rc} : \begin{cases} \dot{\zeta} &= a(\zeta) + b(\zeta)y \\ u &= c(\zeta) \end{cases} \tag{6.42}$$

where $\zeta \in \Re^\nu$, such that the closed-loop system denoted by $\Sigma^{ar} \circ \Sigma^{rc}$ is locally asymptotically-stable and has locally $\mathcal{L}_2$-gain $\leq \gamma$ for all initial conditions in $N \subset \mathcal{X}$ and for all actuator and sensor failures in $\mathcal{Z}_a$ and $\mathcal{Z}_s$ respectively.

To solve the above problem, let $i_a \in \mathcal{Z}_a$, $i_a' \in \bar{\mathcal{Z}}_a$, $j_s \in \mathcal{Z}_s$, $j_s' \in \bar{\mathcal{Z}}_s$ denote the indices of the elements in $\mathcal{Z}_a$, $\bar{\mathcal{Z}}_a$, $\mathcal{Z}_s$ and $\bar{\mathcal{Z}}_s$ respectively, where the "bar" denotes "set complement." Accordingly, introduce the following decomposition

$$
\begin{aligned}
g_2(x) &= g_{2i_a}(x) \oplus g_{2i_a'}(x) \\
u &= u_{i_a} \oplus u_{i_a'} \\
h_2(x) &= h_{2i_s}(x) \oplus h_{2i_s'}(x) \\
y &= y_{i_s} \oplus y_{i_s'} \\
w &= [w_1 \ \dots \ w_m]^T = w_{i_s} \oplus w_{i_s'} \\
b(x) &= [b_1(x) \ b_2(x) \ \dots \ b_m(x)] = b_{i_s}(x) \oplus b_{i_s'}(x)
\end{aligned}
$$

where

$$
\begin{aligned}
g_{2i_a}(x) &= [\delta_{i_a}(1)g_{21}(x) \ \delta_{i_a}(2)g_{22}(x) \ \dots \ \delta_{i_a}(p)g_{2p}(x)] \\
u_{i_a}(x) &= [\delta_{i_a}(1)u_1(x) \ \delta_{i_a}(2)u_2(x) \ \dots \ \delta_{i_a}(p)u_p(x)]^T \\
h_{2i_s}(x) &= [\delta_{i_s}(1)h_{21}(x) \ \delta_{i_s}(2)h_{22}(x) \ \dots \ \delta_{i_s}(m)h_{2m}(x)]^T \\
y_{i_s}(x) &= [\delta_{i_s}(1)y_1 \ \delta_{i_s}(2)y_2 \ \dots \ \delta_{i_s}(m)y_m]^T \\
w_{i_s}(x) &= [\delta_{i_s}(1)w_1 \ \delta_{i_s}(2)w_2 \ \dots \ \delta_{i_a}(m)w_m]^T \\
b_{i_s}(x) &= [\delta_{i_s}(1)b_1(x) \ \delta_{i_s}(2)b_2(x) \ \dots \ \delta_{i_s}(m)b_m(x)]^T
\end{aligned}
$$

and

$$
\delta_{i_a}(i) = \begin{cases} 1 \text{ if } i \in \mathcal{Z}_a \\ 0 \text{ if } i \notin \mathcal{Z}_a \end{cases}
$$

$$
\delta_{i_s}(j) = \begin{cases} 1 \text{ if } j \in \mathcal{Z}_s \\ 0 \text{ if } j \notin \mathcal{Z}_s. \end{cases}
$$

Applying the controller Σ^{rc} to the system when actuator and/or sensor failures corresponding to the indices $i_a \in \mathcal{Z}_a$ and $i_s \in \mathcal{Z}_s$ occur, results in the closed-loop system $\Sigma^{ar}(\mathcal{Z}_s) \circ \Sigma^{rc}(\mathcal{Z}_a)$:

$$
\begin{cases}
\dot{x} &= f(x) + g_1(x)w_0 + g_{2i_a'}(x)c_{i_a'}(\zeta); \ x(0) = x_0 \\
\dot{\zeta} &= a(\zeta) + b_{i_s'}(\zeta)y_{i_s'} \\
&= a(\zeta) + b_{i_s'}(\zeta)h_{2i_s'}(x) + b_{i_s'}(\zeta)w_{i_s'} \\
z_{i_a'} &= \begin{bmatrix} h_1(x) \\ c_{i_a'}(\zeta) \end{bmatrix}.
\end{cases}
\tag{6.43}
$$

The aim now is to find the controller parameters $a(.)$, $b(.)$ and $c(.)$, such that in the event of any actuator or sensor failure for any $i_a \in \mathcal{Z}_a$ and $i_s \in \mathcal{Z}_s$ respectively, the resulting closed-loop system above still has locally $\mathcal{L}_2$-gain $\leq \gamma$ from $w_{i_s'}$ to $z_{i_a'}$ and is locally asymptotically-stable. In this regard, define the Hamiltonian functions in the local coordinates (x, p) on

$T^\star \mathcal{X}$ by

$$
\begin{aligned}
H_s(x,p) &= p^T f(x) + \frac{1}{2}p^T \left[\frac{1}{\gamma^2} g_1(x)g_1^T(x) - g_{2\bar{\mathcal{Z}}_a}(x)g_{2\bar{\mathcal{Z}}_a}^T(x) \right] p + \\
&\quad \frac{1}{2}h_1^T(x)h_1(x) + \frac{1}{2}\gamma^2 h_{2\mathcal{Z}_s}^T(x)h_{2\mathcal{Z}_s}(x), \\
H_0(x,p) &= p^T f(x) + \frac{1}{2\gamma^2}p^T g_1(x)g_1^T(x)p + \frac{1}{2}p^T g_{2\mathcal{Z}_a}(x)g_{2\mathcal{Z}_a}^T(x)p + \\
&\quad \frac{1}{2}h_1^T(x)h_1(x) - \frac{1}{2}\gamma^2 h_{2\bar{\mathcal{Z}}_s}^T(x)h_{2\bar{\mathcal{Z}}_s}(x).
\end{aligned}
$$

Then, the following theorem gives a sufficient condition for the solvability of the reliable control problem.

Theorem 6.3.1 *Consider the nonlinear system (6.41) and assume the following hold:*

(i) *The pair $\{f, h_1\}$ is locally zero-state detectable.*

(ii) *There exists a smooth C^2 function $\psi \geq 0$, $\psi(0) = 0$ and a C^3 positive-definite function $\widetilde{V}$ locally defined in a neighborhood N_0 of $x = 0$ with $\widetilde{V}(0) = 0$ and satisfying the HJIE:*

$$H_s(x, \widetilde{V}_x^T) + \psi(x) = 0. \tag{6.44}$$

(iii) *There exists a C^3 positive-definite function $\widetilde{U}$ locally defined in a neighborhood N_1 of $x = 0$ with $\widetilde{U}(0) = 0$ and satisfying the HJI-inequality:*

$$H_0(x, \widetilde{U}_x^T) + \psi(x) \leq 0, \tag{6.45}$$

and such that $N_0 \cap N_1 \neq \emptyset$, $H_0(x, \widetilde{U}_x^T) + \psi(x)$ has nonsingular Hessian matrix at $x = 0$.

(iv) *$\widetilde{U}(x) - \widetilde{V}(x)$ is positive definite, and there exists an $n \times m$ matrix function $L(.)$ that satisfies the equation*

$$(\widetilde{U}_x - \widetilde{V}_x)(x)L(x) = \gamma^2 h_2^T(x). \tag{6.46}$$

Then, the controller Σ^{rc} solves the $RLMFBNLHICP$ for the system Σ^{ar} with

$$
\begin{aligned}
a(\zeta) &= f(\zeta) + \frac{1}{\gamma^2}g_1(\zeta)g_1^T(\zeta)\widetilde{V}_x^T(\zeta) - g_{2\bar{\mathcal{Z}}_a}(\zeta)g_{2\bar{\mathcal{Z}}_a}^T(\zeta)\widetilde{V}_x^T(\zeta) - L(\zeta)h_2(\zeta) \\
b(\zeta) &= L(\zeta) \\
c(\zeta) &= -g_2^T(\zeta)\widetilde{V}_x^T(\zeta).
\end{aligned}
$$

Proof: The proof is lengthy, but can be found in Reference [285]. $\square$

Remark 6.3.1 *Theorem 6.3.1 provides a sufficient condition for the solvability of reliable controller design problem for the case of the primary contingency problem, where the set of sensors and actuators that are susceptible to failure is known a priori. Nevertheless by enlarging the sets $\mathcal{Z}_a$, $\mathcal{Z}_s$ to $\{1, \ldots, p\}$ and $\{1, \ldots, m\}$ respectively, the scheme can be extended to include all actuators and sensors.*

We consider now an example.

Example 6.3.1 *[285]. Consider the following second-order system.*

$$\begin{bmatrix} \dot{x}_1 \\ \dot{x}_2 \end{bmatrix} = \begin{bmatrix} -2x_1 + x_1 x_2^2 \\ x_2^3 \end{bmatrix} + \begin{bmatrix} 1 \\ x_1 \end{bmatrix} w_0 + \begin{bmatrix} 0 & 1 \\ 1 & 1 \end{bmatrix} \begin{bmatrix} u_1 \\ u_2 \end{bmatrix} \tag{6.47}$$

$$y = 2x_1 + 2x_2 + w_1 \tag{6.48}$$

$$z = [x_1 \ \ x_2^4 \ \ u_1 \ \ u_2]^T. \tag{6.49}$$

The index sets are given by $\mathcal{Z}_a = \{2\}$, $\mathcal{Z}_s = \{\emptyset\}$, and the system is locally zero-state detectable. With $\gamma = 0.81$ and

$$\psi(x) = \frac{1}{2}x_1^2 + \frac{8\gamma^2}{\gamma^2 - x_1^2}x_2^6,$$

approximate solutions to the HJI-inequalities (6.44), (6.45) can be obtained as

$$\tilde{V}(x) = 0.4533x_1^2 + 0.3463x_1^2 x_2^2 + \frac{2\gamma^2}{\gamma^2 - x_1^2}x_2^4 \tag{6.50}$$

$$\tilde{U}(x) = 1.5938x_1^2 + 0.3496x_1 x_2 + 1.2667x_2^2 \tag{6.51}$$

respectively. Finally, equation (6.46) is solved to get

$$L(x) = [1.0132 \ \ 0.8961]^T.$$

The controller can then be realized by computing the values of $a(x)$, $b(x)$ and $c(x)$ accordingly.

6.4   Output Measurement-Feedback $\mathcal{H}_\infty$-Control for a General Class of Nonlinear Systems

In this section, we look at the output measurement-feedback problem for a more general class of nonlinear systems. For this purpose, we consider the following class of nonlinear systems defined on a manifold $\mathcal{X} \subseteq \Re^n$ containing the origin in coordinates $x = (x_1, \ldots, x_n)$:

$$\Sigma^g : \begin{cases} \dot{x} = F(x, w, u); & x(t_0) = x_0 \\ z = Z(x, u) \\ y = Y(x, w) \end{cases} \tag{6.52}$$

where all the variables have their previous meanings, while $F : \mathcal{X} \times \mathcal{W} \times \mathcal{U} \to \mathcal{X}$ is the state dynamics function, $Z : \mathcal{X} \times \mathcal{U} \to \Re^s$ is the controlled output function and $Y : \mathcal{X} \times \mathcal{W} \to \Re^m$ is the measurement output function. Moreover, the functions $F(., ., .)$, $Z(., .)$ and $Y(., .)$ are smooth C^r, $r \geq 1$ functions of their arguments, and the point $x = 0$ is a unique equilibrium-point for the system Σ^g such that $F(0,0,0) = 0$, $Z(0,0) = 0$, $Y(0,0) = 0$. The following assumptions will also be adopted in the sequel.

Assumption 6.4.1 *The linearization of the function $Z(x, u)$ is such that*

$$rank(D_{12}) \triangleq rank\left(\frac{\partial Z}{\partial u}(0,0)\right) = p.$$

The control action to Σ^g is to be provided by a controller of the form (6.3). Motivated by the results of Section 6.1, we can conjecture a dynamic-controller of the form:

$$\Sigma_e^{gc} : \left\{ \begin{array}{rcl} \dot{\zeta} & = & F(\zeta, w, u) + \tilde{G}(\zeta)(y - Y(\zeta, w)) \\ u & = & \tilde{\alpha}_2(\zeta) \end{array} \right. \tag{6.53}$$

where ζ is an estimate of x and $\tilde{G}(.)$ is the output-injection gain matrix which is to be determined.

Again, since w is not directly available, we use its worst-case value in equation (6.53). Denote this value by $w^* = \tilde{\alpha}_1(x)$, where $\tilde{\alpha}_1(.)$ is determined according to the procedure explained in Chapter 5. Then, substituting this in (6.53) we obtain

$$\left\{ \begin{array}{rcl} \dot{\zeta} & = & F(\zeta, \tilde{\alpha}_1(\zeta), \tilde{\alpha}_2(\zeta)) + \tilde{G}(\zeta)(y - Y(\zeta, \tilde{\alpha}_1(\zeta)) \\ u & = & \tilde{\alpha}_2(\zeta). \end{array} \right. \tag{6.54}$$

Now, let $\tilde{x}^e = \left(\begin{array}{c} x \\ \zeta \end{array} \right)$ so that the closed-loop system (6.52) and (6.54) is represented by

$$\left\{ \begin{array}{rcl} \dot{\tilde{x}}^e & = & F^e(\tilde{x}^e, w) \\ z & = & Z^e(\tilde{x}^e) \end{array} \right. \tag{6.55}$$

where

$$\begin{array}{rcl} F^e(\tilde{x}^e, w) & = & \left(\begin{array}{c} F(x, w, \tilde{\alpha}_2(\zeta)) \\ F(\zeta, \tilde{\alpha}_1(\zeta), \tilde{\alpha}_2(\zeta)) + \tilde{G}(\zeta)Y(x, w) - \tilde{G}(\zeta)Y(\zeta, \tilde{\alpha}_1(\zeta)) \end{array} \right), \\ Z^e(\tilde{x}^e) & = & Z(x, \tilde{\alpha}_2(\zeta)). \end{array}$$

The objective then is to render the above closed-loop system dissipative with respect to the supply-rate $s(w, Z) = \frac{1}{2}(\gamma^2 \|w\|^2 - \|Z^e(x^e)\|^2)$ with some suitable storage-function and an appropriate output-injection gain matrix $\tilde{G}(.)$. To this end, we make the following assumption.

Assumption 6.4.2 *Any bounded trajectory $x(t)$ of the system (6.52)*

$$\dot{x}(t) = F(x(t), 0, u(t))$$

satisfying

$$Z(x(t), u(t)) \equiv 0$$

for all $t \geq t_0$, is such that $\lim_{t \to \infty} x(t) = 0$.

Then we have the following proposition.

Proposition 6.4.1 *Consider the system (6.55) and suppose Assumptions 6.4.1, 6.4.2 hold. Suppose also the system*

$$\dot{\zeta} = F(\zeta, \tilde{\alpha}_1(\zeta), 0) - \tilde{G}(\zeta)Y(\zeta, \tilde{\alpha}_1(\zeta)) \tag{6.56}$$

has a locally asymptotically-stable equilibrium-point at $\zeta = 0$, and there exists a locally defined C^1 positive-definite function $\tilde{U} : \mathcal{O} \times \mathcal{O} \to \Re_+$, $\tilde{U}(0) = 0$, $\mathcal{O} \subset \mathcal{X}$, satisfying the dissipation inequality

$$\tilde{U}_{\tilde{x}^e}(\tilde{x}^e)F^e(\tilde{x}^e, w) + \frac{1}{2}\|Z^e(\tilde{x}^e)\|^2 - \frac{1}{2}\gamma^2\|w\|^2 \leq 0 \tag{6.57}$$

for $w = 0$. Then, the system (6.55) has a locally asymptotically-stable equilibrium-point at $\tilde{x}^e = 0$.

Proof: Set $w = 0$ in the dissipation-inequality (6.57) and rewrite it as

$$\dot{\tilde{U}}(\tilde{x}^e(t)) = \tilde{U}_{\tilde{x}^e}(\tilde{x}^e) F^e(\tilde{x}^e, w) = -\frac{1}{2}\|Z(x, \tilde{\alpha}_2(\zeta)) \leq 0,$$

which implies that $\tilde{x}^e = 0$ is stable for (6.55). Further, any bounded trajectory $\tilde{x}^e(t)$ of (6.55) with $w = 0$ resulting in $\dot{\tilde{U}}(\tilde{x}^e(t)) = 0$ for all $t \geq t_s$ for some $t_s \geq t_0$, implies that

$$Z(x(t), \tilde{\alpha}_2(\zeta(t))) \equiv 0 \;\; \forall t \geq t_s.$$

By Assumption 6.4.2 we have $\lim_{t\to\infty} x(t) = 0$. Moreover, Assumption 6.4.1 implies that there exists a smooth function $u = u(x)$ locally defined in a neighborhood of $x = 0$ such that

$$Z(x, u(t)) = 0 \;\; \text{and} \;\; u(0) = 0.$$

Therefore, $\lim_{t\to\infty} x(t) = 0$ and $Z(x(t), \tilde{\alpha}_2(\zeta_2(t))) = 0$ imply $\lim_{t\to\infty} \tilde{\alpha}_2(\zeta(t)) = 0$. Finally, $\lim_{t\to\infty} \tilde{\alpha}_2(\zeta(t)) = 0$ implies $\lim_{t\to\infty} \zeta(t) = 0$ since ζ is a trajectory of (6.56). Hence, by LaSalle's invariance-principle, we conclude that $x^e = 0$ is asymptotically-stable. $\square$

Remark 6.4.1 *As a consequence of the above proposition, the design of the injection-gain matrix $\tilde{G}(\zeta)$ should be such that: (a) the dissipation-inequality (6.57) is satisfied for the closed-loop system (6.55) and for some storage-function $\tilde{U}(\zeta)$; (b) system (6.56) is asymptotically-stable. If this happens, then the controller (6.53) will solve the $MFBNLHICP$.*

In the next several lines, we present the main result for the solution of the $MFBNLHICP$ for the class of systems (6.52). We begin with the following assumption.

Assumption 6.4.3 *The linearization of the output function $Y(x, w)$ is such that*

$$rank(D_{21}) \triangleq rank\left(\frac{\partial Y}{\partial w}(0,0)\right) = m.$$

Define now the Hamiltonian function $K : T^\star \mathcal{X} \times \mathcal{W} \times \Re^m \to \Re$ by

$$K(x, p, w, y) = p^T F(x, w, 0) - y^T Y(x, w) + \frac{1}{2}\|Z(x, 0)\|^2 - \frac{1}{2}\gamma^2\|w\|^2. \tag{6.58}$$

Then, K is concave with respect to w and convex with respect to y by construction. This implies the existence of a smooth maximizing function $\hat{w}(x, p, y)$ and a smooth minimizing function $y_\star(x, p)$ defined in a neighborhood of $(0, 0, 0)$ and $(0, 0)$ respectively, such that

$$\left(\frac{\partial K(x, p, w, y)}{\partial w}\right)_{w=\hat{w}(x,p,y)} = 0, \;\; \hat{w}(0,0,0) = 0, \tag{6.59}$$

$$\left(\frac{\partial^2 K(x, p, w, y)}{\partial w^2}\right)_{(x,p,w,y)=(0,0,0,0)} = -\gamma^2 I, \tag{6.60}$$

$$\left(\frac{\partial K(x, p, \hat{w}(x, p, y), y)}{\partial y}\right)_{y=y_\star(x,p)} = 0, \;\; y_\star(0,0) = 0, \tag{6.61}$$

$$\left(\frac{\partial^2 K(x, p, \hat{w}(x, p, y), y)}{\partial y^2}\right)_{(x,p,w,y)=(0,0,0,0)} = \frac{1}{\gamma^2}D_{21}D_{21}^T. \tag{6.62}$$

Finally, setting

$$w_{**}(x, p) = \hat{w}(x, p, y_*(x, p)),$$

we have the following main result.

Theorem 6.4.1 *Consider the system (6.52) and let the Assumptions 6.4.1, 6.4.2, 6.4.3 hold. Suppose there exists a smooth positive-definite solution $\tilde{V}$ to the HJI-inequality (5.75) and the inequality*

$$K(x, W_x^T(x), w_{**}(x, W_x^T(x)), y_*(x, W_x^T(x))) - \tilde{H}^\star(x, \tilde{V}_x^T(x)) < 0, \qquad (6.63)$$

where $\tilde{H}^\star(.,.)$ is given by (5.72), has a smooth positive-definite solution $W(x)$ defined in the neighborhood of $x = 0$ with $W(0) = 0$. Suppose also in addition that

(i) $W(x) - \tilde{V}(x) > 0 \; \forall x \neq 0$;

(ii) *the Hessian matrix of*

$$K(x, W_x^T(x), w_{**}(x, W_x^T(x)), y_*(x, W_x^T(x)))) - \tilde{H}_\star(x, \tilde{V}_x^T(x))$$

is nonsingular at $x = 0$; and

(iii) *the equation*

$$(W_x(x) - \tilde{V}_x(x))\tilde{G}(x) - y_*^T(r, W_x^T(x)) \qquad (6.64)$$

has a smooth solution $\tilde{G}(x)$.

Then the controller (6.54) solves the $MFBNLHICP$ for the system.

Proof: Let $Q(x) = W(x) - \tilde{V}(x)$ and define

$$\tilde{S}(x, w) \triangleq Q_x[F(x, w, 0) - G(x)Y(x, w)] + \tilde{H}(x, \tilde{V}_x^T, w, 0) - \tilde{H}^\star(x, V_x^T)$$

where $\tilde{H}(., ., ., .)$ is as defined in (5.69). Then,

$$
\begin{aligned}
\tilde{S}(x, w) &= W_x F(x, w, 0) - y_*^T(x, W_x^T)Y(x, w) - \tilde{V}_x F(x, w, 0) + \\
&\quad \tilde{H}(x, \tilde{V}_x^T, w, 0) - \tilde{H}^\star(x, \tilde{V}_x^T) \\
&= W_x F(x, w, 0) - y_*^T(x, W_x^T)Y(x, w) + \frac{1}{2}\|Z(x, 0)\| - \frac{1}{2}\gamma^2\|w\| - \tilde{H}^\star(x, \tilde{V}_x^T) \\
&= K(x, W_x^T, w, y_*(x, W_x^T)) - \tilde{H}^\star(x, \tilde{V}_x^T) \\
&\leq K(x, W_x^T, w_{**}(x, W_x^T), y_*(x, W_x^T)) - \tilde{H}^\star(x, \tilde{V}_x^T) \\
&= x^T \Upsilon(x)x
\end{aligned}
$$

where $\Upsilon(.)$ is some smooth matrix function which is negative-definite at $x = 0$.
    Now, let

$$\tilde{U}(x^e) = Q(x - \zeta) + \tilde{V}(x).$$

Then, by construction, $\tilde{U}$ is such that the conditions (a), (b) of Remark 6.4.1 are satisfied. Moreover, if $\tilde{G}$ is chosen as in (6.64), then

$$
\begin{aligned}
\tilde{U}_{x^e} F(x^e, w) + \frac{1}{2}\|Z^e(x^e)\|^2 - \frac{1}{2}\|w\|^2 \;=\; & Q_x(x - \zeta)[F(x, w, \tilde{\alpha}_2(\zeta)) - \\
& F(\zeta, \tilde{\alpha}_1(\zeta), \tilde{\alpha}_2(\zeta)) - \tilde{G}(\zeta)Y(x, w) + \\
& \tilde{G}(\zeta)Y(\zeta, \tilde{\alpha}_1(\zeta))] + \tilde{V}_x F(x, w, \tilde{\alpha}_2(\zeta)) + \\
& \frac{1}{2}\|Z(x, \tilde{\alpha}_2(\zeta))\|^2 - \frac{1}{2}\gamma^2\|w\|^2 \\
\leq\; & Q_x(x - \zeta)[F(x, w, \tilde{\alpha}_2(\zeta)) - \\
& F(\zeta, \tilde{\alpha}_1(\zeta), \tilde{\alpha}_2(\zeta)) - \tilde{G}(\zeta)Y(x, w) + \\
& \tilde{G}(\zeta)Y(\zeta, \tilde{\alpha}_1(\zeta))] + \tilde{H}(x, \tilde{V}_x^T, w, \tilde{\alpha}_2(\zeta)) - \\
& \tilde{H}^\star(x, V_x^T).
\end{aligned}
$$

If we denote the right-hand-side (RHS) of the above inequality by $L(x, \zeta, w)$, and observe that this function is concave with respect to w, then there exists a unique maximizing function $\tilde{w}(x, \zeta)$ such that

$$
\left(\frac{\partial L(x, \zeta, w)}{\partial w}\right)_{w = \tilde{w}(x, \zeta)} = 0, \quad \tilde{w}(0, 0) = 0,
$$

for all (x, ζ, w) in the neighborhood of $(0, 0, 0)$. Moreover, in this neighbohood, it can be shown (it involves some lengthy calculations) that $L(x, \zeta, \tilde{w}(x, \zeta))$ can be represented as

$$
L(x, \zeta, \tilde{w}(x, \zeta)) = (x - \zeta)^T R(x, \zeta)(x - \zeta)
$$

for some smooth matrix function $R(., .)$ such that $R(0, 0) = \Upsilon(0)$. Hence, $R(., .)$ is locally negative-definite about $(0, 0)$, and thus the dissipation-inequality

$$
\tilde{U}_{x^e} F(x^e, w) + \frac{1}{2}\|Z^e(x^e)\|^2 - \frac{1}{2}\|w\|^2 \leq L(x, \zeta, \tilde{w}(x, \zeta)) \leq 0
$$

is satisfied locally. This guarantees that condition (a) of the remark holds.

Finally, setting $w = \tilde{\alpha}_1(x)$ in the expression for $\tilde{S}(x, w)$, we get

$$
0 > S(x, \tilde{\alpha}_1(x)) \geq Q_x(F(x, \alpha_1(x), 0) - G(x)Y(x, \alpha_1(x))),
$$

which implies that $Q(x)$ is a Lyapunov-function for (6.56), and consequently the condition (b) also holds. $\square$

Remark 6.4.2 *Notice that solving the $MFBNLHICP$ involves the solution of two HJI-inequalities (5.75), (6.63) together with a coupling condition (6.64). This agrees well with the linear theory [92, 101, 292].*

6.4.1   Controller Parametrization

In this subsection, we consider the controller parametrization problem for the general class of nonlinear systems represented by the model Σ^g. For this purpose, we introduce the following additional assumptions.

Assumption 6.4.4 *The matrix $D_{11}^T D_{11} - \gamma^2 I$ is negative-definite for some $\gamma > 0$, where $D_{11} = \frac{\partial Z}{\partial w}(0, 0)$.*

Define similarly the following

$$\tilde{H}(x, p, w, u) \;=\; p^T F(x, w, u) + \frac{1}{2}\|Z(x, w, u)\|^2 - \frac{1}{2}\gamma^2\|w\|^2 \tag{6.65}$$

$$r_{11}(x) \;=\; \left(\frac{\partial^2 \tilde{H}(x, \tilde{V}_x(x), w, u)}{\partial w^2}\right)_{w=\tilde{\alpha}_1(x), u=\tilde{\alpha}_2(x)} \tag{6.66}$$

$$r_{12}(x) \;=\; \left(\frac{\partial^2 \tilde{H}(x, \tilde{V}_x(x), w, u)}{\partial u \partial w}\right)_{w=\tilde{\alpha}_1(x), u=\tilde{\alpha}_2(x)} \tag{6.67}$$

$$r_{21}(x) \;=\; \left(\frac{\partial^2 \tilde{H}(x, \tilde{V}_x(x), w, u)}{\partial w \partial u}\right)_{w=\tilde{\alpha}_1(x), u=\tilde{\alpha}_2(x)} \tag{6.68}$$

$$r_{22}(x) \;=\; \left(\frac{\partial^2 \tilde{H}(x, \tilde{V}_x(x), w, u)}{\partial u^2}\right)_{w=\tilde{\alpha}_1(x), u=\tilde{\alpha}_2(x)}, \tag{6.69}$$

where $\tilde{V}$ solves the HJI-inequality (5.75), while $\tilde{\alpha}_1$ and $\tilde{\alpha}_2$ are the corresponding worst-case disturbance and optimal feedback control as defined by equations (5.73), (5.74) in Section 5.6. Set also

$$\tilde{R}(x) = \begin{bmatrix} (1-\epsilon_1)r_{11}(x) & r_{12}(x) \\ r_{21}(x) & (1+\epsilon_2)r_{22}(x) \end{bmatrix}$$

where $0 < \epsilon_1 < 1$ and $\epsilon_2 > 0$, and define also $\tilde{K} : T^\star \mathcal{X} \times \mathcal{L}_2(\Re_+) \times \Re^m \to \Re$ by

$$\tilde{K}(x, p, w, y) \;=\; p^T F(x, w + \alpha_1(x), 0) - y^T Y(x, w + \alpha_1(x)) +$$
$$\frac{1}{2}\begin{bmatrix} w \\ -\tilde{\alpha}_2(x) \end{bmatrix}^T \tilde{R}(x) \begin{bmatrix} w \\ -\tilde{\alpha}_2(x) \end{bmatrix}.$$

Further, let the functions $w_1(x, p, y)$ and $y_1(x, p)$ defined in the neighborhood of $(0,0,0)$ and $(0,0)$ respectively, be such that

$$\left(\frac{\partial \tilde{K}(x, p, w, y)}{\partial w}\right)_{w=w_1} = 0, \quad w_1(0,0,0) = 0, \tag{6.70}$$

$$\left(\frac{\partial \tilde{K}(x, p, w_1(x, p, y), y)}{\partial w}\right)_{y=y_1} = 0, \quad y_1(0,0) = 0. \tag{6.71}$$

Then we make the following assumption.

Assumption 6.4.5 *There exists a smooth positive-definite function $\tilde{Q}(x)$ locally defined in a neighborhood of $x = 0$ such that the inequality*

$$Y_2(x) = \tilde{K}(x, \tilde{Q}_x^T(x), w_1(x, \tilde{Q}_x^T(x)), y_1(x, \tilde{Q}_x^T(x)), y_1(x, \tilde{Q}_x^T(x))) < 0 \tag{6.72}$$

is satisfied, and the Hessian matrix of the LHS of the inequality is nonsingular at $x = 0$.

Now, consider the following family of controllers defined by

$$\Sigma_q^c : \begin{cases} \dot{\xi} \;=\; F(\xi, \tilde{\alpha}_1(\xi), \tilde{\alpha}_2(\xi) + c(q)) + G(\xi)(y - Y(\xi, \tilde{\alpha}_1(\xi))) + \hat{g}_1(\xi)c(q) + \\ \qquad \hat{g}_2(\xi)d(q) \\ \dot{q} \;=\; a(q, y - Y(\xi, \tilde{\alpha}_1(\xi))) \\ u \;=\; \tilde{\alpha}_2(\xi) + c(q), \end{cases} \tag{6.73}$$

where $\xi \in \mathcal{X}$ and $q \in \Re^v \subset \mathcal{X}$ are defined in the neighborhoods of the origin in $\mathcal{X}$ and $\Re^v$ respectively, while $G(.)$ satisfies the equation

$$\tilde{Q}_x(x)G(x) = y_1(x, \tilde{Q}_x^T(x)) \tag{6.74}$$

with $\tilde{Q}(.)$ satisfying Assumption 6.4.5. The functions $a(.,.)$, $c(.)$ are smooth, with $a(0,0) = 0$, $c(0) = 0$, while $\hat{g}_1(.)$, $\hat{g}_2(.)$ and $d(.)$ are $C^k, k \geq 1$ of compatible dimensions.

Define also the following Hamiltonian function $J : \Re^{2n+v} \times \Re^{2n+v} \times \Re^r \to \Re$ by

$$
\begin{aligned}
J(x_a, p_a, w) \;=\;& p_a^T F_a(x_a, w) + \\
& \frac{1}{2} \left[\begin{array}{c} w - \tilde{\alpha}_1(x) \\ \alpha_2(\xi) + c(q) - \tilde{\alpha}_2(x) \end{array} \right]^T \widetilde{R}(x) \left[\begin{array}{c} w - \tilde{\alpha}_1(x) \\ \tilde{\alpha}_2(\xi) + c(q) - \tilde{\alpha}_2(x) \end{array} \right],
\end{aligned}
$$

where

$$
x_a = \left[\begin{array}{c} x \\ \xi \\ q \end{array} \right], \quad F_a(x_a, w) = \left[\begin{array}{c} F(x, w, \tilde{\alpha}_2(\xi) + c(q)) \\ \tilde{F}(\xi, q) + G(\xi)Y(x, w) + \hat{g}_1(\xi)c(q) + \hat{g}_2(\xi)d(q) \\ a(q, Y(x, w) - Y(\xi, \tilde{\alpha}_1)) \end{array} \right]
$$

and

$$\tilde{F}(\xi, q) = F(\xi, \tilde{\alpha}_1, \tilde{\alpha}_2(\xi) + c(q)) - G(\xi)Y(\xi, \tilde{\alpha}_1(\xi), \tilde{\alpha}_2(\xi) + c(q)).$$

Note also that, since

$$\left(\frac{\partial^2 J(x_a, p_a, w)}{\partial w^2} \right)_{(x_a, p_a, w) = (0,0,0)} = (1 - \epsilon_1)(D_{11}^T D_{11} - \gamma^2 I),$$

there exists a unique smooth solution $w_2(x_a, p_a)$ defined on a neighborhood of $(x_a, p_a) = (0,0)$ satisfying

$$\left(\frac{\partial J(x_a, p_a, w)}{\partial w} \right)_{w = w_2(x_a, p_a)} = 0, \quad w_2(0,0) = 0.$$

The following proposition then gives the parametrization of the set of output measurement-feedback controllers for the system Σ^g.

Proposition 6.4.2 *Consider the system Σ^g and suppose the Assumptions 6.4.1-6.4.5 hold. Suppose the following also hold:*

(H1) *there exists a smooth solution $\tilde{V}$ to the HJI-inequality (5.75), i.e.,*

$$Y_1(x) := \tilde{H}(x, \tilde{V}_x^T(x), \tilde{\alpha}_1(x), \tilde{\alpha}_2(x)) \leq 0$$

for all x about $x = 0$;

(H2) *there exists a smooth real-valued function $M(x_a)$ locally defined in the neighborhood of the origin in $\Re^{2n+v}$ which vanishes at $x_a = col(x, x, 0)$ and is positive everywhere, and is such that*

$$Y_3(x_a) := J(x_a, M_{x_a}^T(x_a), w_2(x_a, M_{x_a}^T(x_a))) < 0$$

and vanishes at $x_a = (x, x, 0)$.

Then the family of controllers Σ_q^g solves the $MFBNLHICP$ for the system Σ^g.

Proof: Consider the Lyapunov-function candidate

$$\Psi_2(x_a) = \tilde{V}(x) + M(x_a)$$

which is positive-definite by construction. Along the trajectories of the closed-loop system (6.1), (6.73):

$$\Sigma_q^{gc} : \begin{cases} \dot{x}_a &= F_a(x_a, w) \\ z &= Z(x, \tilde{\alpha}_2(\xi) + c(q)), \end{cases} \tag{6.75}$$

and by employing Taylor-series approximation of $J(x_a, M_{x_a}^T, w_2(x_a, M_{x_a}^T(x_a))$ about $w = w_2(.,.)$, we have

$$\frac{d\Psi_2}{dt} + \frac{1}{2}\|Z(x, \tilde{\alpha}_2(\xi) + c(q))\|^2 - \frac{1}{2}\gamma^2\|w\|^2 = Y_1(x) + Y_3(x_a) +$$

$$\frac{1}{2}\begin{bmatrix} w - \tilde{\alpha}_1(x) \\ \tilde{\alpha}_2(\xi) + c(q) - \tilde{\alpha}_2(x) \end{bmatrix}^T \begin{bmatrix} \epsilon_1 r_{11}(x) & 0 \\ 0 & -\epsilon_2 r_{22}(x) \end{bmatrix} \begin{bmatrix} w - \tilde{\alpha}_1(x) \\ \tilde{\alpha}_2(\xi) + c(q) - \tilde{\alpha}_2(x) \end{bmatrix} +$$

$$\frac{1}{2}\|w - w_2(x_a, M_{x_a}^T(x_a))\|^2_{\Gamma(x_a)} + o\left(\left\|\begin{matrix} w - \tilde{\alpha}_1(x) \\ \tilde{\alpha}_2(\xi) + c(q) \quad \tilde{\alpha}_2(x) \end{matrix}\right\|^3\right) +$$

$$o(\|w - w_2(x_a, M_{x_a}^T(x_a))\|^3), \tag{6.76}$$

where

$$\Gamma(x_a) = \left(\frac{\partial^2 J(x_a, M_{x_a}^T(x_a), w)}{\partial w^2}\right)_{w=w_2(x_a, M_{x_a}^T(x_a))}$$

and $\Gamma(0) = (1 - \epsilon_1)(D_{11}^T D_{11} - \gamma^2 I)$. Further, setting $w = 0$ in the above equation, we get

$$\frac{d\Psi_2}{dt} = -\frac{1}{2}\|Z(x, \tilde{\alpha}_2(\xi) + c(q))\|^2 + Y_1(x) + Y_3(x_a) +$$

$$\frac{1}{2}\begin{bmatrix} -\tilde{\alpha}_1(x) \\ \tilde{\alpha}_2(\xi) + c(q) - \alpha_2(x) \end{bmatrix}^T \begin{bmatrix} \epsilon_1 r_{11}(x) & 0 \\ 0 & -\epsilon_2 r_{22}(x) \end{bmatrix} \begin{bmatrix} -\tilde{\alpha}_1(x) \\ \tilde{\alpha}_2(\xi) + c(q) - \tilde{\alpha}_2(x) \end{bmatrix} +$$

$$\frac{1}{2}\|w_2(x_a, M_{x_a}^T(x_a))\|^2_{\Gamma(x_a)} + o\left(\left\|\begin{matrix} -\tilde{\alpha}_1(x) \\ \tilde{\alpha}_2(\xi) + c(q) - \tilde{\alpha}_2(x) \end{matrix}\right\|^3\right) +$$

$$o(\|w_2(x_a, M_{x_a}^T(x_a))\|^3)$$

which is negative-semidefinite near $x_a = 0$ by hypothesis and Assumption 6.4.1 as well as the fact that $r_{11}(x) < 0$, $r_{22}(x) > 0$ about $x_a = 0$. This proves that the equilibrium-point $x_a = 0$ of (6.75) is stable.

To conclude asymptotic-stability, observe that any trajectory $(x(t), \xi(t), q(t))$ satisfying

$$\frac{d\Psi_2}{dt}(x(t), \xi(t), q(t)) = 0 \ \forall t \geq t_s$$

(say!), is necessarily a trajectory of

$$\dot{x}(t) = F(x, 0, \tilde{\alpha}_2(\xi) + c(q)),$$

and is such that $x(t)$ is bounded and $Z(x, 0, \tilde{\alpha}_2(\xi) + c(q)) = 0$ for all $t \geq t_s$. Furthermore, by hypotheses H1 and H2 and Assumption 6.4.1, $\dot{\Psi}_2(x(t), \xi(t), q(t)) = 0$ for all $t \geq t_s$, implies $x(t) = \xi(t)$ and $q(t) = 0$ for all $t \geq t_s$. Consequently, by Assumption 6.4.2 we have $\lim_{t\to\infty} x(t) = 0$, $\lim_{t\to\infty} \xi(t) = 0$, and by LaSalle's invariance-principle, we conclude asymptotic-stability. Finally, integrating the expression (6.76) from $t = t_0$ to $t = t_1$, starting at $x(t_0), \xi(t_0), q(t_0)$, and noting that $Y_1(x)$ and $Y(x_a)$ are nonpositive, it can be shown that the closed-loop system has locally $\mathcal{L}_2$-gain $\leq \gamma$ or the disturbance-attenuation property. $\square$

The last step in the parametrization is to show how the functions $\hat{g}_1(.)$, $\hat{g}_2(.)$, $d(.)$ can be selected so that condition $H2$ in the above proposition can be satisfied. This is summarized in the following theorem.

Theorem 6.4.2 *Consider the nonlinear system Σ^g and the family of controllers (6.73). Suppose the Assumptions 6.4.1-6.4.5 and hypothesis $H1$ of Proposition 6.4.2 holds. Suppose the following also holds:*

(H3) *there exists a smooth positive-definite function $L(q)$ locally defined on a neighborhood of $q = 0$ such that the function*

$$Y_4(q, w) = L_q(q)a(q, Y(0, w)) + \frac{1}{2} \begin{bmatrix} w \\ c(q) \end{bmatrix}^T \tilde{R}(0) \begin{bmatrix} w \\ c(q) \end{bmatrix}$$

is negative-definite at $w = w_3(q)$ and its Hessian matrix is nonsingular at $q = 0$, where $w_3(.)$ is defined on a neighborhood of $q = 0$ and is such that

$$\left(\frac{\partial Y_4(q, w)}{\partial w} \right)_{w=w_3(q)} = 0, \quad w_3(0) = 0.$$

Then, if $\hat{g}_1(.)$, $\hat{g}_2(.)$ are selected such that

$$\tilde{Q}_x(x)\hat{g}_1(x) = \tilde{Q}_x(x)\left(\frac{\partial F}{\partial u}(x, 0, 0) - G(x)\frac{\partial Y}{\partial u}(x, 0, 0) \right) + \beta^T(x, 0, 0)r_{12}(x) - $$
$$(1 + \epsilon_2)\tilde{\alpha}_2^T(x)r_{22}(x)$$

and

$$\tilde{Q}_x(x)\hat{g}_2(x) = -Y^T(x, \tilde{\alpha}_1(x) + \beta(x, 0, 0)),$$

respectively, where

$$\beta(x, \xi, q) = w_2(x_a, [\tilde{Q}_x(x - \xi) - \tilde{Q}_x(x - \xi) L_q(q)])$$

and let $d(q) = L_q^T(q)$. Then, each of the family of controllers (6.73) solves the MFBNLHICP for the system (6.52).

Proof: It can easily be shown by direct substitution that the function $M(x_a) = \tilde{Q}(x - \xi) + L(q)$ satisfies the hypothesis $(H2)$ of Proposition 6.4.2. $\square$

6.5   Static Output-Feedback Control for Affine Nonlinear Systems

In this section, we consider the static output-feedback (SOFB) stabilization problem (SOFBP) with disturbance attenuation for affine nonlinear systems. This problem has been extensively considered by many authors for the linear case (see [113, 121, 171] and the references contained therein). However, the nonlinear problem has received very little attention so far [45, 278]. In [45] some sufficient conditions are given in terms of the solution to a Hamilton-Jacobi equation (HJE) with some side conditions, which when specialized to linear systems, reduce to the necessary and sufficient conditions given in [171].

In this section, we present new sufficient conditions for the solvability of the problem in the general affine nonlinear case which we refer to as a "factorization approach" [27]. The sufficient conditions are relatively easy to satisfy, and depend on finding a factorization of the state-feedback solution to yield the output-feedback solution.

We begin with the following smooth model of an affine nonlinear state-space system defined on an open subset $\mathcal{X} \subset \Re^n$ without disturbances:

$$\Sigma^a : \begin{cases} \dot{x} & = & f(x) + g(x)u; \ x(t_0) = x_0 \\ y & = & h(x) \end{cases} \tag{6.77}$$

where $x \in \mathcal{X}$ is the state vector, $u \in \Re^p$ is the control input, and $y \in \Re^m$ is the output of the system. The functions $f : \mathcal{X} \to V^\infty(\mathcal{X})$, and $g : \mathcal{X} \to \mathcal{M}^{n \times p}$, $h : \mathcal{X} \to \Re^m$ are smooth C^∞ functions of x. We also assume that $x = 0$ is an equilibrium point of the system (6.77) such that $f(0) = 0$, $h(0) = 0$.

The problem is to find a SOFB controller of the form

$$u = K(y) \tag{6.78}$$

for the system, such that the closed-loop system (6.1)-(6.78) is locally asymptotically-stable at $x = 0$.

For the nonlinear system (6.77), it is well known that [112, 175], if there exists a C^1 positive-definite local solution $V : \Re^n \to \Re_+$ to the following Hamilton-Jacobi-Bellman equation (HJBE):

$$V_x(x)f(x) - \frac{1}{2}V_x(x)g(x)g^T(x)V_x^T(x) + \frac{1}{2}h^T(x)h(x) = 0, \ \ V(0) = 0 \tag{6.79}$$

(or the inequality with " $\leq$"), then the optimal control law

$$u^* = -g^T(x)V_x^T(x) \tag{6.80}$$

locally asymptotically stabilizes the equilibrium point $x = 0$ of the system and minimizes the cost functional:

$$J_1(u) = \frac{1}{2}\int_{t_0}^\infty (\|y\|^2 + \|u\|^2)dt. \tag{6.81}$$

Moreover, if $V > 0$ is proper (i.e., the set $\Omega_c = \{x | 0 \leq V(x) \leq a\}$ is compact for all $a > 0$), then the above control law (6.80) globally asymptotically stabilizes the equilibrium point $x = 0$. The aim is to replace the state vector in the above control law by the output vector or some function of it, with some possible modifications to the control law. In this regard, consider the system (6.77) and suppose there exists a C^1 positive-definite local solution to the HJE (6.79) and a C^0 function $F : \Re^m \to \Re^p$ such that:

$$F \circ h(x) = -g^T(x)V_x^T(x), \tag{6.82}$$

where "$\circ$" denotes the composition of functions. Then the control law

$$u = F(y) \tag{6.83}$$

solves the SOFBP locally. It is clear that if the function F exists, then

$$u = F(y) = F \circ h(x) = -g^T(x)V_x^T(x).$$

This simple observation leads to the following result.

Proposition 6.5.1 *Consider the nonlinear system (6.77) and the SOFBP. Assume the system is locally zero-state detectable and the state-feedback problem is locally solvable with* *a controller of the form (6.80). Suppose there exist C^0 functions $F_1 : \mathcal{Y} \subset \Re^m \rightarrow \mathcal{U}$, $\phi_1 : X_0 \subset \mathcal{X} \rightarrow \Re^p$, $\eta_1 : X_0 \subset \mathcal{X} \rightarrow \Re_-$ (non-positive), $\eta_1 \in \mathcal{L}[t_0, \infty)$ such that*

$$F_1 \circ h(x) = -g^T(x)V_x^T(x) + \phi_1(x) \tag{6.84}$$

$$V_x(x)g(x)\phi_1(x) = \eta_1(x). \tag{6.85}$$

Then:

(i) *the SOFBP is locally solvable with a feedback of the form*

$$u = F_1(y); \tag{6.86}$$

(ii) *if the optimal cost of the state-feedback control law (6.80) is $J^*_{SFB}(u)$, then the cost $J_{SOFB}(u)$ of the SOFB control law (6.86) is given by*

$$J_{1,SOFB}(u) = J^*_{1,SFB}(u^\star) + \int_{t_0}^{\infty} \eta_1(x)dt.$$

Proof: Consider the closed-loop system (6.77), (6.86):

$$\dot{x} = f(x) + g(x)F_1(h(x)).$$

Differentiating the solution $V > 0$ of (6.79) along the trajectories of this system, we get upon using (6.84), (6.85) and (6.79):

$$
\begin{aligned}
\dot{V} &= V_x(x)(f(x) + g(x)F_1(h(x))) = V_x(x)(f(x) - g(x)g^T(x)V_x^T(x) + g(x)\phi_1(x)) \\
&= V_x(x)f(x) - V_x g(x)g^T(x)V_x^T(x) + V_x(x)g(x)\phi_1(x) \\
&= -\frac{1}{2}V_x(x)g(x)g^T(x)V^T(x) - \frac{1}{2}h^T(x)h(x) + \eta_1(x) \\
&\leq 0. \tag{6.87}
\end{aligned}
$$

This shows that the equilibrium point $x = 0$ of the closed-loop system is Lyapunov-stable. Moreover, by local zero-state detectability of the system, we can conclude local asymptotic-stability. This establishes (i).

To prove (ii), integrate the last expression in (6.87) with respect to t from $t = t_0$ to ∞, and noting that by asymptotic-stability of the closed-loop system, $V(x(\infty)) = 0$, this yields

$$J_1(u^\star) := \frac{1}{2}\int_{t_0}^{\infty}(\|y\|^2 + \|u^\star\|^2)dt = V(x_0) + \int_{t_0}^{\infty}\eta_1(x)dt$$

Since $V(x_0)$ is the total cost of the policy, the result now follows from the fundamental theorem of calculus. $\square$

Remark 6.5.1 *Notice that if Σ^a is globally detectable, $V > 0$ is proper, and the functions $\phi_1(.)$ and $\eta_1(.)$ are globally defined, then the control law (6.86) solves the SOFBP globally.*

A number of interesting corollaries can be drawn from Proposition 6.5.1. In particular, the condition (6.85) in the proposition can be replaced by a less stringent one given in the following corollary.

Corollary 6.5.1 *Take the same assumptions as in Proposition 6.5.1. Then, condition (6.85) can be replaced by the following condition*

$$\phi_1^T(x)F_1 \circ h(x) \geq 0 \tag{6.88}$$

for all $x \in X_1$ a neighborhood of $x = 0$.

Proof: Multiplying equation (6.84) by $\phi_1^T(x)$ from the left and substituting equation (6.85) in it, the result follows by the non-positivity of η_1. $\square$

Example 6.5.1 *We consider the following scalar nonlinear system*

$$\dot{x} = -x^3 + \frac{1}{2}x + u$$
$$y = \sqrt{2}x.$$

Then the HJBE (6.79) corresponding to this system is

$$(-x^3 + \frac{1}{2}x)V_x(x) - \frac{1}{2}V_x^2(x) + x^2 = 0,$$

and it can be checked that the function $V(x) = x^2$ solves the HJE. Moreover, the state-feedback problem is globally solved by the control law $u = -2x$. Thus, clearly, $u = -\sqrt{2}y$, and hence the function $F_1(y) = -y$ solves the SOFBP globally.

In the next section we consider systems with disturbances.

6.5.1   Static Output-Feedback Control with Disturbance-Attenuation

In this section, we extend the simple procedure discussed above to systems with disturbances. For this purpose, consider the system (6.77) with disturbances [113]:

$$\Sigma_2^a : \begin{cases} \dot{x} &= f(x) + g_1(x)w + g_2(x)u \\ y &= h_2(x) \\ z &= \begin{bmatrix} h_1(x) \\ u \end{bmatrix} \end{cases} \tag{6.89}$$

where all the variables are as defined in the previous sections and $w \in \mathcal{W} \subset \mathcal{L}_2[t_0, \infty)$. We recall from Chapter 5 that the state-feedback control (6.80) where $V : N \to \Re_+$ is the smooth positive-definite solution of the Hamilton-Jacobi-Isaacs equation (HJIE):

$$V_x(x)f(x) + \frac{1}{2}V_x(x)[\frac{1}{\gamma^2}g_1(x)g_1^T(x) - g_2(x)g_2^T(x)]V_x^T(x) + \frac{1}{2}h_1^T(x)h_1(x) = 0, \quad V(0) = 0, \tag{6.90}$$

minimizes the cost functional:

$$J_2(u, w) = \frac{1}{2}\int_{t_0}^{\infty}(\|z(t)\|^2 - \gamma^2\|w(t)\|^2)dt$$

and renders the closed-loop system (6.89), (6.80) dissipative with respect to the supply-rate $s(w, z) = \frac{1}{2}(\gamma^2\|w\|^2 - \|z\|^2)$ and hence achieves $\mathcal{L}_2$-gain $\leq \gamma$. Thus, our present problem is to replace the states in the feedback (6.78) with the output y or some function of it. More formally, we define the SOFBP with disturbance attenuation as follows.

Definition 6.5.1 *(Static Output-Feedback Problem with Disturbance Attenuation (SOF-BPDA)). Find (if possible!) a SOFB control law of the form (6.78) such that the closed-loop system (6.89), (6.78) has locally $\mathcal{L}_2$-gain from w to z less or equal $\gamma > 0$, and is locally asymptotically-stable with $w = 0$.*

The following theorem gives sufficient conditions for the solvability of the problem.

Theorem 6.5.1 *Consider the nonlinear system (6.89) and the SOFBPDA. Assume the system is zero-state detectable and the state-feedback problem is locally solvable in $N \subset \mathcal{X}$ with a controller of the form (6.80) and $V > 0$ a smooth solution of the HJIE (6.90). Suppose there exist C^0 functions $F_2 : \mathcal{Y} \subset \Re^m \to \mathcal{U}$, $\phi_2 : X_1 \subset \mathcal{X} \to \Re^p$, $\eta_2 : X_1 \subset \mathcal{X} \to \Re_-$ (non-positive), $\eta_2 \in \mathcal{L}[t_0, \infty)$ such that the conditions*

$$F_2 \circ h_2(x) = -g_2^T(x)V_x^T(x) + \phi_2(x), \tag{6.91}$$

$$V_x(x)g_2(x)\phi_2(x) = \eta_2(x), \tag{6.92}$$

are satisfied. Then:

(i) *the SOFBP is locally solvable with the feedback*

$$u = F_2(y); \tag{6.93}$$

(ii) *if the optimal cost of the state-feedback control law (6.80) is $J_{2,SFB}^*(u^\star, w^\star) = V(x_0)$, then the cost $J_{2,SOFB}(u)$ of the SOFBPDA control law (6.93) is given by*

$$J_{2,SOFB}(u) = J_{2,SFB}^*(u) + \int_{t_0}^{\infty} \eta_2(x)dt.$$

Proof: (i) Differentiating the solution $V > 0$ of (6.90) along a trajectory of the closed-loop system (6.89) with the control law (6.93), we get upon using (6.91), (6.92) and the HJIE (6.90)

$$
\begin{aligned}
\dot{V} &= V_x(x)[f(x) + g_1(x)w + g_2(x)F_2(h_2(x))] \\
&= V_x(x)[f(x) + g_1(x)w - g_2(x)g_2^T(x)V_x^T(x) + g_2(x)\phi_2(x)] \\
&= V_x(x)f(x) + V_x(x)g_1(x)w - V_x(x)g_2(x)g_2^T(x)V^T(x) + V_x(x)g_2(x)\phi_2(x) \\
&\leq -\frac{1}{2}V_x(x)g_2(x)g_2^T(x)V^T(x) - \frac{1}{2}h_1^T(x)h_1(x) + \frac{1}{2}\gamma^2\|w\|^2 - \frac{\gamma^2}{2}\left\|w - \frac{g_1^T(x)V_x^T(x)}{\gamma^2}\right\|^2 \\
&\leq \frac{1}{2}(\gamma^2\|w\|^2 - \|z\|^2). \tag{6.94}
\end{aligned}
$$

Integrating now from $t = t_0$ to $t = T$ we have the dissipation inequality

$$V(x(T)) - V(x_0) \leq \int_{t_0}^{T} \frac{1}{2}(\gamma^2\|w\|^2 - \|z\|^2)dt,$$

and therefore the system has $\mathcal{L}_2$-gain $\leq \gamma$. In addition, with $w \equiv 0$, we get

$$\dot{V} \leq -\frac{1}{2}\|z\|^2,$$

and therefore, the closed-loop system is locally stable. Moreover, the condition $\dot{V} \equiv 0$ for all $t \geq t_c$, for some t_c, implies that $z \equiv 0$ for all $t \geq t_c$, and consequently, $\lim_{t\to\infty} x(t) = 0$ by zero-state detectability. Finally, by the LaSalle's invariance-principle, we conclude asymptotic-stability. This establishes (i).

(ii) Using similar manipulations as in (i) above, it can be shown that

$$\dot{V} = \frac{1}{2}(\gamma^2\|w\|^2 - \|z\|^2) - \frac{\gamma^2}{2}\left\|w - \frac{g_1^T(x)V_x^T(x)}{\gamma^2}\right\|^2 + \eta_2(x)$$

integrating the above expression from $t = t_0$ to ∞, substituting $w = w^\star$ and rearranging, we get

$$J_2^\star(u^\star, w^\star) = V(x_0) + \int_{t_0}^\infty \eta_2(x)dt. \quad \square$$

We consider another example.

Example 6.5.2 *For the second order system*

$$\begin{aligned}
\dot{x}_1 &= x_1 x_2 \\
\dot{x}_2 &= -x_1^2 + x_2 + u + w \\
y &= [x_1 \ \ x_1 + x_2]^T \\
z &= [x_2 \ \ u]^T,
\end{aligned}$$

the HJIE (6.90) corresponding to the above system is given by

$$x_1 x_2 V_{x_1} + (-x_1^2 + x_2)V_{x_2} + \frac{(1-\gamma^2)}{\gamma^2}V_{x_2}^2 + \frac{1}{2}x_2^2 = 0.$$

Then, it can be checked that for $\gamma = \sqrt{\frac{2}{5}}$, the function $V(x) = \frac{1}{2}(x_1^2 + x_2^2)$, solves the HJIE. Consequently, the control law $u = -x_2$ stabilizes the system. Moreover, the function $F_2(y) = y_1 - y_2$ clearly solves the equation (6.91) with $\phi_2(x) = 0$. Thus, the control law $u = y_1 - y_2$ locally stabilizes the system.

We can specialize the result of Theorem 6.5.1 above to the linear system:

$$\Sigma^l : \begin{cases} \dot{x} &= Ax + B_1 w + B_2 u, \quad x(t_0) = 0 \\ y &= C_2 x \\ z &= \begin{bmatrix} C_1 x \\ u \end{bmatrix} \end{cases} \tag{6.95}$$

where $A \in \Re^{n\times n}$, $B_1 \Re^{n\times r}$, $B_2 \in \Re^{n\times p}$, $C_2 \in \Re^{s\times n}$, and $C_1 \in \Re^{(s-p)\times n}$ are constant matrices. We then have the following corollary.

Corollary 6.5.2 *Consider the linear system (6.95) and the SOFBPDA. Assume (A, B_2) is stabilizable, (C_1, A) is detectable and the state-feedback problem is solvable with a controller of the form $u = -B_2^T Px$, where $P > 0$ is the stabilizing solution of the algebraic-Riccati equation (ARE):*

$$A^T P + PA + P[\frac{1}{\gamma^2}B_1 B_1^T - B_2 B_2^T]P + C_1^T C_1 = 0.$$

In addition, suppose there exist constant matrices $\Gamma_2 \in \Re^{p\times m}$, $\Phi_2 \in \Re^{p\times n}$, $0 \geq \Lambda_2 \in \Re^{n\times n}$ such that the conditions

$$\begin{aligned}
\Gamma_2 C_2 &= -B_2^T P + \Phi_2, & (6.96) \\
PB_2 \Phi_2 &= \Lambda_2, & (6.97)
\end{aligned}$$

are satisfied. Then:

(i) *the SOFBP is solvable with the feedback*

$$u = \Gamma_2 y; \tag{6.98}$$

(ii) *if the optimal cost of the state-feedback control law (6.80) is $J^*_{2,SFBl}(u^\star, w^\star) = \frac{1}{2}x_0^T P x_0$, then the cost $J_{SOFB_2}(u, w)$ of the SOFBDA control law (6.93) is given by*

$$J^*_{2,SOFBl}(u^\star, w^\star) = J^*_{2,SFBl}(u^\star, w^\star) - x_0^T \Lambda_2 x_0.$$

In closing, we give a suggestive adhoc approach for solving for the functions F_1, F_2, for of the SOFB synthesis procedures outlined above. Let us represent anyone of the equations (6.84), (6.91) by

$$F \circ h_2(x) = \alpha(x) + \phi(x) \tag{6.99}$$

where $\alpha(.)$ represents the state-feedback control law in each case. Then if we assume for the time-being that ϕ is known, then F can be solved from the above equation as

$$F = \alpha \circ h_2^{-1} + \phi \circ h_2^{-1} \tag{6.100}$$

provided h_2^{-1} exists at least locally. This is the case if h_2 is injective, or h_2 satisfies the conditions of the inverse-function theorem [230]. If neither of these conditions is satisfied, then multiple or no solution may occur.

Example 6.5.3 *Consider the nonlinear system*

$$\begin{aligned}
\dot{x}_1 &= -x_1^3 + x_2 - x_1 + w \\
\dot{x}_2 &= -x_1 - x_2 + u \\
y &= x_2.
\end{aligned}$$

Then, it can be checked that the function $V(x) = \frac{1}{2}(x_1^2 + x_2^2)$ solves the HJIE (6.90) corresponding to the state-feedback problem for the system. Moreover,

$$\alpha = -g_2^T(x)V_x^T(x) = -x_2,$$

$$h_2^{-1}(x) = h_2(x) = x_2.$$

Therefore, with $\phi = 0$, $F(x) = \alpha \circ h_2^{-1}(x) = -x_2$ or $u = F(y) = -y$.

We also remark that the examples presented are really simple, because it is otherwise difficult to find a closed-form solution to the HJIE. It would be necessary to develop a computational procedure for more difficult problems.

Finally, again regarding the existence of the functions F_1, F_2, from equation (6.99) and applying the Composite-function Theorem [230], we have

$$D_x F(h_2(x)) D_x h_2(x) = D_x \alpha(x) + D_x \phi(x). \tag{6.101}$$

This equation represents a system of nonlinear equations in the unknown Jacobian $D_x F$. Thus, if Dh_2 is nonsingular, then the above equation has a unique solution given by

$$D_x F(h_2(x)) = (D_x \alpha(x) + D_x \phi(x)) (D_x h_2(x))^{-1}. \tag{6.102}$$

Moreover, if h_2^{-1} exists, then F can be recovered from the Jacobian, $DF(h_2)$, by composition. However, Dh_2 is seldom nonsingular, and in the absence of this, the generalized-inverse can be used. This may not however lead to a solution. Similarly, h_2^{-1} may not usually exist, and again one can still not rule out the existence of a solution F. More investigation is still necessary at this point. Other methods for finding a solution using for example Groebner basis [85] and techniques from algebraic geometry are also possible.

6.6   Notes and Bibliography

The results of Sections 6.1, 6.3 of the chapter are based mainly on the valuable papers by Isidori et al. [139, 141, 145]. Other approaches to the continuous-time output-feedback problem for affine nonlinear systems can be found in the References [53, 54, 166, 190, 223, 224], while the results for the time-varying and the sampled-data measurement-feedback problems are discussed in [189] and [124, 213, 255, 262] respectively. The discussion on controller parametrization for the affine systems is based on Astolfi [44] and parallels that in [190]. It is also further discussed in [188, 190, 224], while the results for the general case considered in Subsection 6.4.1 are based on the Reference [289]. In addition, the results on the tracking problem are based on [50].

Similarly, the robust control problem is discussed in many references among which are [34, 147, 192, 208, 223, 265, 284]. In particular, the Reference [145] discusses the robust output-regulation (or tracking or servomechanism) problem, while the results on the reliable control problem are based on [285].

Finally, the results of Section 6.5 are based on [27], and a reliable approach using TS-fuzzy models can be found in [278].

7

Discrete-Time Nonlinear $\mathcal{H}_\infty$-Control

In this chapter, we discuss the nonlinear $\mathcal{H}_\infty$ sub-optimal control problem for discrete-time affine nonlinear systems using both state and output measurement-feedbacks. As most control schemes are implemented using a digital controller, studying the discrete-time problem leads to a better understanding of how the control strategy can be implemented on the digital computer, and what potential problems can arise. The study of the discrete-time problem is also important in its own right from a system-theoretic point of view, and delineates many different distinctions from the continuous-time case.

We begin with the derivation of the solution to the full-information problem first, and then specialize it to the state-feedback problem. The parametrization of the set of stabilizing controllers, both static and dynamic, is then discussed. Thereafter, the dynamic output-feedback problem is considered. Sufficient conditions for the solvability of the above problems for the time-invariant discrete-time affine nonlinear systems are given, as well as for a general class of systems.

7.1 Full-Information $\mathcal{H}_\infty$-Control for Affine Nonlinear Discrete-Time Systems

We consider an affine causal discrete-time state-space system defined on $\mathcal{X} \subseteq \Re^n$ in coordinates $x = (x_1, \ldots, x_n)$:

$$\Sigma^{da} : \begin{cases} x_{k+1} &= f(x_k) + g_1(x_k)w_k + g_2(x_k)u_k; \quad x(k_0) = x^0 \\ z_k &= h_1(x_k) + k_{11}(x_k)w_k + k_{12}(x_k)u_k \\ y_k &= h_2(x_k) + k_{21}(x_k)w_k \end{cases} \tag{7.1}$$

where $x \in \mathcal{X}$ is the state vector, $u \in \mathcal{U} \subseteq \Re^p$ is the p-dimensional control input, which belongs to the set of admissible controls $\mathcal{U}$, $w \in \mathcal{W}$ is the disturbance signal, which belongs to the set $\mathcal{W} \subset \ell_2([k_0, K], \Re^r)$ of admissible disturbances, the output $y \in \Re^m$ is the measured output of the system, and $z \in \Re^s$ is the output to be controlled. While the functions $f : \mathcal{X} \to \mathcal{X}$, $g_1 : \mathcal{X} \to \mathcal{M}^{n \times r}(\mathcal{X})$, $g_2 : \mathcal{X} \to \mathcal{M}^{n \times p}(\mathcal{X})$, $h_1 : \mathcal{X} \to \Re^s$, $h_2 : \mathcal{X} \to \Re^m$, and $k_{11} : \mathcal{X} \to \mathcal{M}^{s \times r}(\mathcal{X})$, $k_{12} : \mathcal{X} \to \mathcal{M}^{s \times p}(\mathcal{X})$, $k_{21} : \mathcal{X} \to \mathcal{M}^{m \times r}(\mathcal{X})$ are real C^∞ functions of x. Furthermore, we assume without any loss of generality that the system (7.1) has a unique equilibrium-point at $x = 0$ such that $f(0) = 0$, $h_1(0) = h_2(0) = 0$.

The control action to the system (7.1) can be provided by either pure state-feedback (when only the state of the system is available), or by full-information feedback (when information on both the state and disturbance is available), or by output-feedback (when only the output information is available). Furthermore, as in the previous two chapters, there are two theoretical approaches for solving the nonlinear discrete-time $\mathcal{H}_\infty$-control problem: namely, (i) the theory of differential games, which has been laid out in Chapter 2; and (ii) the theory of dissipative systems the foundations of which are also laid out in Chapter 3. For expository purposes and analytical expediency, we shall alternate between these two approaches. Invariably, the discrete-time problem for the system (7.1) can also be viewed as a two-player zero-sum differential game under the following finite-horizon cost functional:

$$J(u, w) = \frac{1}{2} \sum_{k=k_0}^{K} (\|z_k\| - \gamma^2 \|w_k\|^2), \quad K \in Z \tag{7.2}$$

which is to be maximized by the disturbance function $w_k = \alpha_1(x_k)$ and minimized by the control function $u_k = \alpha_2(x_k)$ which are respectively controlled by the two players. From Chapter 2, a necessary and sufficient condition for the solvability of this problem (which is characterized by a saddle-point) can be summarized in the following theorem.

Theorem 7.1.1 *Consider the discrete-time zero-sum, two-player differential game (7.2), (7.1). Two strategies $u^\star(x_k)$, $w^\star(x_k)$ constitute a feedback saddle-point solution such that*

$$J(u^\star, w) \leq J(u^\star, w^\star) \leq J(u, w^\star) \ \forall u \in \mathcal{U}, \ \forall w \in \mathcal{W}$$

if, and only if, there exist $K - k_0$ functions $V_k(.) : [k_0, K] \times \mathcal{X} \to \Re$ satisfying the discrete-time HJIE (DHJIE):

$$
\begin{aligned}
V_k(x) &= \min_{u_k \in \mathcal{U}} \max_{w_k \in \mathcal{W}} \left\{ \frac{1}{2}(\|z_k\|^2 - \gamma^2 \|w_k\|^2) + V_{k+1}(x_{k+1}) \right\}, \quad V_{K+1}(x) = 0 \\
&= \max_{w_k \in \mathcal{W}} \min_{u_k \in \mathcal{U}} \left\{ \frac{1}{2}(\|z_k\|^2 - \gamma^2 \|w_k\|^2) + V_{k+1}(x_{k+1}) \right\}, \quad V_{K+1}(x) = 0 \\
&= \frac{1}{2}(\|z_k(x, u_k^\star(x), w_k^\star(x))\|^2 - \gamma^2 \|w_k^\star(x)\|^2) + V_{k+1}(f(x) + g_1(x)w_k^\star(x) + \\
&\quad g_2(x)u_k^\star(x)), \quad V_{K+1}(x) = 0.
\end{aligned}
\tag{7.3}
$$

Furthermore, since we are only interested in finding a time-invariant control law, our goal is to seek a time-independent function $V : \mathcal{X} \to \Re$ which satisfies the DHJIE (7.3). Thus, as $k \to \infty$, the DHJIE (7.3) reduces to the time-invariant discrete-time Isaac's equation:

$$
\begin{aligned}
V(f(x) + g_1(x)w^\star(x) + g_2(x)u^\star(x)) - V(x) + \frac{1}{2}(\|z(x, u^\star(x), w^\star(x))\|^2 - \\
\gamma^2 \|w^\star(x)\|^2) = 0, \quad V(0) = 0.
\end{aligned}
\tag{7.4}
$$

Similarly, as in the continuous-time case, the above equation will be the cornerstone for solving the discrete-time sub-optimal nonlinear $\mathcal{H}_\infty$ disturbance-attenuation problem with internal stability by either state-feedback, or full-information feedback, or by dynamic output-feedback. We begin with the full-information and state-feedback problems.

Definition 7.1.1 *(Discrete-Time Full-Information Feedback and State-Feedback (Sub-optimal) Nonlinear $\mathcal{H}_\infty$-Control Problems (respectively DFIFBNLHICP & DSFBNLHICP). Find (if possible!) a static feedback-controller of the form:*

$$u = \alpha_{fif}(x_k, w_k), \quad ---full\text{-}information \ feedback \tag{7.5}$$

or

$$u = \alpha_{sf}(x_k) \quad ----- \ state\text{-}feedback \tag{7.6}$$

$\alpha_{fif} : \mathcal{X} \times \mathcal{W} \to \Re^p$, $\alpha_{sf} : \mathcal{X} \to \Re^p$, $\alpha_{fif}, \alpha_{sf} \in C^2(\mathcal{X})$, $\alpha_{fif}(0, 0) = 0$, $\alpha_{sf}(0) = 0$, *such that the closed-loop system (7.1), (7.5) or (7.1), (7.6) respectively, has locally ℓ_2-gain from the disturbance signal w to the output z less than or equal to some prescribed number $\gamma^\star > 0$ with internal stability.*

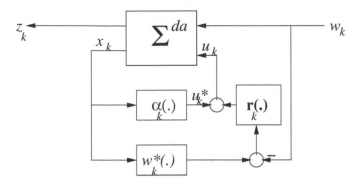

FIGURE 7.1
Full-Information Feedback Configuration for Discrete-Time Nonlinear $\mathcal{H}_\infty$-Control

The set-up for the full-information and state-feedback control law is shown in Figure 7.1 above. We consider the full-information problem first, and then the state-feedback problem as a special case. The following theorem gives sufficient conditions for the solvability of the full-information problem.

Theorem 7.1.2 *Consider the nonlinear discrete-time system (7.1) and suppose for some $\gamma > 0$ there exists a C^2 positive-definite function $V : N \subset \mathcal{X} \to \Re$ locally defined in a neighborhood N of the origin and is such that*

(A1)
$$r_{uu}(0) > 0, \quad r_{ww}(0) - r_{wu}(0)r_{uu}^{-1}(0)r_{uw}(0) < 0$$

where

$$r_{uu}(x) \;\triangleq\; g_2^T(x) \left.\frac{\partial^2 V}{\partial \lambda^2}\right|_{\lambda=F^\star(x)} g_2(x) + k_{12}^T(x)k_{12}(x),$$

$$r_{uw}(x) \;\triangleq\; g_2^T(x) \left.\frac{\partial^2 V}{\partial \lambda^2}\right|_{\lambda=F^\star(x)} g_1(x) + k_{12}^T(x)k_{11}(x) = r_{wu}^T(x),$$

$$r_{ww}(x) \;\triangleq\; g_1^T(x) \left.\frac{\partial^2 V}{\partial \lambda^2}\right|_{\lambda=F^\star(x)} g_1(x) + k_{11}^T(x)k_{11}(x) - \gamma^2 I,$$

$$F^\star(x) = f(x) + g_1(x)w^\star(x) + g_2(x)u^\star(x),$$

for some $u^\star(x)$ and $w^\star(x)$, with $u^\star(0) = 0$, $w^\star(0) = 0$, satisfying

$$\left.\begin{array}{l} \left.\frac{\partial V}{\partial \lambda}\right|_{\lambda=F^\star(x)} g_2(x) + (h_1(x) + k_{11}(x)w^\star(x) + k_{12}(x)u^\star(x))^T k_{12}(x) = 0, \\[2mm] \left.\frac{\partial V}{\partial \lambda}\right|_{\lambda=F^\star(x)} g_1(x) + (h_1(x) + k_{11}(x)w^\star(x) + k_{12}(x)u^\star(x))^T k_{11}(x) - \gamma^2 w^{\star T}(x) = 0; \end{array}\right\}(7.7)$$

(A2) *V satisfies the DHJIE:*

$$V(f(x) + g_1(x)w^\star(x) + g_2(x)u^\star(x)) - V(x) + \frac{1}{2}(\|h_1(x) + k_{11}(x)w^\star(x) +$$
$$k_{12}(x)u^\star(x)\|^2 - \gamma^2\|w^\star(x)\|^2) = 0, \;\; V(0) = 0; \tag{7.8}$$

(A3) *the closed-loop system*

$$x_{k+1} = F^\star(x_k) = f(x_k) + g_1(x_k)w^\star(x_k) + g_2(x_k)u^\star(x_k) \tag{7.9}$$

has $x = 0$ as a locally asymptotically-stable equilibrium-point,

then the full-information feedback law

$$\hat{u}_k = u^\star(x_k) - r_{uu}^{-1}(x_k)r_{uw}(x_k)(w_k - w^\star(x_k)) \tag{7.10}$$

solves the DFIFBNLHICP.

Proof: We consider the Hamiltonian function $H : \mathcal{X} \times \mathcal{U} \times \mathcal{W} \to \Re$ for the system (7.1) under the performance measure

$$J = \frac{1}{2}\sum_{k=k_0}^{\infty}(\|z_k\|^2 - \gamma^2\|w_k\|^2)$$

and defined by

$$H(x, u, w) = V(f(x) + g_1(x)w + g_2(x)u) - V(x) + \frac{1}{2}(\|h_1(x) + k_{11}(x)w + k_{12}u\|^2 - \gamma^2\|w\|^2). \tag{7.11}$$

The Hessian matrix of $H(., ., .)$ with respect to (u, w) is given by

$$\frac{\partial^2 H}{\partial(u, w)^2} = \begin{bmatrix} \frac{\partial^2 H}{\partial u^2} & \frac{\partial^2 H}{\partial w \partial u} \\ \frac{\partial^2 H}{\partial u \partial w} & \frac{\partial^2 H}{\partial w^2} \end{bmatrix}(x) := \begin{bmatrix} r_{uu}(x) & r_{uw}(x) \\ r_{wu}(x) & r_{ww}(x) \end{bmatrix}$$

$$= \begin{bmatrix} g_2^T(x)\frac{\partial^2 V}{\partial \lambda^2}g_2(x) + k_{12}^T(x)k_{12}(x) & g_2^T(x)\frac{\partial^2 V}{\partial \lambda^2}g_1(x) + k_{11}^T(x)k_{12}(x) \\ g_1^T(x)\frac{\partial^2 V}{\partial \lambda^2}g_2(x) + k_{11}^T(x)k_{12}(x) & g_1^T(x)\frac{\partial^2 V}{\partial \lambda^2}g_1(x) + k_{11}^T(x)k_{11}(x) - \gamma^2 I \end{bmatrix}\Bigg|_{\lambda = \mathbf{F}(x)},$$

where

$$\mathbf{F}(x) = f(x) + g_1(x)w(x) + g_2(x)u(x).$$

By assumption $(A1)$ of the theorem and Schur's complement [69], the matrix

$$\frac{\partial^2 H}{\partial(u, w)^2}(0, 0, 0) = \begin{bmatrix} r_{uu}(0) & r_{uw}(0) \\ r_{wu}(0) & r_{ww}(0) \end{bmatrix}$$

is clearly nonsingular. Moreover, since the point $(x, u, w) = (0, 0, 0)$ satisfies (7.7), by the Implicit-function Theorem [157], there exist open neighborhoods X of $x = 0$ in $\mathcal{X}$ and $Y \subset \mathcal{U} \times \mathcal{W}$ of $(u, w) = (0, 0)$ such that there exists a unique smooth solution $\begin{bmatrix} u^\star \\ w^\star \end{bmatrix} : X \to Y$ to (7.7), with $u^\star(0) = 0$ and $w^\star(0) = 0$. Consequently, $(u^\star(x), w^\star(x))$ is a locally critical point of the Hamiltonian function $H(x, u, w)$ about $x = 0$.

Expanding now $H(x, u, w)$ in Taylor-series about $u^\star$ and $w^\star$, we have

$$\begin{aligned} H(x, u, w) &= H(x, u^\star(x), w^\star(x)) + \\ &\quad \frac{1}{2}\begin{bmatrix} u - u^\star(x) \\ w - w^\star(x) \end{bmatrix}^T \begin{bmatrix} r_{uu}(x) & r_{uw}(x) \\ r_{wu}(x) & r_{ww}(x) \end{bmatrix} \begin{bmatrix} u - u^\star(x) \\ w - w^\star(x) \end{bmatrix} + \\ &\quad O\left(\left\|\begin{matrix} u - u^\star(x) \\ w - w^\star(x) \end{matrix}\right\|^3\right), \end{aligned} \tag{7.12}$$

which by Schur's complement can be expressed as

$$\begin{aligned} H(x, u, w) &= H(x, u^\star(x), w^\star(x)) + \frac{1}{2}\begin{bmatrix} u - u^\star(x) + r_{uu}^{-1}(x)r_{uw}(x)(w - w^\star(x)) \\ w - w^\star(x) \end{bmatrix}^T \times \\ &\quad \begin{bmatrix} r_{uu}(x) & 0 \\ 0 & r_{ww}(x) - r_{wu}(x)r_{uu}^{-1}(x)r_{uw}(x) \end{bmatrix} \begin{bmatrix} u - u^\star + r_{uu}^{-1}(x)r_{uw}(x)(w - w^\star(x)) \\ w - w^\star(x) \end{bmatrix} \\ &\quad + O\left(\left\|\begin{matrix} u - u^\star(x) \\ w - w^\star(x) \end{matrix}\right\|^3\right). \end{aligned}$$

Thus, choosing the control law (7.10) yields

$$
\begin{aligned}
H(x, \hat{u}, w) &= H(x, u^*(x), w^*(x)) + \\
&\quad \frac{1}{2}(w - w^*(x))^T[(r_{ww}(x) - r_{wu}(x)r_{uu}^{-1}(x)r_{uw}(x))](w - w^*(x)) + \\
&\quad O(\|w - w^*(x)\|).
\end{aligned}
$$

By hypotheses $(A1)$, $(A2)$ of the theorem, the above equation (7.13) implies that there are open neighborhoods X_0 of $x = 0$ and W_0 of $w = 0$ such that

$$
H(x, \hat{u}, w) \le H(x, u^*(x), w^*(x)) \le 0 \quad \forall x \in X_0, \quad \forall w \in W_0.
$$

Or equivalently

$$
V(f(x) + g_1(x)w + g_2(x)\hat{u}) - V(x) \le \frac{1}{2}\gamma^2\|w\|^2 - \frac{1}{2}\|h_1(x) + k_{11}(x)w + k_{12}(x)\hat{u}\|^2,
$$

for all $w \in W_0$, which is the dissipation-inequality (see Chapter 3). Hence, the closed-loop system under the feedback (7.10) is locally finite-gain dissipative with storage-function V, with respect to the supply-rate $s(w_k, z_k) = \frac{1}{2}(\gamma^2\|w_k\|^2 - \|z_k\|^2)$.

To prove local asymptotic-stability of the closed-loop system with (7.10), we set $w = 0$ resulting in the closed-loop system:

$$
x_{k+1} = f(x_k) + g_2(x_k)(u_k^* - r_{uu}^{-1}(x_k)r_{uw}(x_k)w^*(x_k)). \tag{7.13}
$$

Now, from (7.13) and hypothesis $(A2)$ of the theorem, we deduce that

$$
H(x, \hat{u}, w)|_{w=0} \le \frac{1}{2}w^{*T}(x)[r_{ww}(x) - r_{wu}(x)r_{uu}^{-1}(x)r_{uw}(x)]w^*(x) + O(\|w^*(x)\|).
$$

Since $r_{ww}(x) - r_{wu}(x)r_{uu}^{-1}(x)r_{uw}(x) < 0$ in a neighborhood of $x = 0$ and $w^*(0) = 0$, there exists an open neighborhood X_1 of $x = 0$ such that

$$
H(x, \hat{u}, w)|_{w=0} \le \frac{1}{4}(w^*)^T(r_{ww}(x) - r_{wu}(x)r_{uu}^{-1}(x)r_{uw}(x))w^*(x) \le 0 \quad \forall x \in X_1,
$$

which is equivalent to

$$
\begin{aligned}
V(f(x) + g_2(x)(u^*(x) - r_{uu}^{-1}(x)r_{uw}(x)w^*(x))) - V(x) &\le -\frac{1}{2}\|h_1(x) + k_{12}(x)(u^*(x) - \\
r_{uu}^{-1}(x)r_{uw}(x)w^*(x))\|^2 + \frac{1}{4}w^{*T}(x)(r_{ww}(x) - r_{wu}(x)r_{uu}^{-1}(x)r_{uw}(x))w^*(x) &\le 0.
\end{aligned}
$$

Thus, the closed-loop system is locally stable by Lyapunov's theorem (Chapter 1). In addition, any bounded trajectory of the closed-loop system (7.13) will approach the largest invariant set contained in the set

$$
\Omega_I = \{x \in \mathcal{X} : V(x_{k+1}) = V(x_k) \; \forall k \in Z_+\},
$$

and any trajectory in this set must satisfy

$$
w^{*T}(x)(r_{ww}(x) - r_{wu}(x)r_{uu}^{-1}(x)r_{12}(x))w^*(x) = 0. \tag{7.14}
$$

By hypothesis $(A1)$, there exists a unique nonsingular matrix $T(x)$ such that

$$
r_{ww}(x) - r_{wu}(x)r_{uu}^{-1}(x)r_{12}(x) = -T^T(x)T(x)
$$

for all $x \in \Omega_I$. This, together with the previous condition (7.14), imply that

$$\omega(x_k) \overset{\Delta}{=} T(x_k)w^\star(x_k) = 0 \ \forall k \in Z_+$$

for some function $\omega : \Omega_I \to \Re^\nu$. Consequently, rewriting (7.13) as

$$x_{k+1} = f(x_k) + g_1(x_k)w^\star(x_k) + d(x_k)\omega(x_k) + g_2(x_k)u^\star(x_k) \qquad (7.15)$$

$$d(x_k) \overset{\Delta}{=} -[g_1(x_k) + g_2(x_k)r_{uu}^{-1}(x_k)r_{12}(x_k)]T^{-1}(x_k), \qquad (7.16)$$

then, any bounded trajectory of the above system (7.15), (7.16) such that $V(x_{k+1}) = V(x_k) \ \forall k \in Z_+$, is also a trajectory of (7.9). Hence, by assumption ($A3$) and the LaSalle's invariance-principle, we conclude that $\lim_{k \to \infty} x_k = 0$, or the equilibrium point $x = 0$ is indeed locally asymptotically-stable.

Finally, by the dissipativity of the closed-loop system established above, and the result of Theorem 3.6.1, we also conclude that the system has ℓ_2-gain $\leq \gamma$. $\square$

Remark 7.1.1 *Assumption ($A3$) in Theorem 7.1.2 can be replaced by the following assumption:*

($A4$) *Any bounded trajectory of the system*

$$x_{k+1} = f(x_k) + g_2(x_k)u_k$$

with

$$z_k|_{w_k=0} = h_1(x_k) + k_{12}(x_k)u_k \equiv 0$$

is such that $\lim_{k \to \infty} x_k \equiv 0$.

Thus, any bounded trajectory x_k of the system such that $V(x_{k+1}) = V(x_k) \ \forall k \in Z_+$, is necessarily a bounded trajectory of (7.9) with the constraint $h_1(x_k) + k_{12}(x_k)\hat{u}_k|_{w_k=0} = 0 \ \forall k \in Z_+$. Consequently, by the above assumption ($A4$), this implies $\lim_{k \to \infty} x_k = 0$.

Remark 7.1.2 *The block shown $\mathbf{r}_k(.)$ in Figure 7.1 represents the nonlinear operator $r_{uu}^{-1}(x_k)r_{uw}(x_k)$ in the control law (7.10).*

We now specialize the results of the above theorem to the case of the linear discrete-time system

$$\Sigma^{dl} : \begin{cases} x_{k+1} &= Ax_k + B_1 w_k + B_2 u_k \\ z_k &= C_1 x_k + D_{11}w_k + D_{12}u_k \\ y_k &= C_2 x_k + D_{21}w_k, \end{cases} \qquad (7.17)$$

where $x \in \Re^n$, $z \in \Re^s$, $y \in \Re^m$, $B_1 \in \Re^{n \times r}$, $B_2 \in \Re^{n \times p}$ and the rest of the matrices have compatible dimensions. In this case, if we assume that $rank\{D_{12}\} = p$, then Assumption ($A4$) is equivalent to the following hypothesis [292]:

($A4L$)

$$rank \begin{pmatrix} A - \lambda I & B_2 \\ C_1 & D_{12} \end{pmatrix} = n + p, \ \forall \lambda \in \mathbf{C}, \ |\lambda| < 1.$$

Define now the symmetric matrix

$$R \overset{\Delta}{=} \begin{bmatrix} R_{uu} & R_{uw} \\ R_{wu} & R_{ww} \end{bmatrix}$$

where

$$
\begin{array}{rcl}
R_{uu} & = & B_2^T P B_2 + D_{12}^T D_{12} \\
R_{uw} & = & R_{wu}^T = B_2^T P B_1 + D_{12}^T D_{11} \\
R_{ww} & = & B_1^T P B_1 + D_{11}^T D_{11} - \gamma^2 I.
\end{array}
$$

Then we have the following proposition.

Proposition 7.1.1 *Consider the discrete-time linear system (7.17). Suppose there exists a positive-definite matrix P such that:*

(L1)
$$
R_{uu} > 0, \quad R_{ww} - R_{wu} R_{uu}^{-1} R_{wu} < 0;
$$

(L2) *the discrete-algebraic Riccati equation (DARE)*

$$
P = A^T P A + C^T C - \left(\begin{array}{c} B_2^T P A + D_{12}^T C_1 \\ B_1^T P A + D_{11}^T C_1 \end{array} \right)^T R^{-1} \left(\begin{array}{c} B_2^T P A + D_{12}^T C_1 \\ B_1^T P A + D_{11}^T C_1 \end{array} \right), \tag{7.18}
$$

or equivalently the linear matrix inequality (LMI)

$$
\left[\begin{array}{cc} A^T P A - P + C^T C & \left(\begin{array}{c} B_2^T P A + D_{12}^T C_1 \\ B_1^T P A + D_{11}^T C_1 \end{array} \right)^T \\ \left(\begin{array}{c} B_2^T P A + D_{12}^T C_1 \\ B_1^T P A + D_{11}^T C_1 \end{array} \right) & R \end{array} \right] > 0 \tag{7.19}
$$

holds;

(L3) *the matrix*
$$
A_{cl} \triangleq A - (B_2 \; B_1)^T R^{-1} \left(\begin{array}{c} B_2^T P A + D_{12}^T C_1 \\ B_1^T P A + D_{11}^T C_1 \end{array} \right)
$$

is Hurwitz.

Then, Assumptions $(A1) - (A3)$ of Theorem 7.1.2 are satisfied, with

$$
\left. \begin{array}{rcl}
r_{ij}(0) & = & R_{ij}, \quad i, j = u, w \\
V(x) & = & \frac{1}{2} x^T P x \\
\left[\begin{array}{c} u^\star(x_k) \\ w^\star(x_k) \end{array} \right] & = & -R^{-1} \left(\begin{array}{c} B_2^T P A + D_{12}^T C_1 \\ B_1^T P A + D_{11}^T C_1 \end{array} \right) x_k.
\end{array} \right\} \tag{7.20}
$$

Proof: For the case of the linear system with R_{ij} as defined above, it is clear that Assumptions $(L1)$ and $(A1)$ are identical. So $(A1)$ holds.

To show that $(A2)$ also holds, note that $(L1)$ guarantees that the matrix R is invertible, and (7.7) is therefore explicitly solvable for $u^\star(x)$ and $w^\star(x)$ with $V(x) = \frac{1}{2} x^T P x$. Moreover, these solutions are exactly (7.20). Substituting the above expressions for V, $u^\star$ and $w^\star$ in (7.8), results in the DARE (7.18). Thus, $(A2)$ holds if $(L1)$ holds.

Finally, hypothesis $(L3)$ implies that the system

$$
\begin{array}{rcl}
x_{k+1} & = & A x_k + (B_2 \; B_1)^T R^{-1} \left(\begin{array}{c} B_2^T P A + D_{12}^T C \\ B_1^T P A + D_{11}^T C \end{array} \right) x_k \tag{7.21} \\
& = & A x_k + B_1 w^\star(x_k) + B_2 u^\star(x_k) \tag{7.22}
\end{array}
$$

is asymptotically-stable. Hence, $(A3)$ holds for the linear system. $\square$

The following corollary also follows from the theorem.

Corollary 7.1.1 *Suppose that there exists a symmetric matrix $P > 0$ satisfying conditions $(L1) - (L3)$ of Proposition 7.1.1, then the full-information feedback-control law*

$$u_k \quad = \quad K_1 x_k + K_2 w_k \tag{7.23}$$
$$K_1 \quad \triangleq \quad -(D_{12}^T D_{12} + B_2^T PB)^{-1}(B_2^T PA + D_{12}^T C_1)$$
$$K_2 \quad \triangleq \quad -(D_{12}^T D_{12} + B_2^T PB)^{-1}(B_2^T PE + D_{12}^T D_{11})$$

renders the closed-loop system (7.17),(7.23) internally stable with $\mathcal{H}_\infty$-norm less than or equal to γ.

7.1.1 State-Feedback $\mathcal{H}_\infty$-Control for Affine Nonlinear Discrete-Time Systems

In this subsection, we reduce the full-information static feedback problem to a state-feedback problem in which only the state information is now available for feedback. However, it should be noted that in sharp contrast with the continuous-time problem, the discrete-time problem cannot be solved with $k_{11}(x) = 0$. This follows from (7.10) and Assumption $(A1)$. It is fairly clear that the assumption $k_{11}(x) = 0$ does not imply that $r_{uw}(x) = 0$, and thus $u_k(x)$ defined by (7.5) is independent of w_k.

The following theorem however gives sufficient conditions for the solvability of the *DSFBNLHICP*.

Theorem 7.1.3 *Consider the discrete-time nonlinear system (7.1). Suppose for some $\gamma > 0$ there exists a C^2 positive-definite function $V : N \subset \mathcal{X} \to \Re$ locally defined in a neighborhood of $x = 0$ and which satisfies the following assumption*

(A1s)

$$r_{uu}(0) > 0 \quad and \quad r_{ww}(0) < 0$$

as well as Assumptions $(A2)$ and $(A3)$ of Theorem 7.1.2. Then the static state-feedback control law

$$u_k = u^\star(x_k), \tag{7.24}$$

where $u^\star(.)$ is solved from (7.7), solves the DSFBNLHICP.

Proof: The proof is similar to that of Theorem 7.1.2, therefore we shall provide only a sketch. Obviously, Assumption $(A1s)$ implies $(A1)$. Therefore, we can repeat the arguments in the first part of Theorem 7.1.2 and conclude that there exist unique functions $u^\star(x)$ and $w^\star(x)$ satisfying (7.7), and consequently, the Hamiltonian function $H(x, u, w)$ can be expanded about $(u^\star, w^\star)$ as

$$
\begin{aligned}
H(x, u, w) \quad \triangleq \quad & V(f(x) + g_1(x)w + g_2(x)u) - V(x) + \frac{1}{2}(\|z\|^2 - \gamma^2 \|w\|^2) \\
= \quad & H(x, u^\star(x), w^\star(x)) + \frac{1}{2} \left[\begin{array}{c} u - u^\star(x) \\ w - w^\star(x) \end{array} \right]^T \left[\begin{array}{cc} r_{uu}(x) & r_{uw}(x) \\ r_{wu}(x) & r_{ww}(x) \end{array} \right] \left[\begin{array}{c} u - u^\star(x) \\ w - w^\star(x) \end{array} \right] + \\
& O\left(\left\| \begin{array}{c} u - u^\star(x) \\ w - w^\star(x) \end{array} \right\| \right).
\end{aligned}
\tag{7.25}
$$

Further, Assumption $(A1s)$ implies that there exists a neighborhood U_0 of $x = 0$ such that

$$r_{uu}(x) > 0, \quad r_{ww}(x) < 0 \quad \forall x \in U_0.$$

Substituting the control law (7.24) in (7.25) yields

$$
\begin{aligned}
H(x, u^\star(x), w) \;=\; & H(x, u^\star(x), w^\star(x)) + \frac{1}{2}(w - w^\star(x))^T r_{ww}(x)(w - w^\star(x)) + \\
& O(\|w - w^\star(x)\|),
\end{aligned}
\tag{7.26}
$$

and since $r_{ww}(x) < 0$, it follows that

$$
H(x, u^\star(x), w) \le H(x, u^\star(x), w^\star(x)) \quad \forall x \in U_0, \ \forall w \in W_0.
\tag{7.27}
$$

On the other hand, substituting $w = w^\star$ in (7.25) and using the fact that $r_{uu}(x) > 0 \ \forall x \in U_0$, we have

$$
H(x, u, w^\star(x)) \ge H(x, u^\star(x), w^\star(x)) \quad \forall x \in U_0, \ \forall u \in \mathcal{U}.
\tag{7.28}
$$

This implies that the pair $(u^\star(x), w^\star(x))$ is a local saddle-point of $H(x, u, w)$. Futhermore, by hypothesis $(A2)$ and (7.27) we also have

$$
H(x, u^\star(x), w) \le 0 \quad \forall x \in U_0, \ \forall w \in W_0,
$$

or equivalently

$$
V(f(x) + g_1(x)w + g_2(x)u^\star(x)) - V(x) \le \frac{1}{2}\gamma^2 \|w\|^2 - \frac{1}{2}\|z(x, u^\star(x), w)\|^2 \quad \forall x \in U_0, \ \forall w \in W_0.
\tag{7.29}
$$

Thus, the closed-loop system with $u = u^\star(x)$ given by (7.24) is locally dissipative with storage-function V with respect to the supply-rate $s(w, z) = \frac{1}{2}(\gamma^2 \|w\|^2 - \|z\|^2$, and hence has ℓ_2-gain $\le \gamma$.

Furthermore, setting $w = 0$ in (7.29) we immediately have that the equilibrium-point $x = 0$ of the closed-loop system is Lyapunov-stable. To prove asymptotic-stability, we note from (7.26) and hypothesis $(A1s)$ that there exists a neighborhood N of $x = 0$ such that

$$
H(x, u^\star(x), w) \le H(x, u^\star(x), w^\star(x)) + \frac{1}{4}(w - w^\star(x))^T r_{ww}(x)(w - w^\star(x)) \quad \forall x \in N, \ \forall w \in W_0.
$$

This implies that

$$
V(f(x) + g_2(x)u^\star(x)) - V(x) \le -\frac{1}{2}\|z(x, u^\star(x), w)\|^2 + \frac{1}{4}(w^\star)^T(x)r_{ww}(x)w^\star(x) \le 0.
$$

Hence, the condition

$$
V(f(x) + g_2(x)u^\star(x)) \equiv V(x)
$$

results in

$$
0 = w^{\star T}(x)r_{ww}(x)w^\star(x),
$$

and since $r_{ww}(x) < 0$, there exists a nonsingular matrix $L(x)$ such that

$$
\varphi^T(x)\varphi(x) \overset{\Delta}{=} (w^\star)^T(x)L^T(x)L(x)w^\star(x) = 0 \ \text{ with } \ L^T(x)L(x) = -r_{ww}(x).
$$

Again, representing the closed-loop system

$$
x_{k+1} = f(x_k) + g_2(x_k)u^\star(x_k)
$$

as

$$
x_{k+1} = f(x_k) + g_1(x_k)w^\star(x_k) + g_2(x_k)u^\star(x_k) + \tilde{d}(x_k)\varphi(x_k)
$$

where $\tilde{d}(x_k) = -g_1(x_k)L^{-1}(x_k)$, it is clear that any trajectory of this system resulting in

$\varphi(x_k) = 0 \, \forall k \in \mathbf{Z}_+$, is also a trajectory of (7.9). Therefore, by hypothesis $(A3)$ $\varphi(x_k) = 0$ $\Rightarrow \lim_{k\to\infty} x_k = 0$, and by LaSalle's invariance-principle, we have local asymptotic-stability of the equilibrium-point $x = 0$ of the closed-loop system. $\square$

Remark 7.1.3 *Assumption $(A3)$ in the above Theorem 7.1.3 can similarly be replaced by Assumption $(A4)$ with the same conclusion.*

Again, if we specialize the results of Theorem 7.1.3 to the discrete-time linear system Σ^{dl}, they reduce to the wellknown sufficient conditions for the solvability of the discrete-time linear $\mathcal{H}_\infty$ control problem via state-feedback [252]. This is stated in the following proposition.

Proposition 7.1.2 *Consider the linear system Σ^{dl} and suppose there exists a positive definite matrix P such that*

(L1s)

$$R_{uu} > 0, \quad R_{ww} < 0$$

and Assumptions $(L2)$ and $(L3)$ in Proposition 7.1.1 hold. Then, Assumptions $(A1s)$, $(A2)$ and $(A3)$ of Theorem 7.1.3 hold with $V(x)$, $r_{ij}(0)$, and $\begin{bmatrix} u^\star(x) \\ w^\star(x) \end{bmatrix}$ as in (7.20).

Proof: Proof follows along similar lines as Proposition 7.1.1. $\square$

The following corollary then gives sufficient conditions for the solvability of the discrete linear suboptimal $\mathcal{H}_\infty$-control problem using state-feedback.

Corollary 7.1.2 *Consider the linear discrete-time system Σ_l^d and suppose there exists a matrix $P > 0$ satisfying the hypotheses $(L1s)$, $(L2)$ and $(L3)$ of Proposition 7.1.2. Then the static state-feedback control law*

$$
\begin{aligned}
u_k &= K x_k \\
K &= -(R_{uu} - R_{uw}R_{ww}^{-1}R_{wu})^{-1}(B_2^T PA + D_{12}^T C_1 - R_{uw}R_{ww}^{-1}(B_1^T PA + D_{11}^T C_1)),
\end{aligned}
\tag{7.30}
$$

with $R_{ij}, i, j = u, w$ as defined before, renders the closed-loop system for the linear system asymptotically-stable with $\mathcal{H}_\infty$-norm $\leq \gamma$; in other words, solves the $DSFBNLHICP$ for the system.

Proof: It is shown in [185] that (7.20) is identical to (7.30). $\square$

7.1.2   Controller Parametrization

In this subsection, we discuss the construction of a class of state-feedback controllers that solves the $DSFBNLHICP$. For this, we have the following result.

Theorem 7.1.4 *Consider the nonlinear system (7.1) and suppose for some $\gamma > 0$ there exist a smooth function $\phi : N_1 \to \Re^p$, $\phi(0) = 0$ and a C^2 positive-definite function $V : N_1 \to \Re$ locally defined in a neighborhood $N \subset \mathcal{X}$ of $x = 0$ with $V(0) = 0$ such that Assumption $(A1s)$ of Theorem 7.1.3 is satisfied, as well as the following assumptions:*

(A5) *The DHJIE:*

$$
\begin{aligned}
V(f(x) + g_1(x)w^\star(x) + g_2(x)u^\star(x)) - V(x) + \\
\tfrac{1}{2}(\|h_1(x) + k_{11}(x)w^\star(x) + k_{12}(x)u^\star(x)\|^2 - \gamma^2\|w^\star(x)\|^2) = \\
-\phi^T(x)[r_{uu}(x) - r_{uw}(x)r_{ww}^{-1}(x)r_{wu}(x)]\phi(x), \quad V(0) = 0
\end{aligned}
\tag{7.31}
$$

holds for all $u^\star(x)$, $w^\star(x)$ satisfying (7.7).

(A6) *The system*

$$x_{k+1} = f(x_k) + g_1(x_k)w^\star(x_k) + g_2(x_k)(u^\star(x_k) + \phi(x_k)) \qquad (7.32)$$

is locally asymptotically-stable at $x = 0$.

Then the DSFBNLHICP is solvable by the family of controllers:

$$\mathbf{K}_{SF} = \{u_k | u_k = u^\star(x_k) + \phi(x_k)\}. \qquad (7.33)$$

Proof: Consider again the Hamiltonian function (7.11). Taking the partial-derivatives

$$\frac{\partial H}{\partial u}(x, u, w) = 0, \quad \frac{\partial H}{\partial w}(x, u, w) = 0$$

we get (7.7). By Assumption $(A1s)$ and the Implicit-function Theorem, there exist unique solutions $u^\star(x)$, $w^\star(x)$ locally defined around $x = 0$ with $u^\star(0) = 0$, $w^\star(0) = 0$. Notice also here that

$$\frac{\partial^2 H}{\partial(u,w)^2}(x, u^\star(x), w^\star(x)) \;\triangleq\; \left[\begin{array}{cc} \frac{\partial^2 H}{\partial u^2} & \frac{\partial^2 H}{\partial w \partial u} \\ \frac{\partial^2 H}{\partial u \partial w} & \frac{\partial^2 H}{\partial w^2} \end{array} \right]_{u=u^\star(x), w=w^\star(x)}$$

$$\triangleq \left[\begin{array}{cc} r_{uu}(x) & r_{uw}(x) \\ r_{wu}(x) & r_{ww}(x) \end{array} \right].$$

Further, using the Taylor-series expansion of $H(x, u, w)$ around $u^\star$, $w^\star$ given by (7.12) and substituting (7.33) in it, we get after using Assumption $(A5)$

$$H(x, u^\star + \phi(x), w) = -\phi^T(x)r_{uu}(x)\phi(x) + \phi^T(x)r_{uw}(x)r_{ww}^{-1}(x)r_{wu}(x)\phi(x) +$$

$$\frac{1}{2}\left[\begin{array}{c} \phi(x) \\ w - w^\star(x) \end{array} \right]^T \left\{ \left[\begin{array}{cc} r_{uu}(x) & r_{uw}(x) \\ r_{wu}(x) & r_{ww}(x) \end{array} \right] + O\left(\left\| \begin{array}{c} \phi(x) \\ w - w^\star(x) \end{array} \right\| \right) \right\} \left[\begin{array}{c} \phi(x) \\ w - w^\star(x) \end{array} \right].$$

However, since

$$\frac{1}{2}\phi^T(x)r_{uw}(x)r_{ww}^{-1}(x)r_{wu}(x)\phi(x) - \phi^T(x)r_{uu}(x)\phi(x) =$$

$$\frac{1}{2}\left[\begin{array}{c} \phi(x) \\ w - w^\star(x) \end{array} \right]^T \left[\begin{array}{cc} -2r_{uu}(x) + r_{uw}(x)r_{ww}^{-1}(x)r_{wu}(x) & 0 \\ 0 & 0 \end{array} \right] \left[\begin{array}{c} \phi(x) \\ w - w^\star(x) \end{array} \right],$$

then

$$H(x, u^\star + \phi(x), w) = \frac{1}{2}\|r_{wu}(x)\phi(x)\|^2_{r_{ww}^{-1}(x)} +$$

$$\frac{1}{2}\left[\begin{array}{c} \phi(x) \\ w - w^\star(x) \end{array} \right]^T \left\{ \left[\begin{array}{cc} -r_{uu}(x) + r_{uw}(x)r_{ww}^{-1}(x)r_{wu}(x) & r_{uw}(x) \\ r_{wu}(x) & r_{ww}(x) \end{array} \right] + O\left(\left\| \begin{array}{c} \phi(x) \\ w - w^\star(x) \end{array} \right\| \right) \right\} \times$$

$$\left[\begin{array}{c} \phi(x) \\ w - w^\star(x) \end{array} \right]. \qquad (7.34)$$

Note also that, by Assumption $(A1s)$ the matrix

$$\left[\begin{array}{cc} -r_{uu}(x) + r_{uw}(x)r_{ww}^{-1}(x)r_{wu}(x) & r_{uw}(x) \\ r_{wu}(x) & r_{ww}(x) \end{array} \right] =$$

$$\left[\begin{array}{cc} I & r_{uw}(x)r_{ww}^{-1}(x) \\ 0 & I \end{array} \right] \left[\begin{array}{cc} -r_{uu}(x) & 0 \\ 0 & r_{ww}(x) \end{array} \right] \left[\begin{array}{cc} I & 0 \\ r_{ww}^{-1}(x)r_{wu}(x) & I \end{array} \right]$$

is negative-definite in an open neighborhood of $x = 0$. Therefore, there exists an open neighborhood X_2 of $x = 0$ and W_2 of $w = 0$ such that

$$H(x, u^\star(x) + \phi(x), w) \le \frac{1}{2}\|r_{wu}(x)\phi(x)\|^2_{r_{ww}^{-1}(x)} \le 0 \ \forall x \in X_2, \ \forall w \in W_2.$$

Or equivalently,

$$V(f(x) + g_1(x)w + g_2(x)[u^\star(x) + \phi(x)]) - V(x) \le \frac{1}{2}(\gamma^2\|w\|^2 - \|z\|^2_{u=u^\star(x)+\phi(x)}), \quad (7.35)$$

which implies that the closed-loop system is locally dissipative with storage-function $V(x)$ with respect to the supply-rate $s(w, z) = \frac{1}{2}(\gamma^2\|w\|^2 - \|z\|^2)$, and consequently has ℓ_2-gain $\le \gamma$.

To prove internal-stability, we set $w = 0$ in (7.1) resulting in the closed-loop system

$$x_{k+1} = f(x_k) + g_2(x_k)[u^\star(x_k) + \phi(x_k)]. \quad (7.36)$$

Similarly, subsituting $w = 0$ in (7.35) shows that the closed-loop system is Lyapunov-stable. Furthermore, from (7.34), we have

$$H(x, u^\star + \phi(x), w) \le \frac{1}{2}\|r_{wu}(x)\phi(x)\|^2_{r_{ww}^{-1}(x)} +$$

$$\frac{1}{2}\left[\begin{array}{c} \phi(x) \\ w - w^\star(x) \end{array}\right]^T \left[\begin{array}{cc} -r_{uu}(x) + r_{uw}(x)r_{ww}^{-1}(x)r_{wu}(x) & r_{uw}(x) \\ r_{wu}(x) & r_{ww}(x) \end{array}\right] \left[\begin{array}{c} \phi(x) \\ w - w^\star(x) \end{array}\right]$$

for all $x \in X_3$ and all $w \in W_3$ open neighborhoods of $x = 0$ and $w = 0$ respectively. Again, setting $w = 0$ in the above inequality and using the fact that all the weighting matrices are negative-definite, we get

$$V(f(x) + g_1(x)w + g_2(x)[u^\star(x) + \phi(x)]) - V(x)$$

$$\le \quad -\frac{1}{2}\|z\|^2_{(u=u^\star(x)+\phi(x),w=0)} + \frac{1}{2}\|r_{wu}(x)\phi(x)\|^2_{r_{ww}^{-1}(x)} +$$

$$\frac{1}{4}\left[\begin{array}{c} \phi(x) \\ w^\star(x) \end{array}\right]^T \left[\begin{array}{cc} -r_{uu}(x) + r_{uw}(x)r_{ww}^{-1}(x)r_{wu}(x) & r_{uw}(x) \\ r_{wu}(x) & r_{ww}(x) \end{array}\right] \left[\begin{array}{c} \phi(x) \\ w^\star(x) \end{array}\right]$$

$$= \quad -\frac{1}{2}\|z\|^2_{(u=u^\star(x)+\phi(x),w=0)} + \frac{1}{2}\|r_{wu}(x)\phi(x)\|^2_{r_{ww}^{-1}(x)} +$$

$$\frac{1}{4}\{\|\phi(x)\|^2_{(-r_{uu}(x)+r_{uw}(x)r_{ww}^{-1}(x)r_{wu}(x))} - 2[r_{wu}(x)\phi(x)]^T w^\star(x) + \|w^\star(x)\|^2_{r_{ww}(x)}\}$$

$$\le \quad 0. \quad (7.37)$$

Now define the level-set

$$\Omega_L \stackrel{\Delta}{=} \{x \in X_3 \ : \ V(f(x_k) + g_2(x_k)[u^\star(x_k) + \phi(x_k)]) \equiv V(x_k) \ \forall k\}.$$

Then, from (7.37), any trajectory of the closed-loop system whose Ω-limit set is in Ω_L must be such that

$$w^{\star T}(x)r_{ww}(x)w^\star(x) = 0$$

because of (7.37). Since $r_{ww}(x) < 0$, then there exists a nonsingular matrix $M(x)$ such that $-r_{ww}(x) = M^T(x)M(x)$ and

$$\tilde{\omega}(x_k) \stackrel{\Delta}{=} M(x_k)w^\star(x_k) = 0 \quad \forall k = 0, 1, \ldots.$$

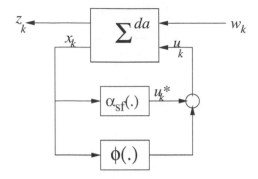

FIGURE 7.2
Controller Parametrization for Discrete-Time State-Feedback Nonlinear $\mathcal{H}_\infty$-Control

Thus, the closed-loop system (7.36) can be represented as

$$x_{k+1} = f(x_k) + g_1(x_k)w^\star(x_k) + \alpha(x_k)\tilde{\omega}(x_k) + g_2(x_k)[u^\star(x_k) + \phi(x_k)],$$

where $\alpha(x_k) = -g_1(x_k)r_{ww}^{-1}(x_k)$. Any trajectory of this system resulting in $\tilde{\omega}(x_k) = 0 \, \forall k = 0, 1, \ldots$, is necessarily a trajectory of the system (7.32). Therefore, by Assumption ($A6$) it implies that $\lim_{k \to \infty} x_k = 0$, and by the LaSalle's invariance-principle, we conclude that the closed-loop system (7.36) is locally asymptotically-stable. $\square$

Remark 7.1.4 *It should be noted that the result of the above theorem also holds if the DHJIE (7.31) is replaced by the corresponding inequality. The results of the theorem will also continue to hold if Assumption ($A6$) is replaced by Assumption ($A4$).*

A block-diagram of the above parametrization is shown in Figure 7.2.
We can also specialize the above result to the linear system Σ^{dl} (7.17). In this regard, recall

$$R = \begin{bmatrix} R_{uu} & R_{uw} \\ R_{wu} & R_{ww} \end{bmatrix} = \begin{bmatrix} B_2^T P B_2 + D_{12}^T D_{12} & B_2^T P B_1 + D_{12}^T D_{11} \\ B_1^T P B_2 + D_{11}^T D_{12} & B_1^T P B1 + D_{11}^T D_{11} - \gamma^2 I \end{bmatrix}.$$

Then, we have the following proposition.

Proposition 7.1.3 *Consider the discrete-time linear system Σ^{dl} and suppose for some $\gamma > 0$ there exists a positive-definite matrix $P \in \Re^{n \times n}$ and a matrix $\Phi \in \Re^{m \times n}$ such that:*

(L1s)

$$R_{uu} > 0 \quad and \quad R_{ww} < 0;$$

(L5) *the DARE:*

$$
\begin{aligned}
P &= A^T P A + C_1^T C_1 - \begin{pmatrix} B_2^T P A + D_{11}^T C_1 \\ B_1^T P A + D_{12}^T C_1 \end{pmatrix}^T R^{-1} \begin{pmatrix} B_2^T P A + D_{11}^T C_1 \\ B_1^T P A + D_{12}^T C_1 \end{pmatrix} + \\
&\quad 2\Phi^T S R^{-1} S^T \Phi
\end{aligned}
\tag{7.38}
$$

holds for all $S \overset{\Delta}{=} [I \ \ 0]$;

(L6) *the matrix*

$$A_{cl} \overset{\Delta}{=} A - (B_2 \ \ B_1)^T \left(R^{-1} \begin{bmatrix} B_2^T P A + D_{11}^T C_1 \\ B_1^T P A + D_{12}^T C_1 \end{bmatrix} - \begin{bmatrix} S R^{-1} S^T \Phi \\ 0 \end{bmatrix} \right)$$

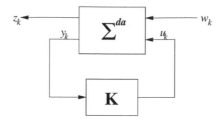

FIGURE 7.3
Configuration for Output Measurement-Feedback Discrete-Time Nonlinear $\mathcal{H}_\infty$-Control

is Hurwitz.

Then, Assumptions (A1s), (A5) and (A6) of Theorem 7.1.4 hold with

$$\left.\begin{array}{rcl}
r_{ij}(0) &=& R_{ij}, \quad i,j = u,w \\
\phi(x_k) &=& SR^{-1}S^T\Phi x_k \\
V(x) &=& \frac{1}{2}x^T P x \\
\left[\begin{array}{c} u^\star(x_k) \\ w^\star(x_k) \end{array}\right] &=& -R^{-1}\left(\begin{array}{c} B_2^T PA + D_{11}^T C_1 \\ B_1^T PA + D_{12}^T C_1 \end{array}\right) x_k.
\end{array}\right\} \tag{7.39}$$

Also, as a consequence of the above proposition, the following corollary gives a parametrization of a class of state-feedback controllers for the linear system Σ_l^d.

Corollary 7.1.3 *Consider the linear discrete-time system Σ_l^d (7.17). Suppose there exists a positive-definite matrix $P \in \Re^{n\times n}$ and a matrix $\Phi \in \Re^{p\times n}$ satisfying the hypotheses (L1s), (L5) and (L6) of Proposition 7.1.3. Then the family of state-feedback controllers:*

$$\mathbf{K}_{SFL} = \left\{ u_k | u_k = -SR^{-1}\left(\left[\begin{array}{c} B_2^T PA + D_{11}^T C_1 \\ B_1^T PA + D_{12}^T C_1 \end{array}\right] - S^T\Phi\right)x_k\right\} \tag{7.40}$$

solves the linear DSFBNLHICP for the system.

7.2 Output Measurement-Feedback Nonlinear $\mathcal{H}_\infty$-Control for Affine Discrete-Time Systems

In this section, we discuss the discrete-time disturbance attenuation problem with internal stability for the nonlinear plant (7.1) using measurement-feedback. This problem has already been discussed for the continuous-time plant in Chapter 5, and in this section we discuss the counterpart solution for the discrete-time case. A block diagram of the discrete-time controller with discrete measurements is shown in Figure 7.3 below.

We begin similarly with the definition of the problem.

Definition 7.2.1 *(Discrete-Time Measurement-Feedback (Sub-Optimal) Nonlinear $\mathcal{H}_\infty$-Control Problem (DMFBNLHICP)). Find (if possible!) a dynamic feedback-controller of the form:*

$$\Sigma_{dyn}^{dc} : \left\{ \begin{array}{rcl} \xi_{k+1} &=& \eta(\xi_k, y_k) \\ u_k &=& \theta(\xi_k, y_k), \end{array}\right. \tag{7.41}$$

where $\xi \in \Xi \subset \mathcal{X}$ a neighborhood of the origin, $\eta : \Xi \times \Re^m \to \Xi$, $\eta(0,0) = 0$, $\theta : \Xi \times \Re^m \to \Re^p$, $\theta(0,0) = 0$ are some smooth functions, which processes the measured variable y of the plant (7.1) and generates the appropriate control action u, such that the closed-loop system (7.1), (7.41) has locally finite ℓ_2-gain from the disturbance signal w to the output z less than or equal to some prescribed number $\gamma^* > 0$ with internal stability, or equivalently, the closed-loop system achieves local disturbance-attenuation less than or equal to γ^* with internal stability.

As in the continuous-time problem, we seek a dynamic compensator with an affine structure

$$\Sigma_{dyn}^{dac} : \begin{cases} \xi_{k+1} &= \tilde{f}(\xi_k) + \tilde{g}(\xi_k)y_k \\ u_k &= \alpha_{dyn}(\xi_k), \end{cases} \tag{7.42}$$

for some smooth function $\alpha_{dyn} : \Xi \to \Re^p$, $\alpha_{dyn}(0) = 0$. Applying this controller to the system (7.1), results in the following closed-loop system:

$$\Sigma_{clp}^{da} : \begin{cases} \dot{x}_{k+1}^e &= f^e(x_k^e) + g^e(x_k^e)w_k \\ z_k &= h^e(x_k^e) + k^e(x_k^e)w_k, \end{cases} \tag{7.43}$$

where $x^e = [x^T \; \xi]^T$,

$$f^e(x^e) = \begin{bmatrix} f(x) + g_2(x)\alpha_{dyn}(\xi) \\ \tilde{f}(\xi) + \tilde{g}(\xi)h_2(x) \end{bmatrix}, \quad g^e(x^e) = \begin{bmatrix} g_1(x) \\ \tilde{g}(\xi)k_{21}(x) \end{bmatrix} \tag{7.44}$$

$$h^e(x^e) = h_2(x) + k_{12}(x)\alpha_{dyn}(x), \quad k^e(x^e) = k_{11}(x). \tag{7.45}$$

The first goal is to achieve finite ℓ_2-gain from w to z for the above closed-loop system to be less than or equal to a prescribed number $\gamma > 0$. A sufficient condition for this can be obtained using the Bounded-real lemma, and is stated in the following lemma.

Lemma 7.2.1 *Consider the system (7.1) and suppose Assumption (A4) in Remark 7.1.1 holds for the system. Further, suppose there exists a C^2 positive-definite function $\Psi : O \times O \subset \mathcal{X} \times \mathcal{X} \to \Re$, locally defined in a neighborhood of $(x, \xi) = (0,0)$, with $\Psi(0,0) = 0$ satisfying*

(S1)

$$(g^e)^T(0,0)\frac{\partial^2 \Psi}{\partial \lambda^2}(0,0)g^e(0,0) + (k^e)^T(0,0)k^e(0,0) - \gamma^2 I < 0. \tag{7.46}$$

(S2)

$$\Psi(f^e(x^e) + g^e(x^e)\alpha_1(x^e)) - \Psi(x^e) + \frac{1}{2}(\|h^e(x^e) + k^e(x^e)\|^2 - \gamma^2 \|\alpha_1(x^e)\|^2) = 0, \tag{7.47}$$

where $w = \alpha_1(x^e)$, with $\alpha_1(0) = 0$, is a locally unique solution of

$$\left.\frac{\partial \Psi}{\partial \lambda}\right|_{\lambda = f^e(x^e) + g^e(x^e)w} g^e(x^e) + w^T(k^{e^T}(x^e)k^e(x^e) - \gamma^2 I) + h^{e^T}(x^e)k^e(x^e) = 0. \tag{7.48}$$

(S3) *The nonlinear discrete-time system*

$$\xi_{k+1} = \tilde{f}(\xi_k) \tag{7.49}$$

is locally asymptotically-stable at the equilibrium $\xi = 0$.

Then the DMFBNLHICP is solvable by the dynamic compensator Σ_{dyn}^{dac}.

Proof: By using Assumptions ($S1$) and ($S2$), it can be shown as in the proof of Theorem 3.6.2 that

$$\Psi(f^e(x^e) + g^e(x^e)w) - \Psi(x^e) \le \frac{1}{2}(\gamma^2\|w\|^2 - \|h^e(x^e) + k^e(x^e)w\|^2) \qquad (7.50)$$

holds for all $x^e \in O_1$ and all $w \in W_1$, where O_1 and W_1 are neighborhoods of $x^e = 0$ and $w = 0$ respectively. Therefore, the closed-loop system Σ_{clp}^{da} has ℓ_2-gain $\le \gamma$. Setting $w = 0$ in the inequality (7.50), implies that Σ_{clp}^{da} is Lyapunov-stable.

Furthermore, any bounded trajectory of Σ_{clp}^{da} corresponding to the condition that

$$\Psi(f^e(x^e)) \equiv \Psi(x^e),$$

is necessarily a trajectory of the system

$$x_{k+1}^e = f^e(x_k^e), \quad k = 0, 1, \ldots$$

under the constraint

$$h^e(x_k^e) \equiv 0, \quad k = 0, 1, \ldots.$$

Thus, we conclude from hypothesis ($A4$) that $\lim_{k\to\infty} x_k^e = 0$. This implies that, every trajectory of the system Σ_{clp}^{da} corresponding to the above condition approaches the trajectory of the system (7.49) which is locally asymptotically-stable at $\xi = 0$. Hence by the LaSalle's invariance-principle, we conclude asymptotic-stability of the equilibrium $x^e = 0$ of the closed-loop system. $\square$

Lemma 7.2.1 gives sufficient conditions for the solvability of the $DMFBNLHICP$ using the Bounded-real lemma. However, the details of how to choose the parameters $\tilde{f}$, $\tilde{g}$ and α_{dyn} are missing. Even though this gives the designer an extra degree-of-freedom to choose the above parameters to achieve additional design objectives, in general a more concrete design procedure is required. Accordingly, a design procedure similar to the continuous-time case discussed in Chapter 6, and in which an observer-based controller which is error-driven is used, can indeed be developed also for the discrete-time case. In this regard, consider the following dynamic-controller

$$\Sigma_{dynobs}^{dac} : \begin{cases} \xi_{k+1} & = & f(\xi_k) + g_1(\xi_k)w^\star(\xi_k) + g_2(\xi_k)u^\star(\xi_k) + G(\xi_k)[y_k - \\ & & h_2(\xi_k) - k_{21}(\xi_k)w^\star(\xi_k)] \\ u_k & = & u^\star(\xi_k), \end{cases} \qquad (7.51)$$

where $u^\star$ and $w^\star$ are the suboptimal feedback control and the worst-case disturbance respectively, determined in Section 7.2, and $G(.) \in \Re^{n\times m}$ is the output-injection gain matrix which has to be specified. Comparing now Σ_{dyn}^{dac} (7.42) with Σ_{dynobs}^{dac} (7.51) we see that

$$\begin{cases} \tilde{f}(\xi) & = & f(\xi) + g_1(\xi)w^\star(\xi) + g_2(\xi)u^\star(\xi) - G(\xi)[h_2(\xi) + k_{21}(\xi)w^\star(\xi)] \\ \tilde{g}(\xi) & = & G(\xi) \\ \alpha_{dyn}(\xi) & = & u^\star(\xi). \end{cases} \qquad (7.52)$$

Consequently, since the control law $\alpha_{dyn}(.)$ has to be obtained by solving the state-feedback problem, Lemma 7.2.1 can be restated in terms of the controller (7.51) as follows.

Lemma 7.2.2 *Consider the system (7.1) and suppose the following assumptions hold:*

(i) *Assumption ($A4$) is satisfied for the system;*

(ii) *there exists a C^2 positive-definite function V locally defined in a neighborhood of $x = 0$ in $\mathcal{X}$, such that Assumptions ($A1s$) and ($A2$) are satisfied;*

(iii) *there exists a C^2 positive-definite function $\Psi(x,\xi)$ locally defined in a neighborhood of $(x,\xi) = (0,0)$ in $\mathcal{X} \times \mathcal{X}$ and an $n \times m$ gain matrix $G(\xi)$ satisfying the hypotheses (S1), (S2) and (S3) of Lemma 7.2.1 for the closed-loop system Σ_{clp}^{da} with the controller (7.51),*

then the DMFBNLHICP is solvable with the observer-based controller Σ_{dynobs}^{dac}.

Remark 7.2.1 *Lemma 7.2.2 can be viewed as a separation principle for dynamic output-feedback controllers, i.e., that the design of the optimal state-feedback control and the output-estimator can be carried out separately.*

The following theorem then is a refinement of Lemma 7.2.2 and is the discrete-time analog of Theorem 6.1.3.

Theorem 7.2.1 *Consider the discrete-time nonlinear system (7.1) and suppose the following are satisfied:*

(I) *Assumption (A4) holds with rank $k_{12}(x) = p$ uniformly.*

(II) *There exists a C^2 positive-definite function V locally defined in a neighborhood of $x = 0$ in $\mathcal{X}$ such that Assumption (A2) is satisfied, and*

$$r_{ww}(0) < 0.$$

(III) *There exists an output-injection gain matrix $G(.)$ and a C^2 real-valued function $W : \tilde{X} \times \tilde{X} \to \Re$ locally defined in a neighborhood $\tilde{X} \times \tilde{X}$ of $(x,\xi) = (0,0)$ with $W(0,0) = 0$, $W(x,\xi) > 0$ for all $x \neq \xi$, and satisfying*

(T1)

$$g^{eT}(0,0)\frac{\partial^2 W}{\partial x^{e2}}(0,0)g^e(0,0) + k^{eT}(0,0)k^e(0,0) + r_{ww}(0) < 0, \tag{7.53}$$

(T2)

$$W(f^e(x^e) + g^e(x^e)\alpha_1(x^e)) - W(x^e) + V(f(x) + g_1(x)\alpha_1(x^e) +$$

$$g_2(x)u^\star(\xi)) - V(x) + \frac{1}{2}(\|h(x) + k_{11}(x)\alpha_1(x^e) + k_{12}(x)u^\star(\xi)\|^2 -$$

$$\gamma^2\|\alpha_1(x^e)\|^2) = 0, \tag{7.54}$$

where $w = \alpha_1(x^e)$, with $\alpha_1(0) = 0$, is a unique solution of

$$\left.\frac{\partial W}{\partial \lambda}\right|_{\lambda=f^e(x^e)+g^e(x^e)w} g^e(x^e) + \left.\frac{\partial V}{\partial \lambda}\right|_{\lambda=f(x)+g_1(x)w+g_2(x)u^\star(\xi)} g_1(x) +$$

$$(h_2(x) + k_{11}(x)w + k_{12}(x)u^\star(\xi))^T k_{12}^T(x) - \gamma^2 w^T = 0. \tag{7.55}$$

(T3) *The nonlinear discrete-time system*

$$\xi_{k+1} = f(\xi_k) + g_1(\xi)w^\star(\xi) - G(\xi)(h_2(\xi) + k_{21}(\xi)w^\star(\xi)) \tag{7.56}$$

is locally asymptotically-stable at the equilibrium point $\xi = 0$.

Then the DMFBNLHICP for the system is solvable with the dynamic compensator Σ_{dynobs}^{da}.

Proof: We sketch the proof as most of the details can be found in the preceding lemmas. We verify that the sufficient conditions given in Lemma 7.2.2 are satisfied. In this respect, we note that Assumption (I) implies that $r_{uu}(0) > 0$, therefore (ii) of Lemma 7.2.2 follows from Assumptions (I) and (II). Similarly, Assumption (III) $\Rightarrow$ (iii) of the lemma. To show this, let

$$\Upsilon(x^e) = V(x) + W(x, \xi).$$

Then it can easily be shown that $(T1) \Rightarrow (S1)$ and $(T2) \Rightarrow (S2)$. Similarly, $(T3) \Rightarrow (S3)$ because by hypothesis $(A4)$ $\lim_{k\to\infty} x_k = 0 \Rightarrow \lim_{k\to\infty} u^\star(\xi_k) = 0$ since $rank\ k_{12}(x) = p$ and

$$\lim_{k\to\infty} [h_1(x_k) + k_{12}(x_k)u^\star(\xi_k)] = 0.$$

Thus, any trajectory of (7.49), where $\tilde{f}$ is given by (7.52), approaches a trajectory of (7.56) which is locally asymptotically-stable at the equilibrium $\xi = 0$. $\square$

Remark 7.2.2 *Note that the DHJIE (7.54) can be replaced by the inequality (≤ 0) and the results of the theorem will still hold.*

The results of Theorem 7.2.1 can similarly be specialized to the linear system Σ^{dl}. If we suppose

$$V(x) = \frac{1}{2}x^T P x, \quad W(x^e) = \frac{1}{2}x^{eT} P^e x^e,$$

then the DHJIE (7.8) is equivalent to the discrete-algebraic-Riccati equation (DARE):

$$A^T PA - P + C_1^T C_1 - \begin{bmatrix} B_2^T PA + D_{12}^T C_1 \\ B_1^T PA + D_{11}^T C_1 \end{bmatrix}^T R^{-1} \begin{bmatrix} B_2^T PA + D_{12}^T C_1 \\ B_1^T PA + D_{11}^T C_1 \end{bmatrix} = 0. \tag{7.57}$$

In addition, the inequality (7.53) and the DHJIE (7.54) reduce respectively to the following inequalities:

$$\Delta \stackrel{\triangle}{=} B^{eT} P^e B^e + R_{ww} < 0; \tag{7.58}$$
$$A^{eT} P^e A^e - P^e - (A^{eT} P^e B^e + C^{eT} R_{uw})\Delta^{-1}(A^{eT} P^e B^e + C^{eT} R_{uw})^T +$$
$$C^{eT} R_{uu} C^e \leq 0 \tag{7.59}$$

where

$$A^e = \begin{bmatrix} \tilde{A} & B_2 F_2 \\ G\tilde{C}_2 & \tilde{A} - G\tilde{C}_2 + B_2 F_2 \end{bmatrix}, \quad B^e = \begin{bmatrix} B_1 \\ GD_{21} \end{bmatrix},$$
$$C^e = [-F_2 \quad F_2], \quad \tilde{A} = A + B_1 F_1, \quad \tilde{C}_2 = C_2 + D_{21} F_1$$
$$\begin{bmatrix} F_1 \\ F_2 \end{bmatrix} = -R^{-1} \begin{bmatrix} B_2^T PA + D_{12}^T C_1 \\ B_1^T PA + D_{11}^T C_1 \end{bmatrix}.$$

Similarly, equation (7.55) can be solved explicitly in this case to get

$$w = \alpha_1(x^e) = w^\star - \Delta^{-1}(B^{eT} P^e A^e + R_{wu} C^e)x^e$$
$$= [F_1 \ 0]x - \Delta^{-1}(B^{eT} P^e A^e + R_{wu} C^e)x^e.$$

Consequently, the following corollary is the linear discrete-time equivalent of Theorem 7.2.1.

Corollary 7.2.1 *Consider the discrete-time linear system Σ_l^d, and suppose the following hold:*

(IL) $rank\{D_{12}\} = p$ and

$$rank \begin{bmatrix} A - \lambda I & B_2 \\ C_1 & D_{12} \end{bmatrix} = n + p \quad \forall \lambda \in \mathbf{C}, \ |\lambda| = 1;$$

(IIL) *there exists a matrix $P > 0$ such that the DARE (7.57) is satisfied with $R_{ww} > 0$;*

(IIIL) *there exists an output injection gain matrix $G \in \Re^{n \times p}$ and a $2n \times 2n$ matrix $P^e \geq 0$ such that the inequalities (7.58), (7.59) are satisfied with the matrix $(A + B_1 F_1 - G(C_2 + D_{21} F_1))$ Hurwitz.*

Then, the linear DMFBNLHICP for Σ^{dl} is solvable using a dynamic compensator of the form:

$$\begin{cases} \xi_{k+1} &= (\tilde{A} - G\tilde{C}_2 + B_2 F_2)\xi_k + Gy_k \\ u_k &= F_2 \xi_k. \end{cases} \tag{7.60}$$

Remark 7.2.3 *The fundamental limitations for the use of the above result in solving the linear DMFBNLHICP are reminiscent of the continuous-time case discussed in Chapter 5 and indeed the discrete-time nonlinear case given in Theorem 7.2.1. These limitations follow from the fact that G and P^e are unknown, and so have to be solved simultaneously from the DARE (7.59) under the constraint (7.58) such that $(A + B_1 F_1 - G(C_2 + D_{21} F_1))$ is Hurwitz. In addition, P^e has twice the dimension of P, thus making the computations more expensive.*

The next lemma presents an approach for circumventing one of the difficulties of solving the inequalities (7.59) and (7.58), namely, the dimension problem.

Lemma 7.2.3 *Suppose there exists an $n \times n$ matrix $S > 0$ satisfying the strict discrete-time Riccati inequality (DRI):*

$$(\tilde{A} - G\tilde{C}_2)^T S(\tilde{A} - G\tilde{C}_2) - S + F_2^T R_{uu} F_2 - [(\tilde{A} - G\tilde{C}_2)^T S(B_1 - GD_{21}) -$$
$$F_2^T R_{uw}]\tilde{\Delta}^{-1}[(\tilde{A} - G\tilde{C}_2)^T S(B_1 - GD_{21}) - F_2^T R_{uw}]^T < 0 \tag{7.61}$$

under the constraint:

$$\tilde{\Delta} := (B_1 - GD_{21})^T S(B_1 - GD_{21}) + R_{ww} < 0. \tag{7.62}$$

Then the semidefinite matrix

$$P^e = \begin{bmatrix} S & -S \\ -S & S \end{bmatrix} \tag{7.63}$$

is a solution to the DARE (7.59) under the constraint (7.58). Moreover, the matrix G satisfying (7.61), (7.58) is such that $(\tilde{A} - G\tilde{C}_2)$ is Hurwitz.

Proof: By direct substitution of P^e given by (7.63) in (7.59), it can be shown that P^e satisfies this inequality together with the constraint (7.58).

Finally, consider the closed-loop system

$$x_{k+1} = (\tilde{A} - G\tilde{C}_2)x_k$$

and the Lyapunov-function candidate

$$Q(x) = x^T S x.$$

Then it can be shown that, along the trajectories of this system,

$$Q(x_{k+1}) - Q(x_k) < -\|F_2 x_k\|_{R_{uu}}^2 + \|[(\tilde{A} - G\tilde{C}_2)x_k]^T S(B_1 - GD_{21}) - (F_2 x_k)^T R_{uw}\|_{\tilde{\Delta}^{-1}}^2 \leq 0.$$

This shows that $(\tilde{A} - G\tilde{C}_2)$ is Hurwitz. $\square$

7.3   Extensions to a General Class of Discrete-Time Nonlinear Systems

In this section, we extend the results of the previous sections to a more general class of discrete-time nonlinear systems that may not necessarily be affine in u and w. We consider the general class of systems described by the state-space equations on $\mathcal{X} \subset \Re^n$ containing the origin $\{0\}$:

$$\Sigma^d \; : \; \begin{cases} x_{k+1} & = \; F(x_k, w_k, u_k); \; x(k_0) = x^0 \\ z_k & = \; Z(x_k, w_k, u_k) \\ y_k & = \; Y(x_k, w_k) \end{cases} \tag{7.64}$$

where all the variables have their usual meanings, and $F : \mathcal{X} \times \mathcal{W} \times \mathcal{U} \to \mathcal{X}$, $F(0,0,0) = 0$, $Z : \mathcal{X} \times \mathcal{W} \times \mathcal{U} \to \Re^s$, $Z(0,0,0) = 0$, $Y : \mathcal{X} \times \mathcal{W} \to \Re^m$. We begin with the full-information and state-feedback problems.

7.3.1   Full-Information $\mathcal{H}_\infty$-Control for a General Class of Discrete-Time Nonlinear Systems

To solve the full-information and state-feedback problems for the system Σ^d (7.64), we consider the Hamiltonian function:

$$\widetilde{H}(x, u, w) = \widetilde{V}(F(x, w, u)) - \widetilde{V}(x) + \frac{1}{2}(\|Z(x, u, w)\|^2 - \gamma^2 \|w\|^2) \tag{7.65}$$

for some positive-definite function $\widetilde{V} : \mathcal{X} \to \Re_+$, and let

$$\frac{\partial^2 \widetilde{H}}{\partial(u, w)^2}(0, 0, 0) \; = \; \begin{bmatrix} \frac{\partial^2 \widetilde{H}}{\partial u^2} & \frac{\partial^2 \widetilde{H}}{\partial w \partial u} \\ \frac{\partial^2 \widetilde{H}}{\partial u \partial w} & \frac{\partial^2 \widetilde{H}}{\partial w^2} \end{bmatrix}(0, 0, 0)$$

$$\stackrel{\Delta}{=} \begin{bmatrix} h_{uu}(0) & h_{uw}(0) \\ h_{wu}(0) & h_{ww}(0) \end{bmatrix},$$

where

$$h_{uu}(0) \; = \; \left[\left(\frac{\partial F}{\partial u}\right)^T \frac{\partial^2 \widetilde{V}}{\partial \lambda^2}(0) \left(\frac{\partial F}{\partial u}\right) + \left(\frac{\partial Z}{\partial u}\right)^T \left(\frac{\partial Z}{\partial u}\right) \right] \Bigg|_{x=0, u=0, w=0},$$

$$h_{ww}(0) \; = \; \left[\left(\frac{\partial F}{\partial w}\right)^T \frac{\partial^2 \widetilde{V}}{\partial \lambda^2}(0) \left(\frac{\partial F}{\partial w}\right) + \left(\frac{\partial Z}{\partial w}\right)^T \left(\frac{\partial Z}{\partial w}\right) - \gamma^2 I \right] \Bigg|_{x=0, u=0, w=0},$$

$$h_{uw}(0) \; = \; \left[\left(\frac{\partial F}{\partial u}\right)^T \frac{\partial^2 \widetilde{V}}{\partial \lambda^2}(0) \left(\frac{\partial F}{\partial w}\right) + \left(\frac{\partial Z}{\partial u}\right)^T \left(\frac{\partial Z}{\partial w}\right) \right] \Bigg|_{x=0, u=0, w=0}$$

$$= \; h_{wu}^T(0).$$

Suppose now that the following assumption holds:

(GN1)

$$h_{uu}(0) > 0, \quad h_{ww}(0) - h_{wu}(0)h_{uu}^{-1}(0)h_{uw}(0) < 0.$$

Then by the Implicit-function Theorem, the above assumption implies that there exist unique smooth functions $\tilde{u}^\star(x)$, $\tilde{w}^\star(x)$ with $\tilde{u}^\star(0) = 0$ and $\tilde{w}^\star(0) = 0$ defined in a neighborhood $\bar{X}$ of $x = 0$ and satisfying the functional equations

$$
\begin{aligned}
0 &= \frac{\partial \tilde{H}}{\partial u}(x, \tilde{u}^\star(x), \tilde{w}^\star(x)) \\
&= \left. \left(\frac{\partial \tilde{V}}{\partial F} \frac{\partial F}{\partial u} + Z^T \frac{\partial Z}{\partial u} \right) \right|_{u = \tilde{u}^\star(x), w = \tilde{w}^\star(x)}
\end{aligned}
\tag{7.66}
$$

$$
\begin{aligned}
0 &= \frac{\partial \tilde{H}}{\partial w}(x, \tilde{u}^\star(x), \tilde{w}^\star(x)) \\
&= \left. \left(\frac{\partial \tilde{V}}{\partial F} \frac{\partial F}{\partial w} + Z^T \frac{\partial Z}{\partial w} - \gamma^2 w^T \right) \right|_{u = \tilde{u}^\star(x), w = \tilde{w}^\star(x)}.
\end{aligned}
\tag{7.67}
$$

Similarly, let

$$
\begin{aligned}
h_{uu}(x) &\triangleq \frac{\partial^2 \tilde{H}}{\partial u^2}(x, \tilde{u}^\star(x), \tilde{w}^\star(x)) \\
h_{ww}(x) &\triangleq \frac{\partial^2 \tilde{H}}{\partial w^2}(x, \tilde{u}^\star(x), \tilde{w}^\star(x)) \\
h_{uw}(x) &\triangleq \frac{\partial^2 \tilde{H}}{\partial w \partial u}(x, \tilde{u}^\star(x), \tilde{w}^\star(x)) = h_{wu}^T(x, \tilde{u}^\star, \tilde{w}^\star)
\end{aligned}
$$

be associated with the optimal solutions $\tilde{u}^\star(x)$, $\tilde{w}^\star(x)$, and let the corresponding DHJIE associated with (7.64) be denoted by

(GN2)

$$
\tilde{V}(F(x, \tilde{w}^\star(x), \tilde{u}^\star(x))) - \tilde{V}(x) + \frac{1}{2}(\|Z(x, \tilde{u}^\star(x), \tilde{w}^\star(x))\|^2 - \gamma^2 \|\tilde{w}^\star(x)\|^2) = 0, \quad \tilde{V}(0) = 0.
\tag{7.68}
$$

Then we have the following result for the solution of the full-information problem.

Theorem 7.3.1 *Consider the discrete-time nonlinear system (7.64), and suppose there exists a C^2 positive-definite function $\tilde{V} : X_1 \subset X \to \Re_+$ locally defined in a neighborhood X_1 of $x = 0$ satisfying the hypotheses (GN1), (GN2). In addition, suppose the following assumption is also satisfied by the system:*

(GN3) *Any bounded trajectory of the free system*

$$
x_{k+1} = F(x_k, 0, u_k),
$$

under the constraint

$$
Z(x_k, 0, u_k) = 0
$$

for all $k \in \mathbf{Z}_+$, is such that, $\lim_{k \to \infty} x_k = 0$.

Then there exists a static full-information feedback control

$$
u_k = \tilde{u}^\star(x_k) - h_{uu}^{-1}(x_k) h_{uw}(x_k)(w_k - \tilde{w}^\star(x_k))
$$

which solves the DFIFBNLHICP for the system.

Proof: The proof can be pursued along similar lines as Theorem 7.1.2. □

The above result can also be easily specialized to the state-feedback case as follows.

Theorem 7.3.2 *Consider the discrete-time nonlinear system (7.64), and suppose there exists a C^2 positive-definite function $\tilde{V} : X_2 \subset \mathcal{X} \to \Re_+$ locally defined in a neighborhood X_2 of $x = 0$ satisfying the hypothesis*

(GN1s)
$$h_{uu}(0) > 0, \quad h_{ww}(0) < 0$$

and hypotheses (GN2), (GN3) above. Then, the static state-feedback control

$$u_k = \tilde{u}^\star(x_k)$$

solves the DSFBNLHICP for the system.

Proof: The theorem can be proven along similar lines as Theorem 7.1.3. □

Moreover, the parametrization of all static state-feedback controllers can also be given in the following theorem.

Theorem 7.3.3 *Consider the discrete-time nonlinear system (7.64), and suppose the following hypothesis holds*

(GN2s) *there exists a C^2 positive-definite function $\tilde{V} : X_3 \subset \mathcal{X} \to \Re_+$ locally defined in a neighborhood X_3 of $x = 0$ satisfying the DHJIE*

$$\tilde{V}(F(x, \tilde{w}^\star(x), \tilde{u}^\star(x))) - \tilde{V}(x) + \frac{1}{2}(\|Z(x, \tilde{u}^\star(x), \tilde{w}^\star(x))\|^2 - \gamma^2\|\tilde{w}^\star(x)\|^2)$$

$$= -\psi(x)[h_{uu}(x) - h_{uw}(x)h_{ww}^{-1}(x)h_{21}(x)]\psi(x), \quad \tilde{V}(0) = 0 \qquad (7.69)$$

for some arbitrary smooth function $\psi : X_3 \to \Re^p$, $\psi(0) = 0$,

as well as the hypotheses (GN3) and (GN1s) with $\tilde{V}$ in place of $\tilde{V}$. Then, the family of controllers

$$\mathbf{K}_{SFg} = \{u_k | u_k = \tilde{u}^\star(x_k) + \psi(x_k)\} \qquad (7.70)$$

is a parametrization of all static state-feedback controllers that solves the DSFBNLHICP for the system.

7.3.2   Output Measurement-Feedback $\mathcal{H}_\infty$-Control for a General Class of Discrete-Time Nonlinear Systems

In this subsection, we discuss briefly the output measurement-feedback problem for the general class of nonlinear systems (7.64). Theorem 7.2.1 can easily be generalized to this class of systems. As in the previous case, we can postulate the existence of a dynamic compensator of the form:

$$\widetilde{\Sigma}_{dynobs}^{dc} : \left\{ \begin{array}{rcl} \xi_{k+1} & = & F(\xi_k, \tilde{w}^\star(\xi_k), \tilde{u}^\star(\xi_k)) + \widetilde{G}(\xi_k)(y_k - Y(\xi_k, \tilde{w}^\star(\xi_k))) \\ u_k & = & \tilde{u}^\star(\xi_k) \end{array} \right. \qquad (7.71)$$

where $\widetilde{G}(.)$ is the output-injection gain matrix, and $\tilde{w}^\star(.)$, $\tilde{u}^\star(.)$, are the solutions to equations (7.66), (7.67). Let the closed-loop system (7.64) with the controller (7.71) be represented as

$$\begin{array}{rcl} x_{k+1}^e & = & F^e(x_k^e, w_k) \\ z_k & = & Z^e(x_k^e, w_k) \end{array}$$

where $x^e = [x^T \ \xi^T]^T$,

$$F^e(x^e, w) = \left[\begin{array}{c} F(x, w, \tilde{u}^\star) \\ F(x, \tilde{w}^\star(\xi), \tilde{u}^\star(\xi)) + \tilde{G}(\xi)(y(x, w) - y(\xi, \tilde{w}^\star(\xi))) \end{array} \right],$$

$$Z^e(x^e, w) = Z(x, w, \tilde{u}^\star(\xi)).$$

Then the following result is a direct extension of Theorem 7.2.1.

Theorem 7.3.4 *Consider the discrete-time nonlinear system (7.64) and assume the following:*

(i) *Assumption (GN3) holds and* $\text{rank}\{\frac{\partial Z}{\partial u}(0,0,0)\} = p$.

(ii) *There exists a C^2 positive-definite function $\tilde{V} : X \to \Re_+$ locally defined in a neighborhood X of $x = 0$ satisfying Assumption (GN2).*

(iii) *There exists an output-injection gain matrix $\tilde{G}(.)$ and a C^2 real-valued function $W : X_4 \times X_4$ locally defined in a neighborhood $X_4 \times X_4$ of $(x, \xi) = (0, 0)$, $X \cap X_4 \neq \emptyset$, with $W(0,0) = 0$, $W(x, \xi) > 0 \, \forall x \neq \xi$ and satisfying*

(GNM1)

$$F^{e^T}(0,0,0) \frac{\partial^2 W}{\partial x^{e2}}(0,0) F^e(0,0,0) + h_{ww}(0) < 0;$$

(GNM2)

$$W(F^e(x^e, \tilde{\alpha}_1(x^e) - W(x^e) + V(F(x, \tilde{\alpha}_1(x^e), u^\star(\xi))) - V(x) +$$

$$\frac{1}{2}(\|Z(x, \tilde{\alpha}_1(x^e), \tilde{u}^\star(\xi))\|^2 - \gamma^2 \|\tilde{\alpha}_1(x^e)\|^2) = 0,$$

where $\tilde{\alpha}_1(x^e)$ with $\tilde{\alpha}_1(0) = 0$ is a locally unique solution of the equation

$$\left. \frac{\partial W}{\partial \beta} \right|_{\beta = F^e(x^e, w)} \frac{\partial F^e}{\partial w}(x^e, w) + \left. \frac{\partial V}{\partial \lambda} \right|_{\lambda = F(x, w, \tilde{u}^\star(\xi))} \frac{\partial F}{\partial w}(x, w, \tilde{u}^\star(\xi)) +$$

$$Z^T(x, w, \tilde{u}^\star(\xi)) \frac{\partial Z}{\partial w}(x, w, \tilde{u}^\star(\xi)) - \gamma^2 w^T = 0.$$

(GNM3) *The discrete-time nonlinear system*

$$x_{k+1} = F(\xi, \tilde{w}^\star(\xi), 0) - \tilde{G}(\xi)Y(\xi, \tilde{w}^\star(\xi))$$

is locally asymptotically-stable at $\xi = 0$.

Then, the DMFBNLHICP for the system (7.64) is solvable with the compensator (7.71).

7.4 Approximate Approach to the Discrete-Time Nonlinear $\mathcal{H}_\infty$-Control Problem

In this section, we discuss alternative approaches to the discrete-time nonlinear $\mathcal{H}_\infty$-Control problem for affine systems. It should have been observed in Sections 7.1, 7.2, that the control laws that were derived are given implicitly in terms of solutions to certain pairs of algebraic equations. This makes the computational burden in using this design method more intensive. Therefore, in this section, we discuss alternative approaches, although approximate, but which can yield explicit solutions to the problem. We begin with the state-feedback problem.

7.4.1   An Approximate Approach to the Discrete-Time State-Feedback Problem

We consider again the nonlinear system (7.1), and assume the following.

Assumption 7.4.1

$$rank\{k_{12}(x)\} = p.$$

Reconsider now the Hamiltonian function (7.11) associated with the problem:

$$H_2(w,u) \;=\; V(f(x)+g_1(x)w+g_2(x)u) - V(x) + \frac{1}{2}(\|h_1(x)+k_{11}(x)w+k_{12}(x)u\|^2 - \gamma^2\|w\|^2) \tag{7.72}$$

for some smooth positive-definite function $V : \mathcal{X} \to \Re_+$. Suppose there exists a smooth real-valued function $\bar{u}(x) \in \Re^p$, $\bar{u}(0) = 0$ such that the HJI-inequality

$$H_2(w,\bar{u}(x)) < 0 \tag{7.73}$$

is satisfied for all $x \in \mathcal{X}$ and $w \in \mathcal{W}$. Then it is clear from the foregoing that the control law

$$u = \bar{u}(x)$$

solves the *DSFBNLHICP* for the system Σ^{da} globally. The bottleneck however, is in getting an explicit form for the function $\bar{u}(x)$. This problem stems from the first term in the HJI-inequality (7.73), i.e.,

$$V(f(x)+g_1(x)w+g_2(x)u),$$

which is a composition of functions and is not necessarily quadratic in u, as in the continuous-time case. Thus, suppose we replace this term by an approximation which is "quadratic" in (w,u), and nonlinear in x, i.e.,

$$V(f(x)+v) = V(f(x)) + V_x(f(x))v + \frac{1}{2}v^T V_{xx}(f(x))v + R_m(x,v)$$

for some vector function $v \in \mathcal{X}$ and where R_m is a remainder term such that

$$\lim_{v \to 0} \frac{R_m(x,v)}{\|v\|^2} = 0.$$

Then, we can seek a saddle-point for the new Hamiltonian function

$$\widehat{H}_2(w,u) \;=\; V(f(x)) + V_x(f(x))(g_1(x)w+g_2(x)u) +$$
$$\frac{1}{2}(g_1(x)w+g_2(x)u)^T V_{xx}(f(x))(g_1(x)+g_2(x)u) - V(x) +$$
$$\frac{1}{2}\|h_1(x)+k_{11}(x)w+k_{12}(x)u\|^2 - \frac{1}{2}\gamma^2\|w\|^2 \tag{7.74}$$

by neglecting the higher-order term $R_m(x,g_1(x)w+g_2(x)u)$. Since $\widehat{H}_2(u,w)$ is quadratic in (w,u), it can be represented as

$$\widehat{H}_2(w,u) = V(f(x)) - V(x) + \frac{1}{2}h_1^T(x)h_1(x) + \widehat{S}(x)\begin{bmatrix} w \\ u \end{bmatrix} + \frac{1}{2}\begin{bmatrix} w \\ u \end{bmatrix}\widehat{R}(x)\begin{bmatrix} w \\ u \end{bmatrix},$$

where

$$\widehat{S}(x) = h_1^T(x)[k_{11}(x)\;\;k_{12}(x)] + V_x(f(x))[g_1(x)\;\;g_2(x)]$$

and

$$\widehat{R}(x) = \begin{pmatrix} k_{11}^T(x)k_{11}(x) - \gamma^2 I & k_{11}^T(x)k_{12}(x) \\ k_{12}^T(x)k_{11}(x) & k_{12}^T(x)k_{12}(x) \end{pmatrix} + \begin{pmatrix} g_1^T(x) \\ g_2^T(x) \end{pmatrix} V_{xx}(f(x))(g_1(x) \ \ g_2(x)).$$

From this, it is easy to determine conditions for the existence of a unique saddle-point and explicit formulas for the coordinates of this point. It can immediately be determined that, if $\widehat{R}(x)$ is nonsingular, then

$$\widehat{H}_2(w,u) = \widehat{H}_2(w^\star(x), u^\star(x)) + \frac{1}{2} \begin{bmatrix} w - \hat{w}^\star(x) \\ u - \hat{u}^\star(x) \end{bmatrix}^T \widehat{R}(x) \begin{bmatrix} w - \hat{w}^\star(x) \\ u - \hat{u}^\star(x) \end{bmatrix} \tag{7.75}$$

where

$$\begin{bmatrix} \hat{w}^\star(x) \\ \hat{u}^\star(x) \end{bmatrix} = -\widehat{R}^{-1}(x)\widehat{S}^T(x). \tag{7.76}$$

One condition that gurantees that $\widehat{R}(x)$ is nonsingular (by Assumption 7.4.1) is that the submatrix

$$\widehat{R}_{11}(x) := k_{11}^T(x)k_{11}(x) - \gamma^2 I + \frac{1}{2}g_1^T(x)V_{xx}(f(x))g_1(x) < 0 \ \ \forall x.$$

If the above condition is satisfied for some $\gamma > 0$, then $\widehat{H}_2(x, w, u)$ has a saddle-point at $(\hat{u}^\star, \hat{w}^\star)$, and

$$\widehat{H}_2(\hat{w}^\star(x), \hat{u}^\star(x)) = V(f(x)) - V(x) - \frac{1}{2}\widehat{S}(x)\widehat{R}^{-1}(x)\widehat{S}^T(x) + \frac{1}{2}h_1^T(x)h_1(x). \tag{7.77}$$

The above development can now be summarized in the following lemma.

Lemma 7.4.1 *Consider the discrete-time nonlinear system (7.1), and suppose there exists a smooth positive-definite function $V : X_0 \subset \mathcal{X} \to \Re_+$, $V(0) = 0$ and a positive number $\delta > 0$ such that*

(i)

$$\widehat{H}_2(\hat{w}^\star(x), \hat{u}^\star(x)) < 0 \ \ \forall 0 \neq x \in X_0, \tag{7.78}$$

(ii)

$$\widehat{H}_2(\hat{w}^\star(x), \hat{u}^\star(x)) < -\frac{1}{2}\delta(\|\hat{w}^\star(x)\|^2 + \|\hat{u}^\star(x)\|^2) \ \ \forall x \in X_0,$$

(iii)

$$\widehat{R}_{11}(x) < 0 \ \ \forall x \in X_0.$$

Then, there exists a neighborhood $X \times W$ of $(w, x) = (0, 0)$ in $\mathcal{X} \times \mathcal{W}$ such that V satisfies the HJI-inequality (7.73) with $\bar{u} = \hat{u}^\star(x)$, $\hat{u}^\star(0) = 0$.

Proof: By construction, $\widehat{H}_2(x, w, u)$ satisfies

$$\widehat{H}_2(x, w, u) = \widehat{H}_2(\hat{w}^\star(x), \hat{u}^\star(x)) + \frac{1}{2}(w - \hat{w}^\star)^T \widehat{R}_{11}(x)(w - \hat{w}^\star(x)).$$

Since $\widetilde{R}_{11}(0)$ is negative-definite by hypothesis (iii), there exists a neighborhood X_1 of $x = 0$ and a positive number $c > 0$ such that

$$(w - \hat{w}^\star)^T \widehat{R}_{11}(x)(w - \hat{w}^\star(x)) \leq c\|w - w^\star(x)\|^2 \ \ \forall x \in X_1, \forall w.$$

Now let $\mu = \min\{\delta, c\}$. Then,

$$(w - \hat{w}^\star)^T \widehat{R}_{11}(x)(w - \hat{w}^\star(x)) \le \mu \|w - \hat{w}^\star(x)\|^2 \quad \forall x \in X_1.$$

Moreover, by hypothesis (ii)

$$\widehat{H}_2(\hat{w}^\star(x), \hat{u}^\star(x)) \le -\frac{\mu}{2}(\|\hat{w}^\star(x)\|^2 + \|u^\star(x)\|^2) \quad \forall x \in X_0.$$

Thus, by the triangle inequality,

$$
\begin{aligned}
\widehat{H}_2(w, \hat{u}^\star(x)) &\le -\frac{\mu}{2}(\|w^\star(x)\|^2 + \frac{1}{2}\|\hat{u}^\star(x)\|^2 + \frac{1}{2}\|w - \hat{w}^\star(x)\|^2) \\
&\le -\frac{\mu}{2}(\|w\|^2 + \|\hat{u}^\star(x)\|^2)
\end{aligned}
\tag{7.79}
$$

for all $x \in X_2$, where $X_2 = X_0 \bigcap X_1$. Notice however that the Hamiltonians $H_2(x, w, u)$ and $\widehat{H}_2(x, w, u)$ defined by (7.72) and (7.74) respectively, are related by

$$H_2(x, w, u) = \widehat{H}_2(x, w, u) + R_m(x, g_1(x)w + g_2(x)u). \tag{7.80}$$

By the result in Section 8.14.3 of reference [94], for all $\kappa > 0$, there exist neighborhoods X_3 of $x = 0$, W_1 of $w = 0$ and U_1 of $u = 0$ such that

$$|R_m(x, g_1(x)w + g_2(x)u)| \le \kappa(\|w\|^2 + \|u\|^2) \quad \forall (x, w, u) \in X_3 \times W_1 \times U_1. \tag{7.81}$$

Finally, combining (7.79), (7.80) and (7.81), one obtains an estimate for $H_2(w, u^\star(x))$, i.e.,

$$H_2(w, u^\star(x)) \le -\frac{\mu}{2}\left(1 - \frac{\kappa}{\mu}\right)(\|w\|^2 + \|\hat{u}^\star(x)\|^2).$$

Choosing $\kappa < \mu$ and $X \subseteq X_3$, $W_1 \subseteq W$, the result follows. $\square$

From the above lemma, one can conclude the following.

Theorem 7.4.1 *Consider the discrete-time nonlinear system (7.1), and assume all the hypotheses (i), (ii), (iii) of Lemma 7.4.1 hold. Then the closed-loop system*

$$
\Sigma^{da} : \begin{cases}
x_{k+1} &= f(x_k) + g_1(x_k)w_k + g_2(x_k)\bar{u}(x_k); \quad x_0 = 0 \\
z_k &= h_1(x_k) + k_{11}(x_k)w_k + k_{12}(x_k)\bar{u}(x_k)
\end{cases}
\tag{7.82}
$$

with $\bar{u}(x) = \hat{u}^\star(x)$ has a locally asymptotically-stable equilibrium-point at $x = 0$, and for every $K \in \mathbf{Z}_+$, there exists a number $\epsilon > 0$ such that the response of the system from the initial state $x_0 = 0$ satisfies

$$\sum_{k=0}^{K} \|z_k\|^2 \le \gamma^2 \sum_{k=0}^{K} \|w_k\|^2$$

for every sequence $w = (w_0, \ldots, w_K)$ such that $\|w_k\| < \epsilon$.

Proof: Since $f(.)$ and $\hat{u}^\star(.)$ are smooth and vanish at $x = 0$, it is easily seen that for every $K > 0$, there exists a number $\epsilon > 0$ such that the response of the closed-loop system to any input sequence $w = (w_0, \ldots, w_K)$ from the initial state $x_0 = 0$ is such that $x_k \in X$ for all $k \le K + 1$ as long as $\|w_k\| < \epsilon$ for all $k \le K$. Without any loss of generality, we may assume that ϵ is such that $w_k \in W$. In this case, using Lemma 7.4.1, we can deduce that the dissipation-inequality

$$V(x_{k+1}) - V(x_k) + \frac{1}{2}(z_k^T z_k - \gamma^2 w_k^T w_k) \le 0 \quad \forall k \le K$$

holds. The result now follows from Chapter 3 and a Lyapunov argument. $\square$

Remark 7.4.1 *Again, in the case of the linear system Σ^{dl} (7.17), the result of Theorem 7.4.1 reduces to wellknown necessary and sufficient conditions for the existence of a solution to the linear $DSFBNLHICP$ [89]. Indeed, setting $B := [B_1 \ B_2]$ and $D := [D_{11} \ D_{12}]$, a quadratic function $V(x) = \frac{1}{2}x^T P x$ with $P = P^T > 0$ satisfies the hypotheses of the theorem if and only if*

$$A^T P A - P + C_1^T C_1 - F_p^T (R + B^T P B) F_p \ < \ 0$$
$$D_{11}^T D_{11} - \gamma^2 I + B_1^T P B_1 \ < \ 0$$

where $F_p = -(R + B^T P B)^{-1}(B^T P A + D^T C_1)$. In this case,

$$\begin{bmatrix} \hat{w}^\star \\ \hat{u}^\star \end{bmatrix} = F_p x.$$

In the next subsection, we consider an approximate approach to the measurement-feedback problem.

7.4.2 An Approximate Approach to the Discrete-Time Output Measurement-Feedback Problem

In this section, we discuss an alternative approximate approach to the discrete-time measurement-feedback problem for affine systems. In this regard, assume similarly a dynamic observer-based controller of the form

$$\bar{\Sigma}_{dynobs}^{dac} \ : \ \begin{cases} \theta_{k+1} & = \ f(\theta_k) + g_1(\theta_k)\hat{w}^\star(\theta_k) + g_2(\theta_k)\hat{u}^\star(\theta_k) + \bar{G}(\theta_k)[y_k - h_2(\theta_k) - \\ & \quad k_{21}(\theta_k)\hat{w}^\star(\theta_k)] \\ u_k & = \ \hat{u}^\star(\theta_k) \end{cases} \tag{7.83}$$

where $\theta \in \mathcal{X}$ is the controller state vector, while $\hat{w}^\star(.)$, $\hat{u}^\star(.)$ are the optimal state-feedback control and worst-case disturbance given by (7.76) respectively, and $\bar{G}(.)$ is the output-injection gain matrix which is to be determined. Accordingly, the corresponding closed-loop system (7.1), (7.83) can be represented by

$$\begin{cases} x_{k+1}^\sharp & = \ f^\sharp(x_k^\sharp) + g^\sharp(x_k^\sharp)w_k \\ z_k^\sharp & = \ h^\sharp(x^\sharp) + k^\sharp(x^\sharp)w_k \end{cases} \tag{7.84}$$

where $x^\sharp = [x^T \ \theta^T]^T$,

$$f^\sharp(x^\sharp) \ = \ \begin{bmatrix} f(x) + g_2(x)\hat{u}^\star(\theta) \\ \begin{pmatrix} f(\theta) + g_1(\theta)\hat{w}^\star(\theta) + g_2(\theta)\hat{u}^\star(\theta) + \\ \bar{G}(\theta)(h_2(x) - h_2(\theta) - k_{21}(\theta)\hat{w}^\star(\theta)) \end{pmatrix} \end{bmatrix},$$

$$g^\sharp(x^\sharp) \ = \ \begin{bmatrix} g_1(x) \\ \bar{G}(\theta)k_{21}(x) \end{bmatrix},$$

and

$$h^\sharp(x^\sharp) = h_1(x) + k_{12}(x)\hat{u}^\star(\theta), \ \ k^\sharp(x^\sharp) = k_{11}(x).$$

The objective is to find sufficient conditions under which the above closed-loop system (7.84) is locally (globally) asymptotically-stable and the estimate $\theta \to x$ as $t \to \infty$. This can be achieved by first rendering the closed-loop system dissipative, and then using some suitable additional conditions to conclude asymptotic-stability.

Thus, we look for a suitable positive-definite function $\Psi_1 : \mathcal{X} \times \mathcal{X} \to \Re_+$, such that the dissipation-inequality

$$\Psi_1(x_{k+1}^\sharp) - \Psi_1(x_k^\sharp) + \frac{1}{2}(\|z_k^\sharp\|^2 - \gamma^2 \|w_k\|^2) \leq 0 \tag{7.85}$$

is satisfied along the trajectories of the system (7.84). To achieve this, we proceed as in the continuous-time case, Chapter 5, and assume the existence of a smooth C^2 function $W : \mathcal{X} \times \mathcal{X} \to \Re$ such that $W(x^\sharp) \geq 0$ for all $x^\sharp \neq 0$ and $W(x^\sharp) > 0$ for all $x \neq \theta$. Further, set

$$H_2^\sharp(w) = W(f^\sharp(x^\sharp) + g^\sharp(x^\sharp)w) - W(x^\sharp) + H_2(w, \hat{u}^\star(\theta)) - \widehat{H}_2(\hat{w}^\star(x), \hat{u}^\star(x)), \tag{7.86}$$

and recall that

$$H_2(w, u) = \widehat{H}_2(w, u) + R_m(x, g_1(x)w + g_2(x)u);$$

so that

$$H_2(w, \hat{u}^\star(\theta)) = \widehat{H}_2(w, \hat{u}^\star(\theta)) + R_m(x, g_1(x)w + g_2(x)\hat{u}^\star(x)).$$

Moreover, by definition

$$
\begin{aligned}
H_2(w, \hat{u}^\star(\theta)) &= V(f(x) + g_1(x)w + g_2(x)\hat{u}^\star(\theta)) - V(x) + \frac{1}{2}\|h_2(x) + \\
&\quad k_{11}(x)w + k_{12}(x)\hat{u}^\star(\theta)\|^2 - \frac{1}{2}\gamma^2\|w\|^2.
\end{aligned} \tag{7.87}
$$

Therefore, subsituting (7.87) in (7.86) and rearranging, we get

$$
\begin{aligned}
H_2^\sharp(w) + \widehat{H}_2(\hat{u}^\star(x), \hat{w}^\star(x)) &= W(f^\sharp(x^\sharp) + g^\sharp(x^\sharp)w) - W(x^\sharp) + \\
V(f(x) + g_1(x)w + g_2(x)\hat{u}^\star(\theta)) - V(x) &+ \tfrac{1}{2}\|h^\sharp(x^\sharp) + k^\sharp(x^\sharp)w\|^2 - \tfrac{1}{2}\gamma^2\|w\|^2 \quad (7.88)
\end{aligned}
$$

From the above identity (7.88), it is clear that, if the right-hand-side is nonpositive, then the positive-definite function

$$\Psi_1(x^\sharp) = W(x^\sharp) + V(x)$$

will indeed have satisfied the dissipation-inequality (7.85) along the trajectories of the closed-loop system (7.84). Moreover, if we assume the hypotheses of Theorem 7.4.1 (respectively Lemma 7.4.1) hold, then the term $\widehat{H}_2(\hat{u}^\star(x), \hat{w}^\star(x))$ is nonpositive. Therefore, it remains to impose on $H_2^\sharp(w)$ to be also nonpositive for all w.

One way to achieve the above objective, is to impose the condition that

$$\max_w H_2^\sharp(w) < 0.$$

However, finding a closed-form expression for $w^{\star\star} = arg\ \max\{H_2^\sharp(w)\}$ is in general not possible as observed in the previous section. Thus, we again resort to an approximate but practical approach. Accordingly, we can replace the term $W(f^\sharp(x^\sharp) + v^\sharp)$ in $H_2^\sharp(.)$ by its second-order Taylor approximation as:

$$W(f^\sharp(x^\sharp) + v^\sharp) = W(f^\sharp(x^\sharp)) + W_{x^\sharp}(f^\sharp(x^\sharp))v^\sharp + \frac{1}{2}v^{\sharp T}W_{x^\sharp x^\sharp}(f^\sharp(x^\sharp))v^\sharp + R_m^\sharp(x^\sharp, v^\sharp)$$

for any $v^\sharp \in \mathcal{X} \times \mathcal{X}$, and where $R_m^\sharp(x^\sharp, v^\sharp)$ is the remainder term. While the last term (recalling from equation (7.75)) can be represented as

$$
\begin{aligned}
H_2(w, \hat{u}^\star(\theta)) - \widehat{H}_2(\hat{w}^\star(x), \hat{u}^\star(x)) &= \frac{1}{2}\begin{bmatrix} w - \hat{w}^\star(x) \\ \hat{u}^\star(\theta) - \hat{u}^\star(x) \end{bmatrix}^T \widehat{R}(x)\begin{bmatrix} w - \hat{w}^\star(x) \\ \hat{u}^\star(\theta) - \hat{u}^\star(x) \end{bmatrix} \\
&\quad + R_m(x, g_1(x)w + g_2(x)\hat{u}^\star(\theta)).
\end{aligned}
$$

Similarly, observe that $R_m(x, g_1(x)w + g_2(x)\hat{u}^\star(\theta))$ can be expanded as a function of $x^\sharp$ with respect to w as

$$R_m(x, g_1(x)w + g_2(x)\hat{u}^\star(\theta)) = R_{m0}(x^\sharp) + R_{m1}(x^\sharp)w + w^T R_{m2}(x^\sharp)w + R_{m3}(x^\sharp, w).$$

Thus, we can now approximate $H_2^\sharp(w)$ with the function

$$
\begin{aligned}
\bar{H}_2^\sharp(w) \;=\; & W(f^\sharp(x^\sharp)) - W(x^\sharp) + R_{m0}(x^\sharp) + (R_{m1}(x^\sharp) + W_{x^\sharp}(f^\sharp(x^\sharp))g^\sharp(x^\sharp))w + \\
& w^T\left(R_{m2}(x^\sharp) + \frac{1}{2}g^{\sharp T}(x^\sharp)W_{x^\sharp x^\sharp}(f^\sharp(x^\sharp))g^\sharp(x^\sharp)\right)w + \\
& \frac{1}{2}\left[\begin{array}{c} w - \hat{w}^\star(x) \\ \hat{u}^\star(\theta) - \hat{u}^\star(x) \end{array}\right]^T \widehat{R}(x)\left[\begin{array}{c} w - \hat{w}^\star(x) \\ \hat{u}^\star(\theta) - \hat{u}^\star(x) \end{array}\right].
\end{aligned}
\tag{7.89}
$$

Moreover, we can now determine an estimate $\hat{w}^{\star\star}$ of $w^{\star\star}$ from the above expression (7.89) for $H_2^\sharp(w)$ by taking derivatives with respect to w and solving the linear equation

$$\frac{\partial \bar{H}_2^\sharp}{\partial w}(\hat{w}^{\star\star}) = 0.$$

It can be shown that, if the matrix

$$\bar{R}(x^\sharp) = \frac{1}{2}g^\sharp(x^\sharp)^T W_{x^\sharp x^\sharp}(f^\sharp(x^\sharp))g^\sharp(x^\sharp) + \widehat{R}_{11}(x) + R_{m2}(x^\sharp)$$

is nonsingular and negative-definite, then $\hat{w}^{\star\star}$ is unique, and is a maximum for $H_2^\sharp(w)$. The design procedure outlined above can now be summarized in the following theorem.

Theorem 7.4.2 *Consider the nonlinear discrete-time system (7.1) and suppose the following hold:*

(i) *there exists a smooth positive-definite function V defined on a neighborhood $X_0 \subset \mathcal{X}$ of $x = 0$, satisfying the hypotheses of Lemma 7.4.1.*

(ii) *there exists a smooth positive-semidefinite function $W(x^\sharp)$, defined on a neighborhood Ξ of $x^\sharp = 0$ in $\mathcal{X} \times \mathcal{X}$, such that $W(x^\sharp) > 0$ for all x, $x \neq 0$, and satisfying*

$$\bar{H}_2^\sharp(\hat{w}^{\star\star}(x^\sharp)) < 0$$

for all $0 \neq x^\sharp \in \Xi$. Moreover, there exists a number $\delta > 0$ such that

$$\bar{H}_2^\sharp(\hat{w}^{\star\star}(x^\sharp)) < -\delta\|\hat{w}^{\star\star}(x^\sharp)\|^2$$
$$\bar{R}(x^\sharp) < 0$$

for all $x^\sharp \in \Xi$.

Then, the controller $\bar{\Sigma}_{dynobs}^{dac}$ given by (7.83) locally asymptotically stabilizes the closed-loop system (7.84), and for every $K \in \mathbf{Z}_+$, there is a number $\epsilon > 0$ such that the response from the initial state $(x_0, \theta_0) = (0,0)$ satisfies

$$\sum_{k=0}^{K} z_k^T z_k \leq \gamma^2 \sum_{k=0}^{K} w_k^T w_k$$

for every sequence $w = (w_0, \ldots, w_K)$ such that $\|w_k\|^2 < \epsilon$.

7.5   Notes and Bibliography

This chapter is entirely based on the papers by Lin and Byrnes [182]-[185]. In particular, the discussion on controller parameterization is from [184]. The results for stable plants can also be found in [51]. Similarly, the results for sampled-data systems have not been discussed here, but can be found in [124, 213, 255].

The alternative and approximate approach for solving the discrete-time problems is mainly from Reference [126], and approximate approaches for solving the DHJIE can also be found in [125]. An information approach to the discrete-time problem can be found in [150], and connections to risk-sensitive control in [151, 150].

8

Nonlinear $\mathcal{H}_\infty$-Filtering

In this chapter, we discuss the nonlinear $\mathcal{H}_\infty$ sub-optimal filtering problem. This problem arises when the states of the system are not available for direct measurement, and so have to be estimated in some way, which in this case is to satisfy an $\mathcal{H}_\infty$-norm requirement. The states information may be required for feedback or other purposes, and the estimator is basically an observer [35] that uses the information on the measured output of the system and sometimes the input, to estimate the states.

It would be seen that the underlying structure of the $\mathcal{H}_\infty$ nonlinear filter is that of the Kalman-filter [35, 48, 119], but differs from it in the fact that (i) the plant is nonlinear; (ii) basic assumptions on the nature of the noise signal (which in the case of the Kalman-filter is Gaussian white noise); and (iii) the cost function to optimize. While the Kalman-filter [35] is a minimum-variance estimator, and is the best unbiased linear optimal filter [35, 48, 119], the $\mathcal{H}_\infty$ filter is derived from a completely deterministic setting, and is the optimal worst-case filter for all $\mathcal{L}_2$-bounded noise signals.

The performance of the Kalman-filter for linear systems has been unmatched, and is still widely applied when the spectra of the noise signal is known. However, in the case when the statistics of the noise or disturbances are not known well, the Kalman-filter can only perform averagely. In addition, the nonlinear enhancement of the Kalman-filter or the *"extended Kalman-filter"* suffers from the usual problem with linearization, i.e., it can only perform well locally around a certain operating point for a nonlinear system, and under the same basic assumptions that the noise inputs are white.

It is therefore reasonable to expect that a filter that is inherently nonlinear and does not make any a priori assumptions on the spectra of the noise input, except that they have bounded energies, would perform better for a nonlinear system. Moreover, previous statistical nonlinear filtering techniques developed using minimum-variance [172] as well as maximum-likelihood [203] criteria are infinite-dimensional and too complicated to solve the filter differential equations. On the other hand, the nonlinear $\mathcal{H}_\infty$ filter is easy to derive, and relies on finding a smooth solution to a HJI-PDE which can be found using polynomial approximations.

The linear $\mathcal{H}_\infty$ filtering problem has been considered by many authors [207, 243, 281]. In [207], a fairly complete theory of linear $\mathcal{H}_\infty$ filtering and smoothing for finite and infinite-horizon problems is given. It is thus the purpose of this chapter to present the nonlinear counterparts of the filtering results. We begin with the continuous-time problem and then the discrete-time problem. Moreover, we also present results on the robust filtering problems.

8.1 Continuous-Time Nonlinear $\mathcal{H}_\infty$-Filtering

The general set-up for this problem is shown in Figure 8.1, where the plant is represented by an affine nonlinear system Σ^a, while $\mathbf{F}$ is the filter. The filter processes the measurement output y from the plant which is also corrupted by the noise signal w, and generates an

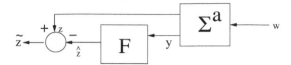

FIGURE 8.1
Configuration for Nonlinear $\mathcal{H}_\infty$-Filtering

estimate $\hat{z}$ of the desired variable z. The plant can be represented by an affine causal state-space system defined on a smooth n-dimensional manifold $\mathcal{X} \subseteq \Re^n$ in coordinates $x = (x_1, \dots, x_n)$ with zero-input:

$$\Sigma^a \; : \; \begin{cases} \dot{x} & = & f(x) + g_1(x)w; \; x(t_0) = x_0 \\ z & = & h_1(x) \\ y & = & h_2(x) + k_{21}(x)w \end{cases} \tag{8.1}$$

where $x \in \mathcal{X}$ is the state vector, $w \in \mathcal{W}$ is an unknown disturbance (or noise) signal, which belongs to the set $\mathcal{W} \subset \mathcal{L}_2([0, \infty), \Re^r)$ of admissible disturbances, the output $y \in \mathcal{Y} \subset \Re^m$ is the measured output (or observation) of the system, and belongs to $\mathcal{Y}$, the set of admissible outputs, and $z \in \Re^s$ is the output to be estimated.

The functions $f : \mathcal{X} \to V^\infty(\mathcal{X})$, $g_1 : \mathcal{X} \to \mathcal{M}^{n \times r}(\mathcal{X})$, $h_1 : \mathcal{X} \to \Re^s$, $h_2 : \mathcal{X} \to \Re^m$, and $k_{21} : \mathcal{X} \to \mathcal{M}^{m \times r}(\mathcal{X})$ are real C^∞ functions of x. Furthermore, we assume without any loss of generality that the system (8.1) has a unique equilibrium-point at $x = 0$ such that $f(0) = 0$, $h_1(0) = h_2(0) = 0$, and we also assume that there exists a unique solution $x(t)$ for the system for all initial conditions x_0 and all $w \in \mathcal{W}$.

The objective is to synthesize a causal filter, $\mathcal{F}$, for estimating the state $x(t)$ (or a function of it $z = h_1(x)$) from observations of $y(\tau)$ up to time t, over a time horizon $[t_0, T]$, i.e., from

$$\mathsf{Y}_t \stackrel{\Delta}{=} \{y(\tau) : \tau \le t\}, \quad t \in [t_0, T],$$

such that the $\mathcal{L}_2$-gain from w to $\tilde{z}$ (the estimation error, to be defined later) is less than or equal to a given number $\gamma > 0$ for all $w \in \mathcal{W}$, for all initial conditions in some subspace $O \subset \mathcal{X}$, i.e., we require that for a given $\gamma > 0$,

$$\int_{t_0}^{T} \|\tilde{z}(\tau)\|^2 d\tau \le \int_{t_0}^{T} \|w(\tau)\|^2 d\tau, \quad T > t_0, \tag{8.2}$$

for all $w \in \mathcal{W}$, and for all $x_0 \in O$. In addition, it is also required that with $w \equiv 0$, the penalty variable or estimation error satisfies $\lim_{t \to \infty} \tilde{z}(t) = 0$.

More formally, we define the local nonlinear $\mathcal{H}_\infty$ suboptimal filtering or state estimation problem as follows.

Definition 8.1.1 *(Nonlinear $\mathcal{H}_\infty$ (Suboptimal) Filtering Problem (NLHIFP)). Given the plant Σ^a and a number $\gamma^\star > 0$. Find a causal filter $\mathcal{F} : \mathcal{Y} \to \mathcal{X}$ which estimates x as $\hat{x}$, such that*

$$\hat{x}(t) = \mathcal{F}(\mathsf{Y}_t)$$

and (8.2) is satisfied for all $\gamma \ge \gamma^\star$, for all $w \in \mathcal{W}$, and for all $x_0 \in O$. In addition, with $w \equiv 0$, we have $\lim_{t \to \infty} \tilde{z}(t) = 0$.

Moreover, if the above conditions are satisfied for all $x(t_0) \in \mathcal{X}$, we say that $\mathcal{F}$ solves the NLHIFP globally.

Remark 8.1.1 *The problem defined above is the finite-horizon filtering problem. We have the infinite-horizon problem if we let $T \to \infty$.*

To solve the above problem, a structure is chosen for the filter $\mathcal{F}$. Based on our experience with linear systems, a *"Kalman"* structure is selected as:

$$\left. \begin{array}{rcl} \dot{\hat{x}} & = & f(\hat{x}) + L(\hat{x}, t)(y(t) - h_2(\hat{x})), \quad \hat{x}(t_0) = \hat{x}_0 \\ \hat{z} & = & h_1(\hat{x}) \end{array} \right\} \tag{8.3}$$

where $\hat{x} \in \mathcal{X}$ is the estimated state, $L(.,.) \in \Re^{n \times m} \times \Re$ is the error-gain matrix which has to be determined, and $\hat{z} \in \Re^s$ is the estimated output function. We can then define the output estimation error as

$$\tilde{z} = z - \hat{z} = h_1(x) - h_1(\hat{x}),$$

and the problem can be formulated as a two-person zero-sum differential game as discussed in Chapter 2. The cost functional is defined as

$$\hat{J}(w, L) \triangleq \frac{1}{2} \int_{t_0}^{T} (\|z(t)\|^2 - \gamma^2 \|w(t)\|^2) dt, \tag{8.4}$$

and we consider the problem of finding $L^\star(.)$ such that $\hat{J}(w, L)$ is minimized subject to the dynamics (8.1), (8.3), for all $w \in \mathcal{L}_2[t_0, T]$, and for all $x_0 \in \mathcal{X}$. To proceed, we augment the system equations (8.1) and (8.3) into the following system:

$$\left\{ \begin{array}{rcl} \dot{x}^e & = & f^e(x^e) + g^e(x^e)w \\ \tilde{z} & = & h_1(x) - h_1(\hat{x}) \end{array} \right. \tag{8.5}$$

where

$$x^e = \left(\begin{array}{c} x \\ \hat{x} \end{array} \right), \quad f^e(x^e) = \left(\begin{array}{c} f(x) \\ f(\hat{x}) + L(\hat{x})(h_2(x) - h_2(\hat{x})) \end{array} \right),$$

$$g^e(x^e) = \left(\begin{array}{c} g_1(x) \\ L(\hat{x})k_{21}(x) \end{array} \right).$$

We then make the following assumption:

Assumption 8.1.1 *The system matrices are such that*

$$\begin{array}{rcl} k_{21}(x)g_1^T(x) & = & 0 \\ k_{21}(x)k_{21}^T(x) & = & I. \end{array}$$

Remark 8.1.2 *The first of the above assumptions means that the measurement-noise and the system-noise are independent; while the second is a normalization to simplify the problem.*

To solve the above problem, we can apply the sufficient conditions given by Theorem 2.3.2 from Chapter 2, i.e., we consider the following HJIE:

$$-Y_t(x^e, t) = \inf_L \sup_w \{Y_{x^e}(x^e, t)(f^e(x^e) + g^e(x^e)w) - \frac{1}{2}\gamma^2 w^T w + \frac{1}{2}z^T z\}, \quad Y(x^e, T) = 0 \tag{8.6}$$

for some smooth C^1 (with respect to both its arguments) function $Y : \widehat{N} \times \widehat{N} \times \Re \to \Re$ locally defined in a neighborhood $\widehat{N}$ of the origin $x^e = 0$. We then have the following lemma which is a restatement of Theorem 5.1.1.

Lemma 8.1.1 *Suppose there exists a pair of strategies $(w, L) = (w^\star, L^\star)$ for which there exists a positive-definite C^1 function $Y : \widehat{N} \times \widehat{N} \times \Re \to \Re_+$, locally defined in a neighborhood $\widehat{N} \times \widehat{N} \subset \mathcal{X} \times \mathcal{X}$ of $x^e = 0$ satisfying the HJIE (8.6). Then the pair $(w^\star, L^\star)$ provides a saddle-point solution for the differential game.*

To find the pair $(w^\star, L^\star)$ that satisfies the HJIE, we proceed as in Chapter 5, by forming the Hamiltonian function $\Lambda : T^\star(\mathcal{X} \times \mathcal{X}) \times \mathcal{W} \times \mathcal{M}^{n \times m} \to \Re$ for the differential game:

$$\Lambda(x^e, w, L, Y_{x^e}) = Y_{x^e}(x^e, t)(f^e(x^e) + g^e(x^e)w) - \frac{1}{2}\gamma^2\|w\|^2 + \frac{1}{2}\|z\|^2. \qquad (8.7)$$

Then we apply the necessary conditions for the unconstrained optimization problem:

$$(w^\star, L^\star) = arg\left\{\sup_w \min_L \Lambda(x^e, w, L, Y_x)\right\}.$$

We summarize the result in the following proposition.

Theorem 8.1.1 *Consider the system (8.5), and suppose there exists a C^1 (with respect to all its arguments) positive-definite function $Y : \widehat{N} \times \widehat{N} \times \Re \to \Re_+$ satisfying the HJIE:*

$$Y_t(x^e, t) + Y_x(x^e, t)f(x) + Y_{\hat{x}}(x^e, t)f(\hat{x}) + \frac{1}{2\gamma^2}Y_x(x^e, t)g_1(x)g_1^T(x)Y_x^T(x^e, t) -$$

$$\frac{\gamma^2}{2}(h_2(x) - h_2(\hat{x}))^T(h_2(x) - h_2(\hat{x})) +$$

$$\frac{1}{2}(h_1(x) - h_1(\hat{x}))^T(h_1(x) - h_1(\hat{x})) = 0, \quad Y(x^e, T) = 0, \qquad (8.8)$$

together with the coupling condition

$$Y_{\hat{x}}(x^e, t)L(\hat{x}, t) = -\gamma^2(h_2(x) - h_2(\hat{x}))^T. \qquad (8.9)$$

Then the matrix $L(\hat{x}, t)$ satisfying (8.9) solves the finite horizon NLHIFP locally in $\widehat{N}$.

Proof: Consider the Hamiltonian function $\Lambda(x^e, w, L, Y_x^e)$. Since it is quadratic in w, we can apply the necessary condition for optimality, i.e.,

$$\left.\frac{\partial\Lambda}{\partial w}\right|_{w=w^\star} = 0$$

to get

$$w^\star := \frac{1}{\gamma^2}(g_1^T(x)Y_x^T(x^e, t) + k_{21}^T(x)L^T(\hat{x}, t)Y_{\hat{x}}^T(x^e, t)).$$

Moreover, it can be checked that the Hessian matrix of $\Lambda(x^e, w^\star, L, Y_x^e)$ is negative-definite, and hence

$$\Lambda(x^e, w^\star, L, Y_x^e) \geq \Lambda(x^e, w, L, Y_x^e) \quad \forall w \in \mathcal{W}.$$

However, Λ is linear in L, so we cannot apply the above technique to obtain $L^\star$. Instead, we use a completion of the squares method. Accordingly, substituting $w^\star$ in (8.7), we get

$$\Lambda(x^e, w^\star, L, Y_x^e) = Y_x(x^e, t)f(x) + Y_{\hat{x}}(x^e, t)f(\hat{x}) + Y_{\hat{x}}(x^e, t)L(\hat{x}, t)(h_2(x) - h_2(\hat{x})) +$$

$$\frac{1}{2\gamma^2}Y_x(x^e, t)g_1(x)g_1^T(x)Y_x^T(x^e, t) + \frac{1}{2}z^Tz +$$

$$\frac{1}{2\gamma^2}Y_{\hat{x}}(x^e, t)L(\hat{x}, t)(x)L^T(\hat{x}, t)Y_{\hat{x}}^T(x^e, t).$$

Now, completing the squares for L in the above expression, we get

$$
\begin{aligned}
\Lambda(x^e, w^\star, L, Y_x^e) \;=\; & Y_x(x^e,t)f(x) + Y_{\hat{x}}(x^e,t)f(\hat{x}) - \frac{1}{2}\gamma^2 \left\| (h_2(x) - h_2(\hat{x})) \right\|^2 + \\
& \frac{1}{2\gamma^2} \left\| L^T(\hat{x},t)Y_{\hat{x}}^T(x^e,t) + \gamma^2(h_2(x) - h_2(\hat{x})) \right\|^2 + \frac{1}{2}z^T z + \\
& \frac{1}{2\gamma^2} Y_x(x^e,t)g_1(x)g_1^T(x)Y_x^T(x^e,t).
\end{aligned}
$$

Thus, taking $L^\star$ as in (8.9) renders the saddle-point conditions

$$
\Lambda(w, L^\star) \leq \Lambda(w^\star, L^\star) \leq \Lambda(w^\star, L)
$$

satisfied for all $(w, L) \in \mathcal{W} \times \Re^{n \times m} \times \Re$, and the HJIE (8.6) reduces to (8.8). By Lemma 8.1.1, we conclude that $(w^\star, L^\star)$ is indeed a saddle-point solution for the game. Finally, it is very easy to show from the HJIE (8.8) that the $\mathcal{L}_2$-gain condition (8.2) is also satisfied. $\square$

Remark 8.1.3 *By virtue of the side-condition (8.9), the HJIE (8.8) can be represented as*

$$
Y_t(x^e,t) + Y_x(x^e,t)f(x) + Y_{\hat{x}}(x^e,t)f(\hat{x}) + \frac{1}{2\gamma^2} Y_x(x^e,t)g_1(x)g_1^T(x)Y_x^T(x^e,t)
$$

$$
-\frac{1}{2\gamma^2} Y_{\hat{x}}(x^e,t)L(\hat{x})L^T(\hat{x})Y_{\hat{x}}^T(x^e,t) + \frac{1}{2}(h_1(x) - h_1(\hat{x}))^T(h_1(x) - h_1(\hat{x})) = 0,
$$

$$
Y(x^e, T) = 0. \tag{8.10}
$$

Remark 8.1.4 *The above result, Theorem 8.1.1, can also be obtained from a dissipative systems perspective. Indeed, it can be checked that a function $Y(.,.)$ satisfying the HJIE (8.8) renders the dissipation-inequality*

$$
Y(x^e(T), T) - Y(x^e(t_0), t_0) \leq \frac{1}{2} \int_{t_0}^{T} (\gamma^2 \|w(t)\|^2 - \|\tilde{z}(t)\|^2)dt \tag{8.11}
$$

satisfied for all $x^e(t_0)$ and all $w \in \mathcal{W}$. Conversely, it can also be shown (as it has been shown in the previous chapters), that a function $Y(.,.)$ satisfying the dissipation-inequality (8.11) also satisfies in more general terms the HJI-inequality (8.8) with "=" replaced by "$\leq$". Thus, this observation allows us to solve a HJI-inequality which is substantially easier and more advantageous to solve.

For the case of the LTI system

$$
\Sigma^l \; : \; \begin{cases} \dot{x} & = \; Ax + B_1 w, \quad x(t_0) = x_0 \\ \tilde{z} & = \; C_1(x - \hat{x}) \\ y & = \; C_2 x + D_{21} w, \end{cases} \tag{8.12}
$$

we have the following corollary.

Corollary 8.1.1 *Consider the LTI system Σ^l (8.12) and the filtering problem for this system. Suppose there exists a symmetric positive-definite solution P to the Riccati-ODE:*

$$
\dot{P}(t) = A^T P(t) + P(t)A + P(t)[\frac{1}{\gamma^2}C_1^T C_1 - C_2^T C_2]P(t) + B_1 B_1^T, \quad P(T) = 0. \tag{8.13}
$$

Then, the filter

$$
\dot{\hat{x}} = A\hat{x} + L(t)(y - C_2\hat{x})
$$

solves the finite-horizon linear $\mathcal{H}_\infty$ filtering problem if the gain-matrix $L(t)$ is taken as

$$L(t) = P(t)C_2^T.$$

Proof: Assume $D_{21}D_{21}^T = I$, $D_{21}B_1 = 0$, $t_0 = 0$. Let $P(t_0) > 0$, and consider the positive-definite function

$$Y(x, \hat{x}, t) = \frac{1}{2}\gamma^2(x - \hat{x})^T P^{-1}(t)(x - \hat{x}), \quad P(t) > 0. \tag{8.14}$$

Taking partial-derivatives and substituting in (8.8), we obtain

$$-\frac{1}{2}\gamma^2(x - \hat{x})^T P^{-1}(t)\dot{P}(t)P^{-1}(t)(x - \hat{x}) + \gamma^2(x - \hat{x})^T P^{-1}(t)A(x - \hat{x}) +$$

$$\frac{\gamma^2}{2}(x - \hat{x})^T P^{-1}(t)B_1 B_1^T P^{-1}(t)(x - \hat{x}) - \frac{\gamma^2}{2}(x - \hat{x})^T C_2^T C_2(x - \hat{x}) +$$

$$\frac{1}{2}(x - \hat{x})^T C_1^T C_1(x - \hat{x}) = 0. \tag{8.15}$$

Splitting the second term in the left-hand-side into two (since it is a scalar):

$$\gamma^2(x - \hat{x})^T P^{-1}(t)A(x - \hat{x}) \quad = \quad \frac{1}{2}\gamma^2(x - \hat{x})^T P^{-1}(t)A(x - \hat{x}) +$$

$$\frac{1}{2}\gamma^2(x - \hat{x})^T A^T P^{-1}(t)(x - \hat{x}),$$

and substituting in the above equation, we get upon cancellation,

$$P^{-1}(t)\dot{P}(t)P^{-1}(t) \quad = \quad P^{-1}(t)A + A^T P^{-1}(t) + P^{-1}(t)B_1 B_1^T P^{-1}(t) -$$

$$C_2^T C_2 + \gamma^{-2}C_1^T C_1. \tag{8.16}$$

Finally, multiplying the above equation from the left and from the right by $P(t)$, we get the Riccati ODE (8.13). The terminal condition is obtained by setting $t = T$ in (8.14) and equating to zero.

Furthermore, substituting in (8.9) we get after cancellation

$$P^{-1}(t)L(t) = C_2^T \quad \text{or} \quad L(t) = P(t)C_2^T. \; \square$$

Remark 8.1.5 *Note that, if we used the HJIE (8.10) instead, or by substituting $C_2 = P^{-1}(t)L(t)$ in (8.16), we get the following Riccati ODE after simplification:*

$$\dot{P}(t) = A^T P(t) + P(t)A + P(t)[\frac{1}{\gamma^2}C_1^T C_1 - L^T(t)L(t)]P(t) + B_1 B_1^T, \quad P(T) = 0.$$

This result is the same as in reference [207]. In addition, cancelling $(x - \hat{x})$ from (8.15), and then multiplying both sides by $P(t)$, followed by the factorization, will result in the following alternative filter Riccati ODE

$$\dot{P}(t) = P(t)A^T + AP(t) + P(t)[\frac{1}{\gamma^2}C_1^T C_1 - C_2^T C_2]P(t) + B_1 B_1^T, \quad P(T) = 0. \tag{8.17}$$

8.1.1 Infinite-Horizon Continuous-Time Nonlinear $\mathcal{H}_\infty$-Filtering

In this subsection, we discuss the infinite-horizon filter in which case we let $T \to \infty$. Since we are interested in finding time-invariant gains for the filter, we seek a time-independent function $Y : \widehat{N}_1 \times \widehat{N}_1 \to \Re_+$ such that the HJIE:

$$Y_x(x^e)f(x) + Y_{\hat{x}}(x^e)f(\hat{x}) + \frac{1}{2\gamma^2}Y_x(x^e)g_1(x)g_1^T(x)Y_x^T(x^e) -$$
$$\frac{\gamma^2}{2}(h_2(x) - h_2(\hat{x}))^T(h_2(x) - h_2(\hat{x})) +$$
$$\frac{1}{2}(h_1(x) - h_1(\hat{x}))^T(h_1(x) - h_1(\hat{x})) = 0, \quad Y(0) = 0, \quad x, \hat{x} \in \widehat{N}_1 \tag{8.18}$$

is satisfied, together with the coupling condition

$$Y_{\hat{x}}(x^e)L(\hat{x}) = -\gamma^2(h_2(x) - h_2(\hat{x}))^T, \quad x, \hat{x} \in \widehat{N}_1. \tag{8.19}$$

Or equivalently, the HJIE:

$$Y_x(x^e)f(x) + Y_{\hat{x}}(x^e)f(\hat{x}) + \frac{1}{2\gamma^2}Y_x(x^e)g_1(x)g_1^T(x)Y_x^T(x^e) -$$
$$\frac{1}{2\gamma^2}Y_{\hat{x}}(x^e)L(\hat{x})L^T(\hat{x})Y_{\hat{x}}^T(x^e) +$$
$$\frac{1}{2}(h_1(x) - h_1(\hat{x}))^T(h_1(x) - h_1(\hat{x})) = 0, \quad Y(0) = 0. \quad x, \hat{x} \in \widehat{N}_1 \tag{8.20}$$

However here, since the estimation is carried over an infinite-horizon, it is necessary to ensure that the interconnected system (8.5) is stable with $w = 0$. This will in turn guarantee that we can find a smooth function $Y(.)$ which satisfies the HJIE (8.18) and provides an optimal gain for the filter. One additional assumption is however required: the system (8.1) must be locally asymptotically-stable. The following theorem summarizes this development.

Proposition 8.1.1 *Consider the nonlinear system (8.1) and the infinite-horizon $NLHIFP$ for it. Suppose the system is locally asymptotically-stable, and there exists a C^1-positive-definite function $Y : \widehat{N}_1 \times \widehat{N}_1 \to \Re_+$ locally defined in a neighborhood of $(x, \hat{x}) = (0,0)$ and satisfying the HJIE (8.18) together with the coupling condition (8.19), or equivalently the HJIE (8.20) for some matrix function $L(.) \in \mathcal{M}^{n \times m}$. Then, the infinite-horizon $NLHIFP$ is locally solvable in $\widehat{N}_1$ and the interconnected system is locally asymptotically-stable.*

Proof: By Remark 8.1.4, any function Y satisfying (8.18)-(8.19) or (8.20) also satisfies the dissipation-inequality

$$Y(x^e(t)) - Y(x^e(t_0)) \le \frac{1}{2}\int_{t_0}^t (\gamma^2\|w(t)\|^2 - \|\tilde{z}(t)\|^2)dt \tag{8.21}$$

for all t and all $w \in \mathcal{W}$. Differentiating this inequality along the trajectories of the interconnected system with $w = 0$, we get

$$\dot{Y}(x^e(t)) = -\frac{1}{2}\|z\|^2.$$

Thus, the interconnected system is stable. In addition, any trajectory $x^e(t)$ of the system starting in $\widehat{N}_1$ neighborhood of the origin $x^e = 0$ such that $\dot{Y}(t) \equiv 0 \ \forall t \ge t_s$, is such that $h_1(x(t)) = h_1(\hat{x}(t))$ and $x(t) = \hat{x}(t) \ \forall t \ge t_s$. This further implies that $h_2(x(t)) = h_2(\hat{x}(t))$, and therefore, it must be a trajectory of the free-system

$$\dot{x}^e = f(x) = \begin{pmatrix} f(x) \\ f(\hat{x}) \end{pmatrix}.$$

By local asymptotic-stability of the free-system $\dot{x} = f(x)$, we have local asymptotic-stability of the interconnected system. $\square$

Remark 8.1.6 *Note that, in the above proposition, it is not necessary to have a stable*

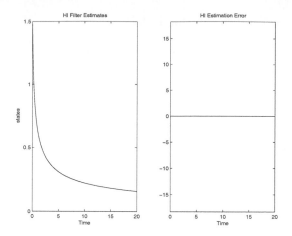

FIGURE 8.2
Nonlinear $\mathcal{H}_\infty$-Filter Performance with Known Initial Condition

system for $\mathcal{H}_\infty$ estimation. However, estimating the states of an unstable system is of no practical benefit.

Example 8.1.1 *Consider a simple scalar example*

$$
\begin{aligned}
\dot{x} &= -x^3 \\
y &= x + w \\
\tilde{z} &= x - \hat{x}.
\end{aligned}
$$

We consider the infinite-horizon problem and the HJIE (8.18) together with (8.19). Substituting in these equations we get

$$
-x^3 Y_x - x^3 Y_{\hat{x}} - \frac{\gamma^2}{2}(x-\hat{x})^2 + \frac{1}{2}(x-\hat{x})^2 = 0, \quad Y(0,0) = 0
$$

$$
Y_{\hat{x}} l_\infty = -2\gamma^2 (x-\hat{x}).
$$

If we let $\gamma = 1$, then

$$
-x^3 Y_x - \hat{x}^3 Y_{\hat{x}} = 0
$$

and it can be checked that

$$
Y_x = x, \quad Y_{\hat{x}} = \hat{x}
$$

solve the HJI-inequality, and result in

$$
Y(x,\hat{x}) = \frac{1}{2}(x^2 + \hat{x}^2) \quad \Longrightarrow \quad l_\infty^\star = \frac{-2(x-\hat{x})}{\hat{x}}.
$$

The results of simulation of the system with this filter are shown on Figures 8.2 and 8.3. A noise signal

$$
w(t) = w_0 + 0.1\sin(t)
$$

where w_0 is a zero-mean Gaussian white-noise with unit variance, is also added to the output.

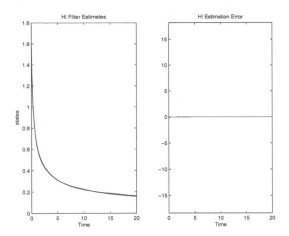

FIGURE 8.3
Nonlinear $\mathcal{H}_\infty$-Filter Performance with Unknown Initial Condition

8.1.2 The Linearized Filter

Because of the difficulty of solving the HJIE in implementation issues, it is sometimes useful to consider the linearized filter and solve the associated Riccati equation. Such a filter will be a variant of the extended-Kalman filter, but is different from it in the sense that, in the extended-Kalman filter, the finite-horizon Riccati equation is solved at every instant, while for this filter, we solve an infinite-horizon Riccati equation. Accordingly, let

$$F = \frac{\partial f}{\partial x}(0), \quad G_1 = g_1(0), \quad H_1 = \frac{\partial h_1}{\partial x}(0), \quad H_2 = \frac{\partial h_2}{\partial x}(0) \tag{8.22}$$

be a linearization of the system about $x = 0$. Then the following result follows trivially.

Proposition 8.1.2 *Consider the nonlinear system (8.1) and its linearization (8.22). Suppose for some $\gamma > 0$ there exists a real positive-definite symmetric solution to the filter algebraic-Riccati equation (FARE):*

$$PF^T + FP + P[\frac{1}{\gamma^2}H_1^T H_1 - H_2^T H_2]P + G_1 G_1^T = 0, \tag{8.23}$$

or

$$PF^T + FP + P[\frac{1}{\gamma^2}H_1^T H_1 - L^T L]P + G_1 G_1^T = 0, \tag{8.24}$$

for some matrix $L = PH_2^T$. Then, the filter

$$\dot{\hat{x}} = F\hat{x} + L(y - H_2\hat{x})$$

solves the infinite-horizon $NLHIFP$ for the system (8.1) locally on a small neighborhood O of $x = 0$ if the gain matrix L is taken as specified above.

Proof: Proof follows trivially from linearization. It can be checked that the function $V(x) = \frac{1}{2}\gamma^2(x - \hat{x})P^{-1}(x - \hat{x})$, P a solution of (8.23) or (8.24) satisfies the HJIE (8.8) together with (8.9) or equivalently the HJIE (8.10) for the linearized system.

Remark 8.1.7 *To guarantee that there exists a positive-definite solution to the ARE (8.23) or (8.24), it is necessary for the linearized system (8.22) or $[F, H_1]$ to be detectable (see [292]).*

8.2   Continuous-Time Robust Nonlinear $\mathcal{H}_\infty$-Filtering

In this section we discuss the continuous-time robust nonlinear $\mathcal{H}_\infty$-filtering problem (*RNLHIF*) in the presence of structured uncertainties in the system. This situation is shown in Figure 8.4 below, and arises when the system model is not known exactly, as is usually the case in practice. For this purpose, we consider the following model of the system with uncertainties:

$$\Sigma^{a,\Delta} : \begin{cases} \dot{x} & = & f(x) + \Delta f(x,t) + g_1(x)w; \quad x(0) = 0 \\ z & = & h_1(x) \\ y & = & h_2(x) + \Delta h_2(x,t) + k_{21}(x)w \end{cases} \tag{8.25}$$

where all the variables have their previous meanings. In addition, $\Delta f : \mathcal{X} \times \Re \to V^\infty \mathcal{X}$, $\Delta f(0,t) = 0$, $\Delta h_2 : \mathcal{X} \times \Re \to \Re^m$, $\Delta h_2(0,t) = 0$ are the uncertainties of the system which belong to the set of admissible uncertainties Ξ_Δ and is defined as follows.

Assumption 8.2.1 *The admissible uncertainties of the system are structured and matched, and they belong to the following set:*

$$\Xi_\Delta = \Big\{ \Delta f, \Delta h_2 \, | \, \Delta f(x,t) = H_1(x)F(x,t)E(x), \quad \Delta h_2(x,t) = H_2(x)F(x,t)E(x),$$

$$E(0) = 0, \quad \int_0^\infty (\|E(x)\|^2 - \|F(x,t)E(x)\|^2)dt \ge 0, \quad and$$

$$[k_{21}(x) \; H_2(x)][k_{21}(x) \; H_2(x)]^T > 0 \; \forall x \in \mathcal{X}, \, t \in \Re \Big\}$$

where $H_1(.), H_2(.), F(.,.), E(.)$ have appropriate dimensions.

The problem is the following.

Definition 8.2.1 *(Robust Nonlinear $\mathcal{H}_\infty$-Filtering Problem (RNLHIFP)). Find a filter of the form*

$$\Sigma^f : \begin{cases} \dot{\xi} & = & a(\xi) + b(\xi)y, \; \xi(0) = 0 \\ \hat{z} & = & c(\xi) \end{cases} \tag{8.26}$$

where $\xi \in \mathcal{X}$ is the state estimate, $y \in \Re^m$ is the system output, $\hat{z}$ is the estimated variable, and the functions $a : \mathcal{X} \to V^\infty \mathcal{X}$, $a(0) = 0$, $b : \mathcal{X} \to \mathcal{M}^{n \times m}$, $c : \mathcal{X} \to \Re^s$, $c(0) = 0$ are smooth C^2 functions, such that the $\mathcal{L}_2$-gain from w to the estimation error $\tilde{z} = z - \hat{z}$ is less than or equal to a given number $\gamma > 0$, i.e.,

$$\int_0^T \|\tilde{z}\|^2 dt \le \gamma^2 \int_0^T \|w(t)\|^2 dt \tag{8.27}$$

for all $T > 0$, all $w \in \mathcal{L}_2[0,T]$, and all admissible uncertainties. In addition with $w \equiv 0$, we have $\lim_{t \to \infty} \tilde{z}(t) = 0$.

To solve the above problem, we first recall the following result [210] which gives sufficient conditions for the solvability of the *NLHIFP* for the nominal system (i.e., without the uncertainties $\Delta f(x,t)$ and $\Delta h_2(x,t)$). Without any loss of generality, we shall also assume for the remainder of this section that $\gamma = 1$ henceforth.

Theorem 8.2.1 *Consider the nominal system (8.25) without the uncertainties $\Delta f(x,t)$ and $\Delta h_2(x,t)$ and the NLHIFP for this system. Suppose there exists a positive-semidefinite*

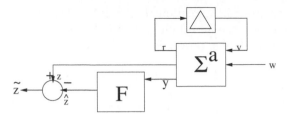

FIGURE 8.4

Configuration for Robust Nonlinear $\mathcal{H}_\infty$-Filtering

function $\Psi : \widetilde{N} \times \widetilde{N} \to \Re$ locally defined in a neighborhood $\widetilde{N} \times \widetilde{N}$ of the origin $(x, \xi) = 0$ such that the following HJIE is satisfied

$$HJI(x, \xi) + \tilde{b}^T(x, \xi)R(x)\tilde{b}(x, \xi) < 0 \tag{8.28}$$

for all $x, \xi \in \widetilde{N} \times \widetilde{N}$, where

$$
\begin{aligned}
HJI(x, \xi) &\triangleq [\Psi_x(x, \xi) \ \Psi_\xi(x, \xi)][\tilde{f}(x, \xi) + \tilde{g}(x, \xi)\Psi_x^T(\xi, \xi)] + \\
&\quad \frac{1}{4}[\Psi_x(x, \xi) \ \Psi_\xi(x, \xi)]\tilde{k}(x, \xi)\tilde{k}^T(x, \xi)\left[\begin{array}{c} \Psi_x^T(x, \xi) \\ \Psi_\xi^T(x, \xi) \end{array} \right] - \\
&\quad \frac{1}{4}\Psi_x(\xi, \xi)g_1(\xi)k_{21}^T(\xi)R^{-1}(x)k_{21}(\xi)g_1^T(\xi)\Psi_x^T(\xi, \xi) - \\
&\quad \tilde{h}_2(x, \xi)R^{-1}(x)\tilde{h}_2(x, \xi) + \tilde{h}_1^T(x, \xi)\tilde{h}_1(x, \xi) + \\
&\quad \frac{1}{2}\Psi_x(\xi, \xi)g_1(\xi)k_{21}^T(\xi)R^{-1}(x)\tilde{h}_2(x, \xi) \\
\tilde{b}(x, \xi) &= \frac{1}{2}b^T(\xi)\Psi_\xi^T(x, \xi) + R^{-1}(x)[\frac{1}{2}k_{21}(x)g_1^T(x)\Psi_x^T(x, \xi) + \tilde{h}_2(x, \xi)],
\end{aligned}
$$

and

$$
\begin{aligned}
\tilde{f}(x, \xi) &= \left[\begin{array}{c} f(x) - g_1(x)k_{21}^T(x)R^{-1}(x)\tilde{h}_2(x, \xi) \\ f(\xi) \end{array} \right], \\
\tilde{g}(x, \xi) &= \left[\begin{array}{c} \frac{1}{2}g_1(\xi)k_{21}^T(\xi)R^{-1}(x)k_{21}(\xi)g_1^T(\xi) \\ \frac{1}{2}g_1(\xi)g^T(\xi) \end{array} \right], \\
\tilde{k}^T(x, \xi) &= [g_1(x)(I - k_{21}^T(x)R^{-1}(x)k_{21}(x))^{\frac{1}{2}} \ 0], \\
\tilde{h}_1(x, \xi) &= h_1(x) - h_1(\xi), \\
\tilde{h}_2(x, \xi) &= h_2(x) - h_2(\xi), \\
R(x) &= k_{21}(x)k_{21}^T(x).
\end{aligned}
$$

Then, the filter (8.26) with

$$
\begin{aligned}
a(\xi) &= f(\xi) - b(\xi)h_2(\xi) \\
c(\xi) &= h_1(\xi)
\end{aligned}
$$

solves the NLHIFP for the system (8.25).

Proof: The proof of this theorem can be found in [210].

Remark 8.2.1 *Note that Theorem 8.2.1 provides alternative sufficient conditions for the solvability of the NLHIFP discussed in Section 8.1.*

Next, before we can present a solution to the $RNLHIFP$, which is a refinement of Theorem 8.2.1, we transform the uncertain system (8.25) into a scaled or auxiliary system (see also [226]) using the matching properties of the uncertainties in Ξ_Δ:

$$\Sigma^{as,\Delta} \; : \; \begin{cases} \dot{x}_s &= f(x_s) + [g_1(x_s) \; \frac{1}{\tau} H_1(x_s)]w_s; \;\; x_s(0) = 0 \\ z_s &= \begin{bmatrix} h_1(x_s) \\ \tau E(x_s) \end{bmatrix} \\ y_s &= h_2(x_s) + [k_{21}(x_s) \; \frac{1}{\tau} H_2(x_s)]w_s \end{cases} \tag{8.29}$$

where $x_s \in \mathcal{X}$ is the state-vector of the scaled system, $w_s \in \mathcal{W} \subset \mathcal{L}_2([0,\infty), \Re^{r+v})$ is the noise input, $\tau > 0$ is a scaling constant, z_s is the new controlled (or estimated) output and has a fictitious component coming from the uncertainties. To estimate z_s, we employ the filter structure (8.26) with an extended output:

$$\Sigma^f \; : \; \begin{cases} \dot{\xi} &= a(\xi) + b(\xi)y_s, \;\; \xi(0) = 0 \\ \hat{z}_s &= \begin{bmatrix} c(\xi) \\ 0 \end{bmatrix}, \end{cases} \tag{8.30}$$

where all the variables have their previous meanings. We then have the following preliminary result which establishes the equivalence between the system (8.25) and the scaled system (8.29).

Theorem 8.2.2 *Consider the nonlinear uncertain system (8.25) and the RNLHIFP for this system. There exists a filter of the form (8.26) which solves the problem for this system for all admissible uncertainties if, and only if, there exists a $\tau > 0$ such that the same filter (8.30) solves the problem for the scaled system (8.29).*

Proof: The proof of this theorem can be found in Appendix B.

We can now present a solution to the $RNLHIFP$ in the following theorem which gives sufficient conditions for the solvability of the problem.

Theorem 8.2.3 *Consider the uncertain system (8.29) and the RNLHIFP for this system. Given a scaling factor $\tau > 0$, suppose there exists a positive-semidefinite function Φ : $\tilde{N}_1 \times \tilde{N}_1 \to \Re$ locally defined in a neighborhood $\tilde{N}_1 \times \tilde{N}_1$ of the origin $(x_s, \xi) = (0, 0)$ such that the following HJIE is satisfied:*

$$\widetilde{HJI}(x_s, \xi) + \hat{b}^T(x_s, \xi)\widehat{R}(x)\hat{b}(x_s, \xi) \le 0 \tag{8.31}$$

for all $x_s, \xi \in \tilde{N}_1 \times \tilde{N}_1$, where

$$
\begin{aligned}
\widetilde{HJI}(x_s, \xi) &\overset{\Delta}{=} \; [\Phi_{x_s}(x_s,\xi) \;\; \Phi_\xi(x_s,\xi)][\hat{f}(x_s,\xi) + \hat{g}(x_s,\xi)\Psi_{x_s}^T(\xi,\xi)] + \\
&\quad \frac{1}{4}[\Phi_{x_s}(x_s,\xi) \;\; \Phi_\xi(x_s,\xi)]\hat{k}(x_s,\xi)\hat{k}^T(x_s,\xi) \begin{bmatrix} \Phi_{x_s}(x_s,\xi) \\ \Psi_\xi(x_s,\xi) \end{bmatrix} \\
&\quad -\frac{1}{4}\Phi_{x_s}(\xi,\xi)\hat{g}_1(\xi)\hat{k}_{21}(\xi)\widehat{R}^{-1}(x_s)\hat{k}_{21}^T(\xi)\hat{g}_1^T(\xi)\Phi_{x_s}^T(\xi,\xi) - \\
&\quad \hat{h}_2^T(x_s,\xi)\widehat{R}^{-1}(x_s)\hat{h}_2(x_s,\xi) + \hat{h}_1^T(x_s,\xi)\hat{h}_1(x_s,\xi) + \\
&\quad \frac{1}{2}\Phi_{x_s}(\xi,\xi)\hat{g}_1(\xi)\hat{k}_{21}^T(\xi)\widehat{R}^{-1}(x_s)\hat{h}_2(x_s,\xi) + \tau^2 E^T(x_s)E(x_s) \\
\hat{b}(x_s,\xi) &= \frac{1}{2}b^T(\xi)\Psi_\xi^T(x_s,\xi) + R^{-1}(x_s)[\frac{1}{2}k_{21}(x_s)g_1^T(x_s)\Psi_{x_s}^T(x,\xi) - \\
&\quad \frac{1}{2}k_{21}(\xi)g_1^T(\xi)\Psi_{x_s}^T(\xi,\xi) + \hat{h}_2(x_s,\xi)],
\end{aligned}
$$

and

$$\hat{f}(x_s, \xi) = \begin{bmatrix} f(x_s) - \hat{g}_1(x)\hat{k}_{21}^T(xs)\widehat{R}^{-1}(x_s)\hat{h}_2(x_s, \xi) \\ f(\xi) \end{bmatrix},$$

$$\hat{g}(x, \xi) = \begin{bmatrix} \frac{1}{2}\hat{g}_1(\xi)\hat{k}_{21}^T(\xi)\widehat{R}^{-1}(x)\hat{k}_{21}(\xi)\hat{g}_1^T(\xi)\Phi_{x_s}^T(\xi, \xi) \\ \frac{1}{2}\hat{g}_1(\xi)\hat{g}^T(\xi)\Phi_{x_s}^T(\xi, \xi) \end{bmatrix}$$

$$\hat{k}^T(x_s, \xi) = [\hat{g}_1(x_s)(I - \hat{k}_{21}^T(x)\widehat{R}^{-1}(x_s)\hat{k}_{21}(x))^{\frac{1}{2}} \quad 0]$$

$$\hat{k}_{21}(x_s) = [k_{21}(x_s) \quad \frac{1}{\tau}H_2(x_s)]$$

$$\hat{g}_1(x_s) = [g_1(x_s) \quad \frac{1}{\tau}H_1(x_s)]$$

$$\hat{h}_1(x_s, \xi) = h_1(x_s) - h_1(\xi)$$

$$\hat{h}_2(x_s, \xi) = h_2(x_s) - h_2(\xi)$$

$$\widehat{R}(x_s) = \hat{k}_{21}(x_s)\hat{k}_{21}^T(x_s).$$

Then the filter (8.30) with

$$a(\xi) = f(\xi) + \frac{1}{2}\hat{g}_1(x)\hat{g}_1^T(x)\Phi_x^T(\xi, \xi) - b(\xi)[h_2(\xi) + \frac{1}{2}\hat{k}_{21}(x)g_1^T(x)\Phi_x^T(\xi, \xi)$$

$$c(\xi) = h_1(\xi)$$

solves the RNLHIFP for the system (8.25).

Proof: The result follows by applying Theorem 8.2.2 for the scaled system (8.25). □

It should be observed in the previous two Sections 8.1, 8.2, that the filters constructed can hardly be implemented in practice, because the gain matrices are functions of the original state of the system, which is to be estimated. Thus, except for the linear case, such filters will be of little practical interest. Based on this observation, in the next section, we present another class of filters which can be implemented in practice.

8.3   Certainty-Equivalent Filters (CEFs)

We discuss in this section a class of estimators for the nonlinear system (8.1) which we refer to as "*certainty-equivalent*" worst-case estimators (see also [139]). The estimator is constructed on the assumption that the asymptotic value of $\hat{x}$ equals x, and the gain matrices are designed so that they are not functions of the state vector x, but of $\hat{x}$ and y only. We begin with the one degree-of-freedom (1-DOF) case, then we discuss the 2-DOF case. Accordingly, we propose the following class of estimators:

$$\Sigma_1^{acef} : \begin{cases} \dot{\hat{x}} = f(\hat{x}) + g_1(\hat{x})\hat{w}^\star + \hat{L}(\hat{x}, y)(y - h_2(\hat{x}) - k_{21}(\hat{x})\hat{w}^\star) \\ \hat{z} = h_2(\hat{x}) \\ \tilde{z} = y - h_2(\hat{x}) \end{cases} \quad (8.32)$$

where $\hat{z} = \hat{y} \in \Re^m$ is the estimated output, $\tilde{z} \in \Re^m$ is the new penalty variable, $\hat{w}^\star$ is the estimated worst-case system noise and $\hat{L} : \mathcal{X} \times \mathcal{Y} \to \mathcal{M}^{n \times m}$ is the gain matrix for the filter.

Again, the problem is to find a suitable gain matrix $\hat{L}(.,.)$ for the above filter, for estimating the state $x(t)$ of the system (8.1) from available observations $\mathsf{Y}_t \overset{\Delta}{=} \{y(\tau), \quad \tau \leq t\}$,

$t \in [t_0, \infty)$, such that the $\mathcal{L}_2$-gain from the disturbance/noise signal w to the error output $\tilde{z}$ (to be defined later) is less than or equal to a desired number $\gamma > 0$, i.e.,

$$\int_{t_0}^{T} \|\tilde{z}(\tau)\|^2 d\tau \le \gamma^2 \int_{t_0}^{T} \|w(\tau)\|^2 d\tau, \ \ T > t_0, \tag{8.33}$$

for all $w \in \mathcal{W}$, for all $x_0 \in O \subset \mathcal{X}$. Moreover, with $w = 0$, we also have $\lim_{t \to \infty} \tilde{z}(t) = 0$.

The above problem can similarly be formulated by considering the cost functional

$$J(\hat{L}, w) \ = \ \frac{1}{2} \int_{t_0}^{\infty} (\|\tilde{z}(\tau)\|^2 - \gamma^2 \|w(\tau)\|^2) d\tau. \tag{8.34}$$

as a differential game with the two contending players controlling L and w respectively. Again by making $J \le 0$, we see that the $\mathcal{L}_2$-gain constraint (8.33) is satisfied, and moreover, we seek for a saddle-point equilibrium solution, $(\hat{L}^\star, w^\star)$, to the above game such that

$$J(\hat{L}^\star, w) \le J(\hat{L}^\star, w^\star) \le J(\hat{L}, w^\star) \ \forall w \in \mathcal{W}, \ \forall \hat{L} \in \mathcal{M}^{n \times m}. \tag{8.35}$$

Before we proceed with the solution to the problem, we introduce the following notion of local zero-input observability.

Definition 8.3.1 *For the nonlinear system Σ^a, we say that it is locally zero-input observable if for all states x_1, $x_2 \in U \subset \mathcal{X}$ and input $w(.) \equiv 0$*

$$y(., x_1, w) = y(., x_2, w) \Longrightarrow x_1 = x_2,$$

where $y(., x_i, w), i = 1, 2$ is the output of the system with the initial condition $x(t_0) = x_i$. Moreover, the system is said to be zero-input observable, if it is locally zero-input observable at each $x_0 \in \mathcal{X}$ or $U = \mathcal{X}$.

Next, we first estimate $\hat{w}^\star$, and for this purpose, define the Hamiltonian function $H : T^\star \mathcal{X} \times \mathcal{W} \times \mathcal{M}^{n \times m} \to \Re$ for the estimator:

$$H(\hat{x}, w, \hat{L}, \hat{V}_{\hat{x}}^T) \ = \ \hat{V}_{\hat{x}}(\hat{x}, y)[f(\hat{x}) + g_1(\hat{x})w + \hat{L}(\hat{x}, y)(y - h_2(\hat{x}) - k_{21}(\hat{x})w)] + \hat{V}_y \dot{y} + \frac{1}{2}(\|\tilde{z}\|^2 - \gamma^2 \|w\|^2) \tag{8.36}$$

for some smooth function $\hat{V} : \mathcal{X} \times \mathcal{Y} \to \Re$ and with the adjoint vector $p = \hat{V}_{\hat{x}}^T(\hat{x}, y)$. Applying now the necessary condition for the worst-case noise

$$\frac{\partial H}{\partial w}\Big|_{w=\hat{w}^\star} = 0 \Longrightarrow \hat{w}^\star = \frac{1}{\gamma^2}[g_1^T(\hat{x})\hat{V}_{\hat{x}}^T(\hat{x}) - k_{21}^T(\hat{x})\hat{L}(\hat{x}, y)\hat{V}_{\hat{x}}^T(\hat{x}, y)].$$

Then substituting $\hat{w}^\star$ into (8.36), we get

$$H(\hat{x}, \hat{w}^\star, \hat{L}, \hat{V}_{\hat{x}}^T) \ = \ V_{\hat{x}} f(\hat{x}) + \hat{V}_y \dot{y} + \frac{1}{2\gamma^2} \hat{V}_{\hat{x}} g_1(\hat{x}) g_1^T(\hat{x}) \hat{V}_{\hat{x}}^T(\hat{x}) + \hat{V}_{\hat{x}} \hat{L}(\hat{x}, y)(y - h_2(\hat{x})) + $$
$$\frac{1}{2\gamma^2} \hat{V}_{\hat{x}} \hat{L}(\hat{x}, y) \hat{L}^T(\hat{x}, y) \hat{V}_{\hat{x}} + \frac{1}{2}(y - h_2(\hat{x}))^T (y - h_2(\hat{x})). \tag{8.37}$$

Completing the squares now for $\hat{L}$ in the above expression for $H(., \hat{w}^\star, L, .)$, we get

$$H(\hat{x}, \hat{w}^\star, \hat{L}, \hat{V}_{\hat{x}}^T) \ = \ V_{\hat{x}} f(\hat{x}) + \hat{V}_y \dot{y} + \frac{1}{2\gamma^2} \hat{V}_{\hat{x}} g_1(\hat{x}) g_1^T(\hat{x}) \hat{V}_{\hat{x}}^T - \frac{\gamma^2}{2}(y - h_2(\hat{x}))^T (y - h_2(\hat{x})) + $$
$$\frac{1}{2\gamma^2} \left\| \hat{L}^T(\hat{x}, y)\hat{V}_{\hat{x}} + \gamma^2(y - h_2(\hat{x})) \right\|^2 + \frac{1}{2}(y - h_2(\hat{x}))^T (y - h_2(\hat{x})).$$

Therefore, choosing $\hat{L}^\star$ as

$$\hat{V}_{\hat{x}}(\hat{x}, y)\hat{L}^\star(\hat{x}, y) = -\gamma^2(y - h_2(\hat{x}))^T \qquad (8.38)$$

minimizes $H(., ., ., ., .)$ and gurantees that the saddle-point conditions (8.35) are satisfied by $(\hat{w}^\star, \hat{L}^\star)$. Finally, substituting this value in (8.37) and setting

$$H(., \hat{w}^\star, \hat{L}^\star, ., .) = 0,$$

yields the HJIE

$$\hat{V}_{\hat{x}}(\hat{x}, y)f(\hat{x}) + \hat{V}_y(\hat{x}, y)\dot{y} + \frac{1}{2\gamma^2}\hat{V}_{\hat{x}}(\hat{x}, y)g_1(\hat{x})g^T(\hat{x})\hat{V}_{\hat{x}}^T(\hat{x}, y) -$$
$$\frac{\gamma^2}{2}(y - h_2(\hat{x}))^T(y - h_2(\hat{x})) +$$
$$\frac{1}{2}(y - h_2(\hat{x}))^T(y - h_2(\hat{x})) = 0, \quad \hat{V}(0,0) = 0, \qquad (8.39)$$

or equivalently the HJIE

$$\hat{V}_{\hat{x}}(\hat{x}, y)f(\hat{x}) + \hat{V}_y(\hat{x}, y)\dot{y} + \frac{1}{2\gamma^2}\hat{V}_{\hat{x}}(\hat{x}, y)g_1(\hat{x})g_1^T(\hat{x})\hat{V}_{\hat{x}}^T(\hat{x}, y) -$$
$$\frac{1}{2\gamma^2}\hat{V}_{\hat{x}}(\hat{x}, y)\hat{L}(\hat{x}, y)\hat{L}^T(\hat{x}, y)\hat{V}_{\hat{x}}^T(\hat{x}, y) +$$
$$\frac{1}{2}(y - h_2(\hat{x}))^T(y - h_2(\hat{x})) = 0, \quad \hat{V}(0,0) = 0. \qquad (8.40)$$

This is summarized in the following result.

Proposition 8.3.1 *Consider the nonlinear system (8.1) and the NLHIFP for it. Suppose Assumption 8.1.1 holds, the plant Σ^a is locally asymptotically-stable about the equilibrium-point $x = 0$ and zero-input observable. Suppose further, there exists a C^1 positive-semidefinite function $\hat{V} : \hat{N} \times \hat{Y} \to \Re_+$ locally defined in a neighborhood $\hat{N} \times \hat{Y} \subset \mathcal{X} \times \mathcal{Y}$ of the origin $(\hat{x}, y) = (0, 0)$, and a matrix function $\hat{L} : \hat{N} \times \hat{Y} \to \mathcal{M}^{n \times m}$, satisfying the HJIE (8.39) or (8.40) together with the coupling condition (8.38). Then the filter Σ_2^{acef} solves the NLHIFP for the system.*

Proof: Let $\hat{V} \geq 0$ be a C^1 solution of the HJIE (8.39) or (8.40). Differentiating $\hat{V}$ along a trajectory of (8.32) with w in place of $\hat{w}^\star$, it can be shown that

$$\dot{\hat{V}} \leq \frac{1}{2}(\gamma^2\|\hat{w}\|^2 - \|\tilde{z}\|^2).$$

Integrating this inequality from $t = t_0$ to $t = \infty$, and since $V \geq 0$ implies that the $\mathcal{L}_2$-gain condition $\int_{t_0}^\infty \|\tilde{z}(t)\|^2 dt \leq \gamma^2 \int_{t_0}^\infty \|w(t)\|^2 dt$ is satisfied. Moreover, setting $w = 0$ in the above inequality implies that $\dot{\hat{V}} \leq -\frac{1}{2}\|\tilde{z}\|^2$ and hence the estimator dynamics is stable. In addition,

$$\dot{\hat{V}} \equiv 0 \; \forall t \geq t_s \implies \tilde{z} \equiv 0 \implies y = h_2(\hat{x}) = \hat{y}.$$

By the zero-input observability of the system Σ^a, this implies that $x = \hat{x}$. $\square$

Remark 8.3.1 *Notice also that for $\gamma = 1$, a control Lyapunov-function $\hat{V}$ [102, 170] for the system (8.1) solves the above HJIE (8.39).*

8.3.1 2-DOF Certainty-Equivalent Filters

Next, we extend the design procedure in the previous section to the 2-DOF case. In this regard, we assume that the time derivative $\dot{y}$ is available as an additional measurement information. This can be obtained or estimated using a differentiator or numerically. Notice also that, with $y = h_2(x) + k_{21}(x)w$ and with $w \in \mathcal{L}_2(0, T)$ for some T sufficiently large, then since the space of continuous functions with compact support C_c is dense in $\mathcal{L}_2$, we may assume that the time derivatives $\dot{y}$, $\ddot{y}$ exist a.e. That is, we may approximate w by piece-wise C^1 functions. However, this assumption would have been difficult to justify if we had assumed w to be some random process, e.g., a white noise process.

Accordingly, consider the following class of filters:

$$
\Sigma_2^{acef} : \begin{cases}
\dot{\hat{x}} = f(\hat{x}) + g_1(\hat{x})\hat{w}^\star + \dot{L}_1(\hat{x}, y, \dot{y})(y - h_2(\hat{x}) - k_{21}(\hat{x})\hat{w}^\star) + \\
\qquad \dot{L}_2(\hat{x}, y, \dot{y})(\dot{y} - \mathcal{L}_{f(\hat{x})}h_2(\hat{x})) \\
\dot{z} = \begin{bmatrix} h_2(\hat{x}) \\ \mathcal{L}_{f(\hat{x})}h_2(\hat{x}) \end{bmatrix} \\
\tilde{z} = \begin{bmatrix} y - h_2(\hat{x}) \\ \dot{y} - \mathcal{L}_{f(\hat{x})}h_2(\hat{x}) \end{bmatrix}
\end{cases}
$$

where $\dot{z} \in \Re^m$ is the estimated output of the filter, $\tilde{z} \in \Re^s$ is the error or penalty variable, $\mathcal{L}_f h_2$ denotes the Lie-derivative of h_2 along f [268], while $\dot{L}_1 : \mathcal{X} \times T\mathcal{Y} \to \mathcal{M}^{n \times m}$, $\dot{L}_2 : \mathcal{X} \times T\mathcal{Y} \to \mathcal{M}^{n \times m}$ are the filter gains, and all the other variables and functions have their corresponding previous meanings and dimensions. As in the previous subsection, we can define the Hamiltonian $\dot{H} : T^\star \mathcal{X} \times \mathcal{W} \times \mathcal{M}^{n \times m} \times \mathcal{M}^{n \times m} \to \Re$ of the system as

$$
\dot{H}(\hat{x}, \hat{w}, \dot{L}_1, \dot{L}_2, \dot{V}_{\hat{x}}^T) = \dot{V}_{\hat{x}}(\hat{x}, y, \dot{y})\Big\{ f(\hat{x}) + g_1(\hat{x})\hat{w} + \dot{L}_1(\hat{x}, y, \dot{y})(y - h_2(\hat{x}) - k_{21}(\hat{x})\hat{w}) +
$$

$$
\dot{L}_2(\hat{x}, y, \dot{y})(\dot{y} - \mathcal{L}_{f(\hat{x})}h_2(\hat{x}))\Big\} + \dot{V}_y \dot{y} + \dot{V}_{\dot{y}}\ddot{y} + \frac{1}{2}(\|\tilde{z}\|^2 - \gamma^2 \|\hat{w}\|^2) \tag{8.41}
$$

for some smooth function $\dot{V} : \mathcal{X} \times T\mathcal{Y} \to \Re$ and by setting the adjoint vector $\dot{p} = \dot{V}_{\hat{x}}^T$. Then proceeding as before, we clearly have

$$
\hat{w}^\star = \frac{1}{\gamma^2}[g_1^T(\hat{x})\dot{V}_{\hat{x}}^T(\hat{x}, y, \dot{y}) - k_{21}^T(\hat{x})\dot{L}_1(\hat{x}, y, \dot{y})\dot{V}_{\hat{x}}^T(\hat{x}, y, \dot{y})].
$$

Substitute now $\hat{w}^\star$ into (8.41) to get

$$
\dot{H}(\hat{x}, \hat{w}^\star, \dot{L}_1, \dot{L}_2, \dot{V}_{\hat{x}}^T) = \dot{V}_{\hat{x}} f(\hat{x}) + \dot{V}_y \dot{y} + \dot{V}_{\dot{y}}\ddot{y} + \frac{1}{2\gamma^2}\dot{V}_{\hat{x}} g_1(\hat{x})g_1^T(\hat{x})\dot{V}_{\hat{x}}^T +
$$

$$
\dot{V}_{\hat{x}}\dot{L}_1(\hat{x}, y, \dot{y})(y - h_2(\hat{x})) + \frac{1}{2\gamma^2}\dot{V}_{\hat{x}}\dot{L}_1(\hat{x}, y, \dot{y})\dot{L}_1^T(\hat{x}, y, \dot{y})\dot{V}_{\hat{x}}^T +
$$

$$
\dot{V}_{\hat{x}}\dot{L}_2(\hat{x}, y, \dot{y})(\dot{y} - \mathcal{L}_f h_2(\hat{x})) + \frac{1}{2}(\dot{y} - \mathcal{L}_f h_2(\hat{x}))^T(\dot{y} - \mathcal{L}_f h_2(\hat{x})) +
$$

$$
\frac{1}{2}(y - h_2(\hat{x}))^T(y - h_2(\hat{x}))
$$

and completing the squares for $\dot{L}_1$ and $\dot{L}_2$, we get

$$
\dot{H}(\hat{x}, \hat{w}^\star, \dot{L}_1, \dot{L}_2, \dot{V}_{\hat{x}}^T) = \dot{V}_{\hat{x}} f(\hat{x}) + \dot{V}_y \dot{y} + \dot{V}_{\dot{y}}\ddot{y} + \frac{1}{2\gamma^2}\dot{V}_{\hat{x}} g_1(\hat{x})g_1^T(\hat{x})\dot{V}_{\hat{x}}^T -
$$

$$
\frac{\gamma^2}{2}(y - h_2(\hat{x}))^T(y - h_2(\hat{x})) + \frac{1}{2\gamma^2}\left\| \dot{L}_1^T(\hat{x}, y, \dot{y})\dot{V}_{\hat{x}}^T + \gamma^2(y - h_2(\hat{x})) \right\|^2 +
$$

$$
\frac{1}{2}\|L_2^T(\hat{x}, y, \dot{y})\dot{V}_{\hat{x}}^T + (\dot{y} - \mathcal{L}_f h_2(\hat{x}))\|^2 - \frac{1}{2}\dot{V}_{\hat{x}} L_2(\hat{x}, y, \dot{y})L_2^T(\hat{x}, y, \dot{y})\dot{V}_{\hat{x}}^T +
$$

$$
\frac{1}{2}(y - h_2(\hat{x}))^T(y - h_2(\hat{x})).
$$

Hence, setting

$$\dot{V}_{\hat{x}}(\hat{x}, y, \dot{y})\dot{L}_1^\star(\hat{x}, y, \dot{y}) \;=\; -\gamma^2(y - h_2(\hat{x}))^T \qquad (8.42)$$

$$\dot{V}_{\hat{x}}(\hat{x}, y, \dot{y})\dot{L}_2^\star(\hat{x}, y, \dot{y}) \;=\; -(\dot{y} - \mathcal{L}_f h_2(\hat{x}))^T \qquad (8.43)$$

minimizes $\dot{H}$ and gurantees that the saddle-point conditions (8.35) are satisfied. Finally, setting

$$\dot{H}(\hat{x}, \dot{w}^\star, \dot{L}_1^\star, \dot{L}_2^\star, \dot{V}_{\hat{x}}^T) = 0$$

yields the HJIE

$$\dot{V}_{\hat{x}}(\hat{x}, y, \dot{y})f(\hat{x}) + \dot{V}_y(\hat{x}, y, \dot{y})\dot{y} + \dot{V}_{\dot{y}}(\hat{x}, y, \dot{y})\ddot{y} +$$
$$\frac{1}{2\gamma^2}\dot{V}_{\hat{x}}(\hat{x}, y, \dot{y})g_1(\hat{x})g_1^T(\hat{x})\dot{V}_{\hat{x}}^T(\hat{x}, y, \dot{y}) -$$
$$\frac{1}{2}(\dot{y} - \mathcal{L}_f h_2(\hat{x}))^T(\dot{y} - \mathcal{L}_f h_2(\hat{x})) +$$
$$\frac{(1-\gamma^2)}{2}(y - h_2(\hat{x}))^T(y - h_2(\hat{x})) = 0, \quad \dot{V}(0,0) = 0. \qquad (8.44)$$

Consequently, we have the following result.

Theorem 8.3.1 *Consider the nonlinear system (8.1) and the NLHIFP for it. Suppose Assumption 8.1.1 holds, the plant Σ^a is locally asymptotically-stable about the equilibrium-point $x = 0$ and zero-input observable. Suppose further, there exists a C^1 positive-semidefinite function $\dot{V} : \dot{N} \times \dot{Y} \to \Re_+$ locally defined in a neighborhood $\dot{N} \times \dot{Y} \subset \mathcal{X} \times T\mathcal{Y}$ of the origin $(\hat{x}, y, \dot{y}) = (0,0,0)$, and matrix functions $L_1 : \dot{N} \times \dot{Y} \to \mathcal{M}^{n\times m}$, $L_2 : \dot{N} \times \dot{Y} \to \mathcal{M}^{n\times m}$, satisfying the H.JIE (8.44) or equivalently the HJIE:*

$$\dot{V}_{\hat{x}}(\hat{x}, y, \dot{y})f(\hat{x}) + \dot{V}_y(\hat{x}, y, \dot{y})\dot{y} + \dot{V}_{\dot{y}}(\hat{x}, y, \dot{y})\ddot{y} + \frac{1}{2\gamma^2}\dot{V}_{\hat{x}}(\hat{x}, y, \dot{y})g_1(\hat{x})g_1^T(\hat{x})\dot{V}_{\hat{x}}^T(\hat{x}, y, \dot{y}) -$$
$$\frac{1}{2\gamma^2}\dot{V}_{\hat{x}}(\hat{x}, y, \dot{y})\dot{L}_1(\hat{x}, y, \dot{y})\dot{L}_1^T(\hat{x}, y, \dot{y})\dot{V}_{\hat{x}}^T(\hat{x}, y, \dot{y}) -$$
$$\frac{1}{2}\dot{V}_{\hat{x}}(\hat{x}, y, \dot{y})\dot{L}_2(\hat{x}, y, \dot{y})\dot{L}_2^T(\hat{x}, y, \dot{y})\dot{V}_{\hat{x}}^T(\hat{x}, y, \dot{y}) + \frac{1}{2}(y - h_2(\hat{x}))^T(y - h_2(\hat{x})) = 0,$$
$$\dot{V}(0,0,0) = 0 \qquad (8.45)$$

together with the coupling condition (8.42), (8.43). Then the filter Σ_2^{acef} solves the HINLFP for the system.

Proof: The first part of the proof has already been established above, that $(\dot{w}^\star, [\dot{L}_1^\star, \dot{L}_2^\star])$ constitute a saddle-point solution for the game (8.34), (8.41). Therefore, we only need to show that the $\mathcal{L}_2$-gain requirement (8.33) is satisfied, and the filter provides asymptotic estimates when $w = 0$.

Let $\dot{V} \geq 0$ be a C^1 solution of the HJIE (8.44), and differentiating it along a trajectory of the filter Σ_3^{acef} with $\dot{w}$ in place of $\dot{w}^\star$, and $\dot{L}_1^\star, \dot{L}_2^\star$, we have

$$\dot{V} = \dot{V}_{\hat{x}}(\hat{x}, y, \dot{y})[f(\hat{x}) + g_1(\hat{x})\dot{w} + \dot{L}_1(\hat{x}, y)(y - h_2(\hat{x}) - k_{21}(\hat{x})\dot{w}) +$$
$$L_2(\hat{x}, y, \dot{y})(\dot{y} - \mathcal{L}_f h_2(\hat{x}))] + \dot{V}_y\dot{y} + \dot{V}_{\dot{y}}\ddot{y}$$

$$= \left\{ \dot{V}_{\hat{x}}f(\hat{x}) + \dot{V}_y\dot{y} + \dot{V}_{\dot{y}}\ddot{y} + \frac{1}{2\gamma^2}\dot{V}_{\hat{x}}g_1(\hat{x})g_1^T(\hat{x})\dot{V}_{\hat{x}}^T + \right.$$
$$\left. \frac{(1-\gamma^2)}{2}(y - h_2(\hat{x}))^T(y - h_2(\hat{x})) - \frac{1}{2}(\dot{y} - \mathcal{L}_f h_2(\hat{x}))^T(\dot{y} - \mathcal{L}_f h_2(\hat{x})) \right\} +$$
$$\dot{V}_{\hat{x}}g_1(\hat{x})\dot{w} - \dot{V}_{\hat{x}}\dot{L}_1^\star(\hat{x}, y)k_{21}(\hat{x})\dot{w} - \frac{1}{2\gamma^2}\dot{V}_{\hat{x}}g_1(\hat{x})g_1^T(\hat{x})\dot{V}_{\hat{x}}^T -$$
$$\frac{\gamma^2}{2}(y - h_2(\hat{x}))^T(y - h_2(\hat{x})) + \dot{V}_{\hat{x}}L_2(\hat{x}, y, \dot{y})(\dot{y} - \mathcal{L}_f h_2(\hat{x})) -$$
$$\frac{1}{2}(y - h_2(\hat{x}))^T(y - h_2(\hat{x})) + \frac{1}{2}(\dot{y} - \mathcal{L}_f h_2(\hat{x}))^T(\dot{y} - \mathcal{L}_f h_2(\hat{x}))$$

$$= -\frac{\gamma^2}{2}\left\|w - \frac{1}{\gamma^2}g_1^T(\hat{x})\hat{V}_{\hat{x}}^T + \frac{1}{\gamma^2}k_{21}^T(\hat{x})\dot{L}_1^{\star^T}(\hat{x},y)\hat{V}_{\hat{x}}^T\right\|^2 + \frac{\gamma^2}{2}\|\hat{w}\|^2 - \frac{1}{2}\|\tilde{z}\|^2$$

$$\leq \frac{1}{2}(\gamma^2\|\hat{w}\|^2 - \|\tilde{z}\|^2).$$

Integrating the above inequality from $t = t_0$ to $t = \infty$, we get that the $\mathcal{L}_2$-gain condition (8.33) is satisfied.

Similarly, setting $\hat{w} = 0$, we have $\dot{V} = -\frac{1}{2}\|\tilde{z}\|^2$. Therefore, the filter dynamics is stable. Moreover, the condition

$$\dot{V} \equiv 0 \; \forall t \geq t_s \implies \tilde{z} \equiv 0 \implies y \equiv h_2(\hat{x}), \;\; \dot{y} \equiv \mathcal{L}_f h_2(\hat{x}).$$

By the zero-input observability of the system Σ^a, this implies that $\hat{x} \equiv x \; \forall t \geq t_s$. $\square$

We consider an example to illustrate and compare the performances of the 1-DOF and the 2-DOF.

Example 8.3.1 *Consider the nonlinear system*

$$\begin{aligned}
\dot{x}_1 &= -x_1^3 - x_2 \\
\dot{x}_2 &= -x_1 - x_2 \\
y &= x_1 + w
\end{aligned}$$

where $w = 5w_0 + 5\sin(t)$ and w_0 is a zero-mean Gaussian white-noise with unit variance. It can be checked that the system is locally zero-input observable and the functions $\hat{V}(x) = \frac{1}{2}(\hat{x}_1^2 + \hat{x}_2^2)$, $\hat{V}(\hat{x}) = \frac{1}{2}(\hat{x}_1^2 + \hat{x}_2^2)$ solve the HJI-inequalities corresponding to (8.39), (8.44) for the 1-DOF and 2-DOF certainty-equivalent filters respectively with $\gamma = 1$. Subsequently, we calculate the gains of the filters as

$$\hat{L}(\hat{x},y) = \begin{bmatrix} 1 \\ -y/\hat{x}_2 \end{bmatrix}, \quad \dot{L}_1(\hat{x},y,\dot{y}) = \begin{bmatrix} 1 \\ -y/\hat{x}_2 \end{bmatrix}, \quad \dot{L}_2(\hat{x},y,\dot{y}] = \begin{bmatrix} -\hat{x}_2^2 \\ -(\dot{y}-\hat{x}_2)/\hat{x}_2 \end{bmatrix}$$

and construct the filters Σ_1^{acef}, Σ_2^{acef} respectively. The results of the individual filter performance with the same initial conditions but unknown system initial conditions are shown in Figure 8.5. It is seen from the simulation that the 2-DOF filter has faster convergence than the 1-DOF filter, though the latter may have better steady-state performance. The estimation errors are also insensitive or robust against a significantly high persistent measurement noise.

8.4 Discrete-Time Nonlinear $\mathcal{H}_\infty$-Filtering

In this section, we consider $\mathcal{H}_\infty$-filtering problem for discrete-time nonlinear systems. The configuration for this problem is similar to the continuous-time problem shown in Figure 8.1 but with discrete-time inputs and output measurements instead, and is shown in Figure 8.7. We consider an affine causal discrete-time state-space system with zero input defined on $\mathcal{X} \subseteq \Re^n$ in coordinates $x = (x_1, \ldots, x_n)$:

$$\Sigma^{da} : \begin{cases} x_{k+1} &= f(x_k) + g_1(x_k)w_k; \;\; x(k_0) = x_0 \\ z_k &= h_1(x_k) \\ y_k &= h_2(x_k) + k_{21}(x_k)w_k \end{cases} \tag{8.46}$$

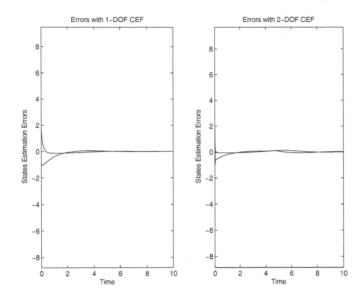

FIGURE 8.5
1-DOF and 2-DOF Nonlinear $\mathcal{H}_\infty$-Filter Performance with Unknown Initial Condition

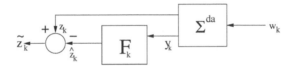

FIGURE 8.6
Configuration for Discrete-Time Nonlinear $\mathcal{H}_\infty$-Filtering

where $x \in \mathcal{X}$ is the state vector, $w \in \mathcal{W} \subset \ell_2[k_0, \infty)$ is the disturbance or noise signal, which belongs to the set $\mathcal{W} \subset \Re^r$ of admissible disturbances or noise signals, the output $y \in \Re^m$ is the measured output of the system, while $z \in \Re^s$ is the output to be estimated. The functions $f : \mathcal{X} \to \mathcal{X}$, $g_1 : \mathcal{X} \to \mathcal{M}^{n \times r}(\mathcal{X})$, $h_1 : \mathcal{X} \to \Re^s$, $h_2 : \mathcal{X} \to \Re^m$, and $k_{12} : \mathcal{X} \to \mathcal{M}^{s \times p}(\mathcal{X})$, $k_{21} : \mathcal{X} \to \mathcal{M}^{m \times r}(\mathcal{X})$ are real C^∞ functions of x. Furthermore, we assume without any loss of generality that the system (8.46) has a unique equilibrium-point at $x = 0$ such that $f(0) = 0$, $h_1(0) = h_2(0) = 0$.

The objective is to synthesize a causal filter, $\mathcal{F}_k$, for estimating the state x_k or a function of it, $z_k = h_1(x_k)$, from observations of y_k up to time $k \in \mathbf{Z}_+$, i.e., from

$$\mathsf{Y}_k \stackrel{\Delta}{=} \{y_i : i \le k\},$$

such that the ℓ_2-gain from w to $\breve{z}$, the error (or penalty variable), of the interconnected system defined as

$$\|\Sigma^{daf} \circ \Sigma^{da}\|_{\ell_\infty} \stackrel{\Delta}{=} \sup_{0 \neq w \in \ell_2} \frac{\|\breve{z}\|_2}{\|w\|_2}, \tag{8.47}$$

where $w := \{w_k\}$, $\breve{z} := \{\breve{z}_k\}$, is rendered less or equal to some given positive number $\gamma > 0$. Moreover, with $w_k \equiv 0$, we have $\lim_{k \to \infty} \breve{z}_k = 0$.

For nonlinear systems, the above condition is interpreted as the ℓ_2-gain constraint and is represented as

$$\sum_{k=k_0}^{K} \|\check{z}_k\|^2 \leq \gamma^2 \sum_{k=k_0}^{K} \|w_k\|^2, \quad K > k_0 \in \mathbf{Z} \tag{8.48}$$

for all $w_k \in \mathcal{W}$, and for all $x^0 \in O \subset \mathcal{X}$.

The discrete-time nonlinear $\mathcal{H}_\infty$ suboptimal filtering or estimation problem can be defined formally as follows.

Definition 8.4.1 *(Discrete-Time Nonlinear $\mathcal{H}_\infty$ (Suboptimal) Filtering/ Prediction Problem (DNLHIFP)). Given the plant Σ^{da} and a number $\gamma^\star > 0$. Find a filter $\mathcal{F}_k : \mathcal{Y} \to \mathcal{X}$ such that*

$$\hat{x}_{k+1} = \mathcal{F}_k(Y_k), \quad k = k_0, \dots, K$$

and the constraint (8.48) is satisfied for all $\gamma \geq \gamma^\star$, for all $w \in \mathcal{W}$, and for all $x^0 \in O$. Moreover, with $w_k \equiv 0$, we have $\lim_{k\to\infty} \check{z}_k = 0$.

Remark 8.4.1 *The problem defined above is the finite-horizon filtering problem. We have the infinite-horizon problem if we let $K \to \infty$.*

To solve the above problem, we consider the following class of estimators:

$$\Sigma^{daf} : \begin{cases} \hat{x}_{k+1} &= f(\hat{x}_k) + L(\hat{x}_k, k)[y_k - h_2(\hat{x}_k)], \quad \hat{x}(k_0) = \hat{x}^0 \\ \hat{z}_k &= h_1(\hat{x}_k) \end{cases} \tag{8.49}$$

where $\hat{x}_k \in \mathcal{X}$ is the estimated state, $L(.,.) \in \mathcal{M}^{n \times m}(\mathcal{X} \times \mathbf{Z})$ is the error-gain matrix which has to be determined, and $\hat{z} \in \Re^s$ is the estimated output of the filter. We can now define the estimation error or penalty variable, $\check{z}$, which has to be controlled as:

$$\check{z}_k := z_k - \hat{z}_k = h_1(x_k) - h_1(\hat{x}_k).$$

Then we combine the plant (8.46) and estimator (8.49) dynamics to obtain the following augmented system:

$$\left. \begin{aligned} \check{x}_{k+1} &= \check{f}(\check{x}_k) + \check{g}(\check{x}_k)w_k, \quad \check{x}(k_0) = (x^{0^T} \ \hat{x}^{0^T})^T \\ \check{z}_k &= \check{h}(\check{x}_k) \end{aligned} \right\}, \tag{8.50}$$

where $\check{x}_k = (x_k^T \ \hat{x}_k^T)^T$,

$$\check{f}(\check{x}) = \begin{pmatrix} f(x_k) \\ f(\hat{x}_k) + L(\hat{x}_k, k)(h_2(x_k) - h_2(\hat{x}_k)) \end{pmatrix}, \quad \check{g}(\check{x}) = \begin{pmatrix} g_1(x_k) \\ L(\hat{x}_k, k)k_{21}(x_k) \end{pmatrix},$$

$$\check{h}(\check{x}_k) = h_1(x_k) - h_1(\hat{x}_k).$$

The problem is then formulated as a two-player zero-sum differential game with the cost functional:

$$\min_L \max_w \tilde{J}(L, w) = \frac{1}{2} \sum_{k=k_0}^{K} \{\|\check{z}_k\|^2 - \gamma^2 \|w_k\|^2\}, \tag{8.51}$$

where $w := \{w_k\}$, $L := L(x_k, k)$. By making $J \leq 0$, we see that the $\mathcal{H}_\infty$ constraint $\|\Sigma^{daf} \circ \Sigma^a\|_{\mathcal{H}_\infty} \leq \gamma$ is satisfied. A saddle-point equilibrium solution to the above game is said to exist if we can find a pair $(L^\star, w^\star)$ such that

$$\tilde{J}(L^\star, w) \leq \tilde{J}(L^\star, w^\star) \leq \tilde{J}(L, w^\star) \quad \forall w \in \mathcal{W}, \forall L \in \mathcal{M}^{n \times m}. \tag{8.52}$$

Sufficient conditions for the solvability of the above game are well known [59]. These are also given as Theorem 2.2.2 which we recall here.

Theorem 8.4.1 *For the two-person discrete-time nonzero-sum game (8.51)-(8.50), under memoryless perfect information structure, a pair of strategies $(L^\star, w^\star)$ provides a feedback saddle-point solution if, and ony if, there exist a set of $K - k_0$ functions $V : \mathcal{X} \times \mathbf{Z} \to \Re$, such that the following recursive equations (or discrete-Hamilton-Jacobi-Isaac's equations (DHJIE)) are satisfied for each $k \in [k_0, K]$:*

$$
\begin{aligned}
V(\breve{x}, k) &= \min_{L \in \mathcal{M}^{n \times m}} \sup_{w \in \mathcal{W}} \left\{ \frac{1}{2}(\|\breve{z}_k\|^2 - \gamma^2 \|w_k\|^2) + V(\breve{x}_{k+1}, k+1) \right\}, \\
&= \sup_{w \in \mathcal{W}} \min_{L \in \mathcal{M}^{n \times m}} \left\{ \frac{1}{2}(\|\breve{z}_k\|^2 - \|w_k\|^2) + V(\breve{x}_{k+1}, k+1) \right\}, \\
&= \frac{1}{2}(\|\breve{z}_k(\breve{x})\|^2 - \gamma^2 \|w_k^\star(\breve{x})\|^2) + V(\breve{x}_{k+1}^\star, k+1), \quad V(\breve{x}, K+1) = 0, \quad (8.53)
\end{aligned}
$$

$k = k_0, \dots, K$, *where $\breve{x} = \breve{x}_k$, and*

$$
x_{k+1}^\star = \breve{f}^\star(\breve{x}_k) + \breve{g}^\star(\breve{x}_k) w_k^\star, \quad \breve{f}^\star(\breve{x}_k) = \breve{f}(\breve{x}_k)\Big|_{L=L^\star}, \quad \breve{g}^\star(\breve{x}_k) = \breve{g}(\breve{x}_k)|_{L=L^\star}.
$$

Equation (8.53) is also known as Isaac's equation.

Thus, we can apply the above theorem to derive sufficient conditions for the solvability of the *DNLHIFP*. To do that, we define the Hamiltonian function $H : (\mathcal{X} \times \mathcal{X}) \times \mathcal{W} \times \mathcal{M}^{n \times m} \times \Re \to \Re$ associated with the cost functional (8.51) and the estimator dynamics as

$$
H(\breve{x}, w_k, L, V) = V(\breve{f}(\breve{x}) + \breve{g}(\breve{x}) w_k(\breve{x}), k+1) - V(\breve{x}, k) + \frac{1}{2}\|\breve{z}_k(\breve{x})\|^2 - \frac{1}{2}\gamma^2 \|w_k(\breve{x})\|^2, \quad (8.54)
$$

where the adjoint variable has been set to $p = V$. Then we have the following result.

Theorem 8.4.2 *Consider the nonlinear system (8.46) and the DNLHIFP for it. Suppose the function h_1 is one-to-one (or injective) and the plant Σ^{da} is locally asymptotically-stable about the equilibrium-point $x = 0$. Further, suppose there exists a C^1 (with respect to both arguments) positive-definite function $V : N \times N \times \mathbf{Z} \to \Re$ locally defined in a neighborhood $N \subset \mathcal{X}$ of the origin $\breve{x} = 0$, and a matrix function $L : N \times \mathbf{Z} \to \mathcal{M}^{n \times m}$ satisfying the following DHJIE:*

$$
V(\breve{x}, k) = V(\breve{f}^\star(\breve{x}) + \breve{g}^\star(\breve{x}) w_k^\star(\breve{x}), k+1) + \frac{1}{2}\|\breve{z}_k(\breve{x})\|^2 - \frac{1}{2}\gamma^2 \|w_k^\star(\breve{x})\|^2, \quad V(\breve{x}, K+1) = 0, \tag{8.55}
$$

$k = k_0, \dots, K$, *together with the side-conditions:*

$$
w_k^\star(\breve{x}) = \frac{1}{\gamma^2} \breve{g}^T(\breve{x}) \frac{\partial^T V(\lambda, k+1)}{\partial \lambda}\bigg|_{\lambda = \breve{f}(\breve{x}) + \breve{g}(\breve{x}) w_k^\star}, \tag{8.56}
$$

$$
L^\star(\hat{x}) = \arg \min_L \{H(\breve{x}, w_k^\star, L, V)\}, \tag{8.57}
$$

$$
\frac{\partial^2 H}{\partial w^2}(\breve{x}, w_k, L^\star, V)\bigg|_{\breve{x}=0} < 0, \tag{8.58}
$$

$$
\frac{\partial^2 H}{\partial L^2}(\breve{x}, w_k^\star, L, V)\bigg|_{\breve{x}=0} > 0. \tag{8.59}
$$

Then: (i) there exists a unique saddle-point equilibrium solution $(w^\star, L^\star)$ for the game

(8.51), (8.50) locally in N; and (ii) the filter Σ^{daf} with the gain-matrix $L(\hat{x}_k, k)$ satisfying (8.57) solves the finite-horizon $DNLHIFP$ for the system locally in N.

Proof: Assume there exists positive-definite solution $V(., k)$ to the DHJIEs (8.55) in $N \subset \mathcal{X}$ for each k, and (i) consider the Hamiltonian function $H(.,.,.,.)$. Apply the necessary condition for optimality, i.e.,

$$\left.\frac{\partial^T H}{\partial w}\right|_{w=w^\star} = \left.\breve{g}^T(\breve{x})\frac{\partial^T V(\lambda, k+1)}{\partial \lambda}\right|_{\lambda=\breve{f}(\breve{x})+\breve{g}(\breve{x})w_k^\star} - \gamma^2 w_k^\star = 0,$$

to get

$$w_k^\star = \left.\frac{1}{\gamma^2}\breve{g}^T(\breve{x})\frac{\partial^T V(\lambda, k+1)}{\partial \lambda}\right|_{\lambda=\breve{f}(\breve{x})+\breve{g}(\breve{x})w_k^\star} := \alpha_0(\breve{x}, w_k^\star). \tag{8.60}$$

Thus, $w^\star$ is expressed implicitly. Moreover, since

$$\frac{\partial^2 H}{\partial w^2} = \left.\breve{g}^T(\breve{x})\frac{\partial^2 V(\lambda, k+1)}{\partial \lambda^2}\right|_{\lambda=\breve{f}(\breve{x})+\breve{g}(\breve{x})w_k^\star} \breve{g}(\breve{x}) - \gamma^2 I$$

is nonsingular about $(\breve{x}, w) = (0, 0)$, the equation (8.60) has a unique solution $\alpha_0(\breve{x})$, $\alpha_0(0) = 0$ in the neighborhood $N \times W$ of $(x, w) = (0, 0)$ by the Implicit-function Theorem [234].

Now substitute $w^\star$ in the expression for $H(.,.,.,.)$ (8.54), to get

$$H(\breve{x}, w_k^\star, L, V) = V(\breve{f}(\breve{x}) + \breve{g}(\breve{x})w_k^\star(\breve{x}), k+1) - V(\breve{x}, k) + \frac{1}{2}\|\breve{z}_k(\breve{x})\|^2 - \frac{1}{2}\gamma^2\|w_k^\star(\breve{x})\|^2$$

and let

$$L^\star = \arg\min_L \{H(\breve{x}, w_k^\star, L, V)\}.$$

Then, by Taylor's theorem, we can expand $H(.,.,.,.)$ about $(L^\star, w^\star)$ [267] as

$$H(\breve{x}, w, L, V) = H(\breve{x}, w^\star, L^\star, V) + \frac{1}{2}(w - w^\star)^T\frac{\partial^2 H}{\partial w^2}(w, L^\star)(w - w^\star) +$$

$$\frac{1}{2}Tr\left\{[I_n \otimes (L - L^\star)^T]\frac{\partial^2 H}{\partial L^2}(w^\star, L)[I_m \otimes (L - L^\star)^T]\right\} +$$

$$O(\|w - w^\star\|^3 + \|L - L^\star\|^3). \tag{8.61}$$

Thus, taking $L^\star$ as in (8.57) and $w^\star = \alpha_0(\breve{x}, w^\star)$ and if the conditions (8.58), (8.59) hold, we see that the saddle-point conditions

$$H(w, L^\star) \leq H(w^\star, L^\star) \leq H(w^\star, L) \quad \forall L \in \mathcal{M}^{n \times m}, \forall w \in \ell_2[k_0, K], \quad k \in [k_0, K]$$

are locally satisfied. Moreover, substituting $(w^\star, L^\star)$ in (8.53) gives the DHJIE (8.55).

(ii) Reconsider equation (8.61). Since the conditions (8.58) and (8.59) are satisfied about $\breve{x} = 0$, by the Inverse-function Theorem [234], there exists a neighborhood $U \subset N \times N$ of $\breve{x} = 0$ for which they are also satisfied. Consequently, we immediately have the important inequality

$$H(\breve{x}, w, L^\star, V) \leq H(\breve{x}, w^\star, L^\star, V) = 0 \quad \forall \breve{x} \in U$$

$$\Longleftrightarrow V(\breve{x}_{k+1}, k+1) - V(\breve{x}, k) \leq \frac{1}{2}\gamma^2\|w_k\|^2 - \frac{1}{2}\|\breve{z}_k\|^2, \quad \forall \breve{x} \in U, \forall w_k \in \mathcal{W}. \tag{8.62}$$

Summing now from $k = k_0$ to $k = K$, we get the dissipation inequality [183]:

$$V(\breve{x}_{K+1}, K+1) - V(x_{k_0}, k_0) \leq \sum_{k_0}^{K}\frac{1}{2}\gamma^2\|w_k\|^2 - \frac{1}{2}\|\breve{z}_k\|^2. \tag{8.63}$$

Thus, the system has ℓ_2-gain from w to $\breve{z}$ less or equal to γ. $\square$

8.4.1 Infinite-Horizon Discrete-Time Nonlinear $\mathcal{H}_\infty$-Filtering

In this subsection, we discuss the infinite-horizon discrete-time filtering problem, in which case we let $K \to \infty$. Since we are interested in finding a time-invariant gain for the filter, we seek a time-independent function $V : \tilde{N} \times \tilde{N} \to \Re$ locally defined in a neighborhood $\tilde{N} \times \tilde{N} \subset \mathcal{X} \times \mathcal{X}$ of $(x, \hat{x}) = (0, 0)$ such that the following stationary DHJIE:

$$\tilde{V}(\breve{f}^\star(\breve{x}) + \breve{g}^\star(\breve{x})\tilde{w}^\star(\breve{x})) - \tilde{V}(\breve{x}) + \frac{1}{2}\|\breve{z}(\breve{x})\|^2 - \frac{1}{2}\gamma^2\|\tilde{w}^\star(\breve{x})\|^2 = 0, \quad \tilde{V}(0) = 0, \quad x, \hat{x} \in \widehat{N} \tag{8.64}$$

is satisfied together with the side-conditions:

$$\tilde{w}^\star(\breve{x}) = \frac{1}{\gamma^2}\breve{g}^T(\breve{x})\frac{\partial^T \tilde{V}(\lambda)}{\partial \lambda}\bigg|_{\lambda=\breve{f}(\breve{x})+\breve{g}(\breve{x})\tilde{w}^\star} := \alpha_1(\breve{x}, \tilde{w}^\star), \tag{8.65}$$

$$\tilde{L}^\star(\hat{x}) = \arg\min_L \left\{\widetilde{H}(\breve{x}, \tilde{w}^\star, \tilde{L}, \tilde{V})\right\} \tag{8.66}$$

$$\frac{\partial^2 \widetilde{H}}{\partial w^2}(\breve{x}, w, \tilde{L}^\star, \tilde{V})\bigg|_{\breve{x}=0} < 0, \tag{8.67}$$

$$\frac{\partial^2 \widetilde{H}}{\partial L^2}(\breve{x}, \tilde{w}^\star, \tilde{L}, \tilde{V})\bigg|_{\breve{x}=0} > 0, \tag{8.68}$$

where

$$\breve{f}^\star(\breve{x}) = \breve{f}(\breve{x})\Big|_{\tilde{L}=\tilde{L}^\star}, \quad \breve{g}^\star(\breve{x}) = \breve{g}(\breve{x})|_{\tilde{L}=\tilde{L}^\star},$$

$$\widetilde{H}(\breve{x}, w_k, \tilde{L}, \tilde{V}) = \tilde{V}(\breve{f}(\breve{x}) + \breve{g}(\breve{x})w) - \tilde{V}(\breve{x}) + \frac{1}{2}\|\breve{z}_k\|^2 - \frac{1}{2}\gamma^2\|w_k\|^2, \tag{8.69}$$

and $\tilde{w}^\star, \tilde{L}^\star$ are the asymptotic values of $w_k^\star, L_k^\star$ respectively. Again here, since the estimation is carried over an infinite-horizon, it is necessary to ensure that the augmented system (8.50) is stable with $w = 0$. However, in this case, we can relax the requirement of asymptotic-stability for the original system (8.46) with a milder requirement of detectability which we define next.

Definition 8.4.2 *The pair $\{f, h\}$ is said to be locally zero-state detectable if there exists a neighborhood $\mathcal{O}$ of $x = 0$ such that, if x_k is a trajectory of $x_{k+1} = f(x_k)$ satisfying $x(k_0) \in \mathcal{O}$, then $h(x_k)$ is defined for all $k \geq k_0$ and $h(x_k) \equiv 0$ for all $k \geq k_s$, implies $\lim_{k\to\infty} x_k = 0$.*

A filter is also required to be stable, so that trajectories do not blow-up for example in an open-loop system. Thus, we define the *"admissibility"* of a filter as follows.

Definition 8.4.3 *A filter $\mathcal{F}$ is admissible if it is (internally) asymptotically (or exponentially) stable for any given initial condition $x(k_0)$ of the plant Σ^{da}, and with $w = 0$*

$$\lim_{k\to\infty} \breve{z}_k = 0.$$

The following proposition can now be proven along the same lines as Theorem 8.4.2.

Proposition 8.4.1 *Consider the nonlinear system (8.46) and the infinite-horizon DNLHIFP for it. Suppose the function h_1 is one-to-one (or injective) and the plant Σ^{da} is zero-state detectable. Further, suppose there exist a C^1 positive-definite function $\tilde{V} : \tilde{N} \times \tilde{N} \to \Re$ locally defined in a neighborhood $\tilde{N} \subset \mathcal{X}$ of the origin $\breve{x} = 0$, $\tilde{V}(0) = 0$ and* *a matrix function $\tilde{L} : \tilde{N} \to \mathcal{M}^{n \times m}$ satisfying the DHJIEs (8.64) together with (8.65)-(8.69). Then: (i) there exists locally a unique saddle-point solution $(\tilde{w}^\star, \tilde{L}^\star)$ for the game; and (ii) the filter Σ^{daf} with the gain matrix $\tilde{L}(\hat{x}) = \tilde{L}^\star(\hat{x})$ satisfying (8.66) solves the infinite-horizon DNLHIFP locally in $\tilde{N}$.*

Proof: (Sketch). We only prove that the filter Σ^{daf} is admissible, as the rest of the proof is similar to the finite-horizon case. Using similar manipulations as in the proof of Theorem 8.4.1, it can be shown that with $w = 0$,

$$\tilde{V}(\breve{x}_{k+1}) - \tilde{V}(\breve{x}_k) \leq -\frac{1}{2}\|\breve{z}_k\|^2.$$

Therefore, the augmented system is locally stable. Further, the condition that $\tilde{V}(\breve{x}_{k+1}) \equiv \tilde{V}(\breve{x}_k) \ \forall k \geq k_c$, for some $k_c \geq k_0$, implies that $\breve{z}_k \equiv 0 \forall k \geq k_c$. Moreover, the zero-state detectability of the system (8.46) implies the zero-state detectability of the system (8.50) since h_1 is injective. Thus, by virtue of this, we have $\lim_{k \to \infty} x_k = 0$, and by LaSalle's invariance-principle, this implies asymptotic-stability of the system (8.50). Hence, the filter Σ^{daf} is admissible. $\square$

8.4.2   Approximate and Explicit Solution

In this subsection, we discuss how the $DNLHIFP$ can be solved approximately to obtain explicit solutions [126]. We consider the infinite-horizon problem, but the approach can also be used for the finite-horizon problem. For simplicity, we make the following assumption on the system matrices.

Assumption 8.4.1 *The system matrices are such that*

$$
\begin{aligned}
k_{21}(x)g_1^T(x) &= 0 \ \forall x \in \mathcal{X}, \\
k_{21}(x)k_{21}^T(x) &= I \ \forall x \in \mathcal{X}.
\end{aligned}
$$

Now, consider the infinite-horizon Hamiltonian function $\widetilde{H}(.,.,.,.)$ defined by (8.69). Expanding it in Taylor-series about $\widehat{f}(\breve{x}) = \begin{pmatrix} f(x) \\ f(\hat{x}) \end{pmatrix}$ up to first-order[1] and denoting this expansion by $\widehat{H}(.,.,.,.)$ and L by $\hat{L}$, we get:

$$
\begin{aligned}
\widehat{H}(\breve{x}, w, \hat{L}, \tilde{V}) &= \Big\{ \tilde{V}(\widehat{f}(\breve{x})) + \tilde{V}_x(\widehat{f}(\breve{x}))g_1(x)w + \tilde{V}_{\hat{x}}(\widehat{f}(\breve{x}))[\hat{L}(\hat{x})(h_2(x) - h_2(\hat{x}) + k_{21}(x)w)] \\
&\quad + O(\|\tilde{v}\|^2) \Big\} - \tilde{V}(\breve{x}) + \frac{1}{2}\|\breve{z}\|^2 - \frac{1}{2}\gamma^2\|w\|^2, \quad \breve{x} \in \widehat{N} \times \widehat{N}, \ w \in \mathcal{W} \quad (8.70)
\end{aligned}
$$

where $\breve{x} = \breve{x}_k$, $\breve{z} = \breve{z}_k$, $w = w_k$, $\tilde{V}_x$, $\tilde{V}_{\hat{x}}$ are the row-vectors of the partial-derivatives of $\tilde{V}$ with respect to x, $\hat{x}$ respectively,

$$
\tilde{v} = \begin{pmatrix} g_1(x)w \\ \hat{L}(\hat{x})[h_2(x) - h_2(\hat{x}) + k_{21}(x)w] \end{pmatrix}
$$

and

$$
\lim_{\tilde{v} \to 0} \frac{O(\|\tilde{v}\|^2)}{\|\tilde{v}\|^2} = 0.
$$

---

[1] A second-order Taylor-series approximation would be more accurate, but the solutions become more complicated. Moreover, the first-order method gives a solution that is close to the continuous-time case [66].

Then, applying the necessary condition for optimality, we get

$$\frac{\partial \widehat{H}}{\partial w_k}\bigg|_{w=\hat{w}^\star} = g_1^T(x)\tilde{V}_x^T(\widehat{f}(\breve{x})) + k_{21}^T(x)\hat{L}^T(\hat{x})\tilde{V}_{\hat{x}}^T(\widehat{f}(\breve{x})) - \gamma^2 \hat{w}^\star = 0,$$

$$\Longrightarrow \hat{w}^\star(\breve{x}) = \frac{1}{\gamma^2}[g_1^T(x)\tilde{V}_x^T(\widehat{f}(\breve{x})) + k_{21}^T(x)\hat{L}^T(\hat{x})\tilde{V}_{\hat{x}}^T(\widehat{f}(\breve{x}))]. \tag{8.71}$$

Now substitute $\hat{w}^\star$ in (8.70) to obtain

$$\begin{aligned}\widehat{H}(\breve{x},\hat{w}_k^\star,\hat{L},\tilde{V}) \approx{}& V(\widehat{f}(\breve{x})) - \tilde{V}(\breve{x}) + \frac{1}{2\gamma^2}\tilde{V}_x(\widehat{f}(\breve{x}))g_1(x)g_1^T(x)\tilde{V}_x^T(\widehat{f}(\breve{x})) + \frac{1}{2}\|\breve{z}\|^2 + \\ & \tilde{V}_{\hat{x}}(\widehat{f}(\breve{x}))\hat{L}(\hat{x})[h_2(x)-h_2(\hat{x})] + \frac{1}{2\gamma^2}\tilde{V}_{\hat{x}}(\widehat{f}(\breve{x}))\hat{L}(\hat{x})\hat{L}^T(\hat{x})\tilde{V}_x^T(\widehat{f}(\breve{x})).\end{aligned}$$

Completing the squares for $\hat{L}$ in the above expression for $\widehat{H}(.,.,.,.)$, we get

$$\begin{aligned}\widehat{H}(\breve{x},\hat{w}^\star,\hat{L},\tilde{V}) \approx{}& \tilde{V}(\widehat{f}(\breve{x})) - \tilde{V}(\breve{x}) + \frac{1}{2\gamma^2}\tilde{V}_x(\widehat{f}(\breve{x}))g_1(x)g_1^T(x)\tilde{V}_x^T(\widehat{f}(\breve{x})) + \frac{1}{2}\|\breve{z}_k\|^2 + \\ & \frac{1}{2\gamma^2}\left\|\hat{L}^T(\hat{x})\tilde{V}_{\hat{x}}^T(\widehat{f}(\breve{x})) + \gamma^2(h_2(x)-h_2(\hat{x}))\right\|^2 - \frac{\gamma^2}{2}\|h_2(x)-h_2(\hat{x})\|^2.\end{aligned}$$

Thus, taking $\hat{L}^\star$ as

$$\tilde{V}_{\hat{x}}(\widehat{f}(\breve{x}))\hat{L}^\star(\hat{x}) = -\gamma^2(h_2(x)-h_2(\hat{x}))^T, \quad x, \hat{x} \in \widehat{N} \tag{8.72}$$

minimizes $\widehat{H}(.,.,.,.)$ and renders the saddle-point condition

$$\widehat{H}(\hat{w}^\star,\hat{L}^\star) \le \widehat{H}(w^\star,\hat{L}) \;\; \forall \hat{L} \in \mathcal{M}^{n\times m}$$

satisfied.

Substitute now $\hat{L}^\star$ as given by (8.72) in the expression for $\widehat{H}(.,.,.,.)$ and complete the squares in w to obtain:

$$\begin{aligned}\widehat{H}(\breve{x},w,\hat{L}^\star,\tilde{V}) ={}& \tilde{V}(\widehat{f}(\breve{x})) - \frac{1}{2\gamma^2}\tilde{V}_{\hat{x}}(\widehat{f}(\breve{x}))\hat{L}(\hat{x})\hat{L}^T(\hat{x})\tilde{V}_{\hat{x}}^T(\widehat{f}(\breve{x}) + \frac{1}{2}\|\breve{z}_k\|^2 + \\ & \frac{1}{2\gamma^2}\tilde{V}_x(\widehat{f}(\breve{x}))g_1(x)g_1^T(x)\tilde{V}_x^T(\widehat{f}(\breve{x})) - \tilde{V}(\breve{x}) \\ & -\frac{\gamma^2}{2}\left\|w - \frac{1}{\gamma^2}g_1^T(x)\tilde{V}_x^T(\widehat{f}(\breve{x})) - \frac{1}{\gamma^2}k_{21}^T(x)\hat{L}^T(\hat{x})\tilde{V}_{\hat{x}}^T(\widehat{f}(\breve{x}))\right\|^2.\end{aligned}$$

Thus, substituting $w = \hat{w}^\star$ as given in (8.71), we see that the second saddle-point condition

$$\widehat{H}(\hat{w}^\star,\hat{L}^\star) \ge H(w,\hat{L}^\star), \;\; \forall w \in \mathcal{W}$$

is also satisfied. Hence, the pair $(\hat{w}^\star,\hat{L}^\star)$ constitutes a unique saddle-point solution to the game corresponding to the Hamiltonian $\widehat{H}(.,.,.,.)$. Finally, substituting $(\hat{w}^\star,\hat{L}^\star)$ in the DHJIE (8.53), we get the following DHJIE:

$$\begin{aligned}& \tilde{V}(\widehat{f}(\breve{x})) - \tilde{V}(\breve{x}) + \tfrac{1}{2\gamma^2}\tilde{V}_x(\widehat{f}(\breve{x}))g_1(x)g_1^T(x)\tilde{V}_x^T(\widehat{f}(\breve{x})) - \\ & \tfrac{1}{2\gamma^2}\tilde{V}_{\hat{x}}(\widehat{f}(\breve{x}))\hat{L}(\hat{x})\hat{L}^T(\hat{x})\tilde{V}_{\hat{x}}^T(\widehat{f}(\breve{x})) + \\ & \tfrac{1}{2}(h_1(x)-h_1(\hat{x}))^T(h_1(x)-h_1(\hat{x})) = 0, \quad \tilde{V}(0) = 0, \quad x, \hat{x} \in \widehat{N}. \tag{8.73}\end{aligned}$$

With the above analysis, we have the following theorem.

Theorem 8.4.3 *Consider the nonlinear system (8.46) and the infinite-horizon DNLHIFP for this system. Suppose the function h_1 is one-to-one (or injective) and the plant Σ^{da} is zero-state detectable. Suppose further there exists a C^1 positive-definite function $\tilde{V}$: $\widehat{N} \times \widehat{N} \to \Re$ locally defined in a neighborhood $\widehat{N} \times \widehat{N} \subset \mathcal{X} \times \mathcal{X}$ of the origin $\breve{x} = 0$, and a matrix function $\hat{L}$: $\widehat{N} \times \widehat{N} \to \mathcal{M}^{n \times m}$ satisfying the DHJIE (8.73) together with the side-condition (8.72). Then:*

(i) *there exists locally in $\widehat{N}$ a unique saddle-point solution $(\hat{w}^\star, \hat{L}^\star)$ for the dynamic game corresponding to (8.51), (8.50);*

(ii) *the filter Σ^{daf} with the gain matrix $\hat{L}(\hat{x}) = \hat{L}^\star(\hat{x})$ satisfying (8.72) solves the infinite-horizon DNLHIFP for the system locally in $\widehat{N}$.*

Proof: Part (i) has already been shown above. For part (ii), consider the time variation of $\tilde{V} > 0$ (a solution to the DHJIE (8.73)) along a trajectory of the augmented system (8.50) with $\hat{L} = \hat{L}^\star$, i.e.,

$$
\begin{aligned}
\tilde{V}(\breve{x}_{k+1}) &= \tilde{V}(\breve{f}(\breve{x}) + \breve{g}(\breve{x})w) \quad \forall \breve{x} \in \widehat{N} \times \widehat{N}, \forall w \in \mathcal{W} \\
&\approx \tilde{V}(\widehat{f}(\breve{x})) + \tilde{V}_x(\widehat{f}(\breve{x}))g_1(x)w + \tilde{V}_{\hat{x}}(\widehat{f}(\breve{x}))[\hat{L}^\star(\hat{x})(h_2(x) - h_2(\hat{x}) + k_{21}(x)w)] \\
&= \tilde{V}(\widehat{f}(\breve{x})) + \frac{1}{2\gamma^2}\tilde{V}_x(\widehat{f}(\breve{x}))g_1(x)g_1^T(x)\tilde{V}_x^T(\widehat{f}(\breve{x})) + \tilde{V}_{\hat{x}}(\widehat{f}(\breve{x}))\hat{L}^\star(\hat{x})[h_2(x) - h_2(\hat{x})] \\
&\quad - \frac{\gamma^2}{2}\|w - \hat{w}^\star\|^2 + \frac{\gamma^2}{2}\|w\|^2 + \frac{1}{2\gamma^2}\tilde{V}_{\hat{x}}(\widehat{f}(\breve{x}))\hat{L}^\star(\hat{x})\hat{L}^{\star^T}(\hat{x})\tilde{V}_{\hat{x}}^T(\widehat{f}(\breve{x})) \\
&= \tilde{V}(\widehat{f}(\breve{x})) + \frac{1}{2\gamma^2}\tilde{V}_x(\widehat{f}(\breve{x}))g_1(x)g_1^T(x)\tilde{V}_x^T(\widehat{f}(\breve{x})) - \frac{\gamma^2}{2}\|w - \hat{w}^\star\|^2 + \frac{\gamma^2}{2}\|w\|^2 - \\
&\quad \frac{1}{2\gamma^2}\tilde{V}_{\hat{x}}(\widehat{f}(\breve{x}))\hat{L}^\star(\hat{x})\hat{L}^{\star^T}(\hat{x})\tilde{V}_{\hat{x}}^T(\widehat{f}(\breve{x})) \\
&\leq \tilde{V}(x) + \frac{\gamma^2}{2}\|w_k\|^2 - \frac{1}{2}\|\breve{z}_k\|^2 \quad \forall \breve{x} \in \widehat{N} \times \widehat{N}, \forall w \in \mathcal{W},
\end{aligned}
$$

where use has been made of the Taylor-series approximation, equation (8.72) and the DHJIE (8.73). Finally, the above inequality clearly implies the infinitesimal dissipation-inequality:

$$
\tilde{V}(\breve{x}_{k+1}) - \tilde{V}(\breve{x}_k) \leq \frac{1}{2}\gamma^2\|w_k\|^2 - \frac{1}{2}\|\breve{z}_k\|^2 \quad \forall \breve{x} \in \widehat{N} \times \widehat{N}, \forall w \in \mathcal{W}.
$$

Therefore, the system (8.50) has locally ℓ_2-gain from w to $\breve{z}$ less or equal to γ. The remaining arguments are the same as in the proof of Proposition 8.4.1. $\square$

We now specialize the result of the above theorem to the linear-time-invariant (LTI) system:

$$
\Sigma^{dl} : \begin{cases} \dot{x}_{k+1} &= Ax_k + B_1 w_k, \quad x(k_0) = x^0 \\ \breve{z}_k &= C_1(x_k - \hat{x}_k) \\ y_k &= C_2 x_k + D_{21} w_k, \quad D_{21}B_1^T = 0, \quad D_{21}^T D_{21} = I, \end{cases} \tag{8.74}
$$

where all the variables have their previous meanings, and $A \in \Re^{n \times n}$, $B_1 \in \Re^{n \times r}$, $C_1 \in \Re^{s \times n}$, $C_2 \in \Re^{m \times n}$ and $D_{21} \in \Re^{m \times r}$ are constant real matrices. We have the following corollary to Theorem 8.4.3.

Corollary 8.4.1 *Consider the discrete-LTI system Σ^{dl} defined by (8.74) and the*

DNLHIFP for it. Suppose C_1 is full column rank and A is Hurwitz. Suppose further, there exists a positive-definite real symmetric solution $\widehat{P}$ to the discrete algebraic-Riccati equation (DARE):

$$A^T\widehat{P}A - \widehat{P} - \frac{1}{2\gamma^2}A^T\widehat{P}B_1B_1^T\widehat{P}A - \frac{1}{2\gamma^2}A^T\widehat{P}\hat{L}\hat{L}^T\widehat{P}A + C_1^TC_1 = 0 \qquad (8.75)$$

together with the coupling condition:

$$A^T\widehat{P}\hat{L}_l^\star = -\gamma^2 C_2^T. \qquad (8.76)$$

Then:

(i) *there exists a unique saddle-point solution $(\hat{w}_l^\star, \hat{L}_l^\star)$ for the game given by*

$$\hat{w}_l^\star = \frac{1}{\gamma^2}(B_1^T + D_{21}^T\hat{L})\widehat{P}A(x - \hat{x}),$$
$$(x - \hat{x})^T A^T\widehat{P}\hat{L}_l^\star = -\gamma^2(x - \hat{x})^T C_2^T;$$

(ii) *the filter $\mathcal{F}$ defined by*

$$\Sigma_{ldf}: \quad \hat{x}_{k+1} = A\hat{x}_k + \hat{L}(y - C_2 x_k), \quad \hat{x}(k_0) = \hat{x}^0$$

with the gain matrix $\hat{L} = \hat{L}_l^\star$ satisfying (8.76) solves the infinite-horizon DNLHIFP for the system.

Proof: Take:

$$\tilde{V}(\breve{x}) = \frac{1}{2}(x - \hat{x})^T\widehat{P}(x - \hat{x}), \quad \widehat{P} = \widehat{P}^T > 0,$$

and apply the result of the theorem. $\square$

Next we consider an example.

Example 8.4.1 *We consider the following scalar example:*

$$\begin{aligned}
x_{k+1} &= x_k^{\frac{1}{3}}\\
y_k &= x_k + w_k\\
z_k &= x_k
\end{aligned}$$

where $w_k = w_0 + 0.1sin(2\pi k)$, and w_0 is zero-mean Gaussian white noise. We compute the solution of the DHJIE (8.73) using the iterative scheme:

$$\begin{aligned}
\tilde{V}^{j+1}(\breve{x}) &= \tilde{V}^k(\hat{f}(\breve{x})) + \frac{1}{2\gamma^2}\tilde{V}_x^j(\hat{f}(\breve{x}))g_1(x)g_1^T(x)\tilde{V}_x^{T,j}(\hat{f}(\breve{x})) - \\
&\quad \frac{1}{2\gamma^2}\tilde{V}_{\hat{x}}^j(\hat{f}(\breve{x}))\hat{L}^j(\hat{x})\hat{L}^{j,T}(\hat{x})\tilde{V}_{\hat{x}}^{T,j}(\hat{f}(\breve{x})) + \frac{1}{2}(h_1(x) - h_1(\hat{x}))^T(h_1(x) - h_1(\hat{x})),\\
&\tilde{V}^j(0) = 0, \quad x, \hat{x} \in \widehat{N}, \quad j = 0,1,\dots \qquad (8.77)
\end{aligned}$$

starting with the initial guess $\tilde{V}^0(\breve{x}) = \frac{1}{2}(x - \hat{x})^2$, $\gamma = 1$ and initial filter-gain $l_0 = 1$, after one iteration, we get $\tilde{V}^1(\breve{x}) = (x^{\frac{1}{3}} - \hat{x}^{\frac{1}{3}})^2 + \frac{1}{2}(x-\hat{x})^2$ and $\tilde{V}_{\hat{x}}^1(\hat{f}(\breve{x})) = -\frac{2}{3}\left(\frac{x^{\frac{1}{9}} - \hat{x}^{\frac{1}{9}}}{\hat{x}^{\frac{2}{9}}}\right) - (x^{\frac{1}{3}} - \hat{x}^{\frac{1}{3}})$. Then we proceed to compute the filter-gain using (8.72). The result of the simulation is shown in Figure 8.7.

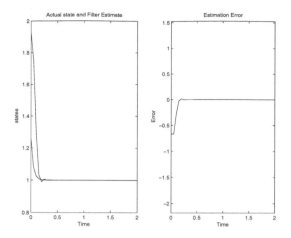

FIGURE 8.7
Discrete-Time Nonlinear $\mathcal{H}_\infty$-Filter Performance with Unknown Initial Condition

8.5   Discrete-Time Certainty-Equivalent Filters (CEFs)

In this section, we present the discrete-time counterpart of the the results of Section 8.3.1 which we referred to as *"certainty-equivalent"* worst-case estimators. It is also similarly apparent that the filter gain derived in equation (8.72) may depend on the estimated state, and this will present a serious stumbling block in implementation. Therefore, in this section we derive results in which the gain matrices are not functions of the state x, but of $\hat{x}$ and y only. The estimator is constructed on the assumption that the asymptotic value of $\hat{x}$ equals x. We first construct the estimator for the 1-DOF case, and then we discuss the 2-DOF case.

We reconsider the nonlinear discrete-time affine causal state-space system defined on a state-space $\mathcal{X} \subseteq \Re^n$ with zero-input:

$$\Sigma^{da} \; : \; \begin{cases} x_{k+1} & = & f(x_k) + g_1(x_k)w_k; \;\; x(k_0) = x^0 \\ y_k & = & h_2(x_k) + k_{21}(x_k)w_k \end{cases} \tag{8.78}$$

where $x \in \mathcal{X}$ is the state vector, $w \in \mathcal{W} \subset \ell_2([k_0, \infty), \Re^r)$ is the disturbance or noise signal, which belongs to the set $\mathcal{W}$ of admissible disturbances and noise signals, the output $y \in \mathcal{Y} \subset \Re^m$ is the measured output of the system, which belongs to the set $\mathcal{Y}$ of measured outputs, while $z \in \Re^s$ is the output to be estimated. The functions $f : \mathcal{X} \to \mathcal{X}$, $g_1 : \mathcal{X} \to \mathcal{M}^{n \times r}$, where $\mathcal{M}^{i \times j}(\mathcal{X})$ is the ring of $i \times j$ matrices over $\mathcal{X}$, $h_2 : \mathcal{X} \to \Re^m$, and $k_{21} : \mathcal{X} \to \mathcal{M}^{m \times r}(\mathcal{X})$ are real C^∞ functions of x. Furthermore, we assume without any loss of generality that the system (8.78) has a unique equilibrium-point at $x = 0$ such that $f(0) = 0$, $h_2(0) = 0$.

Assumption 8.5.1 *The system matrices are such that*

$$\begin{aligned} k_{21}(x)g_1^T(x) & = & 0 \\ k_{21}(x)k_{21}^T(x) & = & I. \end{aligned}$$

Again, the discrete-time $\mathcal{H}_\infty$ nonlinear filtering problem ($DHINLFP$) is to synthesize a filter, $\mathcal{F}_k$, for estimating the state x_k from available observations $Y_k \triangleq \{y_i, \ i \leq k\}$ over a time horizon $[k_0, \infty)$, such that

$$\hat{x}_{k+1} = \mathcal{F}_k(Y_k), \quad k \in [k_0, \infty),$$

and the ℓ_2-gain from the disturbance/noise signal w to the estimation error output $\tilde{z}$ (to be defined later) is less than or equal to a desired number $\gamma > 0$, i.e.,

$$\sum_{k=k_0}^{\infty} \|\tilde{z}_k\|^2 \leq \gamma^2 \sum_{k=k_0}^{\infty} \|w_k\|^2, \quad k \in \mathbf{Z} \tag{8.79}$$

for all $w \in \mathcal{W}$, for all $x_0 \in O \subset \mathcal{X}$.

There are various notions of observability, however for our purpose, we shall adopt the following which also generalizes the notion of *"zero-state observability."*

Definition 8.5.1 *For the nonlinear system (8.78), we say that it is locally zero-input observable if for all states $x_{k_1}, x_{k_2} \in U \subset \mathcal{X}$ and input $w(.) = 0$,*

$$y(., x_{k_1}, w) \equiv y(., x_{k_2}, w) \implies x_{k_1} = x_{k_2},$$

where $y(., x_{k_i}, w), i = 1, 2$ is the output of the system with the initial condition $x(k_0) = x_{k_i}$. Moreover, the system is said to be zero-input observable if it is locally zero-input observable at each $x^0 \in \mathcal{X}$ or $U = \mathcal{X}$.

We now propose the following class of estimators:

$$\Sigma_1^{dacef} : \begin{cases} \dot{\hat{x}}_{k+1} &= f(\hat{x}_k) + g_1(\hat{x})\hat{w}_k^\star + L(\hat{x}_k, y_k)(y_k - h_2(\hat{x}_k) - k_{21}(\hat{x}_k)\hat{w}_k^\star) \\ \hat{z}_k &= h_2(\hat{x}_k) \\ \tilde{z}_k &= y_k - h_2(\hat{x}_k) \end{cases} \tag{8.80}$$

where $\hat{z} = \hat{y} \in \Re^m$ is the estimated output, $\tilde{z}_k \in \Re^m$ is the new estimation error or penalty variable, $\hat{w}^\star$ is the estimated worst-case system noise and $L : \mathcal{X} \times \mathcal{Y} \to \mathcal{M}^{n \times m}$ is the gain-matrix of the filter. We first determine $w^\star$, and accordingly, define the Hamiltonian function $H : \mathcal{X} \times \mathcal{Y} \times \mathcal{W} \times \mathcal{M}^{n \times m} \times \Re \to \Re$ corresponding to (8.80) and the cost functional

$$J = \min_L \max_w \sum_{k=k_0}^{\infty} [\|\tilde{z}_k\|^2 - \gamma^2 \|w_k\|^2], \quad k \in \mathbf{Z} \tag{8.81}$$

by

$$H(\hat{x}, y, w, L, V) = V(f(\hat{x}) + g_1(\hat{x})w + L(\hat{x}, y)(y - h_2(\hat{x}) - k_{21}(\hat{x})w, y) - $$
$$V(\hat{x}, y_{k-1}) + \frac{1}{2}(\|\tilde{z}\|^2 - \gamma^2 \|w\|^2) \tag{8.82}$$

for some smooth function $V : \mathcal{X} \times \mathcal{Y} \to \Re$, where $x = x_k$, $y = y_k$, $w = w_k$ and the adjoint variable p is set as $p = V$. Applying now the necessary condition for the worst-case noise, we get

$$\left. \frac{\partial^T H}{\partial w} \right|_{w=\hat{w}^\star} = (g_1^T(\hat{x}) - k_{21}^T(\hat{x})L^T(\hat{x}, y)) \left. \frac{\partial^T V(\lambda, y)}{\partial \lambda} \right|_{\lambda = f^\star(\hat{x}, y, \hat{w}^\star)}$$
$$-\gamma^2 \hat{w}^\star = 0, \tag{8.83}$$

where

$$f^\star(\hat{x}, y, \hat{w}^\star) = f(\hat{x}) + g_1(\hat{x})\hat{w}^\star + L(\hat{x}, y)(y_k - h_2(\hat{x}) - k_{21}(\hat{x})\hat{w}^\star),$$

and

$$\hat{w}^\star = \frac{1}{\gamma^2}(g_1^T(\hat{x}) - k_{21}^T(\hat{x})L^T(\hat{x},y))\frac{\partial^T V(\lambda,y)}{\partial\lambda}\bigg|_{\lambda=f^\star(\hat{x},y,\hat{w}^\star)} := \alpha_0(\hat{x},\hat{w}^\star,y). \qquad (8.84)$$

Moreover, since

$$\frac{\partial^2 H}{\partial w^2}\bigg|_{w=\hat{w}^\star} = (g_1^T(\hat{x}) - k_{21}^T(\hat{x})L^T(\hat{x},y))\frac{\partial^2 V(\lambda,y)}{\partial\lambda^2}\bigg|_{\lambda=f^\star(\hat{x},y,\hat{w}^\star)}(g_1(\hat{x}) - L(\hat{x},y)k_{21}(\hat{x})) - \gamma^2 I$$

is nonsingular about $(\hat{x},w,y) = (0,0,0)$, equation (8.84) has a unique solution $\hat{w}^\star = \alpha(\hat{x},y)$, $\alpha(0,0) = 0$ in the neighborhood $N \times W \times Y$ of $(x,w,y) = (0,0,0)$ by the Implicit-function Theorem [234].

Now substitute $\hat{w}^\star$ in the expression for $H(.,.,.,.)$ (8.82), to get

$$H(\hat{x},y,\hat{w}^\star,L,V) = V(f(\hat{x}) + g_1(\hat{x})\hat{w}^\star + L(\hat{x},y)(y - h_2(\hat{x}) - k_{21}(\hat{x})\hat{w}^\star),y) -$$
$$V(\hat{x},y_{k-1}) + \frac{1}{2}(\|\tilde{z}\|^2 - \gamma^2\|\hat{w}^\star\|^2)$$

and let

$$L^\star = \arg\min_L\{H(\hat{x},y,\hat{w}^\star,L,V)\}. \qquad (8.85)$$

Then by Taylor's theorem [267], we can expand $H(.,.,.,.)$ about $(L^\star,\hat{w}^\star)$ as

$$H(\hat{x},y,w,L,V) = H(\hat{x},y,\hat{w}^\star,L^\star,V) + \frac{1}{2}(w - \hat{w}^\star)^T\frac{\partial^2 H}{\partial w^2}(w,L^\star)(w - \hat{w}^\star) +$$
$$\frac{1}{2}Tr\left\{[I_n \otimes (L - L^\star)^T]\frac{\partial^2 H}{\partial L^2}(\hat{w}^\star,L)[I_m \otimes (L - L^\star)^T]\right\} +$$
$$O(\|w - \hat{w}^\star\|^3 + \|L - L^\star\|^3). \qquad (8.86)$$

Thus, taking $L^\star$ as in (8.85) and $\hat{w}^\star = \alpha(\hat{x},y)$ and if the conditions

$$\frac{\partial^2 H}{\partial w^2}(\hat{x},y,w,L^\star,V)\bigg|_{(\hat{x}=0,w=0)} < 0, \qquad (8.87)$$

$$\frac{\partial^2 H}{\partial L^2}(\hat{x},y,\hat{w}^\star,L,V)\bigg|_{(\hat{x}=0,w=0)} > 0 \qquad (8.88)$$

hold, we see that the saddle-point conditions

$$H(w,L^\star) \le H(\hat{w}^\star,L^\star) \le H(\hat{w}^\star,L) \quad \forall L \in \mathcal{M}^{n \times m}, \forall w \in \mathcal{W}$$

are locally satisfied. Moreover, setting

$$H(\hat{x},y,\hat{w}^\star,L^\star,V) = 0$$

gives the DHJIE:

$$V(f(\hat{x}) + g_1(\hat{x})\alpha(\hat{x},y) + L^\star(\hat{x},y)(y - h_2(\hat{x}) - k_{21}(\hat{x})\alpha(\hat{x},y)),y) - V(\hat{x},y_{k-1}) +$$
$$\frac{1}{2}\|\tilde{z}(\hat{x})\|^2 - \frac{1}{2}\gamma^2\|\alpha(\hat{x},y)\|^2 = 0, \quad V(0,0) = 0, \quad \hat{x},y \in N \times Y. \qquad (8.89)$$

Consequently, we have the following result.

Proposition 8.5.1 *Consider the discrete-time nonlinear system (8.78) and the DHINLFP for it. Suppose Assumption 8.5.1 holds, the plant Σ^{da} is locally asymptotically-stable about* the equilibrium point $x = 0$ and zero-input observable. Further, suppose there exists a C^2 positive-semidefinite function $V : N \times Y \to \Re_+$ locally defined in a neighborhood $N \times Y \subset \mathcal{X} \times \mathcal{Y}$ of the origin $(\hat{x}, y) = (0, 0)$, and a matrix function $L : N \times Y \to \mathcal{M}^{n \times m}$, satisfying the DHJIE (8.89) together with the side-conditions (8.85), (8.87), (8.88). Then the filter Σ_1^{dacef} solves the DHINLFP for the system locally in N.

Proof: The first part of the theorem on the existence of the saddle-point solutions $(\hat{w}^\star, L^\star)$ has already been shown above. It remains to show that the ℓ_2-gain condition (8.79) is satisfied and the filter provides asymptotic estimates.

For this, let $V \geq 0$ be a C^2 solution of the DHJIE (8.89) and reconsider equation (8.86). Since the conditions (8.87) and (8.88) are satisfied about $\hat{x} = 0$, by the Inverse-function Theorem [234], there exists a neighborhood $U \subset N \times W$ of $(\hat{x}, w) = (0, 0)$ for which they are also satisfied. Consequently, we immediately have the important inequality

$$H(\hat{x}, y, w, L^\star, V) \leq H(\hat{x}, y, \hat{w}^\star, L^\star, V) = 0 \quad \forall \hat{x} \in N, \forall y \in \mathcal{Y}, \forall w \in \mathcal{W}$$
$$\Longleftrightarrow V(\hat{x}_{k+1}, y_k) - V(\hat{x}_k, y_{k-1}) \leq \tfrac{1}{2}\gamma^2 \|w_k\|^2 - \tfrac{1}{2}\|\check{z}_k\|^2. \tag{8.90}$$

Summing now from $k = k_0$ to ∞, we get that the ℓ_2-gain condition (8.79) is satisfied:

$$V(\hat{x}_\infty, y_\infty) + \frac{1}{2} \sum_{k_0}^{\infty} \|\check{z}_k\|^2 \leq \frac{1}{2} \sum_{k=k_0}^{\infty} \gamma^2 \|w_k\|^2 + V(x_{k_0}, y_{k_0-1}). \tag{8.91}$$

Moreover, setting $w_k \equiv 0$ in (8.90), implies that $V(\hat{x}_{k+1}, y_k) - V(\hat{x}_k, y_{k-1}) \leq -\tfrac{1}{2}\|\tilde{z}_k\|^2$ and hence the estimator dynamics is stable. In addition,

$$V(\hat{x}_{k+1}, y_k) - V(\hat{x}_k, y_{k-1}) \equiv 0 \Longrightarrow \tilde{z} \equiv 0 \Longrightarrow y = h_2(\hat{x}) = \hat{y}.$$

By the zero-input observability of the system Σ^{da}, this implies that $x = \hat{x}$. $\square$

8.5.1 2-DOF Proportional-Derivative (PD) CEFs

Next, we extend the above certainty-equivalent design to the 2-DOF case. In this regard, we assume that the time derivative $y_k - y_{k-1}$ is available (or equivalently y_{k-1} is available), and consider the following class of filters:

$$\Sigma_2^{dacef} : \begin{cases} \begin{aligned} \acute{x}_{k+1} &= f(\acute{x}_k) + g_1(\acute{x}_k)\acute{w}^\star(\acute{x}_k) + \acute{L}_1(\acute{x}_k, y_k, y_{k-1})(y_k - h_2(\acute{x}_k) - \\ &\quad k_{21}(\acute{x}_k)\acute{w}^\star(\acute{x}_k)) + \acute{L}_2(\acute{x}_k, y_k, y_{k-1})(y_k - y_{k-1} - h_2(\acute{x}_k) + \\ &\quad h_2(\acute{x}_{k-1})) \\ \acute{z}_k &= \begin{bmatrix} h_2(\acute{x}_k) \\ h_2(\acute{x}_k) - h_2(\acute{x}_{k-1}) \end{bmatrix} \\ \widetilde{z}_k &= \begin{bmatrix} y_k - h_2(\acute{x}_k) \\ (y_k - y_{k-1}) - (h_2(\acute{x}_k) - h_2(\acute{x}_{k-1})) \end{bmatrix} \end{aligned} \end{cases}$$

where $\acute{z} \in \Re^m$ is the estimated output of the filter, $\tilde{z} \in \Re^s$ is the error or penalty variable, while $\acute{L}_1 : \mathcal{X} \times \mathcal{X} \times \mathcal{Y} \times \mathcal{Y} \to \mathcal{M}^{n \times m}$, $\acute{L}_2 : \mathcal{X} \times \mathcal{X} \times \mathcal{Y} \times \mathcal{Y} \to \mathcal{M}^{n \times m}$ are the proportional and derivative gains of the filter respectively, and all the other variables and functions have their corresponding previous meanings and dimensions. As in the previous section, we can define the corresponding Hamiltonian function $\acute{H} : \mathcal{X} \times \mathcal{W} \times \mathcal{M}^{n \times m} \times \mathcal{M}^{n \times m} \times \Re \to \Re$ for the filter as

$$\acute{H}(\acute{x}, \acute{w}, \acute{L}_1, \acute{L}_2, \acute{V}) = \acute{V}(\acute{f}(\acute{x}, \acute{x}_{k-1}, y, y_{k-1}), \acute{x}, y) - \acute{V}(\acute{x}, \acute{x}_{k-1}, y_{k-1}) + \frac{1}{2}(\|\tilde{z}\|^2 - \gamma^2 \|\acute{w}\|^2) \tag{8.92}$$

for some smooth function $\tilde{V} : \mathcal{X} \times \mathcal{X} \times \mathcal{Y} \times \mathcal{Y} \to \Re$ and where

$$\acute{f}(\acute{x}, \acute{x}_{k-1}, y, y_{k-1}) = f(\acute{x}) + g_1(\acute{x})\acute{w} + \acute{L}_1(\acute{x}, \acute{x}_{k-1}, y, y_{k-1})[y - h_2(\acute{x}) - k_{21}(\acute{x})\acute{w}] +$$
$$\acute{L}_2(\acute{x}, \acute{x}_{k-1}, y, y_{k-1})[y - y_{k-1} - (h_2(\acute{x}) - h_2(\acute{x}_{k-1}))].$$

Notice that, in the above and subsequently, we only use the subscripts k, $k-1$ to distinguish the variables, otherwise, the functions are smooth in the variables $\acute{x}$, $\acute{x}_{k-1}$, y, y_{k-1}, $\acute{w}$, etc.

Similarly, applying the necessary condition for the worst-case noise, we have

$$\acute{w}^\star = \frac{1}{\gamma^2}(g_1^T(\acute{x}) - k_{21}^T(\acute{x})\acute{L}_1^T(\acute{x}, \acute{x}_{k-1}, y, y_{k-1}))\frac{\partial^T \acute{V}(\lambda, \acute{x}, y)}{\partial \lambda}\bigg|_{\lambda = \acute{f}(\acute{x}, ., ., .)}$$

$$:= \alpha_1(\acute{x}, \acute{x}_{k-1}, \acute{w}^\star, y, y_{k-1}). \tag{8.93}$$

Morever, since

$$\frac{\partial^2 \acute{H}}{\partial w^2} = (g_1^T(\acute{x}) - k_{21}^T(\acute{x})\acute{L}_1^T(\acute{x}, ., ., .))\frac{\partial^2 \acute{V}(\lambda, \acute{x}, y)}{\partial \lambda^2}\bigg|_{\lambda = \acute{f}(\acute{x}, ., ., .)} (g_1(\acute{x}) - \acute{L}_1(\acute{x}, ., ., .)k_{21}(\acute{x})) - \gamma^2 I$$

is nonsingular about $(\acute{x}, \acute{x}_{k-1}, \acute{w}, y, y_{k-1}) = (0, 0, 0, 0, 0)$, then again by the Impilicit-function Theorem, (8.93) has a unique solution $\acute{w}^\star = \acute{\alpha}(\acute{x}, \acute{x}_{k-1}, y_k, y_{k-1})$, $\acute{\alpha}(0, 0, 0, 0, 0) = 0$ locally about $(\acute{x}, \acute{x}_{k-1}, \acute{w}, y, y_{k-1}) = (0, 0, 0, 0, 0)$.

Substitute now $\acute{w}^\star$ into (8.92) to get

$$\acute{H}(\acute{x}, \acute{w}, \acute{L}_1, \acute{L}_2, \acute{V}) = \acute{V}(\acute{f}(\acute{x}, ., ., .), \acute{x}, y) - \acute{V}(\acute{x}, \acute{x}_{k-1}, y_{k-1}) + \frac{1}{2}(\|\tilde{z}\|^2 - \gamma^2\|\acute{w}^\star\|^2) \tag{8.94}$$

and let

$$[\acute{L}_1^\star \ \acute{L}_2^\star] = \arg\min_{\acute{L}_1, \acute{L}_2}\left\{\acute{H}(\acute{x}, \acute{w}^\star, \acute{L}_1, \acute{L}_2, \acute{V})\right\}. \tag{8.95}$$

Then, it can be shown using Taylor-series expansion similar to (8.86) and if the conditions

$$\frac{\partial^2 \acute{H}}{\partial w^2}(\acute{x}, \acute{w}, \acute{L}_1^\star, \acute{L}_2^\star, \acute{V})\bigg|_{(\acute{x}=0, \acute{w}=0)} < 0 \tag{8.96}$$

$$\frac{\partial^2 \acute{H}}{\partial \acute{L}_1^2}(\acute{x}, \acute{w}^\star, \acute{L}_1, \acute{L}_2^\star, \acute{V})\bigg|_{(\acute{x}=0, \acute{w}=0)} > 0 \tag{8.97}$$

$$\frac{\partial^2 \acute{H}}{\partial \acute{L}_2^2}(\acute{x}, \acute{w}^\star, \acute{L}_1^\star, \acute{L}_2, \acute{V})\bigg|_{(\acute{x}=0, \acute{w}=0)} > 0 \tag{8.98}$$

are locally satisfied, then the saddle-point conditions

$$\acute{H}(\acute{w}, \acute{L}_1^\star, \acute{L}_2^\star) \leq \acute{H}(\acute{w}^\star, \acute{L}_1^\star, \acute{L}_2^\star) \leq \acute{H}(\acute{w}^\star, \acute{L}_1, \acute{L}_2) \ \forall \acute{L}_1, \acute{L}_2 \in \mathcal{M}^{n \times m}, \forall w \in \mathcal{W},$$

are locally satisfied also.

Finally, setting

$$\acute{H}(\acute{x}, \acute{w}^\star, \acute{L}_1^\star, \acute{L}_2^\star, \acute{V}_{\acute{x}}^T) = 0$$

yields the DHJIE

$$\acute{V}(\acute{f}^\star(\acute{x}, \acute{x}_{k-1}, y, y_{k-1}), \acute{x}, y) - \acute{V}(\acute{x}, \acute{x}_{k-1}, y_{k-1}) + \frac{1}{2}(\|\tilde{z}\|^2 - \gamma^2\|\acute{w}^\star\|^2) = 0, \quad \acute{V}(0, 0, 0, 0) = 0, \tag{8.99}$$

where

$$
\begin{aligned}
\acute{f}^\star(\acute{x}, \acute{x}_{k-1}, y, y_{k-1}) &= f(\acute{x}) + g_1(\acute{x})\acute{w}^\star + \acute{L}_1^\star(\acute{x}, x_{k-1}, y, y_{k-1})[y - h_2(\acute{x}) - k_{21}(\acute{x})\acute{w}^\star] + \\
&\quad \acute{L}_2^\star(\acute{x}, x_{k-1}, y, y_{k-1})[y - y_{k-1} - (h_2(\acute{x}) - h_2(\acute{x}_{k-1}))].
\end{aligned}
$$

With the above analysis, we have the following result.

Theorem 8.5.1 *Consider the discrete-time nonlinear system (8.78) and the DHINLFP for it. Suppose Assumption 8.5.1 holds, the plant Σ^{da} is locally asymptotically stable about the equilibrium point $x = 0$ and zero-input observable. Further, suppose there exists a C^2 positive-semidefinite function $\acute{V} : \acute{N} \times \acute{N} \times \acute{Y} \to \Re_+$ locally defined in a neighborhood $\acute{N} \times \acute{N} \times \acute{Y} \times \acute{Y}$ of the origin $(\acute{x}, \acute{x}_{k-1}, y) = (0, 0, 0)$, and matrix functions $\acute{L}_1, \acute{L}_2 \in M^{n \times m}$, satisfying the DHJIE (8.99) together with the side-conditions (8.93), (8.95), (8.97)-(8.98). Then the filter Σ_2^{dacef} solves the DHINLFP for the system locally in $\acute{N}$.*

Proof: We simply repeat the steps of the proof of Proposition 8.5.1. $\square$

8.5.2   Approximate and Explicit Solution

It is hard to appreciate the results of Sections 8.5, 8.5.1, since the filter gains L, $\acute{L}_i, i = 1, 2$, are given implicitly. Therefore, in this subsection, we address this difficulty and derive approximate explicit solutions. More specifically, we shall rederive explicitly the results of Proposition 8.5.1 and Theorem 8.5.1. We begin with the 1-DOF filter Σ_1^{dacef}. Accordingly, consider the Hamiltonian function $H(., ., ., .)$ given by (8.82) and expand it in Taylor-series about $f(\hat{x})$ up to first-order. Denoting this approximation by $\hat{H}(., ., ., .)$ and the corresponding values of L, V and w by $\hat{L}$, $\hat{V}$, and $\hat{w}$ respectively, we get

$$
\begin{aligned}
\hat{H}(\hat{x}, y, \hat{w}, \hat{L}, \hat{V}) &= \left\{ \hat{V}(f(\hat{x}), y) + \hat{V}_{\hat{x}} f(\hat{x}), y)[g_1(\hat{x})\hat{w} + \hat{L}(\hat{x}, y)(y - h_2(\hat{x}) - k_{21}(\hat{x})\hat{w})] \right. \\
&\quad \left. + O(\|\hat{v}\|^2) \right\} - \hat{V}(\hat{x}, y_{k-1}) + \frac{1}{2}(\|\tilde{z}\|^2 - \gamma^2 \|\hat{w}\|^2) \tag{8.100}
\end{aligned}
$$

where

$$
\hat{v} = g_1(\hat{x})\hat{w} + L(\hat{x}, y)(y - h_2(\hat{x}) - k_{21}(\hat{x})\hat{w}),
$$

$$
\lim_{\hat{v} \to 0} \frac{O(\|\hat{v}\|^2)}{\|\hat{v}\|^2} = 0.
$$

Now applying the necessary condition for the worst-case noise, we get

$$
\begin{aligned}
\left. \frac{\partial \hat{H}}{\partial w} \right|_{\hat{w} = \hat{w}^\star} &= g_1^T(\hat{x})\hat{V}_{\hat{x}}^T(\widehat{f}(\hat{x}), y) - k_{21}^T(x)\hat{L}^T(\hat{x})\hat{V}_{\hat{x}}^T(f(\hat{x}), y) - \gamma^2 \hat{w}^\star = 0 \\
&\implies \hat{w}^\star := \frac{1}{\gamma^2}[g_1^T(\hat{x})\hat{V}_{\hat{x}}^T(f(\hat{x}), y) - k_{21}^T(\hat{x})\hat{L}^T(\hat{x})\hat{V}_{\hat{x}}^T(f(\hat{x}), y)]. \tag{8.101}
\end{aligned}
$$

Consequently, substituting $\hat{w}^\star$ into (8.100) and assuming the conditions of Assumption 8.5.1 hold, we get

$$
\begin{aligned}
\hat{H}(\hat{x}, y, \hat{w}^\star, \hat{L}, \hat{V}) &\approx \hat{V}(f(\hat{x}), y) + \frac{1}{2\gamma^2}\hat{V}_{\hat{x}}(f(\hat{x}), y)g_1(\hat{x})g_1^T(\hat{x})\hat{V}_{\hat{x}}^T(f(\hat{x}), y) - \\
&\quad \hat{V}(\hat{x}, y_{k-1}) + \hat{V}_{\hat{x}}(f(\hat{x}), y)\hat{L}(\hat{x}, y)(y - h_2(\hat{x})) + \\
&\quad \frac{1}{2\gamma^2}\hat{V}_{\hat{x}}(f(\hat{x}), y)(\hat{x})\hat{L}(\hat{x}, y)\hat{L}^T(\hat{x}, y)\hat{V}_{\hat{x}}^T(f(\hat{x}), y) + \frac{1}{2}\|\tilde{z}\|^2.
\end{aligned}
$$

Next, we complete the squares with respect to $\hat{L}$ in $\hat{H}(.,.,\hat{w}^\star,.)$ to minimize it, i.e.,

$$\hat{H}(\hat{x}, y, \hat{w}^\star, \hat{L}, \hat{V}) \approx \hat{V}(f(\hat{x}), y) + \frac{1}{2\gamma^2}\hat{V}_{\hat{x}}(f(\hat{x}), y)g_1(\hat{x})g_1^T(\hat{x})\hat{V}_{\hat{x}}^T(f(\hat{x}), y) - \hat{V}(\hat{x}, y_{k-1}) +$$

$$\frac{1}{2\gamma^2}\left\|\hat{L}^T(\hat{x}, y)\hat{V}_{\hat{x}}^T(f(\hat{x})) + \gamma^2(y - h_2(\hat{x}))\right\|^2 - \frac{\gamma^2}{2}\|y - h_2(\hat{x})\|^2 + \frac{1}{2}\|\tilde{z}\|^2.$$

Thus, setting $\hat{L}^\star$ as

$$\hat{V}_{\hat{x}}(f(\hat{x}), y)\hat{L}^\star(\hat{x}) = -\gamma^2(y - h_2(\hat{x}))^T, \quad \hat{x} \in \hat{N} \tag{8.102}$$

minimizes $\hat{H}(.,.,.,.)$ and renders the saddle-point condition

$$\hat{H}(\hat{w}^\star, \hat{L}^\star) \leq \hat{H}(\hat{w}^\star, \hat{L}) \ \ \forall \hat{L} \in \mathcal{M}^{n \times m}$$

satisfied.

Substitute now $\hat{L}^\star$ as given by (8.102) in the expression for $\hat{H}(.,.,.,.)$ and complete the squares in $\hat{w}$ to obtain:

$$\hat{H}(\hat{x}, \hat{w}, \hat{L}^\star, \hat{V}) = \hat{V}(f(\hat{x}), y) - \frac{1}{2\gamma^2}\hat{V}_{\hat{x}}(f(\hat{x}), y)\hat{L}(\hat{x}, y)\hat{L}^T(\hat{x}, y)\hat{V}_{\hat{x}}^T(f(\check{x}), y) - \hat{V}(\hat{x}, y_{k-1}) +$$

$$\frac{1}{2\gamma^2}\hat{V}_{\hat{x}}(f(\hat{x}), y)g_1(\hat{x})g_1^T(\hat{x})\hat{V}_{\hat{x}}^T(f(\hat{x}), y) + \frac{1}{2}\|\tilde{z}\|^2 -$$

$$\frac{\gamma^2}{2}\left\|\hat{w} - \frac{1}{\gamma^2}g_1^T(\hat{x})\hat{V}_{\hat{x}}^T(f(\hat{x}), y) - \frac{1}{\gamma^2}k_{21}^T(\hat{x})\hat{L}^T(\hat{x})\hat{V}_{\hat{x}}^T(f(\hat{x}), y)\right\|^2.$$

Similarly, substituting $\hat{w} = \hat{w}^\star$ as given in (8.101), we see that the second saddle-point condition

$$\hat{H}(\hat{w}^\star, \hat{L}^\star) \geq \hat{H}(w, \hat{L}^\star), \ \ \forall \hat{w} \in \mathcal{W}$$

is also satisfied. Therefore, the pair $(\hat{w}^\star, \hat{L}^\star)$ constitute a unique saddle-point solution to the two-person zero-sum dynamic game corresponding to the Hamiltonian $\hat{H}(.,.,.,.)$. Finally, setting

$$\hat{H}(\hat{x}, \hat{w}^\star, \hat{L}^\star, \hat{V}) = 0$$

yields the following DHJIE:

$$\hat{V}(f(\hat{x}), y) - \hat{V}(\hat{x}, y_{k-1}) + \tfrac{1}{2\gamma^2}\hat{V}_{\hat{x}}(f(\check{x}), y)g_1(\hat{x})g_1^T(\hat{x})\hat{V}_{\hat{x}}^T(f(\hat{x}), y) -$$

$$\tfrac{1}{2\gamma^2}\hat{V}_{\hat{x}}(f(\hat{x}), y)\hat{L}(\hat{x}, y)\hat{L}^T(\hat{x}, y)\hat{V}_{\hat{x}}^T(f(\hat{x}), y) +$$

$$\tfrac{1}{2}(y - h_2(\hat{x}))^T(y - h_2(\hat{x})) = 0, \quad \tilde{V}(0,0) = 0, \ \hat{x} \in \hat{N} \tag{8.103}$$

or equivalently the DHJIE:

$$\hat{V}(f(\hat{x}), y) - \hat{V}(\hat{x}, y) + \tfrac{1}{2\gamma^2}\hat{V}_{\hat{x}}(f(\hat{x}), y)g_1(\hat{x})g_1^T(\hat{x})\hat{V}_{\hat{x}}^T(f(\hat{x}), y) -$$

$$\tfrac{\gamma^2}{2}(y - h_2(\hat{x}))^T(y - h_2(\hat{x})) + \tfrac{1}{2}(y - h_2(\hat{x}))^T(y - h_2(\hat{x})) = 0, \quad \hat{V}(0,0) = 0. \tag{8.104}$$

Consequently, we have the following approximate counterpart of Proposition 8.5.1.

Proposition 8.5.2 *Consider the discrete-time nonlinear system (8.78) and the DHINLFP for it. Suppose Assumption 8.5.1 holds, the plant Σ^{da} is locally asymptotically-stable about the equilibrium point $x = 0$ and zero-input observable. Further, suppose there exists a C^1 positive-semidefinite function $\hat{V} : \hat{N} \times \hat{Y} \to \Re_+$ locally defined in a neighborhood $\hat{N} \times \hat{Y} \subset \mathcal{X} \times \mathcal{Y}$ of the origin $(\hat{x}, y) = (0,0)$, and a matrix function $\hat{L} : \hat{N} \times \hat{Y} \to \mathcal{M}^{n \times m}$,* *satisfying the following DHJIE (8.103) or (8.104) together with the side-conditions (8.102). Then, the filter Σ_1^{dacef} solves the DHINLFP for the system locally in $\hat{N}$.*

Proof: The first part of the theorem on the existence of the saddle-point solutions $(\hat{w}^\star, \hat{L}^\star)$ has already been shown above. It remains to show that the ℓ_2-gain condition (8.79) is satisfied and the filter provides asymptotic estimates.

Accordingly, assume there exists a smooth solution $\hat{V} \geq 0$ to the DHJIE (8.103), and consider the time variation of $\hat{V}$ along the trajectories of the filter Σ_1^{dacef}, (8.80), with $\hat{L} = \hat{L}^\star$, i.e.,

$$
\begin{aligned}
\hat{V}(\hat{x}_{k+1}, y) &= \hat{V}(f(\hat{x}) + \hat{v}), \quad \forall \hat{x} \in \hat{N}, \quad \forall y \in \hat{Y}, \quad \forall \hat{w} \in \mathcal{W} \\
&\approx \hat{V}(f(\hat{x}), y) + \hat{V}_{\hat{x}}(f(\hat{x}), y)g_1(\hat{x})\hat{w} + \hat{V}_{\hat{x}}(f(\hat{x}), y)[\hat{L}^\star(\hat{x}, y)(y - h_2(\hat{x}) - k_{21}(\hat{x})\hat{w})] \\
&= \hat{V}(f(\hat{x}), y) + \frac{1}{2\gamma^2}\hat{V}_{\hat{x}}(f(\hat{x}), y)g_1(\hat{x})g_1^T(\hat{x})\hat{V}_{\hat{x}}^T(f(\hat{x}), y) + \\
&\quad \hat{V}_{\hat{x}}(f(\hat{x}), y)\hat{L}^\star(\hat{x}, y)(y - h_2(\hat{x})) - \frac{\gamma^2}{2}\|\hat{w} - \hat{w}^\star\|^2 + \\
&\quad \frac{\gamma^2}{2}\|\hat{w}\|^2 + \frac{1}{2\gamma^2}\hat{V}_{\hat{x}}(f(\hat{x}), y)\hat{L}^\star(\hat{x}, y)\hat{L}^{\star^T}(\hat{x}, y)\hat{V}_{\hat{x}}^T(f(\hat{x}), y) \\
&= \hat{V}(f(\hat{x}), y) + \frac{1}{2\gamma^2}\hat{V}_{\hat{x}}(f(\hat{x}))g_1(\hat{x})g_1^T(\hat{x})\hat{V}_{\hat{x}}^T(f(\hat{x}), y) + \frac{\gamma^2}{2}\|\hat{w}\|^2 - \\
&\quad \frac{\gamma^2}{2}\|\hat{w} - \hat{w}^\star\|^2 - \frac{1}{2\gamma^2}\hat{V}_{\hat{x}}(f(\hat{x}), y)\hat{L}^\star(\hat{x})\hat{L}^{\star^T}(\hat{x})\hat{V}_{\hat{x}}^T(f(\hat{x}), y) \\
&\leq \hat{V}(\hat{x}, y_{k-1}) + \frac{\gamma^2}{2}\|\hat{w}\|^2 - \frac{1}{2}\|\tilde{z}\|^2 \quad \forall \hat{x} \in \hat{N}, \ y \in \hat{Y}, \ \forall w \in \mathcal{W},
\end{aligned}
$$

where use has been made of the Taylor-series approximation, equation (8.102), and the DHJIE (8.103). Finally, the above inequality clearly implies the infinitesimal dissipation inequality [183]:

$$
\hat{V}(\hat{x}_{k+1}, y_k) - \hat{V}(\hat{x}_k, y_{k-1}) \leq \frac{1}{2}\gamma^2\|\hat{w}\|^2 - \frac{1}{2}\|\tilde{z}\|^2 \quad \forall \hat{x} \in \hat{N}, \ \forall y \in \hat{Y}, \ \forall \hat{w} \subset \mathcal{W}.
$$

Therefore, the filter (8.80) provides locally ℓ_2-gain from $\hat{w}$ to $\tilde{z}$ less or equal to γ. The remaining arguments are the same as in the proof of Proposition 8.5.1. $\square$

Next, we extend the above approximation procedure to the 2-DOF filter Σ_2^{dacef} to arrive at the following result which is the approximate counterpart of Theorem 8.5.1.

Theorem 8.5.2 *Consider the discrete-time nonlinear system (8.78) and the DHINLFP for it. Suppose Assumption 8.5.1 holds, the plant Σ^{da} is locally asymptotically-stable about the equilibrium point $x = 0$ and zero-input observable. Further, suppose there exists a C^1 positive-semidefinite function $\hat{V} : \hat{N} \times \hat{N} \times \hat{Y} \to \Re_+$ locally defined in a neighborhood $\hat{N} \times \hat{N} \times \hat{Y} \times \hat{Y}$ of the origin $(\acute{x}, \acute{x}_{k-1}, y) = (0, 0, 0)$, and matrix functions $\hat{L}_1 \in \mathcal{M}^{n \times m}$, $\hat{L}_2 \in \mathcal{M}^{n \times m}$, satisfying the DHJIE:*

$$
\hat{V}(f(\acute{x}), \acute{x}, y_k) + \frac{1}{2\gamma^2}\hat{V}_{\acute{x}}(f(\acute{x}), ., ., .)g_1(\acute{x})g_1^T(\acute{x})\hat{V}_{\acute{x}}^T(f(\acute{x}), ., ., .) -
$$

$$
\hat{V}(\acute{x}, \acute{x}_{k-1}, y_{k-1}) + \frac{(1-\gamma^2)}{2}(y - h_2(\acute{x}))^T(y - h_2(\acute{x})) -
$$

$$
\frac{1}{2}(\Delta y - \Delta h_2(\acute{x}))^T(\Delta y - \Delta h_2(\acute{x})) = 0, \quad \hat{V}(0, ., 0) = 0 \tag{8.105}
$$

together with the coupling conditions

$$
\hat{V}_{\acute{x}}(\acute{x}, ., ., .)\hat{L}_1^\star(\acute{x}, ., ., .) = -\gamma^2(y - h_2(\acute{x}))^T \tag{8.106}
$$

$$
\hat{V}_{\acute{x}}(\acute{x}, ., ., .)\hat{L}_2^\star(\acute{x}, ., ., .) = -(\Delta y - \Delta h_2(\acute{x}))^T, \tag{8.107}
$$

where $\Delta y = y_k - y_{k-1}$, $\Delta h_2(x) = h_2(x_k) - h_2(x_{k-1})$. Then the filter Σ_2^{dacef} solves the DHINLFP for the system locally in $\widehat{N}$.

Proof: (Sketch) We can similarly write the first-order Taylor-series approximation of $\acute{H}$ as

$$
\begin{aligned}
\widehat{H}(\acute{x}, \widehat{w}, \widehat{L}_1, \widehat{L}_2, \widehat{V}) &= \widehat{V}(f(\acute{x}), \acute{x}, y) + \widehat{V}_{\acute{x}}(f(\acute{x}), ., .)g_1(\acute{x})\acute{w} + \\
&\quad \widehat{V}_{\acute{x}}(f(\acute{x}), ., .)\widehat{L}_1(\acute{x}, ., ., .)(y - h_2(\acute{x}) - k_{21}(\acute{x})\acute{w}) + \\
&\quad \widehat{V}_{\acute{x}}(f(\acute{x}), ., .)\widehat{L}_2(\acute{x}, ., ., .)(\Delta y - \Delta h_2(\hat{x})) - \\
&\quad \widehat{V}(\acute{x}, \acute{x}_{k-1}, y_{k-1}) + \frac{1}{2}(\|\tilde{z}_k\|^2 - \gamma^2\|\acute{w}\|^2)
\end{aligned}
$$

where $\widehat{w}$, $\widehat{L}_1$ and $\widehat{L}_2$ are the corresponding approximate values of $\acute{w}$, $\acute{L}_1$ and $\acute{L}_2$ respectively. Then we can also calculate the approximate estimated worst-case system noise, $\widehat{w}^\star \approx \acute{w}^\star$, as

$$
\widehat{w}^\star = \frac{1}{\gamma^2}[g_1^T(\acute{x})\widehat{V}_{\acute{x}}^T(f(\acute{x}), ., .) - k_{21}^T(\acute{x})\widehat{L}_1^T(\acute{x}, ., ., .)\widehat{V}_{\acute{x}}^T(f(\acute{x}), ., .)].
$$

Subsequently, going through the rest of the steps of the proof as in Proposition 8.5.2, we get (8.106), (8.107), and setting

$$
\widehat{H}(\acute{x}, \widehat{w}^\star, \widehat{L}_1^\star, \widehat{L}_2^\star, \acute{V}_{\acute{x}}^T) = 0
$$

yields the DHJIE (8.105). $\square$

We consider a simple example.

Example 8.5.1 *We consider the following scalar system*

$$
\begin{aligned}
x_{k+1} &= x_k^{\frac{1}{5}} + x_k^{\frac{1}{3}} \\
y_k &= x_k + w_k
\end{aligned}
$$

where $w_k = w_{0k} + 0.1\sin(20\pi k)$ and w_0 is a zero-mean Gaussian white-noise.

We compute the approximate solutions of the DHJIEs (8.104) and (8.105) using an iterative process and then calculate the filter gains respectively. We outline each case below.

    1-DOF Filter
Let $\gamma = 1$ and since $g_1(x) = 0$, we assume $\hat{V}^0(\hat{x}, y) = \frac{1}{2}(\hat{x}^2 + y^2)$ and compute

$$
\begin{aligned}
\hat{V}^1(x, y) &= \frac{1}{2}(\hat{x}^{\frac{1}{5}} + \hat{x}^{\frac{1}{3}})^2 + \frac{1}{2}y^2 \\
\hat{V}_x^1(x_k, y_k) &= (\hat{x}_k^{\frac{1}{5}} + \hat{x}_k^{\frac{1}{3}})(\frac{1}{5}\hat{x}_k^{\frac{-4}{5}} + \frac{1}{3}\hat{x}_k^{\frac{-2}{3}})
\end{aligned}
$$

Therefore,

$$
L(\hat{x}_k, y_k) = -\frac{y_k - h_2(\hat{x}_k)}{(\hat{x}_k^{\frac{1}{2}} + \hat{x}_k^{\frac{1}{3}})(\frac{1}{5}\hat{x}_k^{\frac{-4}{5}} + \frac{1}{3}\hat{x}_k^{\frac{-2}{3}})}.
$$

The filter is then simulated with a different initial condition from the system, and the results of the simulation are shown in Figure 8.8.

    2-DOF Filter
Similarly, we compute an approximate solution of the DHJIE (8.105) starting with the initial guess $\widehat{V}(\acute{x}, y) = \frac{1}{2}(\acute{x}^2 + y^2)$ and $\gamma = 1$. Moreover, we can neglect the last term in (8.105)

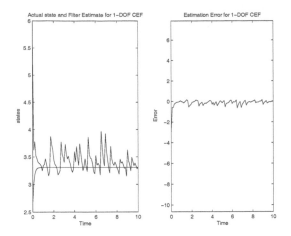

FIGURE 8.8
1-DOF Discrete-Time $\mathcal{H}_\infty$-Filter Performance with Unknown Initial Condition; Reprinted from Int. J. of Robust and Nonlinear Control, vol. 20, no. 7, pp. 818-833, © 2010, "2-DOF Discrete-time nonlinear $\mathcal{H}_\infty$ filtering," by M. D. S. Aliyu and E. K. Boukas, with permission from Wiley Blackwell.

since it is negative; hence the approximate solution we obtain will correspond to the solution of the DHJI-inequality corresponding to (8.105):

$$\widehat{V}^1(\acute{x},\acute{x}_{k-1},y) = \frac{1}{2}(\acute{x}^{\frac{1}{2}}+\acute{x}^{\frac{1}{3}})^2+\frac{1}{2}\acute{x}_{k-1}^2+\frac{1}{2}y_{k-1}^2$$

$$\Longrightarrow V_x^1(\acute{x}_k,\acute{x}_{k-1},y_{k-1}) = (\acute{x}_k^{\frac{1}{2}}+\acute{x}_k^{\frac{1}{3}})(\frac{1}{2}\acute{x}_k^{\frac{-1}{2}}+\frac{1}{3}\acute{x}_k^{\frac{-2}{3}})$$

using an iterative procedure [23], and compute the filter gains as

$$L_1(\acute{x}_k,\acute{x}_{k-1},y_k,y_{k-1}) = -\frac{y_k-h_2(\acute{x}_k)}{(\acute{x}_k^{\frac{1}{2}}+\acute{x}_k^{\frac{1}{3}})(\frac{1}{2}\acute{x}_k^{\frac{-1}{2}}+\frac{1}{3}\acute{x}_k^{\frac{-2}{3}})} \qquad (8.108)$$

$$L_2(\acute{x}_k,\acute{x}_{k-1},y_k,y_{k-1}) = -\frac{(\Delta y_k-\Delta h_2(\acute{x}_k))}{(\acute{x}_k^{\frac{1}{2}}+\acute{x}_k^{\frac{1}{3}})(\frac{1}{2}\acute{x}_k^{\frac{-1}{2}}+\frac{1}{3}\acute{x}_k^{\frac{-2}{3}})}. \qquad (8.109)$$

This filter is simulated with the same initial condition as the 1-DOF filter above and the results of the simulation are shown similarly on Figure 8.9. The results of the simulations show that the 2-DOF filter has slightly improved performance over the 1-DOF filter.

8.6   Robust Discrete-Time Nonlinear $\mathcal{H}_\infty$-Filtering

In this section, we discuss the robust $\mathcal{H}_\infty$-filtering problem for a class of uncertain nonlinear discrete-time systems described by the following model, and defined on $\mathcal{X} \subset \Re^n$:

$$\Sigma_\Delta^{ad} : \begin{cases} x_{k+1} &= [A+\Delta A_k]x_k+Gg(x_k)+Bw_k; \quad x_{k_0}=x^0, \quad k \in \mathbf{Z} \\ z_k &= C_1 x_k \\ y_k &= [C_2+\Delta C_{2,k}]x_k+Hh(x_k)+Dw_k \end{cases} \qquad (8.110)$$

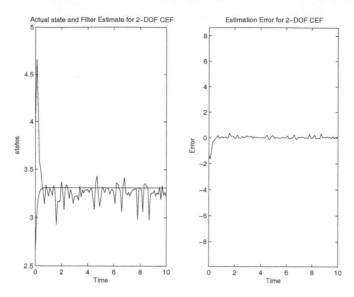

FIGURE 8.9
2-DOF Discrete-Time $\mathcal{H}_\infty$-Filter Performance with Unknown Initial Condition; Reprinted from Int. J. of Robust and Nonlinear Control, vol. 20, no. 7, pp. 818-833, © 2010, "2-DOF Discrete-time nonlinear $\mathcal{H}_\infty$ filtering," by M. D. S. Aliyu and E. K. Boukas, with permission from Wiley Blackwell.

where all the variables have their previous meanings. In addition, A, $\Delta A_k \in \Re^{n \times n}$, $G \in \Re^{n \times n_1}$, $g : \mathcal{X} \to \Re^{n_1}$, $B \in \Re^{n \times r}$, $C_1 \in \Re^{s \times n}$, $C_2, \Delta C_{2,k} \in \Re^{m \times n}$, and $H \in \Re^{m \times n_2}$, $h : \mathcal{X} \to \Re^{n_1}$. ΔA_k is the uncertainty in the system matrix A while $\Delta C_{2,k}$ is the uncertainty in the system output matrix C_2 which are both time-varying. Moreover, the uncertainties are matched, and belong to the following set of admissible uncertainties.

Assumption 8.6.1 *The admissible uncertainties of the system are structured and matched, and they belong to the following set:*

$$\Xi_{d,\Delta} \;=\; \left\{ \Delta A_k, \Delta C_{2,k} \,|\, \Delta A_k = H_1 F_k E, \;\; \Delta C_{2,k} = H_2 F_k E, \;\; F_k^T F_k \le I \right\}$$

where H_1, H_2, F_k, E are real constant matrices of appropriate dimensions, and F_k is an unknown time-varying matrix.

Whereas the nonlinearities $g(.)$ and $h(.)$ satisfy the following assumption.

Assumption 8.6.2 *The nonlinearies $g(.)$ and $h(.)$ are Lipschitz continuous, i.e., for any $x_1, x_2 \in \mathcal{X}$, there exist constant matrices Γ_1, Γ_2 such that:*

$$g(0) \;=\; 0, \tag{8.111}$$
$$\|g(x_1) - g(x_2)\| \;\le\; \|\Gamma_1(x_1 - x_2)\|, \tag{8.112}$$
$$\|h(x_1) - h(x_2)\| \;\le\; \|\Gamma_2(x_1 - x_2)\|. \tag{8.113}$$

for some constant matrices Γ_1, Γ_2.

The problem is then the following.

Definition 8.6.1 *(Robust Discrete-Time Nonlinear $\mathcal{H}_\infty$ (Suboptimal) Filtering Problem (RDNLHIFP)). Given the system (8.110) and a prescribed level of noise attenuation $\gamma > 0$, find a causal filter $\mathcal{F}_k$ such that the ℓ_2-gain from (w, x^0) to the filtering error (to be defined later), $\tilde{z}$, is attenuated by γ, i.e.,*

$$\|\tilde{z}\|_{\ell_2} \leq \gamma^2 \left\{ \|w\|_{\ell_2} + x^0 \tilde{R} x^0 \right\},$$

and the error-dynamics (to be defined) is globally exponentially-stable for all $(0,0) \neq (w, x^0) \in \ell_2[k_0, \infty) \oplus \mathcal{X}$ and all $\Delta A_k, \Delta C_{2,k} \in \Xi_\Delta^d$, $k \in \mathbf{Z}$ where $\tilde{R} = \tilde{R}^T > 0$ is some suitable weighting matrix.

Before we present a solution to the above filtering problem, we establish the following bounded-real-lemmas for discrete-time-varying systems (see also Chapter 3) that will be required in the proof of the main result of this section.

Consider the linear discrete-time-varying system:

$$\Sigma^{dl} : \begin{cases} \dot{\check{x}}_{k+1} &= \check{A}_k \check{x}_k + \check{B}_k w_k, \quad x_{k_0} = \check{x}^0 \\ \check{z}_k &= \check{C}_k \check{x}_k \end{cases} \tag{8.114}$$

where $\check{x}_k \in \mathcal{X}$ is the state vector, $\check{w}_k \in \ell_2([k_0, \infty), \mathfrak{R}^r)$ is the input vector, $\check{z}_k \in \mathfrak{R}^s$ is the controlled output, and $\check{A}_k, \check{B}_k, \check{C}_k$ are bounded time-varying matrices. The induced ℓ_∞-norm (or $\mathcal{H}_\infty(j\mathfrak{R})$-norm in the context of our discussion) from $(\check{w}, \check{x}^0)$ to $\check{z}_k$ for the above system is defined by:

$$\|\Sigma^{dl}\|_{l_\infty} \triangleq \sup_{(\check{w}, \check{x}_0) \in \ell_2 \oplus \mathcal{X}} \frac{\|\check{z}\|^2}{\|\check{w}\|^2 + \check{x}^{0T} \tilde{R} \check{x}^0}. \tag{8.115}$$

Then, we have the following lemma.

Lemma 8.6.1 *For the linear time-varying discrete-time system (8.114) and a given $\gamma > 0$, the following statements are equivalent:*

(a) *the system is exponentially stable and $\|\Sigma^{dl}\|_{l_\infty} < \gamma$;*

(b) *there exists a bounded time-varying matrix function $Q_k = Q_k^T \geq 0$, $\forall k \geq k_0$ satisfying:*

$$\check{A}_k Q_k \check{A}_k^T - Q_{k+1} + \gamma^{-2} \check{A}_k Q_k \check{C}_k^T (I - \gamma^{-2} \check{C}_k Q_k \check{C}_k^T)^{-1} \check{C}_k Q_k \check{A}_k^T + \check{B}_k \check{B}_k^T = 0; \quad Q_{k_0} = \tilde{R}^{-1},$$
$$I - \gamma^{-2} \check{C}_k Q_k \check{C}_k^T > 0 \quad \forall k \geq k_0$$

and the closed-loop system

$$\check{x}_{k+1} = [\check{A}_k + \gamma^{-2} \check{A}_k Q_k \check{C}_k^T (I - \gamma^{-2} \check{C}_k Q_k \check{C}_k^T)^{-1} \check{C}_k] \check{x}_k$$

is exponentially stable;

(c) *there exists a scalar $\delta_1 > 0$ and a bounded time-varying matrix function $\check{P}_k = \check{P}_k^T$, $\forall k \geq k_0$ satisfying:*

$$\check{A}_k \check{P}_{k+1} \check{A}_k^T - \check{P}_k + \gamma^{-2} \check{A}_k^T \check{P}_{k+1} \check{B}_k (I - \gamma^{-2} \check{B}_k^T \check{P}_{k+1} \check{B}_k)^{-1} \check{B}_k^T \check{P}_{k+1} \check{A}_k + \check{C}_k^T \check{C}_k + \delta_1 I < 0;$$
$$\check{P}_{k_0} < \gamma^2 \tilde{R},$$
$$I - \gamma^{-2} \check{B}_k^T \check{P}_{k+1} \check{B}_k > 0 \quad \forall k \geq k_0.$$

Proof: $(a) \Leftrightarrow (b) \Rightarrow (c)$ has been shown in Reference [280], Theorem 3.1. To show that $(a) \Rightarrow (c)$, we consider the extended output $z^b = \begin{bmatrix} \check{C}_k \\ \sqrt{\delta_1} I \end{bmatrix} x_k$. By exponential stability of the system, and the fact that $\|\Sigma_0^{dl}\|_{l_\infty} < \gamma$, there exists a sufficiently small number $\delta_1 > 0$ such that $\|\Sigma^{dl}\|_{l_\infty} < \gamma$ for all $(w, x_0) \in \ell_2 \oplus \mathcal{X}$ to $z^\flat$. The result then follows again from Theorem 3.1, Reference [280]. $\square$

The following lemma gives the bounded-real conditions for the system (8.110).

Lemma 8.6.2 *Consider the nonlinear discrete-time system Σ_Δ^{ad} (8.110) satisfying Assumptions 8.6.1, 8.6.2. For a given $\gamma > 0$ and a weighting matrix $\tilde{R} = \tilde{R}^T > 0$, the system is globally exponentially-stable and*

$$\|z\|_{l_2}^2 < \gamma\{\|w\|_{\ell_2}^2 + x^{0^T}\tilde{R}x^0\}$$

for any non-zero $(x_0, w) \in \mathcal{X} \oplus \ell_2$ and for all ΔA_k if there exists scalars $\epsilon > 0$, $\delta_1 > 0$ and a bounded time-varying matrix function $Q_k = Q_k^T > 0$, $\forall k \geq k_0$ satisfying:

$$A^T Q_{k+1} A - Q_k + \gamma^{-2} A^T Q_{k+1} B_1 (I - \gamma^{-2} B_1^T Q_{k+1} B_1)^{-1} B_1^T Q_{k+1} A + C_1^T C_1 +$$
$$\epsilon^2 E^T E + \Gamma_1^T \Gamma_1 + \delta_1 I < 0;$$
$$Q_{k_0} < \gamma^2 \tilde{R}, \tag{8.116}$$
$$I - \gamma^{-2} B_1^T Q_{k+1} B_1 > 0 \quad \forall k \geq k_0,$$

where

$$B_1 = \begin{bmatrix} B & \dfrac{\gamma}{\epsilon} H_1 & \gamma G \end{bmatrix}.$$

Proof: The inequality (8.116) implies that $Q_k > \delta_1 I \forall k \geq k_0$. Moreover, since Q_k is bounded, there exists a scalar $\delta_2 > 0$ such that $Q_k \leq \delta_2 I \forall k \geq k_0$. Now consider the Lyapunov-function candidate:

$$V(x, k) = x_k^T Q_k x_k$$

such that

$$\delta_1 \|x\|^2 \leq V(x, k) \leq \delta_2 \|x\|^2.$$

Then, it can be shown (using similar arguments as in the proof of Theorem 4.1, Reference [279]) that along any trajectory of the free-system (8.110) with $w_k = 0 \forall k \geq k_0$,

$$\Delta V(x, k) = V(k + 1, x) - V(x, k) \leq -\delta_1 \|x_k\|^2.$$

Therefore, by Lyapunov's theorem [157], the free-system is globally exponentially-stable. $\square$

We now present the solution to $RDNLHIFP$ for the class of nonlinear discrete-time systems Σ_Δ^{dl}. For this, we need some additional assumptions on the system matrices.

Assumption 8.6.3 *The system (8.110) matrices are such that*

(a1) (C_2, A) *is detectable.*

(a2) $[D \ H_2 \ H]$ *has full row-rank.*

(a3) *The matrix A is nonsingular.*

Theorem 8.6.1 *Consider the uncertain nonlinear discrete-time system (8.110) satisfying the Assumptions 8.6.1-8.6.3. Given $\gamma > 0$ and $\tilde{R} = \tilde{R}^T > 0$, let $\nu > 0$ be a small number and suppose the following conditions hold:*

(a) *for some constant number $\epsilon > 0$, there exists a stabilizing solution $P = P^T > 0$ to the stationary DARE:*

$$A^T P A - P + \gamma^{-2} A^T P \tilde{B} (I - \gamma^{-2} \tilde{B}^T P \tilde{B})^{-1} \tilde{B}^T P A + E_1^T E_1 + \nu I = 0 \tag{8.117}$$

such that $P < \gamma^2 \tilde{R}$ and $I - \gamma^{-2} \tilde{B}^T P \tilde{B} > 0$, where

$$E_1 = (\epsilon^2 E^T E + \Gamma_1^T \Gamma_1)^{\frac{1}{2}}, \quad \tilde{B} = \begin{bmatrix} B & \dfrac{\gamma}{\epsilon} H_1 & \gamma G \end{bmatrix}.$$

(b) *there exists a bounded time-varying matrix $S_k = S_k^T \geq 0 \; \forall k \geq k_0$, satisfying*

$$
\begin{aligned}
S_{k+1} &= \hat{A} S_k \hat{A}^T - (\hat{A} S_k \hat{C}_1^T + \hat{B} \hat{D}_1^T)(\hat{C}_1 S_k \hat{C}_1^T + \hat{R})^{-1}(\hat{C}_1 S_k \hat{A}^T + \hat{D}_1 \hat{B}^T) + \hat{B} \hat{B}^T; \\
S_0 &= (\tilde{R} - \gamma^{-2} P)^{-1}, \\
&\quad I - \gamma^{-2} \hat{M}^T S_k \hat{M}^T > 0 \quad \forall k \geq k_0,
\end{aligned}
\tag{8.118}
$$

and the system

$$\rho_{k+1} := A_{2k}\rho_k = [\hat{A} - (\hat{A} S_k \hat{C}_1^T + \hat{B}\hat{D}_1^T)(\hat{C}_1 S_k \hat{C}_1^T + \hat{R})^{-1}\hat{C}_1]\rho_k \tag{8.119}$$

is exponentially-stable, where

$$
\begin{aligned}
\hat{A} &= A + \delta A_e := A + \gamma^{-2}\bar{B}\bar{B}^T(P^{-1} - \gamma^{-2}\tilde{B}\tilde{B}^T)^{-1}A \\
\hat{C}_2 &= C_2 + \delta C_{2e} := C_2 + \gamma^{-2}\bar{D}\bar{B}^T(P^{-1} - \gamma^{-2}\tilde{B}\tilde{B}^T)^{-1}A
\end{aligned}
$$

$$\hat{B} = [\bar{B}Z \;\; \gamma G \;\; 0], \quad \hat{D} = [\bar{D}Z \;\; 0 \;\; \gamma H]$$

$$\hat{C}_1 = \begin{bmatrix} \gamma^{-1}\hat{M} \\ \hat{C}_2 \end{bmatrix}, \quad \hat{D}_1 = \begin{bmatrix} 0 \\ \hat{D} \end{bmatrix}, \quad \hat{R} = \begin{bmatrix} -I & 0 \\ 0 & \hat{D}\hat{D}^T \end{bmatrix}$$

$$\bar{B} = \begin{bmatrix} B & \dfrac{\gamma}{\epsilon} H_1 \end{bmatrix}, \quad \bar{D} = \begin{bmatrix} D & \dfrac{\gamma}{\epsilon} H_2 \end{bmatrix}$$

$$\hat{M} = [C_1^T C_1 + \Gamma^T \Gamma]^{\frac{1}{2}}, \quad \Gamma = [\Gamma_1^T \;\; \Gamma_2^T]^T$$

$$Z = [I + \gamma^{-2}\bar{B}^T(P^{-1} - \gamma^{-2}\tilde{B}\tilde{B}^T)^{-1}\bar{B}]^{\frac{1}{2}}.$$

Then, the RNLHIFP for the system is solvable with a finite-dimensional filter. Moreover, if the above conditions are satisfied, a suitable filter is given by

$$\Sigma^{daf} : \begin{cases} \hat{x}_{k+1} &= \hat{A}\hat{x} + Gg(\hat{x}) + \hat{L}_k[y_k - \hat{C}_2\hat{x} - Hh(\hat{x})], \quad \hat{x}_{k_0} = 0 \\ \hat{z} &= C_1\hat{x}, \end{cases} \tag{8.120}$$

where $\hat{L}_k$ is the gain-matrix and is given by

$$
\begin{aligned}
\hat{L}_k &= (\hat{A}\hat{S}_k\hat{C}_2^T + \hat{B}\hat{D}^T)(\hat{C}_2\hat{S}_k\hat{C}_2^T + \hat{D}\hat{D}^T)^{-1} \\
\hat{S}_k &= S_k + \gamma^{-2}Sk\hat{M}^T(I - \gamma^{-2}\hat{M}S_k\hat{M}^T)^{-1}\hat{M}S_k.
\end{aligned}
\tag{8.121}
$$

Proof: We note that, $P^{-1} - \gamma^{-2}\tilde{B}\tilde{B}^T$ is positive-definite since $P > 0$ and $I - \gamma^{-2}\tilde{B}^T P \tilde{B} > 0$. Thus, Z is welldefined. Similarly, $I - \gamma^{-2}\hat{M}S_k\hat{M}^T > 0 \forall k \geq k_0$ and together with Assumption 8.6.3:(a2), imply that $\hat{C}_1 S_k \hat{C}_1^T + \hat{R}$ is nonsingular for all $k \geq k_0$. Consequently, equation (8.118) is welldefined.

Next, consider the filter Σ^{daf} and rewrite its equation as:

$$
\begin{aligned}
\hat{x}_{k+1} &= (A + \delta A_e)\hat{x} + Gg(\hat{x}) + \hat{L}_k[y_k - (C_2 + \delta C_{2e})\hat{x} - Hh(\hat{x})], \quad \hat{x}_{k_0} = 0 \\
\hat{z}_k &= C_1\hat{x}
\end{aligned}
$$

where δA_e and δC_{2e} are defined above, and represent the uncertain and time-varying components of A and C_2 (i.e., ΔA_k and ΔC_{2k}) respectively, that are compensated in the estimator. Then the dynamics of the state estimation error $\tilde{x}_k := x_k - \hat{x}_k$ is given by

$$
\begin{cases}
\tilde{x}_{k+1} &= [A + \delta A_e - \hat{L}_k(C_2 + \delta C_{2e})]\tilde{x} + [(\Delta A - \delta A_e) - \hat{L}_k(\Delta C_2 - \delta C_{2e})]x_k \\
&\quad + (B - \hat{L}_k D)w_k + G[g(x_k) - g(\hat{x})] - \hat{L}_k H[h(x_k) - h(\hat{x})], \quad \tilde{x}_{k_0} = x_0 \quad (8.122) \\
\tilde{z} &= C_1\tilde{x}
\end{cases}
$$

where $\tilde{z} := z - \hat{z}$ is the output estimation error. Now combine the system (8.110) and the error-dynamics (8.122) into the following augmented system:

$$\eta_{k+1} = (A_a + H_a F_k E_a)\eta_k + G_a g_a(x_k, \hat{x}) + B_a w_k; \quad \eta_{k_0} = [x^{0^T} \ x^{0^T}]^T$$
$$e_k = C_a \eta_k$$

where

$$\eta = [x^T \ \tilde{x}^T]^T$$

$$A_a = \begin{bmatrix} A & 0 \\ -(\delta A_e - \hat{L}_k \delta C_{2e}) & A + \delta A_e - \hat{L}_k(C_2 + \delta C_{2e}) \end{bmatrix}$$

$$B_a = \begin{bmatrix} B \\ B - \hat{L}_k D \end{bmatrix}, \quad H_a = \begin{bmatrix} H_1 \\ H_1 - \hat{L}_k H_2 \end{bmatrix}$$

$$G_a = \begin{bmatrix} G & 0 & 0 \\ 0 & G & -\hat{L}_k H \end{bmatrix}, \quad g_a(x_k, \hat{x}_k) = \begin{bmatrix} g(x_k) \\ g(x_k) - g(\hat{x}_k) \\ h(x_k) - h(\hat{x}_k) \end{bmatrix}$$

$$C_a = [0 \ C_1], \quad E_a = [E \ 0].$$

Then, by Assumption 8.6.2

$$\|g_a(x_k, \hat{x}_k)\| \le \|\widehat{\Gamma}\eta_k\|, \quad \text{with} \ \widehat{\Gamma} = Blockdiag\{\Gamma_1, \ \Gamma\}.$$

Further, define

$$\Pi = \begin{bmatrix} \Pi_{11} & \Pi_{12} \\ \Pi_{12}^T & \Pi_{22} \end{bmatrix} = A_a X_k A_a^T - X_{k+1} + A_a X_k \hat{C}_k^T (I - \hat{C}_a X_k \hat{C}_a^T)^{-1} \hat{C}_a X_k A_a^T +$$
$$\hat{B}_a \hat{B}_a^T \tag{8.123}$$

where

$$\hat{B}_a = [\gamma^{-1} B_a \ \epsilon^{-1} H_a \ G_a], \quad \hat{C}_a = \begin{bmatrix} \hat{E}_1 & 0 \\ 0 & \hat{M} \end{bmatrix}$$

and $\hat{E}_1$ is such that

$$\hat{E}_1^T \hat{E}_1 = \epsilon^2 E^T E + V_1^T V_1 + \nu I.$$

Also, let $Q_k = \gamma^{-2} S_k$ and

$$X_k = \begin{bmatrix} P^{-1} & 0 \\ 0 & Q_k \end{bmatrix}.$$

Then by standard matrix manipulations, it can be shown that

$$\Pi_{11} = AP^{-1}A^T - P^{-1} + AP^{-1}\hat{E}_1(I - \hat{E}_1 P^{-1}\hat{E}_1^T)^{-1}\hat{E}_1 P^{-1}A^T + \gamma^{-2}\tilde{B}\tilde{B}^T$$

and

$$\Pi_{12} = -A(P - \hat{E}_1^T \hat{E}_1)^{-1}(\delta A_e - \hat{L}_k \delta C_{2e})^T + \gamma^{-2}BB^T + \epsilon^{-2}H_1 H_1^T - (\gamma^{-2}BD^T + \epsilon^{-2}H_1 H_1^T)\hat{L}_k^T.$$

Moreover, since A is nonsingular, in view of (8.117) and the definition of δA_e, δC_{2e}, it implies that

$$\Pi_{11} = 0, \quad \Pi_{12} = 0.$$

It remains to show that $\Pi_{22} = 0$. Using similar arguments as in Reference [89] (Theorem 3.1), it follows from (8.123) that Q_k satisfies the DRE:

$$
\begin{aligned}
Q_{k+1} &= \hat{A}\hat{Q}_k\hat{A}^T - (\hat{A}\hat{Q}_k\hat{C}^T + \gamma^{-2}\hat{B}\hat{D}^T)(\hat{C}\hat{Q}_k\hat{C}^T + \gamma^{-2}\hat{D}\hat{D}^T)^{-1}(\hat{C}\hat{Q}_k\hat{A}^T + \\
&\quad \gamma^{-2}\hat{D}\hat{B}^T) + \gamma^{-2}\hat{B}\hat{B}^T; \quad Q_{k_0} = \gamma^2 R - P
\end{aligned}
\tag{8.124}
$$

where

$$
\hat{Q}_k = Q_k + Q_k\hat{M}^T(I - \hat{M}Q_k\hat{M}^T)^{-1}\hat{M}Q_k.
$$

Now, from (8.123) using some matrix manipulations we get

$$
\begin{aligned}
\Pi_{22} &= \hat{A}\hat{Q}_k\hat{A}^T - (\hat{A}\hat{Q}_k\hat{C}^T + \gamma^{-2}\hat{B}\hat{D}^T)\hat{L}_k - \hat{L}_k(\hat{C}\hat{Q}_k\hat{A}^T + \gamma^{-2}\hat{D}\hat{B}^T) + \\
&\quad \hat{L}_k(\hat{C}\hat{Q}\hat{C}^T + \gamma^{-2}\hat{D}\hat{D}^T)\hat{L}_k^T + \gamma^{-2}\hat{B}\hat{B}^T
\end{aligned}
$$

and the gain matrix $\hat{L}_k$ from (8.121) can be rewritten as

$$
\hat{L}_k = (\hat{A}\hat{Q}_k\hat{C}^T + \gamma^{-2}\hat{B}\hat{D}^T)(\hat{C}\hat{Q}_k\hat{C}^T + \gamma^{-2}\hat{D}\hat{D}^T)^{-1}.
\tag{8.125}
$$

Thus, from (8.124) and (8.125), it follows that $T_{22} = 0$, and hence we conclude from (8.123) that

$$
A_a X_k A_a^T - X_{k+1} + A_a X_k \hat{C}_a^T (I - \hat{C}_a X_k \hat{C}_a^T)^{-1} \hat{C}_a X_k A_a^T + \hat{B}_a \hat{B}_a^T = 0; \quad X_0 = \hat{R}^{-1}
\tag{8.126}
$$

where

$$
\hat{R} = \begin{bmatrix} P & 0 \\ 0 & \gamma^2\tilde{R} - P \end{bmatrix} > 0.
$$

Next, we show that, X_k is such that the time-varying system

$$
\rho_{k+1} = \hat{A}_a\rho_k = [\hat{A}_a + (\hat{A}_a X_k \hat{C}_a^T (I - \hat{C}_a X_k \hat{C}_a^T)^{-1}\hat{C}_a]\rho_k
\tag{8.127}
$$

is exponentially-stable. Let

$$
\hat{A}_a := \begin{bmatrix} \bar{A} & 0 \\ * & A_{2k} \end{bmatrix}
$$

where A_{2k} is as defined in (8.119), '$*$' denotes a bounded but otherwise unimportant term, and

$$
\bar{A} = A + \gamma^{-2}\tilde{B}(I - \gamma^{-2}\tilde{B}P\tilde{B})^{-1}\tilde{B}^T PA.
$$

$\bar{A}$ is Schur-stable[2] since P is a stabilizing solution of (8.117). Moreover, by exponential-stability of the system (8.119), it follows that (8.127) is also exponentially-stable. Therefore, X_k is the stabilizing solution of (8.124). Consequently, by Lemma 8.6.1 there exists a scalar $\delta_1 > 0$ and a bounded time-varying matrix $Y_k = Y_k^T > 0 \, \forall k \geq k_0$ such that

$$
A_a^T Y_k A_a - Y_k + A_a^T Y_{k+1}\hat{B}_a(I - \hat{B}_a^T Y_{k+1}\hat{B}_a)^{-1}\hat{B}_a^T Y_{k+1}A_a + \hat{C}_a^T\hat{C}_a + \delta_1 I < 0; \quad Y_{k_0} < \hat{R}.
$$

Noting that

$$
\hat{C}_a^T\hat{C}_a = \hat{C}_a^T\hat{C}_a + \epsilon^2 E_a^T E_a + \hat{\Gamma}^T\hat{\Gamma} + \begin{bmatrix} \nu I & 0 \\ 0 & 0 \end{bmatrix},
$$

we see that Y_k satisfies the following inequality:

$$
\begin{aligned}
&A_a^T Y_k A_a - Y_k + A_a^T Y_{k+1}\hat{B}_a(I - \hat{B}_a^T Y_{k+1}\hat{B}_a)^{-1}\hat{B}_a^T Y_{k+1}A_a + \hat{C}_a^T\hat{C}_a + \epsilon^2 E_a^T E_a + \\
&\hat{\Gamma}^T\hat{\Gamma} + \delta_1 I < 0, \quad Y_{k_0} < \hat{R}.
\end{aligned}
$$

---

[2]Eigenvalues of $\bar{A}$ are inside the unit circle.

In addition,

$$\eta_0^T \widehat{R}\eta_0 = \gamma^2 x_0^T \tilde{R} x_0.$$

Finally, application of Lemma 8.6.2 and using the definition of $\hat{B}_a$ imply that the error-dynamics (8.122) are exponentially-stable and

$$\|\tilde{z}\|_{\ell_2}^2 \leq \gamma\{\|w\|_{\ell_2}^2\| + x_0^T \tilde{R} x_0\}$$

for all $(0,0) \neq (x^0, w) \in \mathcal{X} \oplus \ell_2$ and all ΔA_k, $\Delta C_{2,k} \in \Xi_{d,\Delta}$. $\square$

8.7   Notes and Bibliography

The material of Section 8.1 is based on the Reference [66], while the material in Section 8.4 is based on the Reference [15]. An alternative to the solution of the discrete-time problem is also presented in Reference [244] under some simplifying assumptions. The materials of Sections 8.3 and 8.5 on 2-DOF and certainty equivalent filters are based on the References [22, 23, 24]. In particular continuous-time and discrete-time 2-DOF proportional-integral (PI) filters which are the counterpart of the PD filters presented in the chapter, are discussed in [22] and [24] respectively.

Furthermore, the results on $RNLHIFP$ - Section 8.2 are based on the reference [211], while the discrete-time case in Section 8.6 is based on [279]. Lastly, comparison of simulation results between the $\mathcal{H}_\infty$ filter and the extended-Kalman-filter can be found in the same references.

9

Singular Nonlinear $\mathcal{H}_\infty$-Control and $\mathcal{H}_\infty$-Control for Singularly-Perturbed Nonlinear Systems

In this chapter, we discuss the singular nonlinear $\mathcal{H}_\infty$-control problem. This problem arises when the full control signal is not available in the penalty variable due to some rank-deficiency of the gain matrix, and hence the problem is not well posed. The problem also arises when studying certain robustness issues in parametric or multiplicative uncertain systems.

Two approaches for solving the above problem are: (i) using the regular nonlinear $\mathcal{H}_\infty$-control techniques discussed in Chapters 5 and 6; and (ii) using high-gain feedback and/or converting the problem to the problem of "almost-disturbance-decoupling" discussed also in Chapter 5. We shall discuss both approaches in the chapter. We shall also discuss the measurement-feedback problem. However, only the continuous-time problem for affine non-linear systems will be presented. Moreover, the approaches are extensions to similar techniques used for linear systems as discussed in References [237, 252, 253].

Another problem that is in the class of "singular" problems is that of $\mathcal{H}_\infty$-control of singularly-perturbed systems. These class of systems possess fast and slow modes which are weakly coupled. Such models are also used to represent systems with algebraic constraints, and the solution to the system with constraint is found as the asymptotic limit of the solution to an extended system without constraints. We shall study this problem in the later part of the chapter.

9.1 Singular Nonlinear $\mathcal{H}_\infty$-Control with State-Feedback

At the outset, we consider affine nonlinear systems defined on a state-space manifold $\mathcal{X} \subset \Re^n$ defined in local coordinates $(x_1, \ldots, x_n)$:

$$\tilde{\Sigma}^a : \begin{cases} \dot{x} &= f(x) + g_1(x)w + \tilde{g}_2(x)\tilde{u}; \quad x(t_0) = x_0 \\ z &= h_1(x) + \tilde{k}_{12}(x)\tilde{u} \\ y &= x \end{cases} \tag{9.1}$$

where $u \in \mathcal{U} \subset \Re^p$ is the p-dimensional control input, which belongs to the set of admissible controls $\mathcal{U} \subset \mathcal{L}_2([t_0, T], \Re^p)$, $w \in \mathcal{W}$ is the disturbance signal, which belongs to the set $\mathcal{W} \subset \mathcal{L}_2([t_0, T], \Re^r)$ of admissible disturbances, the output $y \in \Re^m$ is the states-vector of the system which are measured directly, and $z \in \Re^s$ is the output to be controlled. The functions $f : \mathcal{X} \to V^\infty \mathcal{X}$, $g_1 : \mathcal{X} \to \mathcal{M}^{n \times r}(\mathcal{X})$, $\tilde{g}_2 : \mathcal{X} \to \mathcal{M}^{n \times p}(\mathcal{X})$, $h_1 : \mathcal{X} \to \Re^s$, and $\tilde{k}_{12} : \mathcal{X} \to \mathcal{M}^{p \times m}(\mathcal{X})$ are assumed to be real C^∞-functions of x. Furthermore, we assume that $x = 0$ is the only equilibrium of the system and is such that $f(0) = 0$, $h_1(0) = 0$. We also assume that the system is well defined, i.e., for any initial state $x(t_0) \in \mathcal{X}$ and any admissible input $u \in \mathcal{U}$, there exists a unique solution $x(t, t_0, x_0, u)$ to (9.1) on $[t_0, \infty)$ which continuously depends on the initial conditions, or the system satisfies the local existence and uniqueness theorem for ordinary differential equations [157].

The objective is to find a static state-feedback control law of the form

$$\tilde{u} = \tilde{\alpha}(x), \quad \tilde{\alpha}(0) = 0 \tag{9.2}$$

which achieves locally $\mathcal{L}_2$-gain from w to z less than or equal to $\gamma^\star > 0$ for the closed-loop system (9.1), (9.2) and asymptotic-stability with $w = 0$.

The problem could have been solved by the techniques of Chapters 5 and 6, if not for the fact that the coefficient matrix in the penalty variable k_{12} is not full-rank, and this creates the *"singularity"* to the problem. To proceed, let $\kappa = \min\ rank\{k_{21}(x)\} < p$ assumed to be constant over the neighborhood $M \subset X$ of $x = 0$. Then, it is possible to find a local diffeomorphism[1] (a local coordinate-transformation)

$$u = \varphi(x)\tilde{u}$$

which transforms the system (9.1) into

$$\Sigma^a : \begin{cases} \dot{x} &=& f(x) + g_1(x)w + g_2(x)u; \quad x(t_0) = x_0 \\ z &=& h_1(x) + k_{12}(x)u \\ y &=& x \end{cases} \tag{9.3}$$

with

$$g_2(x) = \tilde{g}_2(x)\varphi^{-1}(x), \quad k_{12}(x) = \tilde{k}_{12}(x)\varphi^{-1}(x) = [D_{12}\ 0] \tag{9.4}$$

and $D_{12} \in \Re^{s\times\kappa}$, $D_{12}^T D_{12} = I$. The control vector can now be partitioned conformably with the partition (9.4) so that $u = \begin{bmatrix} u_1 \\ u_2 \end{bmatrix}$, where $u_1 \in \Re^\kappa$, $u_2 \in \Re^{p-\kappa}$, and the system (9.3) is represented as

$$\Sigma^a : \begin{cases} \dot{x} &=& f(x) + g_1(x)w + g_{21}(x)u_1 + g_{22}(x)u_2; \quad x(t_0) = x_0 \\ z &=& h_1(x) + D_{12}u_1 \\ y &=& x, \end{cases} \tag{9.5}$$

where $g_2(x) = [g_{21}(x)\ g_{22}(x)]$, and g_{21}, g_{22} have compatible dimensions. The problem can now be more formally defined as follows.

Definition 9.1.1 *(State-Feedback Singular (Suboptimal) Nonlinear $\mathcal{H}_\infty$-Control Problem (SFBSNLHICP)). For a given number $\gamma^\star > 0$, find (if possible!) a state-feedbback control law of the form*

$$u = \begin{bmatrix} u_1 \\ u_2 \end{bmatrix} = \begin{bmatrix} \alpha_{21}(x) \\ \alpha_{22}(x) \end{bmatrix}, \quad \alpha_{2j}(0) = 0, j = 1, 2, \tag{9.6}$$

such that the closed-loop system (9.5), (9.6) is locally asymptotically-stable with $w = 0$, and has locally finite $\mathcal{L}_2$-gain from w to z less than or equal to $\gamma^\star$.

The following theorem gives sufficient conditions under which the $SFBSNLHICP$ can be solved using the techniques discussed in Chapter 5.

---

[1] A coordinate transformation that is bijective (therefore invertible) and smooth (or a smooth homeomorphism). A global diffeomorphism can be found if there exists a minor of $\tilde{k}_{12}$ with constant rank κ for all $x \in X$.

Theorem 9.1.1 *Suppose the state-feedback $\mathcal{H}_\infty$-control problem for the subsystem*

$$\left.\begin{aligned}
\dot{x} &= f(x) + g_1(x)w + g_{21}(x)u_1; \quad x(t_0) = x_0 \\
y &= x \\
z &= h_1(x) + k_{12}(x)u_1, \quad k_{12}^T(x)k_{12}(x) = I, \quad h_1^T(x)k_{12}(x) = 0
\end{aligned}\right\} \tag{9.7}$$

is solvable for a given $\gamma > 0$, with the control law

$$u_1^\star = -g_{21}^T(x)V_x^T(x),$$

where $V : M \to \Re$ is a smooth positive-definite solution of the HJIE:

$$V_x(x)f(x) + \frac{1}{2}V_x(x)[\frac{1}{\gamma^2}g_1(x)g_1^T(x) - g_{21}(x)g_{21}^T(x)]V_x^T(x) + \frac{1}{2}h_1^T(x)h_1(x) = 0, \quad V(0) = 0. \tag{9.8}$$

In addition, suppose there exists a function $\alpha_{22} : M \to \Re^{p-\kappa}$ such that

$$V_x(x)g_{22}(x)\alpha_{22}(x) \le 0 \quad \forall x \in M \tag{9.9}$$

and the pair $\{f(x) + g_{22}(x)\alpha_{22}(x), h_1(x)\}$ is zero-state detectable. Then, the state-feedback control law

$$u = \left[\begin{array}{c} u_1^\star(x) \\ u_2^\star(x) \end{array}\right] = \left[\begin{array}{c} -g_{21}^T(x)V_x^T \\ \alpha_{22}(x) \end{array}\right] \tag{9.10}$$

solves the SFBSNLHICP for the system (9.1) in M.

Proof: Differentiating the solution $V > 0$ of (9.8) along a trajectory of the closed-loop system (9.5) with the control law (9.10), we get upon using the HJIE (9.8):

$$\begin{aligned}
\dot{V} &= V_x(x)[f(x) + g_1(x)w + g_{21}(x)u_1 + g_{22}(x)u_2] \\
&= V_x(x)[f(x) + g_1(x)w - g_{21}(x)g_{21}^T(x)V_x^T(x) + g_{22}(x)\alpha_{22}(x)] \\
&= V_x(x)f(x) + V_x(x)g_1(x)w - V_x(x)g_{21}(x)g_{21}^T(x)V^T(x) + V_x(x)g_{22}(x)\alpha_{22}(x) \\
&\le -\frac{1}{2}V_x(x)g_{21}(x)g_{21}^T(x)V_x^T(x) - \frac{1}{2}h^T(x)h(x) + \frac{1}{2}\gamma^2\|w\|^2 - \\
&\quad \frac{\gamma^2}{2}\left\|w - \frac{g_1^T(x)V_x^T(x)}{\gamma^2}\right\|^2 \\
&\le \frac{1}{2}(\gamma^2\|w\|^2 - \|z\|^2). \tag{9.11}
\end{aligned}$$

Integrating now from $t = t_0$ to $t = T$ we have the dissipation-inequality

$$V(x(T)) - V(x_0) \le \int_{t_0}^{T} \frac{1}{2}(\gamma^2\|w\|^2 - \|z\|^2)dt$$

and therefore the system has $\mathcal{L}_2$-gain $\le \gamma$. Further, with $w = 0$, we get

$$\dot{V} \le -\frac{1}{2}\|z\|^2,$$

and thus, the closed-loop system is locally stable. Moreover, the condition $\dot{V} \equiv 0$ for all $t \ge t_c$, for some $t_c \ge t_0$, implies that $z \equiv 0$ and $h_1(x) \equiv 0$, $u_1^\star(x) \equiv 0$ for all $t \ge t_c$. Consequently, it is a trajectory of $\dot{x} = f(x) + g_{22}(x)\alpha_{22}(x)$. By the zero-state detectability of $\{f(x) + g_{22}(x)\alpha_{22}(x), h_1(x)\}$, this implies $\lim_{t\to\infty} x(t) = 0$. Finally, by LaSalle's invariance-principle, we conclude asymptotic-stability. $\square$

Remark 9.1.1 *The above theorem gives sufficient conditions for the solvability of the SFBSNLHICP for the case in which the regular SFBHICP for the subsystem (9.5) is solvable. Thus, it follows that the control input u_1, which is referred to as the "regular control," has enough power to provide disturbance attenuation and stability for the system; while the remaining input u_2, which is referred to as the "singular control" can be utilized to achieve additional objectives such as transient performance. If however, this is not the case, then some other approach must be used for solving the singular problem. One approach converts the problem to the "almost-disturbance-decoupling problem" discussed in Chapter 5 [42], while another approach uses high-gain feedback. The latter approach will be discussed in the next subsection.*

Remark 9.1.2 *Note that the condition (9.9) is fulfilled with $u_2 = 0$. However, for any other function $u_2 = \alpha_{22}(x) \neq 0$ which satisfies (9.9), we have that*

$$\|\Sigma^a(u_2 = \alpha_{22}(x) \neq 0)\|_{\mathcal{L}_2} < \|\Sigma^a(u_2 = 0)\|_{\mathcal{L}_2}.$$

In fact, a better choice of u_2 is

$$u_2 = -R^{-1}g_{22}^T(x)V_x^T(x)$$

for some weighting matrix $R > 0$.

9.1.1 State-Feedback Singular Nonlinear $\mathcal{H}_\infty$-Control Using High-Gain Feedback

In this subsection, we discuss an alternative approach to the $SFBSNLHICP$. For this purpose, rewrite the system (9.3) in the form

$$\Sigma^a : \begin{cases} \dot{x} &= f(x) + g_1(x)w + g_{21}(x)u_1 + g_{22}(x)u_2; \quad x(t_0) = x_0 \\ z &= \begin{bmatrix} h_1(x) \\ u_1 \end{bmatrix} \\ y &= x, \end{cases} \tag{9.12}$$

where all the variables and functions have their previous meanings, and consider the following auxiliary ε-perturbed version of it:

$$\Sigma_\varepsilon^a : \begin{cases} \dot{x} &= f(x) + g_1(x)w + g_{21}(x)u_1 + g_{22}(x)u_2; \quad x(t_0) = x_0 \\ z_\varepsilon &= \begin{bmatrix} h_1(x) \\ u_1 \\ \sqrt{\varepsilon}u_2 \end{bmatrix} \\ y &= x. \end{cases} \tag{9.13}$$

Then we have the following proposition.

Proposition 9.1.1 *Suppose there exists a feedback of the form (9.6) which solves the $SFBSNLHICP$ for the system (9.12) with finite $\mathcal{L}_2$-gain from w to α_{22}. Then, the closed-loop system (9.12), (9.6) has $\mathcal{L}_2$-gain $\leq \gamma$ if, and only if, the closed-loop system (9.13), (9.6) has $\mathcal{L}_2$-gain $\leq \gamma$ for some sufficiently small $\varepsilon > 0$.*

Proof: ($\Rightarrow$) By assumption, the $\mathcal{L}_2$-gain from w to α_{22} is finite (assume $\leq \rho < \infty$). Thus,

$$\int_{t_0}^T \varepsilon\|\alpha_{22}(x(t))\|^2 \leq \varepsilon\rho^2 \int_{t_0}^T \|w(t)\|^2 dt \quad \forall T > 0.$$

Moreover, the closed-loop system (9.12), (9.6) has $\mathcal{L}_2$-gain $\leq \gamma$, therefore

$$\int_{t_0}^{T} \|z(t)\|^2 \leq \gamma^2 \int_{t_0}^{T} \|w(t)\|^2 dt \quad \forall T > 0.$$

Adding the above two inequalities, we get

$$\int_{t_0}^{T} \|z(t)\|^2 dt + \varepsilon \|\alpha_{22}(x(t))\|^2 \leq \gamma^2 \int_{t_0}^{T} \|w(t)\|^2 dt + \varepsilon \rho^2 \int_{t_0}^{T} \|w(t)\|^2 dt \quad \forall T > 0.$$

Since $w \in \mathcal{L}_2[0, \infty)$ and ε is small, there exists a constant $\beta < \infty$ such that for all $w \in \mathcal{L}_2[0, \infty)$, $\varepsilon \rho^2 \int_{t_0}^{T} \|w(t)\|^2 dt < \beta$. Therefore,

$$\int_{t_0}^{T} \|z_\varepsilon(t)\|^2 dt \leq \gamma^2 \int_{t_0}^{T} \|w(t)\|^2 dt + \beta,$$

and hence the result.

($\Leftarrow$) This is straight-forward. Clearly

$$\int_{t_0}^{T} \|z_\varepsilon(t)\|^2 dt \quad \leq \quad \gamma^2 \int_{t_0}^{T} \|w(t)\|^2 dt$$

$$\Longrightarrow \int_{t_0}^{T} \|z(t)\|^2 dt \quad \leq \quad \gamma^2 \int_{t_0}^{T} \|w(t)\|^2 dt \quad \square$$

Based on the above proposition, we can proceed to design a regular stabilizing feedback for the system (9.12) using the system (9.13). For this, we have the following theorem.

Theorem 9.1.2 *Consider the nonlinear system Σ^a defined by (9.12) and the SFBSNLHICP for it. Suppose there exists a C^1 solution $V \geq 0$ to the HJIE:*

$$V_x(x)f(x) + \frac{1}{2}V_x(x)[\frac{1}{\gamma^2}g_1(x)g_1^T(x) - g_{21}(x)g_{21}^T(x) - \frac{1}{\varepsilon}g_{22}(x)g_{22}^T(x)]V_x^T(x) +$$

$$\frac{1}{2}h_1^T(x)h_1(x) = 0, \quad V(0) = 0, \tag{9.14}$$

for some $\gamma > 0$ and some sufficiently small $\varepsilon > 0$. Then the state-feedback

$$u = \begin{bmatrix} u_1^\star(x) \\ u_2^\star(x) \end{bmatrix} = -\begin{bmatrix} g_{21}^T(x) \\ \frac{1}{\varepsilon}g_{22}^T(x) \end{bmatrix} V_x^T(x) \tag{9.15}$$

solves the SFBSNLHICP for the system.

Proof: It can be shown as in the proof of Theorem 9.1.1 that the state-feedback (9.15) when applied to the system Σ_ε^a leads to a closed-loop system which is locally asmptotically-stable and has $\mathcal{L}_2$-gain $\leq \gamma$ from w to z_ε. The result then follows by application of Proposition 9.1.1. $\square$

We can specialize the result of Theorem 9.1.2 to the linear system

$$\Sigma^l : \begin{cases} \dot{x} & = & Ax + B_1 w + B_{21}u_1 + B_{22}u_2, \quad x(t_0) = x_0 \\ z & = & \begin{bmatrix} C_1 x \\ u_1 \end{bmatrix} \\ y & = & x, \end{cases} \tag{9.16}$$

where all the variables have their previous meanings, and $A \in \Re^{n \times n}$, $B_1 \in \Re^{n \times r}$, $B_{21} \in \Re^{n \times \kappa}$, $B_{22} \in \Re^{n \times p-\kappa}$, $C_1 \in \Re^{s \times n}$ are real constant matrices. We have the following corollary to Theorem 9.1.2.

Corollary 9.1.1 *Consider the linear system Σ^l defined by (9.16) and the SFBSNLHICP for it. Suppose there exists a symmetric solution $P > 0$ to the ARE:*

$$A^T P + PA + P[\frac{1}{\gamma^2}B_1 B_1^T - B_{21}B_{21}^T - \frac{1}{\varepsilon}B_{22}B_{22}^T]P + C_1^T C_1 + \epsilon Q = 0, \tag{9.17}$$

for some matrix $Q > 0$, $\gamma > 0$ and some sufficiently small $\epsilon > 0$. Then, the state-feedback

$$u_l = \begin{bmatrix} u_{1,l}^\star \\ u_{2,l}^\star \end{bmatrix} = -\begin{bmatrix} B_{21}^T \\ \frac{1}{\varepsilon}B_{22}^T \end{bmatrix} Px \tag{9.18}$$

solves the SFBSNLHICP for the system.

Remark 9.1.3 *Note that there is no assumption of detectability of (C_1, A) in the above corollary. This is because, since $\varepsilon Q + C_1^T C_1 > 0$, there exists a C such that $\epsilon Q + C_1^T C_1 = C^T C$ and (C, A) is always detectable.*

9.2   Output Measurement-Feedback Singular Nonlinear $\mathcal{H}_\infty$-Control

In this section, we discuss a solution to the singular nonlinear $\mathcal{H}_\infty$-control problem for the system Σ^a using a dynamic measurement-feedback controller of the form:

$$\Sigma_{dyn}^c : \begin{cases} \dot{\xi} = \eta(\xi) + \theta(\xi)y \\ u = \begin{bmatrix} u_1 \\ u_2 \end{bmatrix} = \begin{bmatrix} \alpha_{21}(\xi) \\ \alpha_{22}(\xi) \end{bmatrix}, \end{cases} \tag{9.19}$$

where $\xi \in \Xi$ is the state of the compensator with $\Xi \subset \mathcal{X}$ a neighborhood of the origin, $\eta : \Xi \to V^\infty(\Xi)$, $\eta(0) = 0$, and $\theta : \Xi \to \mathcal{M}^{n \times m}$ are some smooth functions. The controller processes the measured variable y of the plant (9.1) and generates the appropriate control action u, such that the closed-loop system $\Sigma^a \circ \Sigma_{dyn}^c$ has locally $\mathcal{L}_2$-gain from the disturbance signal w to the output z less than or equal to some prescribed number $\gamma^\star > 0$ with internal stability. This problem will be abbreviated as $MFBSNLHICP$.

To solve the above problem, we consider the following representation of the system (9.3) with disturbances and measurement noise:

$$\tilde{\Sigma}^a : \begin{cases} \dot{x} = f(x) + g_1(x)w_1 + g_{21}(x)u_1 + g_{22}(x)u_2; \quad x(t_0) = x_0 \\ z = \begin{bmatrix} h_1(x) \\ u_1 \end{bmatrix} \\ y = h_2(x) + w_2, \end{cases} \tag{9.20}$$

where $h_2 : \mathcal{X} \to \Re^m$ is a smooth matrix while all the other functions and variables have their previous meanings. As in the previous section, we can proceed to design the controller for the following auxiliary ε-perturbed version of the plant $\tilde{\Sigma}^a$:

$$\tilde{\Sigma}_\varepsilon^a : \begin{cases} \dot{x} = f(x) + g_1(x)w_1 + g_{21}(x)u_1 + g_{22}(x)u_2; \quad x(t_0) = x_0 \\ z = \begin{bmatrix} h_1(x) \\ u_1 \\ \sqrt{\varepsilon}u_2 \end{bmatrix} \\ y = h_2(x) + w_2. \end{cases} \tag{9.21}$$

Moreover, it can be shown similarly to Proposition 9.1.1, that under the assumption that the $\mathcal{L}_2$-gain from w to u_2 is finite, the closed-loop system (9.20), (9.19) has $\mathcal{L}_2$-gain $\leq \gamma$ if and only if the closed-loop system (9.21), (9.19) has $\mathcal{L}_2$-gain $\leq \gamma$. We can then employ the techniques of Chapter 6 to design a *"certainty-equivalent worst-case"* dynamic controller Σ^c_{dyn} for the system. The following theorem gives sufficient conditions for the solvability of the problem.

Theorem 9.2.1 *Consider the nonlinear system $\tilde{\Sigma}^a$ and the MFBSNLHICP. Suppose for some sufficiently small $\varepsilon > 0$ there exist smooth C^2 solutions $V \geq 0$ to the HJIE (9.14) and $W \geq 0$ to the HJIE:*

$$W_x(x)f(x) + \frac{1}{2\gamma^2}W_x(x)g_1(x)g_1^T(x)W_x^T(x) + \frac{1}{2}h_1(x)h_1^T(x) - \frac{1}{2}\gamma^2 h_2^T(x)h_2(x) = 0, \quad W(0) = 0,$$
$$(9.22)$$

on $M \subset \mathcal{X}$ such that

$$f - g_{21}g_{21}^T V_x^T - \frac{1}{\varepsilon}g_{22}(x)g_{22}^T(x)V_x^T + \frac{1}{\gamma^2}g_1 g_1^T V_x^T \quad \text{is exponentially stable,}$$

$$-\left(f + \frac{1}{\gamma^2}g_1 g_1^T W_x^T\right) \quad \text{is exponentially stable,}$$

$$W_{xx}(x) > V_{xx}(x) \; \forall x \in M.$$

Then, the dynamic controller defined by

$$\Sigma^c_{dyn} : \begin{cases} \dot{\xi} &= f(\xi) - g_{21}(\xi)g_{21}^T(\xi)V_\xi^T(\xi) - \frac{1}{\varepsilon}g_{22}(\xi)g_{22}^T(\xi)V_\xi^T(\xi) + \\ & \quad \frac{1}{\gamma^2}g_1(\xi)g_1^T(\xi)V_\xi^T(\xi) + \gamma^2[W_{\xi\xi}(\xi) - V_{\xi\xi}(\xi)]^{-1}\frac{\partial h_2}{\partial \xi}(\xi)(y - h_2(\xi)) \\ u &= \begin{bmatrix} u_1 \\ u_2 \end{bmatrix} = -\begin{bmatrix} g_{21}^T(\xi) \\ \frac{1}{\varepsilon}g_{22}^T(\xi) \end{bmatrix} V_\xi^T(\xi) \end{cases} \quad (9.23)$$

solves the MFBSNLHICP for the system locally on M.

Proof: (Sketch). For the purpose of the proof, we consider the finite-horizon problem on the interval $[t_0, T]$, and derive the solution of the infinite-horizon problem by letting $T \to \infty$. Accordingly, we consider the cost-functional:

$$\min_{u \in \mathcal{U}} \max_{w \in \mathcal{W}} J_m(w, u) = \frac{1}{2}\int_{t_0}^T \{\|u_1\|^2 + \varepsilon\|u_2\|^2 + \|h_1(x)\|^2 - \gamma^2\|w_1\|^2 - \gamma^2\|w_2\|^2\}dt$$

where $u(\tau)$ depends on $y(\tau)$, $\tau \leq t$. As in Chapter 6, the problem can then be split into two subproblems: (i) the state-feedback subproblem; and (ii) the state-estimation subproblem.

Subproblem (i) has already been dealt with in the previous section leading to the feedbacks (9.15). For (ii), we consider the certainty-equivalent worst-case estimator:

$$\dot{\xi} = f(\xi) + g_1(\xi)\alpha_1(\xi) + g_{21}(\xi)\alpha_{21}(\xi) + g_{22}(\xi)\alpha_{22}(\xi) + G_s(\xi)(y - h_2(\xi)),$$

where $\alpha_1(x) = w^\star = \frac{1}{\gamma^2}g_1(x)g_1^T(x)V_x^T(x)$ is the worst-case disturbance, $\alpha_{21}(x) = u_1^\star(x)$, $\alpha_{22}(x) = u_2^\star(x)$ and $G_s(.)$ is the output-injection gain matrix. Then, we can design the gain matrix $G_s(.)$ to minimize the estimation error $e = y - h_2(\xi)$ and render the closed-loop system asymptotically-stable.

Acordingly, let $\tilde{u}_1(t)$, $\tilde{u}_2(t)$ and $\tilde{y}(t)$, $t \in [t_0, \tau]$ be a given pair of inputs and corresponding measured output. Then, we consider the problem of maximizing the cost functional

$$\tilde{J}_m(w, u) = V(x(\tau), \tau) + \frac{1}{2}\int_{t_0}^\tau \{\|\tilde{u}_1\|^2 + \varepsilon\|\tilde{u}_2\|^2 + \|h_1(x)\|^2 - \gamma^2\|w_1\|^2 - \gamma^2\|w_2\|^2\}dt,$$

with respect to $x(t)$, $w_1(t)$, $w_2(t)$, where $V(x,\tau)$ is the value-function for the first optimization subproblem to determine the state-feedbacks, and subject to the constraint that the output of the system (9.21) equals $\tilde{y}(t)$. Moreover, since w_2 directly affects the observation y, we can substitute $w_2 = h_2(x) - \tilde{y}$ into the cost functional $\tilde{J}_m(.,.)$ such that the above constraint is automatically satisfied. The resulting value-function for this maximization subproblem is then given by $S(x,\tau) = V(x,\tau) - W(x,\tau)$, where $W \geq 0$ satisfies the HJIE

$$W_t(x,t) + W_x(x,t)[f(x) + g_{21}(x)\tilde{u}_1(t) + g_{22}(x)\tilde{u}_2(t)] + \tfrac{1}{2}h_1^T(x)h_1(x)$$
$$\tfrac{1}{2\gamma^2}W_x(x,t)g_1(x)g_1^T(x)W_x^T(x,t) + -\tfrac{1}{2}\gamma^2 h_2^T(x)h_2(x) + \gamma^2 h_2^T(x)\tilde{y}(t) -$$
$$\tfrac{1}{2}\gamma^2\|\tilde{y}(t)\|^2 + \tfrac{1}{2}\|\tilde{u}_1(t)\|^2 + \tfrac{1}{2}\varepsilon\|\tilde{u}_2(t)\|^2 = 0, \quad W(x,t_0) = 0. \tag{9.24}$$

Assuming that the maximum of $S(.,.)$ is determined by the condition $S_x(\xi(t),t) = 0$ and that the Hessian is nondegenerate,[2] then the corresponding state-equation for ξ can be found by differentiation of $S_x(\xi(t),t) = 0$.

Finally, we obtain the controller which solves the infinite-horizon problem by letting $T \to \infty$ while imposing that $x(t) \to 0$, and $t_0 \to -\infty$ while $x(t_0) \to 0$. A finite-dimensional approximation to this controller is given by (9.23). $\square$

9.3   Singular Nonlinear $\mathcal{H}_\infty$-Control with Static Output-Feedback

In this section, we briefly extend the static output-feedback approach developed in Section 6.5 to the case of singular control for the affine nonlinear system. In this regard, consider the following model of the system with the penalty variable defined in the following form

$$\Sigma_1^a : \begin{cases} \dot{x} &= f(x) + g_1(x)w + \tilde{g}_2(x)\tilde{u} \\ y &= h_2(x) \\ z &= \begin{bmatrix} h_1(x) \\ \tilde{k}_{12}(x)\tilde{u} \end{bmatrix}, \end{cases} \tag{9.25}$$

where all the variables and functions have their previous meanings and dimensions, and $\min \, rank\{k_{12}(x)\} = \kappa < p$ (assuming it is constant for all x). Then as discussed in Section 9.1, in this case, there exists a coordinate transformation φ under which the system can be represented in the following form [42]:

$$\Sigma_2^a : \begin{cases} \dot{x} &= f(x) + g_1(x)w + g_{21}(x)u_1 + g_{22}(x)u_2 \\ y &= h_2(x) \\ z &= \begin{bmatrix} h_1(x) \\ D_{12}u_1 \end{bmatrix}, \quad D_{12}^T D_{12} = I, \end{cases} \tag{9.26}$$

where $g_2 = [g_{21}(x) \ \ g_{22}]$, $D_{12} \in \Re^{s-\kappa\times\kappa}$, $u_1 \in \Re^\kappa$ is the regular control, and $u_2 \in \Re^{p-\kappa}$, is the singular control. Furthermore, it has been shown that the state-feedback control law

$$u = \begin{bmatrix} u_1 \\ u_2 \end{bmatrix} = \begin{bmatrix} -g_{21}^T(x)\tilde{V}_x^T(x) \\ \alpha_{22}(x) \end{bmatrix}, \tag{9.27}$$

where $\alpha_{22}(x)$ is such that

$$\tilde{V}_x(x)g_{22}(x)\alpha_{22}(x) \leq 0$$

---
[2]The Hessian is nondegenerate if it is not identically zero.

and $\tilde{V} > 0$ is a smooth local solution of the HJIE (9.8), solves the state-feedback singular $\mathcal{H}_\infty$-control problem $(SFBSNLHICP)$ for the system (9.25) if the pair $\{f + g_{22}\alpha, h_1\}$ is locally zero-state detectable. The following theorem then shows that, if the state-feedback singular $\mathcal{H}_\infty$-control problem is locally solvable with a regular control, then a set of conditions similar to (6.91), (6.92) are sufficient to guarantee the solvability of the SOFBP.

Theorem 9.3.1 *Consider the nonlinear system (9.25) and the singular $\mathcal{H}_\infty$-control problem for this system. Suppose the $SFBSNLHICP$ is locally solvable with a regular control and there exist C^0 functions $F_3 : \mathcal{Y} \subset \Re^m \to \Re^\kappa$, $\phi_3 : X_2 \subset \mathcal{X} \to \Re^p$, $\psi_3 : \mathcal{Y} \to \Re^{p-\kappa}$ such that the conditions*

$$F_3 \circ h_2(x) = -g_{21}^T(x)\tilde{V}_x^T(x) + \phi_3(x), \tag{9.28}$$

$$\tilde{V}_x(x)[g_{21}(x)\phi_3(x) + g_{22}(x)\psi_3(h_2(x))] \leq 0, \tag{9.29}$$

are satisfied. In addition, suppose also the pair $\{f(x) + g_{22}(x)\psi_3 \circ h_2(x),\ h_1(x)\}$ is zero-state detectable. Then, the static output-feedback control law

$$u = \begin{bmatrix} u_1 \\ u_2 \end{bmatrix} = \begin{bmatrix} F_3(y) \\ \psi_3(y) \end{bmatrix} \tag{9.30}$$

solves the singular $\mathcal{H}_\infty$-control problem for the system locally.

Proof: Differentiating the solution $\tilde{V} > 0$ of (9.8) along a trajectory of the closed-loop system (9.25) with the control law (9.30), we get upon using (9.28), (9.29) and the HJIE (9.8):

$$
\begin{aligned}
\dot{\tilde{V}} &= \tilde{V}_x(x)[f(x) + g_1(x)w + g_{21}(x)(F_3(h_2(x)) + g_{22}(x)\psi_3(y)] \\
&= \tilde{V}_x(x)[f(x) + g_1(x)w - g_{21}(x)g_{21}^T(x)\tilde{V}_x^T(x) + g_{21}(x)\phi_3(x) + g_{22}(x)\psi_3(y)] \\
&= \tilde{V}_x(x)f(x) + \tilde{V}_x(x)g_1(x)w - \tilde{V}_x(x)g_{21}(x)g_{21}^T(x)\tilde{V}^T(x) + \\
&\quad \tilde{V}_x(x)(g_{21}(x)\phi_3(x) + \psi_3(h_2(x))) \\
&\leq -\frac{1}{2}\tilde{V}_x(x)g_{21}(x)g_{21}^T(x)\tilde{V}^T(x) - \frac{1}{2}h_1^T(x)h_1(x) + \frac{1}{2}\gamma^2\|w\|^2 - \\
&\quad \frac{\gamma^2}{2}\left\| w - \frac{g_1^T(x)\tilde{V}_x^T(x)}{\gamma^2} \right\|^2 \\
&\leq \frac{1}{2}(\gamma^2\|w\|^2 - \|z\|^2). \tag{9.31}
\end{aligned}
$$

Integrating now the above inequality from $t = t_0$ to $t = T$ we have the dissipation inequality

$$\tilde{V}(x(T)) - \tilde{V}(x_0) \leq \int_{t_0}^{T} \frac{1}{2}(\gamma^2\|w\|^2 - \|z\|^2)dt$$

and therefore the system has $\mathcal{L}_2$-gain $\leq \gamma$. Moreover, with $w = 0$, we get

$$\dot{\tilde{V}} \leq -\frac{1}{2}\|z\|^2,$$

and thus, the closed-loop system is locally stable. Moreover, the condition $\dot{\tilde{V}} \equiv 0$ for all $t \geq t_c$, for some $t_c \geq t_0$, implies that $z \equiv 0$ and $h_1(x) = 0$, $u^\star(x) = 0$ for all $t \geq t_c$. Consequently, it is a trajectory of $\dot{x} = f(x) + g_{22}(x)\psi_3(h_2(x))$. By the zero-state detectability of $\{f(x) + g_{22}(x)\psi_3(h_2(x)),\ h_1(x)\}$, this implies $\lim_{t \to \infty} x(t) = 0$, and using LaSalle's invariance-principle we conclude asymptotic-stability. $\square$

On the other hand, if the singular problem is not solvable using the regular control above, then it has been shown in Section 9.1.1 that a high-gain feedback can be used to solve it. Thus, similarly, the results of Theorem 9.1.2 can be extended straightforwardly to the static-output feedback design. It can easily be guessed that if there exist C^0 functions $F_4 : \mathcal{Y} \subset \Re^m \to \mathcal{U}$, $\phi_4 : X_4 \subset \mathcal{X} \to \Re^p$, $\psi_4 : \mathcal{Y} \to \Re^{p-\kappa}$, such that the conditions

$$F_4 \circ h_2(x) = - \left[\begin{array}{c} g_{21}^T(x) \\ \frac{1}{\varepsilon} g_{22}^T(x) \end{array} \right] \tilde{V}_x^T(x) + \phi_4(x), \tag{9.32}$$

$$\tilde{V}_x(x)[g_{21}(x)\phi_4(x) + g_{22}(x)\psi_4(h_2(x))] \leq 0, \tag{9.33}$$

where $\tilde{V}$ is a local solution of the HJIE (9.8), are satisfied, then the output-feedback control law

$$u = \left[\begin{array}{c} F_4(y) \\ \psi_4(y) \end{array} \right] \tag{9.34}$$

solves the singular $\mathcal{H}_\infty$-control problem for the system locally.

9.4 Singular Nonlinear $\mathcal{H}_\infty$-Control for Cascaded Nonlinear Systems

In this section, we discuss the $SFBSNLHICP$ for a fairly large class of cascaded nonlinear systems that are totally singular, and we extend some of the results presented in Section 9.1 to this case. This class of systems can be represented by an aggregate state-space model defined on a manifold $\mathcal{X} \subset \Re^n$:

$$\Sigma_{agg}^a : \left\{ \begin{array}{rcl} \dot{x} & = & f(x) + g_1(x)w + g_2(x)u; \quad x(t_0) = x_0 \\ z & = & h(x) \\ y & = & x, \end{array} \right. \tag{9.35}$$

where all the variables have their previous meanings, while $g_1 : \mathcal{X} \to \mathcal{M}^{n \times r}$, $g_2 : \mathcal{X} \to \Re^{n \times p}$, $h : \mathcal{X} \to \Re^s$ are C^∞ function of x. We assume that the system has a unique equilibrium-point $x = 0$, and $f(0) = 0$, $h(0) = 0$. In addition, we also assume that there exist local coordinates $[x_1^T \; x_2^T] = [x^1, \ldots, x^q, x^{q+1}, \ldots, x^n]$, $1 \leq q < n$ such that the system Σ_{agg}^a can be decomposed into

$$\Sigma_{dec}^a : \left\{ \begin{array}{rcl} \dot{x}_1 & = & f_1(x_1) + g_{11}(x_1)w + g_{21}(x_1)h_2(x_1, x_2) \quad x_1(t_0) = x_{10} \\ \dot{x}_2 & = & f_2(x_1, x_2) + g_{12}(x_1, x_2)w + g_{22}(x_1, x_2)u; \quad x_2(t_0) = x_{20} \\ z & = & \left[\begin{array}{c} z_1 \\ z_2 \end{array} \right] = \left[\begin{array}{c} h_1(x_1) \\ h_2(x_1, x_2) \end{array} \right] \\ y & = & x \end{array} \right. \tag{9.36}$$

where $z_1 \in \Re^{s_1}$, $z_2 \in \Re^{s_2}$, $s_1 \geq 1$, $s = s_1 + s_2$. Moreover, in accordance with the representation (9.35), we have

$$f(x) = \left[\begin{array}{c} f_1(x_1) + g_{21}(x_1)z_2 \\ f_2(x_1, x_2) \end{array} \right], \quad g_1(x) = \left[\begin{array}{c} g_{11}(x_1) \\ g_{12}(x_1, x_2) \end{array} \right],$$

$$g_2(x) = \left[\begin{array}{c} 0 \\ g_{22}(x_1, x_2) \end{array} \right], \quad h(x) = \left[\begin{array}{c} h_1(x_1) \\ h_2(x_1, x_2) \end{array} \right].$$

The decomposition (9.36) can also be viewed as a cascade of two subsystems:

$$\Sigma_1^a \quad : \quad \begin{cases} \dot{x} & = & f_1(x_1) + g_{11}(x_1)w + g_{21}(x_1)z_2; \\ z_1 & = & h_1(x_1) \end{cases}$$

$$\Sigma_2^a \quad : \quad \begin{cases} \dot{x}_2 & = & f_2(x_1, x_2) + g_{12}(x_1, x_2)w + g_{22}(x_1, x_2)u \\ z_2 & = & h_2(x_1, x_2). \end{cases}$$

Such a model represents many physical dynamical systems we encouter everyday; for example, in mechnical systems, the subsystem Σ_1^a can represent the kinematic subsystem, while Σ_2^a represents the dynamic subsystem.

To solve the $SFBSNLHICP$ for the above system, we make the following assumption.

Assumption 9.4.1 *The $SFBHICP$ for the subsystem Σ_1^a is solvable, i.e., there exists a smooth solution $P : M_1 \to \Re,\ P \geq 0,\ M_1 \subset \mathcal{X}$ a neighborhood of $x_1 = 0$ to the HJI-inequality*

$$P_{x_1}(x_1)f_1(x_1) + \frac{1}{2}P_{x_1}(x_1)[\frac{1}{\gamma^2}g_{11}(x_1)g_{11}^T(x_1) - g_{21}(x_1)g_{21}^T(x_1)]P_{x_1}^T(x_1) +$$

$$\frac{1}{2}h_1^T(x_1)h_1(x_1) \leq 0, \quad P(0) = 0, \tag{9.37}$$

such that $z_2(x)$ viewed as the control is given by

$$z_2^\star = -g_{21}^T(x_1)P_{x_1}^T(x_1),$$

and the worst-case disturbance affecting the subsystem is also given by

$$w_1^\star = \frac{1}{\gamma^2}g_{11}^T(x_1)P_{x_1}^T(x_1).$$

We can now define the following auxiliary system

$$\bar{\Sigma}_{dec}^a \quad : \quad \begin{cases} \dot{x}_1 & = & f_1(x_1) + g_{11}(x_1)\bar{w} + g_{21}(x_1)z_2 + \frac{1}{\gamma^2}g_{11}(x_1)g_{11}^T(x_1)P_{x_1}^T(x_1) \\ \dot{x}_2 & = & f_2(x) + g_{12}(x)\bar{w} + g_{22}(x)u + \frac{1}{\gamma^2}g_{12}(x)g_{11}^T(x_1)P_{x_1}^T(x_1) \\ \bar{z} & = & z_2 - z_2^\star = h_2(x) + g_{11}^T(x_1)P_{x_1}^T(x_1). \end{cases} \tag{9.38}$$

where $\bar{z} \in \Re^{s-s_1}$ and

$$\bar{w} = w - \frac{1}{\gamma^2}g_{11}^T(x_1)P_{x_1}^T(x_1).$$

Now define

$$\bar{h}(x) \quad = \quad \bar{z} = z_2 - z_2^\star = h_2(x) + g_{21}^T(x_1)P_{x_1}^T(x_1)$$

$$\bar{f}(x) \quad = \quad \begin{bmatrix} f_1(x_1) + g_1(x_1)z_2 + \frac{1}{\gamma^2}g_{11}(x_1)g_{11}^T(x_1)P_{x_1}^T(x_1) \\ f_2(x) + \frac{1}{\gamma^2}g_{12}^T(x)g_{11}^T(x_1)P_{x_1}^T(x_1) \end{bmatrix}.$$

Then, (9.38) can be represented in the aggregate form

$$\bar{\Sigma}_{agg}^a \quad : \quad \begin{cases} \dot{x} & = & \bar{f}(x) + g_1(x)\bar{w} + g_2(x)u; \quad \bar{f}(0) = 0 \\ \bar{z} & = & \bar{h}(x), \quad \tilde{h}(0) = 0, \end{cases} \tag{9.39}$$

and we have the following lemmas.

Lemma 9.4.1 *For the system representations Σ_{agg}^a (9.35) and $\bar{\Sigma}_{agg}^a$ (9.39), the following inequality holds:*

$$\frac{1}{2}\|\bar{z}\|^2 \geq P_{x_1}(x_1)(f_1(x_1) + g_{21}(x_1)z_2) + \frac{1}{2\gamma^2}\|g_{11}^T(x_1)P_{x_1}^T(x_1)\|^2 + \frac{1}{2}\|z\|^2. \tag{9.40}$$

Proof: Note,

$$\frac{1}{2}\|\bar{z}\|^2 = \frac{1}{2}\|z_2 - z_2^\star\|^2 = P_{x_1}(x_1)g_{21}(x_1)z_2 + \frac{1}{2}\|z_2\|^2 + \frac{1}{2}\|g_{21}^T(x_1)P_{x_1}^T(x_1)\|^2,$$

and from the HJI-inequality (9.37), we get

$$\frac{1}{2}\|g_{21}^T(x_1)P_{x_1}^T(x_1)\|^2 \geq P_{x_1}(x_1)f_1(x_1) + \frac{1}{2\gamma^2}\|g_{11}^T(x_1)P_{x_1}^T(x_1)\|^2 + \frac{1}{2}\|z_1\|^2.$$

Upon inserting this in the previous equation, the result follows. □

Lemma 9.4.2 *Let $K_\gamma : T^\star\mathcal{X} \times \mathcal{U} \times \mathcal{W} \to \Re$ be the pre-Hamiltonian for the system Σ_{agg}^a defined by*

$$K_\gamma(x, p, u, w) = p^T(f(x) + g_1(x)w + g_2(x)u) + \frac{1}{2}\|z\|^2 - \frac{1}{2}\gamma^2\|w\|^2,$$

where $(x^1, \ldots, x^n)^T$, $(p^1, \ldots, p^n)^T$ are local coordinates for $T^\star\mathcal{X}$. Similarly, let $\bar{K}_\gamma : T^\star\mathcal{X} \times \mathcal{U} \times \mathcal{W} \to \Re$ be the pre-Hamiltonian for $\bar{\Sigma}_{agg}^a$. Then

$$K_\gamma(x, p + P_x^T, u, w) \leq \bar{K}_\gamma(x, p, u, \bar{w}). \tag{9.41}$$

Proof: We note that $\bar{f}(x) + g_1(x)\bar{w} = f(x) + g_1(x)w$. Then

$$
\begin{aligned}
\bar{K}_\gamma(x, p, u, \bar{w}) &= p^T(\bar{f}(x) + g_1(x)\bar{w} + g_2(x)u) + \frac{1}{2}\|\bar{z}\|^2 - \frac{1}{2}\gamma^2\|\bar{w}\|^2 \\
&= p^T(f(x) + g_1(x)w + g_2(x)u) + \frac{1}{2}\|\bar{z}\|^2 - \frac{1}{2}\gamma^2\|\bar{w}\|^2 \\
&\geq p^T(f(x) + g_1(x)w + g_2(x)u) + P_{x_1}(x_1)(f_1(x_1) + g_1(x_1)z_2) + \\
&\quad \frac{1}{2\gamma^2}\|g_{11}^T(x_1)P_{x_1}(x_1)\|^2 + \frac{1}{2}\|z\|^2 - \frac{1}{2}\gamma^2\|\bar{w}\|^2, \tag{9.42}
\end{aligned}
$$

where the last inequality follows from Lemma 9.4.1. Noting that

$$
\begin{aligned}
\frac{1}{2}\gamma^2\|\bar{w}\|^2 &= \frac{1}{2}\gamma^2 \left(w - \frac{1}{2\gamma^2}g_{11}^T(x_1)P_{x_1}^T(x_1)\right)^T \left(w - \frac{1}{2\gamma^2}g_{11}^T(x_1)P_{x_1}^T(x_1)\right) \\
&= \frac{1}{2}\gamma^2\|w\|^2 + \frac{1}{2\gamma^2}\|g_{11}^T(x_1)P_{x_1}^T(x_1)\|^2 - P_{x_1}^T(x_1)g_{11}(x_1)w,
\end{aligned}
$$

then substituting the above expression in the inequality (9.42), gives

$$
\begin{aligned}
\bar{K}_\gamma(x, p, u, \bar{w}) &\geq p^T(f(x) + g_1(x)w + g_2(x)u) + P_{x_1}(x_1)(f_1(x_1) + g_1(x_1)z_2 + g_{11}(x_1)w) + \\
&\quad \frac{1}{2}\|z\|^2 - \frac{1}{2}\gamma^2\|w\|^2 \\
&= K_\gamma(x, p + P_{x_1}(x_1), u, w). \quad □
\end{aligned}
$$

As a consequence of the above lemma, we have the following theorem.

Theorem 9.4.1 *Let $\gamma > 0$ be given, and suppose there exists a static state-feedback control*

$$u = \alpha(x), \quad \alpha(0) = 0$$

that solves the SFBSNLHICP for the system $\bar{\Sigma}_{agg}^a$ such that there exists a C^1 solution $W : M \to \Re$, $W \geq 0$, $M \subset \mathcal{X}$ to the dissipation-inequality

$$\bar{K}_\gamma(x, W_x^T(x), \alpha(x), \bar{w}) \leq 0, \quad W(0) = 0, \forall \bar{w} \in \mathcal{W}.$$

Then, the same control law also solves the SFBSNLHICP for the system Σ_{agg}^a and the dissipation-inequality

$$K_\gamma(x, V_x^T(x), \alpha(x), w) \le 0, \quad V(0) = 0, \forall w \in \mathcal{W}$$

is satisfied as well for a nonnegative C^1 function $V : M \to \Re$, such that $V = W + P$.

Proof: Substituting $p = W_x^T$ and $u = \alpha(x)$ in (9.41), we get

$$K_\gamma(x, W_x^T(x) + P_x^T, \alpha(x), w) \le \bar{K}_\gamma(x, W_x^T(x), \alpha(x), \bar{w}) \le 0 \ \forall \bar{w} \in \mathcal{W} \tag{9.43}$$

$$\implies \bar{K}_\gamma(x, V_x^T(x), \alpha(x), w) \le 0 \ \forall w \in \mathcal{W}. \tag{9.44}$$

Moreover, $V(0) = 0$ also, and a solution to the HJIE for the *SFBHICP* in Chapter 5. $\square$

Remark 9.4.1 *Again as a consequence of Theorem (9.4.1), we can consider the SFBSNLHICP for the auxiliary system $\bar{\Sigma}_{agg}^a$ instead of that of Σ_{agg}^a. The benefit of solving the former is that the penalty variable $\bar{z}$ has lower dimension than z. This implies that, while the $s_2 \times p$ matrix $L_{g_2}\bar{h}(x)$ may have full-rank, the $s \times p$ matrix $L_{g_2}h(x)$ is always rank deficient. Consequently, the former can be strongly input-output decoupled by the feedback*

$$u = L_{g_2}^{-1}\bar{h}(x)(v - L_f\bar{h}(x)).$$

This feature will be used in solving the SFBSNLHICP.

The following theorem then gives a solution to the *SFBSNLHICP* for the system Σ_{agg}^a.

Theorem 9.4.2 *Consider the system $\bar{\Sigma}_{agg}^a$ and the SFBSNLHICP. Suppose $s_2 = p$ and the $s_2 \times s_2$ matrix $L_{g_2}\bar{h}(x)$ is invertible for all $x \in M$. Let*

$$C(x) = I + \frac{s_2}{4\gamma^2}D(x),$$

where

$$D(x) = diag\left\{\|L_{g_1}^T\bar{h}_1(x)\|^2, \dots, \|L_{g_1}^T\bar{h}_{s_2}(x)\|^2\right\}.$$

Then, the state-feedback

$$u = \alpha(x) = -L_{g_2}\bar{h}(x)[L_{\bar{f}}\bar{h}(x) + C(x)\bar{h}(x)] \tag{9.45}$$

renders the differential dissipation-inequality

$$\bar{K}_\gamma(x, W_x^T(x), \alpha(x), \bar{w}) \le 0, \quad W(0) = 0, \ \forall \bar{w} \in \mathcal{W}$$

satisfied for

$$W(x) = \frac{1}{4}\bar{h}^T(x)\bar{h}(x).$$

Consequently, by Theorem 9.4.1, the same feedback control also solves the SFBSNLHICP for the system Σ_{agg}^a with

$$V(x) = \frac{1}{4}\bar{h}^T(x)\bar{h}(x) + P(x)$$

satisfying

$$K_\gamma(x, V_x^T(x), \alpha(x), w) \le 0, \quad V(0) = 0 \ \forall w \in \mathcal{W}.$$

Proof: Consider the closed-loop system (9.39), (9.45). Then using completion of squares (see also [199]), we have

$$
\begin{aligned}
\bar{K}_\gamma(x, W_x^T(x), \alpha(x), \bar{w}) &= W_x(x)[\bar{f}(x) + g_1(x)\bar{w} + g_2(x)\alpha(x)] - \frac{1}{2}\gamma^2\|\bar{w}\|^2 + \frac{1}{2}\|\bar{z}\|^2 \\
&= \frac{1}{2}\bar{h}^T(x)[L_{\bar{f}}\bar{h}(x) + L_{g_1}\bar{h}(x)\bar{w} + L_{g_2}\bar{h}(x)\alpha(x)] - \frac{1}{2}\gamma^2\|\bar{w}\|^2 + \frac{1}{2}\|\bar{z}\|^2 \\
&= \frac{1}{2}\bar{h}^T(x)[-C(x)\bar{h}(x) + L_{g_1}\bar{h}(x)\bar{w}] - \frac{1}{2}\gamma^2\|\bar{w}\|^2 + \frac{1}{2}\|\bar{z}\|^2 \\
&= -\frac{1}{2}\frac{s_2}{4\gamma^2}\bar{h}^T(x)D(x)\bar{h}(x) + \frac{1}{2}\bar{h}^T(x)L_{g_1}\bar{h}(x)\bar{w} - \frac{1}{2}\gamma^2\|\bar{w}\|^2 \\
&= -\frac{1}{2}\frac{s_2}{\gamma^2}\sum_{i=1}^{s_2}\left[\frac{1}{4}\bar{h}_i^2(x)\|L_{g_1}^T\bar{h}(x)\|^2 - \frac{\gamma^2}{s_2}\bar{h}_i(x)L_{g_1}\bar{h}(x)\bar{w} + \frac{\gamma^4}{s_2^2}\|\bar{w}\|^2\right] \\
&= -\frac{1}{2}\frac{s_2}{\gamma^2}\sum_{i=1}^{s_2}\left[\left\|\frac{1}{2}\bar{h}_i(x)L_{g_1}^T\bar{h}(x) - \frac{\gamma^2}{s_2}\bar{w}\right\|^2\right] \le 0.
\end{aligned}
$$

Moreover, $W(0) = 0$, since $\bar{h}(0) = 0$, and the result follows. $\square$

We apply the results developed above to a couple of examples.

Example 9.4.1 *[88]. Consider the system Σ_{agg}^a, and suppose the subsystem Σ_1^a is passive (Chapter 3), i.e., it satisfies the KYP property:*

$$V_{1,x_1}(x_1)f_1(x_1) \le 0, \quad V_{1,x_1}(x_1)g_{21}(x_1) = h_1^T(x_1), \quad V_1(0) = 0$$

for some storage-function $V_1 \ge 0$. Let also

$$g_{11}(x_1) = cg_{21}(x_1), \quad c \in \Re, \quad and \quad \gamma > \sqrt{|c|}.$$

Then the function

$$P_1(x_1) = \frac{\gamma}{\sqrt{\gamma^2 - c^2}}V_1(x_1)$$

solves the HJI-inequality (9.37) and

$$z_2^\star = -\frac{\gamma}{\sqrt{\gamma^2 - c^2}}h_1(x_1).$$

Therefore,

$$
\bar{f}(x) = \begin{bmatrix} f_1(x_1) + g_{21}(x_1)\left(z_2 + \frac{c^2}{\gamma\sqrt{\gamma^2-c^2}}h_1(x_1)\right) \\ f_2(x) + \frac{c^2}{\gamma\sqrt{\gamma^2-c^2}}g_{12}(x)h_1(x_1) \end{bmatrix} \tag{9.46}
$$

$$
\bar{h}(x) = z_2 + \frac{\gamma}{\sqrt{\gamma^2 - c^2}}. \tag{9.47}
$$

Example 9.4.2 *[88]. Consider a rigid robot dynamics given by the following state-space model:*

$$
\Sigma_r^a : \begin{cases} \dot{x}_1 &= x_2 \\ \dot{x}_2 &= -\bar{M}^{-1}[C(x_1, x_2)x_2 + e(x_1)] + \bar{M}^{-1}(x_1)w + \bar{M}^{-1}(x_1)u \\ z &= \begin{bmatrix} x_1 \\ x_2 \end{bmatrix}, \end{cases} \tag{9.48}
$$

where $x_1, x_2 \in \Re^n$ represent joint positions and velocities, $\bar{M}(x_1)$ is the positive-definite inertia matrix, $C(x_1, x_2)x_2$ is the coriolis and centrifugal forces, while $e(x_1)$ is the gravity load.

The above model Σ_r^a of the robot can be viewed as a cascade system with a kinematic subsystem Σ_{r1}^a represented by the state x_1, and a dynamic subsystem represented by the state x_2. Moreover, $g_{11}(x_1) = 0$, $c = 0$, and applying the results of Theorem 9.4.2, we get

$$\bar{f}(x) = \left[\begin{array}{c} x_2 \\ -\bar{M}^{-1}(x_1)(C(x_1, x_2)x_2 + e(x_1)) \end{array} \right]$$

$$g_2(x) = g_1(x) = \left[\begin{array}{c} 0 \\ M^{-1}(x_1) \end{array} \right], \quad \bar{h}(x) = x_1 + x_2.$$

Furthermore, $s_2 = m = n$ and $L_{g_2}\bar{h}(x) = \bar{M}^{-1}(x_1) > 0$ is nonsingular. Thus,

$$L_f \bar{h}(x) = x_2 - \bar{M}^{-1}(x_1)(C(x_1, x_2)x_2 + e(x_1)),$$

$$L_{g_1} \bar{h}(x) = \bar{M}^{-1}(x_1),$$

and

$$D(x_1) = diag\{\|\bar{M}_{.1}^{-1}(x_1)\|^2, \dots, \|\bar{M}_{.m}^{-1}(x_1)\|^2\}.$$

Therefore, by Theorem 9.4.2, the state-feedback

$$\begin{aligned} u &= -\bar{M}(x_1)x_2 + C(x_1, x_2)x_2 + e(x_1) - \bar{M}(x_1)\left(I + \frac{n}{4\gamma^2}D(x_1)\right)(x_1 + x_2) \\ &= -k_1(x_1)x_1 - k_2(x_1, x_2)x_2 + e(x_1) \end{aligned}$$

where the feedback gain matrices k_1, k_2 are given by

$$k_1(x_1) = \bar{M}(x_1)\left(I + \frac{n}{4\gamma^2}D(x_1)\right),$$

$$k_2(x_1, x_2) = -C(x_1, x_2) + \bar{M}(x_1)\left(2I + \frac{n}{4\gamma^2}D(x_1)\right),$$

solves the SFBSNLHICP for the robot system. Moreover, the function

$$V(x) = \frac{1}{2}x_1^T x_1 + \frac{1}{4}(x_1 + x_2)^T(x_1 + x_2)$$

is positive-definite and proper. Consequently, the closed-loop system is globally asymptotically-stable at $x = 0$ with $w = 0$.

9.5 $\mathcal{H}_\infty$-Control for Singularly-Perturbed Nonlinear Systems

In this section, we discuss the state-feedback $\mathcal{H}_\infty$-control problem for nonlinear singularly-perturbed systems. Singularly-perturbed systems are those class of systems that are characterized by a discontinuous dependence of the system properties on a small perturbation parameter ε. They arise in many physical systems such as electrical power systems and electrical machines (e.g., an asynchronous generator, a dc motor, electrical converters), electronic systems (e.g., oscillators) mechanical systems (e.g., fighter aircrafts), biological systems (e.g., bacterial-yeast cultures, heart) and also economic systems with various competing sectors. This class of systems has multiple time-scales, namely, a "fast" and a "slow" dynamics. This makes their analysis and control more complicated than regular systems. Nevertheless, they have been studied extensively [157, 165]. The control problem for this class of systems is also closely related to the control problems that were discussed in the previous sections.

We consider the following affine class of singularly-perturbed systems defined on $\mathcal{X} \subset \Re^n$:

$$\Sigma_{sp}^a \; : \; \begin{cases} \dot{x}_1 &= f_1(x_1, x_2) + g_{11}(x_1, x_2)w + g_{21}(x_1, x_2)u; \;\; x(t_0) = x_0 \\ \varepsilon \dot{x}_2 &= f_2((x_1, x_2) + g_{12}(x_1, x_2)w + g_{22}(x_1, x_2)u; \\ z &= \begin{bmatrix} h_1(x_1, x_2) \\ u \end{bmatrix} \\ y &= x, \end{cases} \tag{9.49}$$

where $x_1 \in \Re^{n_1} \subset \mathcal{X}$ is the slow state, $x_2 \in \Re^{n_2} \subset \mathcal{X}$ is the fast state, $u \in \mathcal{U} \subset \mathcal{L}_2([t_0, T], \Re^p)$ is the control input, $w \in \mathcal{W} \subset \mathcal{L}_2([t_0, T], \Re^r)$ is the disturbance input, $z \in \Re^s$ is the controlled output, while $y \in \Re^m$ is the measured output and ε is a small parameter. The functions $f_i : \mathcal{X} \times \mathcal{X} \to V^\infty \mathcal{X}$, $g_{ij} : \mathcal{X} \times \mathcal{X} \to M^{n \times *}, i, j = 1, 2, * = r$ if $j = 1$ or $* = p$ if $j = 2$, and $h_1 : \mathcal{X} \to \Re^s$ are smooth C^∞ functions of $x = (x_1^T, x_2^T)^T$. We also assume that $f_1(0, 0) = 0$ and $h_1(0, 0) = 0$.

The problem is to find a static state-feedback control of the form

$$u = \beta(x), \;\; \beta(0) = 0, \tag{9.50}$$

such that the closed-loop system (9.49), (9.50) has locally $\mathcal{L}_2$-gain from w to z less than or equal to a given number $\gamma^\star > 0$ with closed-loop local asymptotic-stability. The following result gives a solution to this problem, and is similar to the result in Chapter 6 for the regular problem.

Proposition 9.5.1 *Consider the system (9.49) and the state-feedback $\mathcal{H}_\infty$-control problem (SFBNLHICP) for it. Assume the system is zero-state detectable, and suppose for some $\gamma > 0$ and each $\varepsilon > 0$, there exists a C^2 solution $V : \tilde{M} \to \Re, V \geq 0, \tilde{M} \subset \mathcal{X}$ to the HJIE*

$$V_{x_1}(x)f_1(x_1, x_2) + V_{x_2}(x)f_2(x) + \frac{1}{2}(V_{x_1}^T(x) \;\; \varepsilon^{-1}V_{x_2}^T(x)) \begin{pmatrix} S_{11}(x) & S_{12}(x) \\ S_{21}(x) & S_{22}(x) \end{pmatrix} \times$$

$$\begin{pmatrix} V_{x_1}(x) \\ \varepsilon^{-1}V_{x_2}(x) \end{pmatrix} + \frac{1}{2}h_1^T(x)h_1(x) = 0, \;\; V(0) = 0 \tag{9.51}$$

where

$$S_{ij}(x) = \gamma^{-2}g_{1i}(x)g_{1j}^T(x) - g_{2i}(x)g_{2j}^T(x), \;\; i, j = 1, 2.$$

Then the state-feedback control

$$u^\star = -[g_{21}^T(x) \;\; \varepsilon^{-1}g_{22}^T(x)]V_x^T(x) \tag{9.52}$$

solves the SFBNLHICP for the system on $\tilde{M}$, i.e., the closed-loop system (9.49), (9.52) has $\mathcal{L}_2$-gain from w to z less than or equal to γ and is asymptotically-stable with $w = 0$.

Proof: Proof follows along similar lines as in Chapter 5 with the slight modification for the presence of ε. $\square$

The controller constructed above depends on ε which may present some computational difficulties. A composite controller that does not depend on ε can be constructed as an asymptotic approximation to the above controller as $\varepsilon \to 0$. To proceed, define the Hamiltonian system corresponding to the *SFBNLHICP* for the system as defined in Section 5.1:

$$
\left.
\begin{array}{ccl}
\dot{x}_1 & = & \frac{\partial H_\gamma}{\partial p_1} = F_1(x_1, p_1, x_2, p_2) \\
\dot{p}_1 & = & -\frac{\partial H_\gamma}{\partial x_1} = F_2(x_1, p_1, x_2, p_2) \\
\varepsilon \dot{x}_2 & = & \frac{\partial H_\gamma}{\partial p_2} = F_3(x_1, p_1, x_2, p_2) \\
\varepsilon \dot{p}_2 & = & -\frac{\partial H_\gamma}{\partial x_2} = F_4(x_1, p_1, x_2, p_2),
\end{array}
\right\}
\tag{9.53}
$$

with Hamiltonian function $H_\gamma : T^*\mathcal{X} \to \Re$ defined by

$$
\begin{aligned}
H_\gamma(x_1, x_2, p_1, p_2) & = & p_1^T f_1(x_1, x_2) + p_2^T f_2(x) + \\
& & \frac{1}{2}(p_1^T \; p_2^T) \begin{pmatrix} S_{11}(x) & S_{12}(x) \\ S_{21}(x) & S_{22}(x) \end{pmatrix} \begin{pmatrix} p_1 \\ p_2 \end{pmatrix} + \frac{1}{2} h_1^T(x) h_1(x).
\end{aligned}
$$

Let now $\varepsilon \to 0$ in the above equations (9.53), and consider the resulting algebraic equations:

$$
F_3(x_1, p_1, x_2, p_2) = 0, \quad F_4(x_1, p_1, x_2, p_2) = 0.
$$

If we assume that H_γ is nondegenerate at $(x, p) = (0, 0)$, then by the Implicit-function Theorem [157] , there exist nontrivial solutions to the above equations

$$
x_2 = \phi(x_1, p_1), \quad p_2 = \psi(x_1, x_2).
$$

Substituting these solutions in (9.53) results in the following reduced Hamiltonian system:

$$
\left\{
\begin{array}{ccl}
\dot{x}_1 & = & F_1(x_1, p_1, \phi(x_1, p_1), \psi(x_1, p_1)) \\
\dot{p}_1 & = & F_2(x_1, p_1, \phi(x_1, p_1), \psi(x_1, p_1)).
\end{array}
\right.
\tag{9.54}
$$

To be able to analyze the asymptotic behavior of the above system and its invariant-manifold, we consider its linearization of (9.49) about $x = 0$. Let $A_{ij} = \frac{\partial f_i}{\partial x_j}(0, 0)$, $B_{ij} = g_{ij}(0, 0)$, $C_i = \frac{\partial h_1}{\partial x_j}(0)$, $i, j = 1, 2$, so that we have the following linearization

$$
\Sigma_{sp}^l : \left\{
\begin{array}{ccl}
\dot{x}_1 & = & A_{11}x_1 + A_{12}x_2 + B_{11}w + B_{21}u; \quad x(t_0) = x_0 \\
\varepsilon \dot{x}_2 & = & A_{21}x_1 + A_{22}x_2 + B_{12}w + B_{22}u; \\
z & = & \begin{bmatrix} C_1 x_1 + C_2 x_2 \\ u \end{bmatrix} \\
y & = & x.
\end{array}
\right.
\tag{9.55}
$$

Similarly, the linear Hamiltonian system corresponding to this linearization is given by

$$
\begin{bmatrix} \dot{x} \\ \dot{p} \end{bmatrix} = \bar{H}_\gamma \begin{bmatrix} x \\ p \end{bmatrix},
\tag{9.56}
$$

where $\bar{H}_\gamma$ is the linear Hamiltonian matrix corresponding to (9.55) which is defined as

$$
\bar{H}_\gamma = \begin{bmatrix} H_{11} & H_{12} \\ \varepsilon^{-1} H_{21} & \varepsilon^{-1} H_{22} \end{bmatrix},
$$

and H_{ij} are the sub-Hamiltonian matrices:

$$
H_{ij} = \begin{bmatrix} A_{ij} & -S_{ij}(0) \\ -C_i^T C_i & -A_{ji}^T \end{bmatrix},
$$

with $S_{ij} = \gamma^{-2}B_{1i}B_{1j} - B_{2i}B_{2j}$.

The Riccati equations corresponding to the fast and slow dynamics are respectively given by

$$A_{22}^T X_f + X_f A_{22} - X_f S_{22}(0)X_f + C_2^T C_2 = 0, \tag{9.57}$$
$$A_0^T X_s + X_s A_0 - X_0 S_0 X_0 + Q_0 = 0, \tag{9.58}$$

where

$$\begin{bmatrix} A_0 & -S_0 \\ -Q_0 & -A_0^T \end{bmatrix} = H_{11} - H_{12}H_{22}^{-1}H_{21}.$$

Then it is wellknown [216] that, if the system (9.55) is stabilizable, detectable and does not have invariant-zeros on the imaginary axis (this holds if the Hamiltonian matrix $\bar{H}_\gamma$ is hyperbolic [194]), then the $\mathcal{H}_\infty$-control problem for the system is solvable for all small ε. In fact, the control

$$u_{cl} = -B_{21}^T X_s x_1 - B_{22}^T (X_f x_2 + X_c x_1), \tag{9.59}$$

$$X_c = [X_f \quad -I]H_{22}^{-1}H_{21}\begin{bmatrix} I \\ X_s \end{bmatrix}$$

which is ε-independent, is one such state-feedback. It is also well known that this feedback controller also stabilizes the nonlinear system (9.49) locally about $x = 0$. The question is: how big is the domain of validity of the controller, and how far can it be extended for the nonlinear system (9.49)?

To answer the above question, we make the following assumptions.

Assumption 9.5.1 *For a given $\gamma > 0$, the ARE (9.57) has a symmetric solution $X_f \geq 0$ such that $A_{cf} = A_{22} - S_{22}X_f$ is Hurwitz.*

Assumption 9.5.2 *For a given $\gamma > 0$, the ARE (9.58) has a symmetric solution $X_s \geq 0$ such that $A_{cs} = A_0 - S_0 X_s$ is Hurwitz.*

Then, under Assumption 9.5.2 and the theory of ordinary differential-equations, the system (9.54) has a stable invariant-manifold

$$M_s^{*^-} = \{x_1, p_1 = \sigma_s(x_1) = X_s x_1 + O(\|x_1\|^2)\}, \tag{9.60}$$

and the dynamics of the system (which is asymptotically-stable) restricted to this manifold is given by

$$\dot{x}_1 = F_1(x_1, \sigma_s(x_1), \phi(x_1, \sigma_s(x_1)), \psi(x_1, \sigma_s(x_1))), \tag{9.61}$$

x_1 in a small neighborhood of $x = 0$. Moreover, from (9.54), the function σ satisfies the partial-differential equation (PDE)

$$\frac{\partial \sigma_s}{\partial x_1} F_1(x_1, \sigma_s(x_1), \phi(x_1, \sigma_s(x_1)), \psi(x_1, \sigma_s(x_1))) = F_2(x_1, \sigma_s(x_1), \phi(x_1, \sigma_s(x_1)),$$

$$\psi(x_1, \sigma_s(x_1))) \tag{9.62}$$

which could be solved approximately using power-series expansion [193]. Such a solution can be represented as

$$\sigma_s(x_1) = X_s x_1 + \sigma_{s2}(x_1) + \sigma_{s3}(x_1) + \ldots \tag{9.63}$$

where $\sigma_{si}(.), i = 2, \ldots,$ are the higher-order terms, and $\sigma_{s1} = X_s$. In fact, substituting

(9.63) in (9.62) and equating terms of the same powers in x_1, the higher-order terms could be determined recursively.

Now, substituting the solution (9.60) in the fast dynamic subsystem, i.e., in (9.53), we get for all x_1 such that $\tilde{M}$ exists,

$$\dot{x}_2 = \bar{F}_3(x_1, x_2, p_2), \quad \dot{p}_2 = \bar{F}_4(x_1, x_2, p_2),$$

where

$$\bar{F}_3(x_1, x_2, p_2) = F_3(x_1, \sigma_s(x_1), x_2 + \phi(x_1, \sigma_s(x_1), p_2 + \sigma_s(x_1))),$$
$$\bar{F}_4(x_1, x_2, p_2) = F_4(x_1, \sigma_s(x_1), x_2 + \phi(x_1, \sigma_s(x_1), p_2 + \sigma_s(x_1))).$$

The above fast Hamiltonian subsystem has a stable invariant-manifold

$$M_f^{*^-} = \{x_2, p_2 = \vartheta_f(x_1, x_2)\}$$

with asymptotically-stable dynamics

$$\dot{x}_2 = \bar{F}_3(x_1, \sigma_s(x_1), x_2, \vartheta_f(x_1, x_2))$$

for all $x_1, x_2 \in \Omega$ neighborhood of $x = 0$. The function ϑ_f similarly has a representation about $x = 0$ as

$$\vartheta_f(x_1, x_2) = X_f x_2 + O((\|x_1\| + \|x_2\|)\|x_2\|)$$

and satisfies a similar PDE:

$$\frac{\partial \vartheta}{\partial x_2} \bar{F}_3(x_1, x_2, \vartheta_f(x_1, x_2)) = \bar{F}_4(x_1, x_2, \vartheta_f(x_1, x_2)).$$

Consequently, $\vartheta_f(.,.)$ has the following power-series expansion in powers of x_2

$$\vartheta_f(x_1, x_2)) = X_f x_2 + \vartheta_{f2}(x_1, x_2) + \vartheta_{f3}(x_1, x_2) + \dots$$

where $\vartheta_{fi}, i - 2, 3, \dots$ denote the higher-order terms.

Finally, we can define the following composite controller

$$u_c = -g_{21}^T(x)\sigma_s(x_1) - g_{22}^T(x)[\psi(x_1, \sigma_s(x_1)) + \vartheta(x_1, x_2 - \phi(x_1, \sigma_s(x_1)))] \tag{9.64}$$

which is related to the linear composite controller (9.59) in the following way:

$$u_c = u_{cl} + O(\|x_1\| + \|x_2\|).$$

Thus, (9.64) solves the nonlinear problem locally in $\tilde{M}$. Summing up the above analysis, we have the following theorem.

Theorem 9.5.1 *Consider the system (9.49) and the SFBNLHICP for it. Suppose Assumptions 9.5.1, 9.5.2 hold. Then there exist indices $m_1, m_2 \in \mathbf{Z}_+$ and an $\varepsilon_0 > 0$ such that, for all $\varepsilon \in [0, \varepsilon_0)$ the following hold:*

(i) *there exists a C^2 solution $V : \Omega_{m_1} \times \Omega_{m_2} \to \Re_+$, $\Omega_{m_1} \times \Omega_{m_2} \subset \mathcal{X}$ to the HJIE (9.51) whose approximation is given by*

$$V(x_1, x_2) = V_0(x_1) + O(\varepsilon)$$

for some function $V_0 = \int \sigma_s(x_1)dx_1$. Consequently, the control law (9.52) also has the approximation:

$$u = u_c + O(\varepsilon);$$

(ii) *the SFBNLHICP for the system is locally solvable on $\Omega_{m_1} \times \Omega_{m_2}$ by the composite control (9.64);*

(iii) *the SFBNLHICP for the system is locally solvable on $\Omega_{m_1} \times \Omega_{m_2}$ by the linear composite control (9.59).*

Proof: (i) The existence of a C^2 solution to the HJIE (9.51) follows from linearization. Indeed, $V_0 \approx x^T X_s x \geq 0$ is C^2.
(ii) Consider the closed-loop system (9.49), (9.64). We have to prove that Assumptions 9.5.1, 9.5.2 hold for the linearized system:

$$\dot{x}_1 = \tilde{A}_{11}x_1 + \tilde{A}_{12}x_2 + B_{11}w, \quad \dot{x}_2 = \tilde{A}_{21}x_1 + \tilde{A}_{22}x_2 + B_{12}w, \quad z = [\tilde{C}_1 \ \tilde{C}_2]x,$$

where $\tilde{C}_1 = [C_1 \ B_{21}^T X_s - B_{22}^T X_c]$, $\tilde{C}_2 = [C_2, -B_{22}^T X_f]$. This system corresponds to the closed-loop system (9.55), (9.59) whose Hamiltonian system is also given similarly by (9.56) with Hamiltonian matrix $\tilde{H}_\gamma$, and where all matrices are replaced by "tilde" superscript, with $\tilde{S}_{ij} = -\gamma^2 B_{1i}B_{1j}$, $i,j = 1,2$.

It should be noted that Assumptions 9.5.1, 9.5.2 imply that the fast and reduced subsystems have stable invariant-manifolds. Setting $\varepsilon = 0$ and substituting $p_1 = X_s x_1$, $p_2 = X_c x_1 + X_f x_2$ leads to the Hamiltonian system (9.56) with $\varepsilon = 0$. Therefore, $p_2 = X_f x_2$ and $p_1 = X_s x_1$ are the desired stable manifolds.
(iii) This proof is similar to (ii). $\square$

9.6   Notes and Bibliography

The results of Sections 9.1-9.2 are based on the References [42, 194]. A detailed proof of Theorem 9.2.1 can be found in [194]. Moreover, an alternative approach to the $MFBSNLHICP$ can be found in [176]. In addition, the full-information problem for a class of nonlinear systems is discussed in [43].

The material of Section 9.4 is based on the Reference [88]. Application of singular $\mathcal{H}_\infty$-control to the control of a rigid spacecraft can also be found in the same reference.

The results presented in Section 9.5 are based on References [104]-[106]. Results on descriptor systems can also be found in the same references. In addition, robust control of nonlinear singularly-perturbed systems is discussed in Reference [218].

Finally, the results of Section 9.3 are based on [27].

10

$\mathcal{H}_\infty$-Filtering for Singularly-Perturbed Nonlinear Systems

In this chapter, we discuss the counterpart filtering results for continuous-time singularly-perturbed affine nonlinear systems. The linear $\mathcal{H}_\infty$ filtering problem has been considered by a number of authors [178, 246], and the nonlinear problem for fuzzy T-S models has also been considered by a number of authors [39, 40, 41, 283]. This represents an interpolation of a number of linear models for the aggregate nonlinear system. Thus, the filter equations can be solved using linear-matrix inequalities (LMI) which makes the approach computationally attractive.

However, the general affine nonlinear filtering problem has only been recently considered by the authors [25, 26]. The results we present in this chapter represent a generalization of the results of Chapter 9. But in addition, various classes of filter configuration are possible in this case, ranging from decomposition to aggregate and to reduced-order filters.

10.1 Problem Definition and Preliminaries

We consider the following affine nonlinear causal state-space model of the plant which is defined on $\mathcal{X} \subseteq \Re^{n_1+n_2}$ with zero control input:

$$\mathbf{P}_{sp}^a : \begin{cases} \dot{x}_1 &= f_1(x_1, x_2) + g_{11}(x_1, x_2)w; \quad x_1(t_0, \varepsilon) = x_{10} \\ \varepsilon\dot{x}_2 &= f_2(x_1, x_2) + g_{21}(x_1, x_2)w; \quad x_2(t_0, \varepsilon) = x_{20} \\ y &= h_{21}(x_1) + h_{22}(x_2) + k_{21}(x_1, x_2)w \end{cases} \tag{10.1}$$

where $x = \begin{pmatrix} x_1 \\ x_2 \end{pmatrix} \in \mathcal{X}$ is the state vector with x_1 the slow state which is n_1-dimensional, and x_2 the fast, which is n_2-dimensional; $w \in \mathcal{W} \subseteq \mathcal{L}_2([t_0, \infty), \Re^r)$ is an unknown disturbance (or noise) signal, which belongs to the set $\mathcal{W}$ of admissible exogenous inputs; $y \in \mathcal{Y} \subset \Re^m$ is the measured output (or observation) of the system, and belongs to $\mathcal{Y}$, the set of admissible measured-outputs; while $\varepsilon > 0$ is a small perturbation parameter.

The functions $\begin{pmatrix} f_1 \\ f_2 \end{pmatrix} : \mathcal{X} \to V^\infty(\mathcal{X}) \subseteq \Re^{2(n_1+n_2)}$, $g_{11} : \mathcal{X} \to \mathcal{M}^{n_1 \times r}(\mathcal{X})$, $g_{21} : \mathcal{X} \to \mathcal{M}^{n_2 \times r}(\mathcal{X})$, where $\mathcal{M}^{i \times j}$ is the ring of $i \times j$ smooth matrices over $\mathcal{X}$, $h_{21}, h_{22} : \mathcal{X} \to \Re^m$, and $k_{21} : \mathcal{X} \to \mathcal{M}^{m \times r}(\mathcal{X})$ are real C^∞ functions of x. Furthermore, we assume without any loss of generality that the system (10.1) has an isolated equilibrium-point at $(x_1, x_2) = (0, 0)$ and such that $f_1(0, 0) = 0$, $f_2(0, 0) = 0$, $h_{21}(0, 0) = h_{22}(0, 0) = 0$. We also assume that there exists a unique solution $x(t, t_0, x_0, w, \varepsilon) \; \forall t \in \Re$ for the system for all initial conditions x_0, for all $w \in \mathcal{W}$, and all $\varepsilon \in \Re$.

In addition, to guarantee local asymptotic-stability of the system (10.1) with $w = 0$, we assume that (10.1) satisfies the conditions of Theorem 8.2, [157], i.e., there exists an $\varepsilon^\star > 0$ such that (10.1) is locally asymptotically-stable about $x = 0$ for all $\varepsilon \in [0, \varepsilon^\star)$.

The $\mathcal{H}_\infty$-suboptimal local filtering/state estimation problem is redefined as in Chapter 8.

Definition 10.1.1 *(Nonlinear $\mathcal{H}_\infty$ (Suboptimal) Filtering Problem (NLHIFP)). Find a filter, $\mathbf{F}$, for estimating the state $x(t)$ from observations $Y_t \triangleq \{y(\tau) : \tau \leq t\}$ of $y(\tau)$ up to time t, to obtain the estimate*

$$\hat{x}(t) = \mathbf{F}(Y_t),$$

such that the $\mathcal{L}_2$-gain from the input w to some suitable penalty variable $\tilde{z}$ is rendered less or equal to a given number $\gamma > 0$, i.e.,

$$\int_{t_0}^\infty \|\tilde{z}(\tau)\|^2 dt \leq \gamma^2 \int_{t_0}^\infty \|w(\tau)\|^2 dt, \quad \forall w \in \mathcal{W}, \tag{10.2}$$

for all initial conditions $x_0 \in \mathcal{O} \subset \mathcal{X}$. In addition, with $w \equiv 0$, we also have $\lim_{t \to \infty} \tilde{z} = 0$. Moreover, if the filter solves the problem for all $x_0 \in \mathcal{X}$, we say the problem is solved globally.

We shall also adopt the following definition of zero-input observability.

Definition 10.1.2 *For the nonlinear system $\mathbf{P}_{sp}^a$, we say that it is locally zero-input observable, if for all states x_1, $x_2 \in U \subset \mathcal{X}$ and input $w(.) \equiv 0$*

$$y(t; x_1, w) \equiv y(t; x_2, w) \implies x_1 = x_2,$$

where $y(., x_i, w), i = 1, 2$ is the output of the system with the initial condition $x(t_0) = x_i$. Moreover, the system is said to be zero-input observable, if it is locally observable at each $x_0 \in \mathcal{X}$ or $U = \mathcal{X}$.

10.2 Decomposition Filters

In this section, we present a decomposition approach to the $\mathcal{H}_\infty$ state estimation problem. As in the linear case [79, 178], we assume that there exists locally a smooth invertible coordinate transformation (a diffeomorphism)

$$\xi_1 = \varphi_1(x), \quad \varphi_1(0) = 0, \quad \xi_2 = \varphi_2(x), \quad \varphi_2(0) = 0, \quad \xi_1 \in \Re^{n_1}, \xi_2 \in \Re^{n_2}, \tag{10.3}$$

such that the system (10.1) can be decomposed into the form

$$\tilde{\mathbf{P}}_{sp}^a : \begin{cases} \dot{\xi}_1 &= \tilde{f}_1(\xi_1) + \tilde{g}_{11}(\xi)w; \quad \xi_1(t_0) = \varphi_1(x_0) \\ \varepsilon\dot{\xi}_2 &= \tilde{f}_2(\xi_2) + \tilde{g}_{21}(\xi)w; \quad \xi_2(t_0) = \varphi_2(x_0) \\ y &= \tilde{h}_{21}(\xi_1) + \tilde{h}_{22}(\xi_2) + \tilde{k}_{21}(\xi)w. \end{cases} \tag{10.4}$$

The necessary conditions that guarantee the existence of such a transformation are given in [26]. Subsequently, we can proceed to design the filter based on this transformed model (10.4) with the systems states partially decoupled. Accordingly, we propose the following composite filter

$$\mathbf{F}_{1c}^a : \begin{cases} \dot{\hat{\xi}}_1 &= \tilde{f}_1(\hat{\xi}_1) + \tilde{g}_{11}(\hat{\xi})\hat{w}^\star + \hat{L}_1(\hat{\xi}, y)(y - \tilde{h}_{21}(\hat{\xi}_1) - \tilde{h}_{22}(\hat{\xi}_2)); \\ &\quad \hat{\xi}_1(t_0) = \varphi_1(0) \\ \varepsilon\dot{\hat{\xi}}_2 &= \tilde{f}_2(\hat{\xi}_2) + \tilde{g}_{21}(\hat{\xi})\hat{w}^\star + \hat{L}_2(\hat{\xi}, y)(y - \tilde{h}_{21}(\hat{\xi}_1) - \tilde{h}_{22}(\hat{\xi}_2)); \\ &\quad \hat{\xi}_2(t_0) = \varphi_2(0). \end{cases} \tag{10.5}$$

where $\hat{\xi} \in \mathcal{X}$ is the filter state, $\hat{w}^\star$ is the certainty-equivalent worst-case noise, $\hat{L}_1 \in \Re^{n_1 \times m}$, $\hat{L}_2 \in \Re^{n_2 \times m}$ are the filter gains, while all the other variables have their corresponding previous meanings and dimensions. We can then define the penalty variable or estimation error as

$$z = y - \tilde{h}_{21}(\hat{\xi}_1) - \tilde{h}_{22}(\hat{\xi}_2). \tag{10.6}$$

Similarly, the problem can then be formulated as a zero-sum differential game with the following cost functional (Chapter 2):

$$\min_{\substack{\hat{L}_1 \in \Re^{n_1 \times m}, \\ \hat{L}_2 \in \Re^{n_2 \times m}}} \sup_{w \in \mathcal{W}} \hat{J}_1(\hat{L}_1, \hat{L}_2, w) = \frac{1}{2}\int_{t_0}^{\infty}[\|z\|^2 - \gamma^2\|w\|^2]dt,$$

$$s.t.\ (10.5)\quad \text{and with } w \equiv 0, \lim_{t \to \infty}\{\hat{\xi}(t) - \xi(t)\} = 0. \tag{10.7}$$

A pair of strategies $(\hat{L}^\star, w^\star)$, $\hat{L}^\star = [\hat{L}_1^\star, \hat{L}_2^\star]$ is said to constitute a *"saddle-point solution"* to the above game, if the following conditions are satisfied [57]:

$$\hat{J}_1(\hat{L}^\star, w) \le \hat{J}_1(\hat{L}^\star, w^\star) \le \hat{J}_1(\hat{L}, w^\star); \quad \forall \hat{L}_1 \in \Re^{n_1 \times m}, \hat{L}_2 \in \Re^{n_2 \times m}, \forall w \in \mathcal{W}. \tag{10.8}$$

To proceed to find the above saddle-points, we form the Hamiltonian function $\hat{H}$: $T^\star\mathcal{X} \times T^\star\mathcal{Y} \times \mathcal{W} \times \Re^{n_1 \times m} \times \Re^{n_2 \times m} \to \Re$ corresponding to the above dynamic game:

$$
\begin{aligned}
\hat{H}(\hat{\xi}, y, w, \hat{L}_1, \hat{L}_2, \hat{V}_{\hat{\xi}}^T, \hat{V}_y^T) &= \hat{V}_{\hat{\xi}_1}(\hat{\xi}, y)[\tilde{f}_1(\hat{\xi}_1) + \tilde{g}_{11}(\hat{\xi})w + \hat{L}_1(\hat{\xi}, y)(y - \tilde{h}_{21}(\hat{\xi}_1) - \\
&\quad h_{22}(\hat{\xi}_2))] + \frac{1}{\varepsilon}\hat{V}_{\hat{\xi}_2}(\hat{\xi}, y)[\tilde{f}_2(\hat{\xi}_2) + \tilde{g}_{21}(\xi)w + \\
&\quad \hat{L}_2(\hat{\xi}, y)(y - \tilde{h}_{21}(\hat{\xi}_1) - h_{22}(\hat{\xi}_2))] + \hat{V}_y(\hat{\xi}, y)\dot{y} + \\
&\quad \frac{1}{2}(\|z\|^2 - \gamma^2\|w\|^2) \tag{10.9}
\end{aligned}
$$

for some C^1 function $\hat{V} : \mathcal{X} \times \mathcal{Y} \to \Re$. Then, applying the necessary condition for the worst-case noise, we have

$$\left.\frac{\partial \hat{H}}{\partial w}\right|_{w=\hat{w}^\star} = 0 \implies \hat{w}^\star = \frac{1}{\gamma^2}[\tilde{g}_{11}^T(\hat{\xi})\hat{V}_{\hat{\xi}_1}^T(\hat{\xi}, y) + \frac{1}{\varepsilon}\tilde{g}_{21}^T(\hat{\xi})\hat{V}_{\hat{\xi}_2}^T(\hat{\xi}, y)]. \tag{10.10}$$

Moreover, we note that

$$\frac{\partial^2 \hat{H}}{\partial w^2} = -\gamma^2 I \implies \hat{H}(\hat{\xi}, y, w, \hat{L}_1, \hat{L}_2, \hat{V}_{\hat{\xi}}^T, \hat{V}_y^T) \le \hat{H}(\hat{\xi}, y, \hat{w}^\star, \hat{L}_1, \hat{L}_2, \hat{V}_{\hat{\xi}}^T, \hat{V}_y^T)\ \forall w \in \mathcal{W}.$$

Substituting now $\hat{w}^\star$ in (10.9) and completing the squares for $\hat{L}_1$ and $\hat{L}_2$, we have

$$
\begin{aligned}
\hat{H}(\hat{\xi}, y, \hat{w}^\star, \hat{L}_1, \hat{L}_2, \hat{V}_{\hat{\xi}}^T, \hat{V}_y^T) &= \hat{V}_{\hat{\xi}_1}(\hat{\xi}, y)\tilde{f}_1(\hat{\xi}_1) + \frac{1}{\varepsilon}\hat{V}_{\hat{\xi}_2}(\hat{\xi}, y)\tilde{f}_2(\hat{\xi}_2) + \hat{V}_y(\hat{\xi}, y)\dot{y} + \\
&\quad \frac{1}{2\gamma^2}[\hat{V}_{\hat{\xi}_1}(\hat{\xi}, y)\tilde{g}_{11}(\hat{\xi})\tilde{g}_{11}^T(\hat{\xi})\hat{V}_{\hat{\xi}_1}^T(\hat{\xi}, y) + \frac{1}{\varepsilon}\hat{V}_{\hat{\xi}_1}(\hat{\xi}, y)\tilde{g}_{11}(\xi)\tilde{g}_{21}^T(\hat{\xi})\hat{V}_{\hat{\xi}_1}^T(\hat{\xi}, y) + \\
&\quad \frac{1}{\varepsilon}\hat{V}_{\hat{\xi}_2}(\hat{\xi}, y)\tilde{g}_{21}(\xi)\tilde{g}_{11}^T(\hat{\xi})\hat{V}_{\hat{\xi}_1}^T(\hat{\xi}, y) + \frac{1}{\varepsilon^2}\hat{V}_{\hat{\xi}_2}(\hat{\xi}, y)\tilde{g}_{21}(\xi)\tilde{g}_{21}^T(\hat{\xi})\hat{V}_{\hat{\xi}_2}^T(\hat{\xi}, y)] + \\
&\quad \frac{1}{2}\left\|\hat{L}_1^T(\hat{\xi}, y)\hat{V}_{\hat{\xi}_1}^T(\hat{\xi}, y) + (y - \tilde{h}_{21}(\hat{\xi}_1) - \tilde{h}_{22}(\hat{\xi}_2))\right\|^2 - \frac{1}{2}\|(y - \tilde{h}_{21}(\hat{\xi}_1) - \tilde{h}_{22}(\hat{\xi}_2))\|^2 - \\
&\quad \frac{1}{2}\hat{V}_{\hat{\xi}_1}(\hat{\xi}, y)\hat{L}_1(\hat{\xi}, y)\hat{L}_1^T(\hat{\xi}, y)V_{\hat{\xi}_1}^T(\hat{\xi}, y) + \frac{1}{2}\left\|\frac{1}{\varepsilon}\hat{L}_2^T(\hat{\xi}, y)\hat{V}_{\hat{\xi}_2}^T(\hat{\xi}, y) + (y - \tilde{h}_{21}(\hat{\xi}_1) - \tilde{h}_{22}(\hat{\xi}_2))\right\|^2 - \\
&\quad \frac{1}{2}\|(y - \tilde{h}_{21}(\hat{\xi}_1) - \tilde{h}_{22}(\hat{\xi}_2))\|^2 - \frac{1}{2\varepsilon^2}\hat{V}_{\hat{\xi}_2}(\hat{\xi}, y)\hat{L}_2(\hat{\xi}, y)\hat{L}_2^T(\hat{\xi}, y)\hat{V}_{\hat{\xi}_2}^T(\hat{\xi}, y) + \frac{1}{2}\|z\|^2.
\end{aligned}
$$

Therefore, setting the optimal gains $\hat{L}_1^\star(\hat{\xi}, y)$, $\hat{L}_2^\star(\hat{\xi}, y)$ as

$$\hat{V}_{\hat{\xi}_1}(\hat{\xi}, y)\hat{L}_1^\star(\hat{\xi}, y) = -(y - \tilde{h}_{21}(\hat{\xi}_1) - \tilde{h}_{22}(\hat{\xi}_2))^T \tag{10.11}$$

$$\frac{1}{\varepsilon}\hat{V}_{\hat{\xi}_2}(\hat{\xi}, y)\hat{L}_2^\star(\hat{\xi}, y) = -(y - \tilde{h}_{21}(\hat{\xi}_1) - \tilde{h}_{22}(\hat{\xi}_2))^T \tag{10.12}$$

minimizes the Hamiltonian (10.9) and implies that the saddle-point condition

$$\hat{H}(\hat{\xi}, y, \hat{w}^\star, \hat{L}_1^\star, \hat{L}_2^\star, \hat{V}_{\hat{\xi}}^T, \hat{V}_y^T) \leq \hat{H}(\hat{\xi}, y, \hat{w}^\star, \hat{L}_1, \hat{L}_2, \hat{V}_{\hat{\xi}}^T) \tag{10.13}$$

is satisfied. Finally, setting

$$\hat{H}(\hat{\xi}, y, \hat{w}^\star, \hat{L}_1^\star, \hat{L}_2^\star, \hat{V}_{\hat{\xi}}^T, \hat{V}_y^T) = 0$$

results in the following Hamilton-Jacobi-Isaacs equation (HJIE):

$$\hat{V}_{\hat{\xi}_1}(\hat{\xi}, y)\tilde{f}_1(\hat{\xi}_1) + \frac{1}{\varepsilon}\hat{V}_{\hat{\xi}_2}(\hat{\xi}, y)\tilde{f}_2(\hat{\xi}_2) + \hat{V}_y(\hat{\xi}, y)\dot{y} +$$

$$\frac{1}{2\gamma^2}[\hat{V}_{\hat{\xi}_1}(\hat{\xi}, y)\ \hat{V}_{\hat{\xi}_2}(\hat{\xi}, y)] \begin{bmatrix} \tilde{g}_{11}(\hat{\xi})\tilde{g}_{11}^T(\hat{\xi}) & \frac{1}{\varepsilon}\tilde{g}_{11}(\hat{\xi})\tilde{g}_{21}^T(\hat{\xi}) \\ \frac{1}{\varepsilon}\tilde{g}_{21}(\hat{\xi})\tilde{g}_{11}^T(\hat{\xi}) & \frac{1}{\varepsilon^2}\tilde{g}_{21}(\hat{\xi})\tilde{g}_{21}^T(\hat{\xi}) \end{bmatrix} \begin{bmatrix} \hat{V}_{\hat{\xi}_1}^T(\hat{\xi}, y) \\ \hat{V}_{\hat{\xi}_2}^T(\hat{\xi}, y) \end{bmatrix} -$$

$$\frac{1}{2}\hat{V}_{\hat{\xi}_1}(\hat{\xi}, y)\hat{L}_1(\hat{\xi}, y)\hat{L}_1^T(\hat{\xi}, y)\hat{V}_{\hat{\xi}_1}^T(\hat{\xi}, y) - \frac{1}{2\varepsilon^2}\hat{V}_{\hat{\xi}_2}(\hat{\xi}, y)\hat{L}_2(\hat{\xi}, y)\hat{L}_2^T(\hat{\xi}, y)\hat{V}_{\hat{\xi}_2}^T(\hat{\xi}, y) -$$

$$\frac{1}{2}(y - \tilde{h}_{21}(\hat{\xi}_1) - \tilde{h}_{22}(\hat{\xi}_2))^T(y - \tilde{h}_{21}(\hat{\xi}_1) - \tilde{h}_{22}(\hat{\xi}_2)) = 0, \quad \hat{V}(0,0) = 0, \tag{10.14}$$

or equivalently the HJIE

$$\hat{V}_{\hat{\xi}_1}(\hat{\xi}, y)\tilde{f}_1(\hat{\xi}_1) + \frac{1}{\varepsilon}\hat{V}_{\hat{\xi}_2}(\hat{\xi}, y)\tilde{f}_2(\hat{\xi}_2) + \hat{V}_y(\hat{\xi}, y)\dot{y} +$$

$$\frac{1}{2\gamma^2}[\hat{V}_{\hat{\xi}_1}(\hat{\xi}, y)\ \hat{V}_{\hat{\xi}_2}(\hat{\xi}, y)] \begin{bmatrix} \tilde{g}_{11}(\hat{\xi})\tilde{g}_{11}^T(\hat{\xi}) & \frac{1}{\varepsilon}\tilde{g}_{11}(\hat{\xi})\tilde{g}_{21}^T(\hat{\xi}) \\ \frac{1}{\varepsilon}\tilde{g}_{21}(\hat{\xi})\tilde{g}_{11}^T(\hat{\xi}) & \frac{1}{\varepsilon^2}\tilde{g}_{21}(\hat{\xi})\tilde{g}_{21}^T(\hat{\xi}) \end{bmatrix} \begin{bmatrix} \hat{V}_{\hat{\xi}_1}^T(\hat{\xi}, y) \\ \hat{V}_{\hat{\xi}_2}^T(\hat{\xi}, y) \end{bmatrix} -$$

$$\frac{3}{2}(y - \tilde{h}_{21}(\hat{\xi}_1) - \tilde{h}_{22}(\hat{\xi}_2))^T(y - \tilde{h}_{21}(\hat{\xi}_1) - \tilde{h}_{22}(\hat{\xi}_2)) = 0, \quad \hat{V}(0,0) = 0. \tag{10.15}$$

In addition, from (10.9), we have

$$\hat{H}(\hat{\xi}, y, w, \hat{L}_1^\star, \hat{L}_2^\star, \hat{V}_{\hat{\xi}}^T, \hat{V}_y^T) = \hat{V}_{\hat{\xi}_1}(\hat{\xi}, y)\tilde{f}_1(\hat{\xi}_1) + \frac{1}{\varepsilon}\hat{V}_{\hat{\xi}_2}(\hat{\xi}, y)\tilde{f}_2(\hat{\xi}_2) + \hat{V}_y(\hat{\xi}, y)\dot{y} -$$

$$\frac{\gamma^2}{2}\|w - w^\star\|^2 + \frac{\gamma^2}{2}\|w^\star\|^2 - \frac{3}{2}\|z\|^2$$

$$= \hat{H}(\hat{\xi}, y, w^\star, \hat{L}_1^\star, \hat{L}_2^\star, \hat{V}_{\hat{\xi}}^T, \hat{V}_y^T) - \frac{\gamma^2}{2}\|w - w^\star\|^2.$$

Therefore,

$$\hat{H}(\hat{\xi}, y, w, \hat{L}_1^\star, \hat{L}_2^\star, \hat{V}_{\hat{\xi}}^T, \hat{V}_y^T) \leq \hat{H}(\hat{\xi}, y, w^\star, \hat{L}_1^\star, \hat{L}_2^\star, \hat{V}_{\hat{\xi}}^T, \hat{V}_y^T). \tag{10.16}$$

Combining now (10.13) and (10.16), we have that the saddle-point conditions (10.8) are satisfied and the pair $([\hat{L}_1^\star, \hat{L}_2^\star], w^\star)$ constitutes a saddle-point solution to the game (10.7). Consequently, we have the following result.

Proposition 10.2.1 *Consider the nonlinear system (10.1) and the NLHIFP for it. Suppose the plant $\mathbf{P}_{sp}^a$ is locally asymptotically-stable about the equilibrium-point $x = 0$ and zero-input observable for all $\varepsilon \in [0, \varepsilon^\star)$. Further, suppose also there exist a local diffeomorphism φ*

that transforms the system to the partially decoupled form (10.4), a C^1 positive-semidefinite function $\hat{V} : \hat{N} \times \hat{\Upsilon} \to \Re_+$ locally defined in a neighborhood $\hat{N} \times \hat{\Upsilon} \subset \mathcal{X} \times \mathcal{Y}$ of the origin $(\hat{\xi}, y) = (0, 0)$, and matrix functions $\hat{L}_i : \hat{N} \times \hat{\Upsilon} \to \Re^{n_i \times m}$, $i = 1, 2$, satisfying the HJIE (10.14) together with the coupling-conditions (10.11), (10.12) for some $\gamma > 0$ and $\varepsilon < \varepsilon^\star$. Then, the filter $\mathbf{F}_{1c}^a$ solves the NLHIFP for the system locally in $\varphi^{-1}(\hat{N})$.

Proof: The first part of the proof regarding the optimality of the filter gains $\hat{L}_1^\star$, $\hat{L}_2^\star$ has already been shown above. It remains to prove asymptotic convergence of the estimation error vector. For this, let $\hat{V} \geq 0$ be a C^1 solution of the HJIE (10.14) or equivalently (10.15). Then, differentiating this solution along a trajectory of (10.5) with the optimal gains $\hat{L}_1^\star$, $\hat{L}_2^\star$, and for some $w \in \mathcal{W}$ inplace of $\hat{w}^\star$, we get

$$
\begin{aligned}
\dot{\hat{V}} &= \hat{V}_{\hat{\xi}_1}(\hat{\xi}, y)[\tilde{f}_1(\hat{\xi}_1) + \tilde{g}_{11}(\hat{\xi})w + \hat{L}_1^\star(\hat{\xi}, y)(y - \tilde{h}_{21}(\hat{\xi}_1) - \tilde{h}_{22}(\hat{\xi}_2))] + \\
&\quad \frac{1}{\varepsilon}\hat{V}_{\hat{\xi}_2}(\hat{\xi}, y)[\tilde{f}_2(\hat{\xi}_2) + \tilde{g}_{21}(\hat{\xi})w + \hat{L}_2^\star(\hat{\xi}, y)(y - \tilde{h}_{21}(\hat{\xi}_1) - \tilde{h}_{22}(\hat{\xi}_2))] + \hat{V}_y(\hat{\xi}, y)\dot{y} \\
&= -\frac{\gamma^2}{2}\|w - \hat{w}^\star\|^2 + \frac{1}{2}\gamma^2\|w\|^2 - \frac{1}{2}\|z\|^2 \\
&\leq \frac{1}{2}\gamma^2\|w\|^2 - \frac{1}{2}\|z\|^2,
\end{aligned}
$$

where the last equality follows from using the HJIE (10.14). Integrating the above inequality from $t = t_0$ to $t = \infty$ and since the system is asymptotically-stable, implies that the $\mathcal{L}_2$-gain condition (10.2) is satisfied.

In addition, setting $w = 0$ in the above inequality implies that $\dot{\hat{V}}(\hat{\xi}(t), y(t)) \leq -\frac{1}{2}\|z\|^2$. Therefore, the filter dynamics is stable, and $\hat{V}(\hat{\xi}(t), y(t))$ is non-increasing along a trajectory of (10.5). Moreover, the condition that $\dot{\hat{V}}(\hat{\xi}(t), y(t)) \equiv 0 \; \forall t \geq t_s$, for some $t_s \geq t_0$, implies that $z \equiv 0$, which further implies that $y = \tilde{h}_{21}(\hat{\xi}_1) + \tilde{h}_{22}(\hat{\xi}_2) \; \forall t \geq t_s$. By the zero-input observability of the system, this implies that $\hat{\xi} = \xi$. Finally, since φ is a diffeomorphism and $\varphi(0) = 0$, $\hat{\xi} = \xi$ implies $\hat{x} = \varphi^{-1}(\hat{\xi}) = \varphi^{-1}(\xi) = x$. $\square$

The above result can be specialized to the linear singularly-perturbed system (LSPS)[178, 246]

$$
\mathbf{P}_{sp}^l : \begin{cases} \dot{x}_1 = A_1 x_1 + A_{12} x_2 + B_{11} w; & x_1(t_0) = x_{10} \\ \varepsilon \dot{x}_2 = A_{21} x_1 + A_2 x_2 + B_{21} w; & x_2(t_0) = x_{20} \\ y = C_{21} x_1 + C_{22} x_2 + w \end{cases} \tag{10.17}
$$

where $A_1 \in \Re^{n_1 \times n_1}$, $A_{12} \in \Re^{n_1 \times n_2}$, $A_{21} \in \Re^{n_2 \times n_1}$, $A_2 \in \Re^{n_2 \times n_2}$, $B_{11} \in \Re^{n_1 \times s}$, and $B_{21} \in \Re^{n_2 \times s}$, while the other matrices have compatible dimensions. Then, an explicit form of the required transformation φ above is given by the Chang transformation [79]:

$$
\begin{bmatrix} \xi_1 \\ \xi_2 \end{bmatrix} = \begin{bmatrix} I_{n_1} - \varepsilon \mathsf{HL} & -\varepsilon \mathsf{H} \\ \mathsf{L} & I_{n_2} \end{bmatrix} \begin{bmatrix} x_1 \\ x_2 \end{bmatrix} \tag{10.18}
$$

where the matrices L and H satisfy the equations

$$
\begin{aligned}
0 &= A_2 \mathsf{L} - A_{21} - \varepsilon \mathsf{L}(A_1 - A_{12}\mathsf{L}) \\
0 &= -\mathsf{H}(A_2 + \varepsilon \mathsf{L}A_{12}) + A_{12} + \varepsilon(A_1 - A_{12}\mathsf{L})\mathsf{H}.
\end{aligned}
$$

The system is then represented in the new coordinates by

$$
\tilde{\mathbf{P}}_{sp}^l : \begin{cases} \dot{\xi}_1 = \tilde{A}_1 \xi_1 + \tilde{B}_{11} w; & \xi_1(t_0) = \xi_{10} \\ \varepsilon \dot{\xi}_2 = \tilde{A}_2 \xi_2 + \tilde{B}_{21} w; & \xi_2(t_0) = \xi_{20} \\ y = \tilde{C}_{21} \xi_1 + \tilde{C}_{22} \xi_2 + w \end{cases} \tag{10.19}
$$

where

$$
\begin{aligned}
\tilde{A}_1 &= A_1 - A_{12}\mathsf{L} = A_1 - A_{12}A_2^{-1}A_{21} + O(\varepsilon) \\
\tilde{B}_{11} &= B_{11} - \varepsilon\mathsf{HL}B_{11} - \mathsf{H}B_{21} = B_{11} - A_{12}A_2^{-1}B_{21} + O(\varepsilon) \\
\tilde{A}_2 &= A_2 + \varepsilon\mathsf{L}A_{12} = A_2 + O(\varepsilon) \\
\tilde{B}_{21} &= B_{21} + \varepsilon\mathsf{L}B_{11} = B_{21} + O(\varepsilon) \\
\tilde{C}_{21} &= C_{21} - C_{22}\mathsf{L} = C_{21} - C_{22}A_2^{-1}A_{21} + O(\varepsilon) \\
\tilde{C}_{22} &= C_{22} + \varepsilon(C_{21} - C_{22})\mathsf{H} = C_{22} + O(\varepsilon).
\end{aligned}
$$

Specializing the filter (10.5) to the LSPS (10.19) results in the following filter

$$
\mathbf{F}_{1c}^l : \begin{cases}
\dot{\hat{\xi}}_1 = (\tilde{A}_1 + \frac{1}{\gamma^2}\tilde{B}_{11}\tilde{B}_{11}^T\hat{P}_1)\hat{\xi}_1 + \frac{1}{\gamma^2\varepsilon}\tilde{B}_{11}\tilde{B}_{21}^T\hat{P}_2\hat{\xi}_2 + \\
\quad \hat{L}_1(y - \tilde{C}_{21}\hat{\xi}_1 - \tilde{C}_{22}\hat{\xi}_2), \quad \hat{\xi}_1(t_0) = 0 \\
\varepsilon\dot{\hat{\xi}}_2 = (\tilde{A}_2 + \frac{1}{\gamma^2\varepsilon}\tilde{B}_{21}\tilde{B}_{21}^T\hat{P}_2)\hat{\xi}_2 + \frac{1}{\gamma^2}\tilde{B}_{21}\tilde{B}_{11}^T\hat{P}_1\hat{\xi}_1 + \\
\quad \hat{L}_2(y - \tilde{C}_{21}\hat{\xi}_1 - \tilde{C}_{22}\hat{\xi}_2), \quad \hat{\xi}_2(t_0) = 0,
\end{cases} \tag{10.20}
$$

where $\hat{P}_1, \hat{P}_2, \hat{L}_1, \hat{L}_2$ satisfy the following matrix inequalities:

$$
\begin{bmatrix}
\tilde{A}_1^T\hat{P}_1 + \hat{P}_1\tilde{A}_1 + \frac{1}{\gamma^2}\hat{P}_1\tilde{B}_{11}\tilde{B}_{11}^T\hat{P}_1 - 3\tilde{C}_{21}^T\tilde{C}_{21} & \frac{1}{\gamma^2\varepsilon}\hat{P}_1\tilde{B}_{11}\tilde{B}_{21}^T\hat{P}_2 + 3\tilde{C}_{21}^T\tilde{C}_{22} & 3\tilde{C}_{21}^T & 0 \\
\frac{1}{\gamma^2\varepsilon}\hat{P}_2\tilde{B}_{21}\tilde{B}_{11}^T\hat{P}_1 + 3\tilde{C}_{22}^T\tilde{C}_{21} & \tilde{A}_2^T\hat{P}_2 + \hat{P}_2\tilde{A}_2 + \frac{1}{\gamma^2\varepsilon}\hat{P}_2\tilde{B}_{21}\tilde{B}_{21}^T\hat{P}_2 - 3\tilde{C}_{22}^T\tilde{C}_{22} & 3\tilde{C}_{22}^T & 0 \\
3\tilde{C}_{21} & 3\tilde{C}_{22} & -3I & \frac{1}{2}Q \\
0 & 0 & \frac{1}{2}Q & 0
\end{bmatrix} \le 0 \tag{10.21}
$$

$$
\begin{bmatrix}
0 & 0 & \frac{1}{2}(\hat{P}_1\hat{L}_1 - \tilde{C}_{21}^T) \\
0 & 0 & -\frac{1}{2}\tilde{C}_{22}^T \\
\frac{1}{2}(\hat{P}_1\hat{L}_1 - \tilde{C}_{21}^T)^T & -\frac{1}{2}\tilde{C}_{22}^T & (1 - \mu_1)I
\end{bmatrix} \le 0 \tag{10.22}
$$

$$
\begin{bmatrix}
0 & 0 & -\frac{1}{2}\tilde{C}_{21}^T \\
0 & 0 & \frac{1}{2\varepsilon}(\hat{P}_2\hat{L}_2 - \tilde{C}_{22}^T) \\
-\frac{1}{2}\tilde{C}_{21} & \frac{1}{2\varepsilon}(\hat{P}_2\hat{L}_2 - \tilde{C}_{22}^T)^T & (1 - \mu_2)I
\end{bmatrix} \le 0 \tag{10.23}
$$

for some symmetric matrix $Q \in \Re^{m \times m} \ge 0$, and numbers $\mu_1, \mu_2 \ge 1$. Consequently, we have the following corollary to Proposition 10.2.1.

Corollary 10.2.1 *Consider the LSPS (10.17) and the $\mathcal{H}_\infty$-filtering problem for it. Suppose the plant $\mathbf{P}_{sp}^l$ is asymptotically-stable about the equilibrium-point $x = 0$ and observable for all $\varepsilon \in [0, \varepsilon^\star)$. Suppose further, the system is transformable to the form (10.19), and there exist positive-semidefinite matrices $\hat{P}_1 \in \Re^{n_1 \times n_1}$, $\hat{P}_2 \in \Re^{n_2 \times n_2}$, $Q \in \Re^{m \times m}$, together with matrices $\hat{L}_1, \hat{L}_2 \in \Re^{n \times m}$, satisfying the matrix-inequalities (MIs) (10.21)-(10.23) for some $\gamma > 0$ and $\varepsilon < \varepsilon^\star$. Then, the filter $\mathbf{F}_{1c}^l$ solves the $\mathcal{H}_\infty$-filtering problem for the system.*

Proof: Take

$$
V(\hat{\xi}, y) = \frac{1}{2}(\hat{\xi}_1^T P_1 \hat{\xi}_1 + \hat{\xi}_2^T P_2 \hat{\xi}_2 + y^T Q y)
$$

and apply the result of the proposition. $\square$

So far we have not exploited the benefit of the coordinate transformation φ in designing the filter (10.5) for the system (10.4). Moreover, for the linear system (10.17), the resulting governing equations (10.21)-(10.23) are not linear in the unknown variables $\hat{P}_1$, $\hat{P}_2$. Therefore, we shall now consider the design of separate reduced-order filters for the two decomposed subsystems. Accordingly, let $\varepsilon \downarrow 0$ in (10.4) and obtain the following reduced system model:

$$\tilde{\mathbf{P}}_{\mathbf{r}}^{\mathbf{a}} : \begin{cases} \dot{\xi}_1 &= \tilde{f}_1(\xi_1) + \tilde{g}_{11}(\xi)w \\ 0 &= \tilde{f}_2(\xi_2) + \tilde{g}_{21}(\xi)w \\ y &= \tilde{h}_{21}(\xi_1) + \tilde{h}_{22}(\xi_2) + \tilde{k}_{21}(\xi)w. \end{cases} \tag{10.24}$$

Then we assume the following.

Assumption 10.2.1 *The system (10.1), (10.24) is in the "standard form," i.e., the equation*

$$0 = \tilde{f}_2(\xi_2) + \tilde{g}_{21}(\xi)w \tag{10.25}$$

has $l \geq 1$ isolated roots, we can denote any one of these solutions by

$$\bar{\xi}_2 = q(\xi_1, w) \tag{10.26}$$

for some smooth function $q : \mathcal{X} \times \mathcal{W} \to \mathcal{X}$.

Using Assumption 10.2.1 results in the reduced-order slow subsystem

$$\mathbf{P}_{\mathbf{r}}^{\mathbf{a}} : \begin{cases} \dot{\xi}_1 &= \tilde{f}_1(\xi_1) + \tilde{g}_{11}(\xi_1, q(\xi_1, w))w + O(\varepsilon) \\ y &= \tilde{h}_{21}(\xi_1) + \tilde{h}_{22}(q(\xi_1, w)) + \tilde{k}_{21}(\xi_1, q(\xi_1, w))w + O(\varepsilon) \end{cases} \tag{10.27}$$

and a boundary-layer (or quasi-steady-state) subsystem

$$\frac{d\bar{\xi}_2}{d\tau} = \tilde{f}_2(\bar{\xi}_2(\tau)) + \tilde{g}_{21}(\xi_1, \bar{\xi}_2(\tau))w \tag{10.28}$$

where $\tau = t/\varepsilon$ is a stretched-time parameter. This subsystem is guaranteed to be asymptotically-stable for $0 < \varepsilon < \varepsilon^\star$ (see Theorem 8.2 in Ref. [157]) if the original system (10.1) is asymptotically-stable.

Then, we can design separate filters for the above two subsystems (10.27), (10.28) respectively as

$$\tilde{\mathbf{F}}_{2c}^{a} : \begin{cases} \dot{\breve{\xi}}_1 &= \tilde{f}_1(\breve{\xi}_1) + \tilde{g}_{11}(\breve{\xi}_1, q(\breve{\xi}_1, \breve{w}_1^\star))\breve{w}_1^\star + \breve{L}_1(\breve{\xi}_1, y)(y - \tilde{h}_{21}(\breve{\xi}_1) - \\ &\quad h_{22}(q(\breve{\xi}_1, \breve{w}_1^\star))); \quad \breve{\xi}_1(t_0) = 0 \\ \varepsilon\dot{\breve{\xi}}_2 &= \tilde{f}_2(\breve{\xi}_2) + \tilde{g}_{21}(\breve{\xi})w_2^\star + \breve{L}_2(\breve{\xi}_2, y)(y - \tilde{h}_{21}(\breve{\xi}_1) - \tilde{h}_{22}(\breve{\xi}_2)), \\ &\quad \breve{\xi}_2(t_0) = 0. \\ \breve{z} &= y - \tilde{h}_{21}(\breve{\xi}_1) - \tilde{h}_{22}(\breve{\xi}_2) \end{cases} \tag{10.29}$$

where we have decomposed w into two components w_1 and w_2 for convenience, and $\breve{w}_i^\star$ is predetermined with $\breve{\xi}_j$ constant [82], $i \neq j$, $i, j = 1, 2$.

The following theorem summarizes the design.

Theorem 10.2.1 *Consider the nonlinear system (10.1) and the $\mathcal{H}_\infty$ local state estimation problem for it. Suppose the plant $\mathbf{P}_{\mathbf{sp}}^{\mathbf{a}}$ is locally asymptotically-stable about the equilibrium-point $x = 0$ and zero-input observable for all $\varepsilon \in [0, \varepsilon^\star)$. Suppose further, there exists a local diffeomorphism φ that transforms the system to the partially decoupled form (10.4), and Assumption 10.2.1 holds. In addition, suppose for some $\gamma > 0$ and $\varepsilon \in [0, \varepsilon^\star)$, there* *exist C^1 positive-semidefinite functions $\breve{V}_i : \breve{N}_i \times \breve{\Upsilon}_i \to \Re_+$, $i = 1, 2$, locally defined in neighborhoods $\breve{N}_i \times \breve{\Upsilon}_i \subset \mathcal{X} \times \mathcal{Y}$ of the origin $(\breve{\xi}_i, y) = (0, 0)$, $i = 1, 2$ respectively, and matrix functions $\breve{L}_i : \breve{N}_i \times \breve{\Upsilon}_i \to \Re^{n_i \times m}$, $i = 1, 2$ satisfying the HJIEs:*

$$\breve{V}_{1\breve{\xi}_1}(\breve{\xi}_1, y)\tilde{f}_1(\breve{\xi}_1) + \frac{1}{2\gamma^2}\breve{V}_{1\breve{\xi}_1}(\breve{\xi}_1, y)\tilde{g}_{11}(\breve{\xi}_1, q(\breve{\xi}_1, \breve{w}_1^\star))\tilde{g}_{11}^T(\breve{\xi}_1, q(\breve{\xi}_1, \breve{w}_1^\star))\breve{V}_{1\breve{\xi}_1}^T(\breve{\xi}_1, y) +$$

$$\breve{V}_{1y}(\breve{\xi}_1, y)\dot{y} - \frac{1}{2}(y - \tilde{h}_{21}(\breve{\xi}_1) - \tilde{h}_{22}(q(\breve{\xi}_1, \breve{w}_1^\star)))^T(y - \tilde{h}_{21}(\breve{\xi}_1) - \tilde{h}_{22}(q(\breve{\xi}_1, \breve{w}_1^\star))) = 0,$$

$$\breve{V}_1(0, 0) = 0 \tag{10.30}$$

$$\frac{1}{\varepsilon}\breve{V}_{2\breve{\xi}_2}(\breve{\xi}, y)\tilde{f}_2(\breve{\xi}_2) + \frac{1}{2\gamma^2\varepsilon^2}\breve{V}_{2\breve{\xi}_2}(\breve{\xi}, y)\tilde{g}_{21}(\breve{\xi})\tilde{g}_{21}^T(\breve{\xi})\breve{V}_{2\breve{\xi}_2}^T(\breve{\xi}, y) + \breve{V}_{2y}(\breve{\xi}, y)\dot{y} -$$

$$\frac{1}{2}(y - \tilde{h}_{21}(\breve{\xi}_1) - \tilde{h}_{22}(\breve{\xi}_2))^T(y - \tilde{h}_{21}(\breve{\xi}_1) - \tilde{h}_{22}(\breve{\xi}_2)) = 0, \quad \breve{V}_2(0, 0) = 0 \tag{10.31}$$

$$\breve{w}_1^\star = \frac{1}{\gamma^2}\tilde{g}_{11}^T(\breve{\xi}_1, \bar{\xi}_2)\breve{V}_{1\breve{\xi}_1}^T(\breve{\xi}_1, y) \tag{10.32}$$

together with the coupling conditions

$$\breve{V}_{1\breve{\xi}_1}(\breve{\xi}_1, y)\breve{L}_1(\breve{\xi}_1, y) = -(y - \tilde{h}_{21}(\breve{\xi}_1) - \tilde{h}_{22}(q(\breve{\xi}_1, \breve{w}_1^\star)))^T \tag{10.33}$$

$$\frac{1}{\varepsilon}\breve{V}_{2\breve{\xi}_2}(\breve{\xi}, y)\breve{L}_2(\breve{\xi}, y) = -(y - \tilde{h}_{21}(\breve{\xi}_1) - \tilde{h}_{22}(\breve{\xi}_2))^T. \tag{10.34}$$

Then, the filter $\widetilde{\mathbf{F}}_{2c}^a$ solves the NLHIFP for the system locally in $\varphi^{-1}(\breve{N}_1 \times \breve{N}_2)$.

Proof: (Sketch). Define the two separate Hamiltonian functions $\breve{H}_i : T^\star\mathcal{X} \times \mathcal{W} \times \Re^{n_i \times m} \to \Re$, $i = 1, 2$ with respect to the cost-functional (10.7) for the two filters (10.29) as

$$\breve{H}_1(\breve{\xi}_1, y, w_1, \breve{L}_1, \breve{L}_2, \breve{V}_\xi^T, \breve{V}_y^T) = \breve{V}_{1\breve{\xi}}(\breve{\xi}_1, y)[\tilde{f}_1(\breve{\xi}_1) + \tilde{g}_{11}(\breve{\xi}_1, \bar{\xi}_2)w_1 +$$

$$\breve{L}_1(\breve{\xi}_1, y)(y - \tilde{h}_{21}(\breve{\xi}_1) - h_{22}(\bar{\xi}_2))] + \frac{1}{2}(\|z\|^2 - \gamma^2\|w_1\|^2)$$

$$\breve{H}_2(\breve{\xi}, y, w_2, \breve{L}_1, \breve{L}_2, \breve{V}_\xi^T, \breve{V}_y^T) = \frac{1}{\varepsilon}\breve{V}_{2\breve{\xi}}(\breve{\xi}, y)[\tilde{f}_2(\breve{\xi}_2) + \tilde{g}_{21}(\breve{\xi})w_2 +$$

$$\breve{L}_2(\breve{\xi}, y)(y - \tilde{h}_{21}(\breve{\xi}_1) - \tilde{h}_{22}(\breve{\xi}_2))] + \frac{1}{2}(\|z\|^2 - \gamma^2\|w_2\|^2)$$

for some smooth functions $\breve{V}_i : \mathcal{X} \times \mathcal{Y} \to \Re$, $i = 1, 2$. Then, we can determine $\breve{w}_1^\star$, $\breve{w}_2^\star$ by applying the necessary conditions for the worst-case noise as

$$\breve{w}_1^\star = \frac{1}{\gamma^2}\tilde{g}_{11}^T(\breve{\xi}_1, \bar{\xi}_2)\breve{V}_{1\breve{\xi}_1}^T(\breve{\xi}_1, y)$$

$$\breve{w}_2^\star = \frac{1}{\varepsilon\gamma^2}\tilde{g}_{12}^T(\breve{\xi})\breve{V}_{2\breve{\xi}_2}^T(\breve{\xi}, y)$$

where $\breve{w}_1^\star$ is determined with $\bar{\xi}_2$ fixed. The rest of the proof follows along the same lines as Proposition 10.2.1. $\square$

The limiting behavior of the filter (10.29) as $\varepsilon \downarrow 0$ results in the following reduced-order filter

$$\widetilde{\mathbf{F}}_{2r}^a : \begin{cases} \dot{\breve{\xi}}_1 = \tilde{f}_1(\breve{\xi}_1) + \tilde{g}_{11}(\breve{\xi}_1, q(\breve{\xi}_1, \breve{w}_1^\star))\breve{w}_1^\star + \breve{L}_1(\breve{\xi}_1, y)(y - \tilde{h}_{21}(\breve{\xi}_1) - h_{22}(q(\breve{\xi}_1, \breve{w}_1^\star))) \\ \breve{\xi}_1(t_0) = 0 \end{cases} \tag{10.35}$$

which is governed by the HJIE (10.30).

The result of Theorem 10.2.1 can similarly be specialized to the LSPS (10.17). Assuming A_2 is nonsingular (i.e., Assumption 10.2.1 is satisfied), then we can solve for

$$\bar{\xi}_2 = -A_2^{-1}B_{21}w,$$

and we substitute in the slow-subsystem to obtain the composite filter

$$\mathbf{F}_{2c}^l : \begin{cases} \dot{\check{\xi}}_1 = \tilde{A}_1\check{\xi}_1 + \frac{1}{\gamma^2}\tilde{B}_{11}\tilde{B}_{11}^T\check{P}_1\check{\xi}_1 + \check{L}_1(y - \tilde{C}_{21}\check{\xi}_1 + \frac{1}{\gamma^2}\tilde{C}_{22}\tilde{A}_2^{-1}\tilde{B}_{21}\tilde{B}_{11}^T\check{P}_1\check{\xi}_1), \\ \check{\xi}_1(t_0) = 0 \\ \varepsilon\dot{\check{\xi}}_2 = \tilde{A}_2\check{\xi}_2 + \frac{1}{\gamma^2\varepsilon}\tilde{B}_{21}\tilde{B}_{21}^T\check{P}_2\check{\xi}_2 + \check{L}_2(y - \tilde{C}_{21}\check{\xi}_1 - \tilde{C}_{22}\check{\xi}_2), \quad \check{\xi}_2(t_0) = 0. \end{cases}$$

Thus, the following corollary can be deduced.

Corollary 10.2.2 *Consider the LSPS (10.17) and the $\mathcal{H}_\infty$-filtering problem for it. Suppose the plant $\mathbf{P}_{sp}^l$ is asymptotically-stable about the equilibrium-point $x = 0$ and observable for all $\varepsilon \in [0, \varepsilon^\star)$. Further, suppose it is transformable to the form (10.19) and Assumption 10.2.1 holds, or A_2 is nonsingular. In addition, suppose for some $\gamma > 0$ and $\varepsilon \in [0, \varepsilon^\star)$, there exist positive-semidefinite matrices $\check{P}_1 \in \Re^{n_1 \times n_1}$, $\check{P}_2 \in \Re^{n_2 \times n_2}$, $\check{Q}_1, \check{Q}_2 \in \Re^{m \times m}$ and matrices $\check{L}_1 \in \Re^{n_1 \times m}$, $\check{L}_2 \in \Re^{n_2 \times m}$, satisfying the linear-matrix-inequalities (LMIs)*

$$\left[\begin{array}{cc} \left(\begin{array}{c} \tilde{A}_1^T\check{P}_1 + \check{P}_1\tilde{A}_1 - \tilde{C}_{21}^T\tilde{C}_{21} + \frac{1}{\gamma^2}\tilde{C}_{21}^T\tilde{C}_{22}\tilde{A}_2^{-1}\tilde{B}_{21}\tilde{B}_{11}^T\check{P}_1+ \\ \frac{1}{\gamma^2}\check{P}_1\tilde{B}_{11}\tilde{B}_{21}^T\tilde{A}_2^{-T}\tilde{C}_{22}^T\tilde{C}_{21} \end{array} \right) & \check{P}_1\tilde{B}_{11} \\ \tilde{B}_{11}^T\check{P}_1 & -\gamma^{-2}I \\ \check{P}_1\tilde{B}_{11}\tilde{B}_{21}^T\tilde{A}_2^{-T}\tilde{C}_{22}^T & 0 \\ \tilde{C}_{21} - \frac{1}{\gamma^2}\tilde{C}_{22}\tilde{A}_2^{-1}\tilde{B}_{21}\tilde{B}_{11}^T\check{P}_1 & 0 \\ 0 & 0 \end{array} \right.$$

$$\left. \begin{array}{cccc} \frac{1}{\gamma^2}\tilde{C}_{22}\tilde{A}_2^{-1}\tilde{B}_{21}\tilde{B}_{11}^T\check{P}_1 & \tilde{C}_{21}^T - \frac{1}{\gamma^2}\check{P}_1\tilde{B}_{11}\tilde{B}_{21}^T\tilde{A}_2^{-T}\tilde{C}_{22}^T & 0 \\ 0 & 0 & 0 \\ -\gamma^{-2}I & 0 & 0 \\ 0 & -I & \check{Q}_1 \\ 0 & \check{Q}_1 & 0 \end{array} \right] \leq 0 \qquad (10.36)$$

$$\left[\begin{array}{ccccc} -\tilde{C}_{21}^T\tilde{C}_{21} & -\tilde{C}_{21}^T\tilde{C}_{22} & 0 & \tilde{C}_{21}^T & 0 \\ -\tilde{C}_{22}^T\tilde{C}_{21} & \frac{1}{\varepsilon}(\tilde{A}_2^T\check{P}_2 + \check{P}_2\tilde{A}_2) - \tilde{C}_{22}^T\tilde{C}_{21} & \check{P}_2\tilde{B}_{21} & \tilde{C}_{22}^T & 0 \\ 0 & \tilde{B}_{21}^T\check{P}_2 & -\varepsilon^{-2}\gamma^{-2}I & 0 & 0 \\ \tilde{C}_{21} & \tilde{C}_{22} & 0 & -I & \check{Q}_2 \\ 0 & 0 & 0 & \check{Q}_2 & 0 \end{array} \right] \leq 0 \qquad (10.37)$$

$$\left[\begin{array}{cc} 0 & \frac{1}{2}\left(\begin{array}{c} \check{P}_1\check{L}_1 - \tilde{C}_{21}^T + \\ \frac{1}{\gamma^2}\check{P}_1\tilde{B}_{11}\tilde{B}_{21}\tilde{A}_2^{-T}\tilde{C}_{22}^T \end{array} \right) \\ \frac{1}{2}\left(\begin{array}{c} \check{P}_1\check{L}_1 - \tilde{C}_{21}^T \\ +\frac{1}{\gamma^2}\check{P}_1\tilde{B}_{11}\tilde{B}_{21}\tilde{A}_2^{-T}\tilde{C}_{22}^T \end{array} \right)^T & (1 - \delta_1)I \end{array} \right] \leq 0 \qquad (10.38)$$

$$\left[\begin{array}{ccc} 0 & 0 & -\frac{1}{\varepsilon}\tilde{C}_{21}^T \\ 0 & 0 & \frac{1}{2}(\frac{1}{\varepsilon}\check{P}_2\check{L}_2 - \tilde{C}_{22}^T) \\ -\frac{1}{2}\tilde{C}_{21} & \frac{1}{2}(\frac{1}{\varepsilon}\check{P}_2\check{L}_2 - \tilde{C}_{22}^T)^T & (1 - \delta_2)I \end{array} \right] \leq 0 \qquad (10.39)$$

for some numbers $\delta_1, \delta_2 \geq 1$. Then the filter $\mathbf{F}_{2c}^l$ solves the $\mathcal{H}_\infty$-filtering problem for the system.

Proof: Take

$$\begin{aligned}
\breve{V}_1(\breve{\xi}_1, y) &= \frac{1}{2}(\breve{\xi}_1^T \breve{P}_1 \breve{\xi}_1 + y^T \breve{Q}_1 y) \\
\breve{V}_2(\breve{\xi}_2, y) &= \frac{1}{2}(\breve{\xi}_2^T \breve{P}_2 \breve{\xi}_2 + y^T \breve{Q}_2 y)
\end{aligned}$$

and apply the result of the theorem. Moreover, the nonsingularity of A_2 guarantees that a reduced-order subsystem exists. $\square$

10.3   Aggregate Filters

If the coordinate transformation, φ, discussed in the previous section cannot be found, then an aggregate filter for the system (10.1) must be designed. Accordingly, consider the following class of filters:

$$\mathbf{F}_{3ag}^a : \begin{cases}
\dot{\hat{x}}_1 = f_1(\hat{x}_1, \hat{x}_2) + g_{11}(\hat{x}_1, \hat{x}_2)\hat{w}^\star + \grave{L}_1(\hat{x}, y)(y - h_{21}(\hat{x}_1) + h_{22}(\hat{x}_2)); \\
\quad\quad \hat{x}_1(t_0) = 0 \\
\varepsilon\dot{\hat{x}}_2 = f_2(\hat{x}_1, \hat{x}_2) + g_{12}(\hat{x}_1, \hat{x}_2)\hat{w}^\star + \grave{L}_2(\hat{x}, y)(y - h_{21}(\hat{x}_1) + h_{22}(\hat{x}_2)); \\
\quad\quad \hat{x}_2(t_0) = 0 \\
\grave{z} = y - h_{21}(\hat{x}_1) + h_{22}(\hat{x}_2)
\end{cases}$$

where $\grave{L}_1 \in \Re^{n_1 \times m}$, $\grave{L}_2 \in \Re^{n_2 \times m}$ are the filter gains, and $\grave{z}$ is the new penalty variable. Then the following result can be derived using similar steps as outlined in the previous section.

Theorem 10.3.1 *Consider the nonlinear system (10.1) and the NLHIFP for it. Suppose the plant $\mathbf{P}_{sp}^a$ is locally asymptotically-stable about the equilibrium-point $x = 0$ and observable for all $\varepsilon \in [0, \varepsilon^\star)$. Further, suppose for some $\gamma > 0$ and $\varepsilon \in [0, \varepsilon^\star)$, there exists a C^1 positive-semidefinite function $\grave{V} : \grave{N} \times \grave{\Upsilon} \to \Re_+$, locally defined in a neighborhood $\grave{N} \times \grave{\Upsilon} \subset \mathcal{X} \times \mathcal{Y}$ of the origin $(\hat{x}_1, \hat{x}_2, y) = (0, 0, 0)$, and matrix functions $\grave{L}_i : \grave{N} \times \grave{\Upsilon} \to \Re^{n_i \times m}$, $i = 1, 2$, satisfying the HJIE:*

$$\grave{V}_{\hat{x}_1}(\hat{x}, y) f_1(\hat{x}_1, \hat{x}_2) + \frac{1}{\varepsilon}\grave{V}_{\hat{x}_2}(\hat{x}, y) f_2(\hat{x}_1, \hat{x}_2) + \grave{V}_y(\hat{x}, y)\dot{y} +$$

$$\frac{1}{2\gamma^2}[\grave{V}_{\hat{x}_1}(\hat{x}, y) \;\; \grave{V}_{\hat{x}_2}(\hat{x}, y)] \begin{bmatrix} g_{11}(\hat{x})g_{11}^T(\hat{x}) & \frac{1}{\varepsilon}g_{11}(\hat{x})g_{21}^T(\hat{x}) \\ \frac{1}{\varepsilon}g_{21}(\hat{x})g_{11}^T(\hat{x}) & \frac{1}{\varepsilon^2}g_{21}(\hat{x})g_{21}^T(\hat{x}) \end{bmatrix} \begin{bmatrix} \grave{V}_{\hat{x}_1}^T(\hat{x}, y) \\ \grave{V}_{\hat{x}_2}^T(\hat{x}, y) \end{bmatrix} -$$

$$\frac{3}{2}(y - h_{21}(\hat{x}_1) - h_{22}(\hat{x}_2))^T (y - h_{21}(\hat{x}_1) - h_{22}(\hat{x}_2)) = 0, \;\; \grave{V}(0, 0) = 0. \quad\quad (10.40)$$

together with the side-conditions

$$\grave{V}_{\hat{x}_1}(\hat{x}, y)\grave{L}_1(\hat{x}, y) = -(y - h_{21}(\hat{x}_1) - h_{22}(\hat{x}_2))^T \quad\quad (10.41)$$

$$\frac{1}{\varepsilon}\grave{V}_{\hat{x}_2}(\hat{x}, y)\grave{L}_2(\hat{x}, y) = -(y - h_{21}(\hat{x}_1) - h_{22}(\hat{x}_2))^T. \quad\quad (10.42)$$

Then, the filter $\mathbf{F}_{3ag}^a$ with

$$\hat{w}^\star = \frac{1}{\gamma^2}[g_{11}^T(\hat{x})\grave{V}_{\hat{x}_1}^T(\hat{x}, y) + \frac{1}{\varepsilon}g_{21}^T(\hat{x})\grave{V}_{\hat{x}_2}^T(\hat{x}, y)]$$

solves the NLHIFP for the system locally in $\grave{N}$.

Proof: Proof follows along the same lines as Proposition 10.2.1. □

The above result, Theorem 10.3.1, can similarly be specialized to the LSPS $\mathbf{P}^l_{sp}$.

To obtain the limiting filter (10.40) as $\varepsilon \downarrow 0$, we use Assumption 10.2.1 to obtain a reduced-order model of the system (10.1). If we assume the equation

$$0 = f_2(x_1, x_2) + \tilde{g}_{21}(x_1, x_2)w \tag{10.43}$$

has $k \geq 1$ isolated roots, we can denote any one of these roots by

$$\bar{x}_2 = p(x_1, w), \tag{10.44}$$

for some smooth function $p : \mathcal{X} \times \mathcal{W} \to \mathcal{X}$. The resulting reduced-order system is given by

$$\mathbf{P}^a_{spr} : \begin{cases} \dot{x}_1 &= f_1(x_1, \bar{x}_2) + g_{11}(x_1, \bar{x}_2)w; \quad x_1(t_0) = x_{10} \\ y &= h_{21}(x_1) + h_{22}(\bar{x}_2) + k_{21}(x_1, \bar{x}_2)w, \end{cases} \tag{10.45}$$

and the corresponding reduced-order filter is then given by

$$\mathbf{F}^a_{3agr} : \begin{cases} \acute{x}_1 &= f_1(\acute{x}_1, p(\acute{x}_1, \acute{w}^\star)) + g_{11}(\acute{x}_1, p(\acute{x}_1, \acute{w}^\star))\acute{w}^\star + \\ &\quad \acute{L}_1(\acute{x}, y)(y - h_{21}(\acute{x}_1) + h_{22}(p(\acute{x}_2, \acute{w}^\star))); \quad \acute{x}_1(t_0) = 0 \\ \acute{z} &= y - h_{21}(\acute{x}_1) + h_{22}(p(\acute{x}_1, \acute{w}^\star)), \end{cases}$$

where all the variables have their corresponding previous meanings and dimensions, while

$$\acute{w}^\star = \frac{1}{\gamma^2} g_{11}^T(\acute{x})\acute{V}^T_{\acute{x}_1}(\acute{x}, y)$$

$$\acute{V}_{\acute{x}_1}(\acute{x}, y)\acute{L}_1(\acute{x}, y) = -(y - h_{21}(\acute{x}_1) - h_{22}(p(\acute{x}_1, \acute{w}^\star)))^T$$

and $\acute{V}$ satisfies the following HJIE:

$$\acute{V}_{\acute{x}_1}(\acute{x}, y)f_1(\acute{x}_1, p(\acute{x}_1, \acute{w}^\star)) + \acute{V}_y(\acute{x}_1, y)\dot{y} +$$

$$\frac{1}{2\gamma^2}\acute{V}_{\acute{x}_1}(\acute{x}_1, y)g_{11}(\acute{x}, p(\acute{x}_1, \acute{w}^\star))g_{11}^T(\acute{x}, p(\acute{x}_1, \acute{w}^\star))\acute{V}^T_{\acute{x}_1}(\acute{x}_1, y) -$$

$$\frac{1}{2}(y - h_{21}(\acute{x}_1) - h_{22}(p(\acute{x}_1, \acute{w}^\star)))^T(y - h_{21}(\acute{x}_1) - h_{22}(p(\acute{x}_1, \acute{w}^\star))) = 0, \tag{10.46}$$

with $\acute{V}(0, 0) = 0$. In the next section, we consider some examples.

10.4   Examples

Consider the following singularly-perturbed nonlinear system

$$\begin{aligned} \dot{x}_1 &= -x_1^3 + x_2 \\ \varepsilon\dot{x}_2 &= -x_1 - x_2 + w \\ y &= x_1 + x_2 + w, \end{aligned}$$

where $w \in \mathcal{L}_2[0, \infty)$, $\varepsilon \geq 0$. We construct the aggregate filter $\mathbf{F}^a_{3ag}$ presented in the previous section for the above system. It can be checked that the system is locally zero-input observable, and the function $\acute{V}(\acute{x}) = \frac{1}{2}(\acute{x}_1^2 + \varepsilon\acute{x}_2^2)$, solves the inequality form of the HJIE (10.40) corresponding to the system. Subsequently, we calculate the gains of the filter as

$$\acute{L}_1(\acute{x}, y) = -\frac{(y - \acute{x}_1 - \acute{x}_2)}{\acute{x}_1}, \quad \acute{L}_2(\acute{x}, y) = -\frac{\varepsilon(y - \acute{x}_1 - \acute{x}_2)}{\acute{x}_2}, \tag{10.47}$$

where $\acute{L}_1(\acute{x}, y)$, $\acute{L}_2(\acute{x}, y)$ are set equal to zero if $\|\acute{x}\| < \epsilon$ (small) to avoid the singularity at $\acute{x} = 0$.

10.5   Notes and Bibliography

This chapter is mainly based on [26]. Similar results for the $\mathcal{H}_2$ filtering problem can be found in [25]. Results for fuzzy TS nonlinear models can be found in [40, 41, 283].

11

Mixed $\mathcal{H}_2/\mathcal{H}_\infty$ Nonlinear Control

In this chapter, we discuss the mixed $\mathcal{H}_2/\mathcal{H}_\infty$-control problem for nonlinear systems. This problem arises when a higher-degree of performance for the system is desired, and two criteria are minimized to derive the controller that enjoys both the properties of an $\mathcal{H}_2$ (or LQG [174]) and $\mathcal{H}_\infty$-controller. A stronger motivation for this problem though is that, because the solution to the $\mathcal{H}_\infty$-control problem is nonunique (if it is not optimal, it can hardly be unique) and only the suboptimal problem could be solved easily, then is it possible to formulate another problem that could be solved optimally and obtain a unique solution?

The problem for linear systems was first considered by Bernstein and Haddad [07], where a solution for the output-feedback problem in terms of three coupled algebraic-Riccati-equations (AREs) was obtained by formulating it as an LQG problem with an $\mathcal{H}_\infty$ constraint. The dual to this problem has also been considered by Doyle, Zhou and Glover [93, 293]. While Mustapha and Glover [205, 206] have considered entropy minimization which provides an upper bound on the $\mathcal{H}_2$-cost under an $\mathcal{H}_\infty$-constraint.

Another contribution to the linear literature was from Khargonekhar and Rotea [161] and Scherer et al. [238, 239], who considered more general multi-objective problems using convex optimization and/or linear-matrix-inequalities (LMI). And more lately, by Limebeer et al. [179] and Chen and Zhou [81] who considered a two-person nonzero-sum differential game approach with a multi-objective flavor (for the latter). This approach is very transparent and is reminiscent of the minimax approach to $\mathcal{H}_\infty$-control by Basar and Bernhard [57]. The state-feedback problem is solved in Limebeer et al. [179], while the output-feedback problem is solved in Chen and Zhou [81]. By-and-large, the outcome of the above endeavors are a parametrization of the solution to the mixed $\mathcal{H}_2/\mathcal{H}_\infty$-control problem in terms of two cross-coupled nonstandard Riccati equations for the state-feedback problem, and an additional standard Riccati equation for the output-feedback problem.

Similarly, the nonlinear control problem has also been considered more recently by Lin [180]. He extended the results of Limebeer et al. [179], and derived a solution to the state-feedback problem in terms of a pair of cross-coupled Hamilton-Jacobi-Isaac's equations. In this chapter, we discuss mainly this approach to the problem for both continuous-time and discrete-time nonlinear systems.

11.1 Continuous-Time Mixed $\mathcal{H}_2/\mathcal{H}_\infty$ Nonlinear Control

In this section, we discuss the mixed $\mathcal{H}_2/\mathcal{H}_\infty$ nonlinear control problem using state-feedback. The general set-up for this problem is shown in Figure 11.1 with the plant represented by an affine nonlinear system Σ^a, while the static controller is represented by $\mathbf{K}$. The disturbance/noise signal $w = \begin{pmatrix} w_0 \\ w_1 \end{pmatrix}$, is comprised of two components: (i) a bounded-spectral signal (e.g., a white Gaussian-noise signal) $w_0 \in \mathcal{S}$ (the space of bounded-spectral signals), and (ii) a bounded-power signal or $\mathcal{L}_2$ signal $w_1 \in \mathcal{P}$ (the space of bounded power

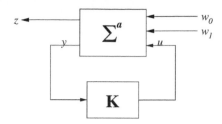

FIGURE 11.1
Set-Up for Nonlinear Mixed $\mathcal{H}_2/\mathcal{H}_\infty$-Control

signals). Thus, the induced norm from the input w_0 to z is the $\mathcal{L}_2$-norm of the closed-loop system $\mathbf{K} \circ \Sigma^a$, i.e.,

$$\|\mathbf{K} \circ \Sigma^a\|_{\mathcal{L}_2} \triangleq \sup_{0 \neq w_0 \in \mathcal{S}} \frac{\|z\|_{\mathcal{P}}}{\|w_0\|_{\mathcal{S}}}, \tag{11.1}$$

while the induced norm from w_1 to z is the $\mathcal{L}_\infty$-norm of the closed-loop system $\mathbf{K} \circ \Sigma^a$, i.e.,

$$\|\mathbf{K} \circ \Sigma^a\|_{\mathcal{L}_\infty} \triangleq \sup_{0 \neq w_1 \in \mathcal{P}} \frac{\|z\|_2}{\|w_1\|_2} \tag{11.2}$$

where

$$\mathcal{P} \triangleq \{w(t): \ w \in \mathcal{L}_\infty, \ R_{ww}(\tau), \ S_{ww}(j\omega) \text{ exist for all } \tau \text{ and all } \omega \text{ resp.},$$
$$\|w\|_{\mathcal{P}} < \infty\},$$

$$\mathcal{S} \triangleq \{w(t): \ w \in \mathcal{L}_\infty, \ R_{ww}(\tau), \ S_{ww}(j\omega) \text{ exist for all } \tau \text{ and all } \omega \text{ resp.},$$
$$\|S_{ww}(j\omega)\|_\infty < \infty\},$$

$$\|z\|_{\mathcal{P}}^2 \triangleq \lim_{T \to \infty} \frac{1}{2T} \int_{-T}^{T} \|z(t)\|^2 dt,$$

$$\|w_0\|_{\mathcal{S}}^2 = \|S_{w_0 w_0}(j\omega)\|_\infty,$$

and $R_{ww}(\tau)$, $S_{ww}(j\omega)$ are the autocorrelation and power-spectral density matrices of $w(t)$ respectively [152]. Notice also that, $\|(.)\|_{\mathcal{P}}$ and $\|(.)\|_{\mathcal{S}}$ are seminorms. In addition, if the plant is stable, we replace the induced $\mathcal{L}$-norms above by their equivalent $\mathcal{H}$-subspace norms. The standard optimal mixed $\mathcal{H}_2/\mathcal{H}_\infty$ state-feedback control problem is to synthesize a feedback control of the form

$$u = \alpha(x), \quad \alpha(0) = 0 \tag{11.3}$$

such that the above induced-norms (11.1), (11.2) of the closed-loop system are minimized, and the closed-loop system is also locally asymptotically-stable. However, in this chapter, we do not solve the above problem, instead we solve an associated suboptimal $\mathcal{H}_2/\mathcal{H}_\infty$-control problem which involves a single disturbance $w \in \mathcal{L}_2$ entering the plant, and where the objective is to minimize the output energy $\|z\|_{\mathcal{H}_2}$ of the closed-loop system while rendering $\|\mathbf{K} \circ \Sigma^a\|_{\mathcal{H}_\infty} \leq \gamma^\star$.

The plant is represented by an affine nonlinear causal state-space system defined on a manifold $\mathcal{X} \subset \Re^n$ containing the origin $x = 0$:

$$\Sigma^a: \begin{cases} \dot{x} &= f(x) + g_1(x)w + g_2(x)u; \ x(t_0) = x_0 \\ z &= h_1(x) + k_{12}(x)u \\ y &= x, \end{cases} \tag{11.4}$$

where $x \in \mathcal{X}$ is the state vector, $u \in \mathcal{U} \subseteq \Re^p$ is the p-dimensional control input, which belongs to the set of admissible controls $\mathcal{U} \subset \mathcal{L}_2([t_0, T], \Re^p)$, $w \in \mathcal{W}$ is the disturbance signal, which belongs to the set $\mathcal{W} \subset \Re^r$ of admissible disturbances (to be defined more precisely later), the output $y \in \Re^m$ is the measured output of the system, and $z \in \Re^s$ is the output to be controlled. The functions $f : \mathcal{X} \to V^\infty(\mathcal{X})$, $g_1 : \mathcal{X} \to \mathcal{M}^{n \times r}(\mathcal{X})$, $g_2 : \mathcal{X} \to \mathcal{M}^{n \times p}(\mathcal{X})$, $h_1 : \mathcal{X} \to \Re^s$ and $k_{12} : \mathcal{X} \to \mathcal{M}^{s \times p}(\mathcal{X})$, are real C^∞ functions of x.

Furthermore, we assume without any loss of generality that the system (11.4) has a unique equilibrium-point at $x = 0$ such that $f(0) = 0$, $h_1(0) = h_2(0) = 0$, and for simplicity, we also make the following assumption.

Assumption 11.1.1 *The system matrices are such that*

$$\left. \begin{array}{rcl} h_1^T(x)k_{12}(x) & = & 0 \\ k_{12}^T(x)k_{12}(x) & = & I. \end{array} \right\} \tag{11.5}$$

The problem can now be formally defined as follows.

Definition 11.1.1 *(State-Feedback Mixed $\mathcal{H}_2/\mathcal{H}_\infty$ Nonlinear Control Problem (SF-BMH2HINLCP)).*

(A) *Finite-Horizon Problem ($T < \infty$): Find (if possible!) a time-varying static state-feedback control law of the form:*

$$u - \tilde{\alpha}_2(x, t), \quad \tilde{\alpha}_2(t, 0) = 0, \quad t \in \Re,$$

such that:

(a) *the closed-loop system*

$$\Sigma^{cla} : \left\{ \begin{array}{rcl} \dot{x} & = & f(x) + g_1 w + g_2(x)\tilde{\alpha}_2(x, t) \\ z & = & h_1(x) + k_{21}(x)\tilde{\alpha}_2(x, t) \end{array} \right. \tag{11.6}$$

is stable with $w = 0$ and has locally finite $\mathcal{L}_2$-gain from w to z less or equal to $\gamma^\star$, starting from $x_0 = 0$, for all $t \in [0, T]$ and all $w \in \mathcal{W} \subseteq \mathcal{L}_2[0, T]$;

(b) *the output energy $\|z\|_{\mathcal{H}_2}$ of the system is minimized.*

(B) *Infinite-Horizon Problem ($T \to \infty$): In addition to the items (a) and (b) above, it is also required that*

(c) *the closed-loop system Σ^{cla} defined above with $w \equiv 0$ is locally asymptotically-stable about the equilibrium-point $x = 0$.*

Such a problem can be formulated as a two-player nonzero-sum differential game (Chapter 2) with two cost functionals:

$$\min_{u \in \mathcal{U}, w \in \mathcal{W}} J_1(u, w) = \frac{1}{2} \int_{t_0}^T (\gamma^2 \|w(\tau)\|^2 - \|z(\tau)\|^2) d\tau \tag{11.7}$$

$$\min_{u \in \mathcal{U}, w \in \mathcal{W}} J_2(u, w) = \frac{1}{2} \int_{t_0}^T \|z(\tau)\|^2 d\tau \tag{11.8}$$

for the finite-horizon problem, with $T \geq t_0$. Here, the first functional is associated with the $\mathcal{H}_\infty$-constraint criterion, while the second functional is related to the output energy of the system or $\mathcal{H}_2$-criterion. It can easily be seen that, by making $J_1 \geq 0$, the $\mathcal{H}_\infty$-constraint $\|\mathbf{K} \circ \Sigma^a\|_{\mathcal{H}_\infty} = \|\Sigma^{cla}\|_{\mathcal{H}_\infty} \leq \gamma$ is satisfied. Subsequently, minimizing J_2 will achieve the

$\mathcal{H}_2/\mathcal{H}_\infty$ design objective. A Nash-equilibrium solution to the above game is said to exist if we can find a pair of strategies $(u^\star, w^\star)$ such that

$$J_1(u^\star, w^\star) \leq J_1(u^\star, w) \quad \forall w \in \mathcal{W}, \tag{11.9}$$

$$J_2(u^\star, w^\star) \leq J_2(u, w^\star) \quad \forall u \in \mathcal{U}. \tag{11.10}$$

Furthermore, by minimizing the first objective with respect to w and substituting in the second objective which is then minimized with respect to u, the above pair of Nash-equilibrium strategies could be found. A sufficient condition for the solvability of the above differential game is provided from Theorem 2.3.1, Chapter 2, by the following pair of cross-coupled HJIEs for the finite-horizon state-feedback problem:

$$-Y_t(x, t) = \inf_{w \in \mathcal{W}} \left\{ Y_x(x, t) f(x, u^\star(x), w(x)) + \gamma^2 \|w(x)\|^2 - \|z^\star(x)\|^2 \right\}, \quad Y(x, T) = 0,$$

$$-V_t(x, t) = \min_{u \in \mathcal{U}} \left\{ V_x(x, t) f(x, u(x), w^\star(x)) + \|z^\star(x)\|^2 \right\} = 0, \quad V(x, T) = 0,$$

for some negative-definite function $Y : \mathcal{X} \to \Re$ and positive-definite function $V : \mathcal{X} \to \Re$, where $z^\star(x) = h_1(x) + k_{12}(x)u^\star(x)$.

In view of the above result, the following theorem gives sufficient conditions for the solvability of the finite-horizon problem.

Theorem 11.1.1 *Consider the nonlinear system Σ^a defined by (11.4) and the finite-horizon SFBMH2HINLCP with cost functionals (11.7), (11.8). Suppose there exists a pair of negative and positive-definite C^1-functions (with respect to both arguments) $Y, V : N \times [0, T] \to \Re$ locally defined in a neighborhood N of the origin $x = 0$, such that $Y(0, t) = 0$ and $V(0, t) = 0$, and satisfying the coupled HJIEs:*

$$-Y_t(x, t) = Y_x(x, t) f(x) - \frac{1}{2} V_x(x, t) g_2(x) g_2^T(x) V_x^T(x, t) -$$

$$\frac{1}{2\gamma^2} Y_x(x, t) g_1(x) g_1^T(x) Y_x^T(x, t) - Y_x(x, t) g_2(x) g_2^T(x) V_x^T(x, t) -$$

$$\frac{1}{2} h_1^T(x) h_1(x), \quad Y(x, T) = 0 \tag{11.11}$$

$$-V_t(x, t) = V_x(x, t) f(x) - \frac{1}{2} V_x(x, t) g_2(x) g_2^T(x) V_x^T(x, t) -$$

$$\frac{1}{\gamma^2} V_x(x, t) g_1(x) g_1^T(x) Y_x^T(x, t) + \frac{1}{2} h_1^T(x) h_1(x), \quad V(x, T) = 0. \tag{11.12}$$

Then the state-feedback controls

$$u^\star(x, t) = -g_2^T(x) V_x^T(x, t) \tag{11.13}$$

$$w^\star(x, t) = -\frac{1}{\gamma^2} g_1^T(x) Y_x^T(x, t) \tag{11.14}$$

solve the finite-horizon SFBMH2HINLCP for the system. Moreover, the optimal costs are given by

$$J_1^\star(u^\star, w^\star) = Y(t_0, x_0) \tag{11.15}$$

$$J_2^\star(u^\star, w^\star) = V(t_0, x_0). \tag{11.16}$$

Proof: Assume there exists locally solutions $Y < 0$, $V > 0$ to the HJIEs (11.11), (11.12) in $N \subset \mathcal{X}$. We prove item (a) of Definition 11.1.1 first. Rearranging the HJIE (11.11) and completing the squares, we have

$$
\begin{aligned}
Y_t + Y_x(f(x) + g_1(x)w - g_2(x)g^T(x)V_x^T(x)) &= \frac{1}{2}\|h_1(x)\|^2 + \frac{\gamma^2}{2}\|w - w^\star\|^2 + \\
&\quad \frac{1}{2}\|u^\star\|^2 - \frac{1}{2}\gamma^2\|w\|^2 \\
\Longleftrightarrow \dot{Y}(x,t) &= \frac{1}{2}\|z\|^2 - \frac{1}{2}\gamma^2\|w\|^2 + \frac{\gamma^2}{2}\|w - w^\star\|^2 \\
\Longrightarrow \dot{Y}(x,t) &\leq \frac{1}{2}\gamma^2\|w\|^2 - \frac{1}{2}\|z\|^2
\end{aligned}
$$

for some function $\tilde{Y} = -Y > 0$. Integrating now the above expression from t_0 and $x(t_0)$ to $t = T$ and $x(T)$, we get the dissipation-inequality

$$
\dot{Y}(x(T), T') - \tilde{Y}(x(t_0, t_0)) \leq \int_{t_0}^T \frac{1}{2}(\gamma^2\|w\|^2 - \|z\|^2)dt.
$$

Therefore, the system has locally $\mathcal{L}_2$-gain $\leq \gamma$ from w to z. Furthermore, the closed-loop system with $u = u^\star$ and $w = 0$ is given by

$$
\dot{x} = f(x) - g_2(x)g_2^T(x)V_x^T(x).
$$

Differentiating $\tilde{Y}$ from above along a trajectory of this system, we have

$$
\begin{aligned}
\dot{\tilde{Y}} &= \tilde{Y}_t(x,t) + \tilde{Y}_x(x,t)(f(x) - g_2(x)g_2^T(x)V_x^T) \quad \forall x \in N \\
&\leq -\frac{1}{2}\|z\|^2 \leq 0.
\end{aligned}
$$

Hence, the closed-loop system is Lyapunov-stable.

Next we prove item (b). Consider the cost functional $J_1(u, w)$ first. For any $T \geq t_0$, the following holds

$$
J_1(u, w) + Y(x(T), T) - Y(x(t_0), t_0) = \int_{t_0}^T \{\frac{1}{2}(\gamma^2\|w\|^2 - \|z\|^2) + \dot{Y}(x,t)\}dt
$$

$$
= \int_{t_0}^T \{\frac{1}{2}(\gamma^2\|w\|^2 - \|z\|^2) + Y_t + Y_x(f(x) + g_1(x)w + g_2(x)u)\}dt
$$

$$
= \int_{t_0}^T \left\{Y_t + Y_x f(x) + Y_x g_2(x)u + \frac{\gamma^2}{2}\left\|w + \frac{1}{\gamma^2}g_1^T(x)Y_x^T\right\|^2 - \frac{\gamma^2}{2}\|\frac{1}{\gamma^2}g_1^T(x)Y_x^T\|^2 \right.
$$

$$
\left. - \frac{1}{2}\|u\|^2 - \frac{1}{2}\|h_1(x)\|^2\right\}dt.
$$

Using the HJIE (11.11), we have

$$
\begin{aligned}
J_1(u, w) + Y(x(T), T) - Y(x(t_0), t_0) &= \int_{t_0}^T \left\{\frac{\gamma^2}{2}\left\|w + \frac{1}{\gamma^2}g_1^T(x)Y_x^T\right\|^2 - \right. \\
&\quad \frac{1}{2}\|u + g_2^T(x)V_x^T\|^2 + \\
&\quad \|g_2^T(x)V_x^T\|^2 + V_x(x)g_2(x)u + \\
&\quad \left. Y_x g_2(x)(u - g_2^T(x)V_x^T(x))\right\}dt,
\end{aligned}
$$

and substituting $u = u^\star$, we have

$$J_1(u^\star, w) + Y(x(T), T) - Y(x(t_0), t_0) = \int_{t_0}^T \frac{\gamma^2}{2} \|w - w^\star\|^2 dt \geq 0.$$

Therefore

$$J_1(u^\star, w^\star) \leq J_1(u^\star, w)$$

with

$$J_1(u^\star, w^\star) = Y(x(t_0), t_0)$$

since $Y(x(T), T) = 0$.

Similarly, considering the cost functional $J_2(u, w)$, the following holds for any $T > 0$

$$
\begin{aligned}
J_2(u, w) + V(x(T), T) - V(x(t_0), t_0) &= \int_{t_0}^T \left\{ \frac{1}{2}\|z\|^2 + \dot{V}(x, t) \right\} dt \\
&= \int_{t_0}^T \left\{ \frac{1}{2}\|z\|^2 + V_t(x, t) + V_x(f(x) + \right. \\
&\quad \left. g_1(x)w + g_2(x)u) \right\} dt \\
&= \int_{t_0}^T \left\{ V_t + V_x f(x) + V_x g_1(x) w + \right. \\
&\quad \left. \frac{1}{2}\|u + g_2^T V_x^T\|^2 - \|g_2^T(x)V_x^T\|^2 + \frac{1}{2}\|h_1^T(x)\|^2 \right\} dt.
\end{aligned}
$$

Using the HJIE (11.12) in the above, we get

$$J_2(u, w) + V(x(T), T) - V(x(t_0), t_0) = \int_{t_0}^T \left\{ \frac{1}{2}\|u + g_2^T V_x^T\|^2 + V_x g_1(x)(w - \frac{1}{\gamma^2}g_1^T Y_x^T) \right\} dt.$$

Substituting now $w = w^\star$ we get

$$J_2(u, w^\star) + V(x(T), T) - V(x(t_0), t_0) = \int_{t_0}^T \frac{1}{2}\|u + g_2^T V_x^T\|^2 dt \geq 0,$$

and therefore

$$J_2(u^\star, w^\star) \leq J_2(u, w^\star)$$

with

$$J_2(u^\star, w^\star) = V(x(t_0), t_0).\;\square$$

We can specialize the results of the above theorem to the linear system

$$\Sigma^l : \begin{cases} \dot{x} &= Ax + B_1 w + B_2 u \\ z &= C_1 x + D_{12} w \\ y &= x \end{cases} \tag{11.17}$$

under the following assumption.

Assumption 11.1.2

$$C_1^T D_{12} = 0, \quad D_{12}^T D_{12} = I.$$

Then we have the following corollary.

Corollary 11.1.1 *Consider the linear system Σ^l under the Assumption 11.1.2. Suppose* *there exist $P_1(t) \leq 0$ and $P_2(t) \geq 0$ solutions of the cross-coupled Riccati ordinary-differential-equations (ODEs):*

$$-\dot{P}_1(t) = A^T P_1 + P_1(t)A - [P_1(t) \ P_2(t)] \begin{bmatrix} \gamma^{-2}B_1 B_1^T & B_2 B_2^T \\ B_2 B_2^T & B_2 B_2^T \end{bmatrix} \begin{bmatrix} P_1(t) \\ P_2(t) \end{bmatrix} - C_1^T C_1,$$

$$P_1(T) = 0$$

$$-\dot{P}_2(t) = A^T P_2 + P_2(t)A - [P_1(t) \ P_2(t)] \begin{bmatrix} 0 & \gamma^{-2}B_1 B_1^T \\ \gamma^{-2}B_1 B_1^T & B_2 B_2^T \end{bmatrix} \begin{bmatrix} P_1(t) \\ P_2(t) \end{bmatrix} + C_1^T C_1,$$

$$P_2(T) = 0$$

on $[0,T]$. Then, the Nash-equilibrium strategies uniquely specified by

$$u^\star = -B_2^T P_2(t)x(t)$$

$$w^\star = -\frac{1}{\gamma^2}B_1^T P_1(t)x(t)$$

solve the finite-horizon $SFBMH2HICP$ for the system. Moreover, the optimal costs for the game associated with the system are given by

$$J_1(u^\star, w^\star) = \frac{1}{2}x^T(t_0)P_1(t_0)x(t_0),$$

$$J_2(u^\star, w^\star) = \frac{1}{2}x^T(t_0)P_2(t_0)x(t_0).$$

Proof: Take

$$Y(x(t),t) = \frac{1}{2}x^T(t)P_1(t)x(t), \quad P_1(t) \leq 0$$

$$V(x(t),t) = \frac{1}{2}x^T(t)P_2(t)x(t), \quad P_2(t) \geq 0$$

and apply the results of the Theorem. $\square$

Remark 11.1.1 *In the above corollary, we considered negative and positive-(semi)definite solutions of the Riccati (ODEs), while in Theorem 11.1.1 we considered strict definite solutions of the HJIEs. However, it is generally sufficient to consider semidefinite solutions of the HJIEs.*

11.1.1 The Infinite-Horizon Problem

In this subsection, we consider the infinite-horizon $SFBMH2HINLCP$ for the affine nonlinear system Σ^a. In this case, we let $T \to \infty$, and seek time-invariant functions and feedback gains that solve the HJIEs. Because of this, it is necessary to require that the closed-loop system is locally asymptotically-stable as stated in item (c) of the definition. However, to achieve this, some additional assumptions on the system might be necessary. The following theorem gives sufficient conditions for the solvability of this problem. We recall the definition of detectability first.

Definition 11.1.2 *The pair $\{f,h\}$ is said to be locally zero-state detectable if there exists a neighborhood $\mathcal{O}$ of $x = 0$ such that, if $x(t)$ is a trajectory of $\dot{x}(t) = f(x)$ satisfying $x(t_0) \in \mathcal{O}$, then $h(x(t))$ is defined for all $t \geq t_0$, and $h(x(t)) \equiv 0$, for all $t \geq t_s$, implies $\lim_{t \to \infty} x(t) = 0$. Moreover, $\{f,h\}$ is detectable if $\mathcal{O} = \mathcal{X}$.*

Theorem 11.1.2 *Consider the nonlinear system Σ^a defined by (11.4) and the infinite-horizon $SFBMH2HINLCP$ with cost functions (11.7), (11.8). Suppose*

(H1) *the pair $\{f, h_1\}$ is zero-state detectable;*

(H2) *there exists a pair of negative and positive-definite C^1-functions $\tilde{Y}, \tilde{V} : \tilde{N} \to \Re$ locally defined in a neighborhood $\tilde{N}$ of the origin $x = 0$, and satisfying the coupled HJIEs:*

$$\tilde{Y}_x(x)f(x) - \frac{1}{2}\tilde{V}_x(x)g_2(x)g_2^T(x)\tilde{V}_x^T(x) - \frac{1}{2\gamma^2}\tilde{Y}_x(x)g_1(x)g_1^T(x)\tilde{Y}_x^T(x) -$$

$$\tilde{Y}_x(x)g_2(x)g_2^T(x)\tilde{V}^T(x) - \frac{1}{2}h_1^T(x)h_1(x) = 0, \quad \tilde{Y}(0) = 0 \qquad (11.18)$$

$$\tilde{V}_x(x)f(x) - \frac{1}{2}\tilde{V}_x(x)g_2(x)g_2^T(x)\tilde{V}_x^T(x) - \frac{1}{\gamma^2}\tilde{V}_x(x)g_1(x)g_1^T(x)\tilde{Y}_x^T(x) +$$

$$\frac{1}{2}h_1^T(x)h_1(x) = 0, \quad \tilde{V}(0) = 0. \qquad (11.19)$$

Then the state-feedback controls

$$u^\star(x) = -g_2^T(x)\tilde{V}_x^T(x) \qquad (11.20)$$

$$w^\star(x) = -\frac{1}{\gamma^2}g_1^T(x)\tilde{Y}_x^T(x) \qquad (11.21)$$

solve the infinite-horizon $SFBMH2HINLCP$ for the system. Moreover, the optimal costs are given by

$$J_1^\star(u^\star, w^\star) = \tilde{Y}(x_0) \qquad (11.22)$$

$$J_2^\star(u^\star, w^\star) = \tilde{V}(x_0). \qquad (11.23)$$

Proof: We only prove item (c) in the definition, since the proofs of items (a) and (b) are similar to the finite-horizon case. Using similar manipulations as in the proof of item (a) of Theorem 11.1.1, it can be shown that with $w \equiv 0$,

$$\dot{\tilde{Y}} = -\frac{1}{2}\|z\|^2,$$

for some function $\check{Y} = -\tilde{Y} > 0$. Therefore the closed-loop system is Lyapunov-stable. Further, the condition $\dot{\check{Y}} \equiv 0 \forall t \geq t_s$, for some $t_s \geq t_0$, implies that $u^\star \equiv 0$, $h_1(x) \equiv 0$. By hypothesis $(H1)$, this implies $\lim_{t\to\infty} x(t) = 0$, and we can conclude asymptotic-stability by LaSalle's invariance-principle. $\square$

The above theorem can again be specialized to the linear system Σ^l in the following corollary.

Corollary 11.1.2 *Consider the linear system Σ^l under the Assumption 11.1.2. Suppose (C_1, A) is detectable and there exists symmetric solutions $\bar{P}_1 \leq 0$ and $\bar{P}_2 \geq 0$ of the cross-coupled AREs:*

$$A^T\bar{P}_1 + \bar{P}_1 A - [\bar{P}_1 \ \ \bar{P}_2]\begin{bmatrix} \gamma^{-2}B_1B_1^T & B_2B_2^T \\ B_2B_2^T & B_2B_2^T \end{bmatrix}\begin{bmatrix} \bar{P}_1 \\ \bar{P}_2 \end{bmatrix} - C_1^TC_1 = 0,$$

$$A^T\bar{P}_2 + \bar{P}_2 A - [\bar{P}_1 \ \ \bar{P}_2]\begin{bmatrix} 0 & \gamma^{-2}B_1B_1^T \\ \gamma^{-2}B_1B_1^T & B_2B_2^T \end{bmatrix}\begin{bmatrix} \bar{P}_1 \\ \bar{P}_2 \end{bmatrix} + C_1^TC_1 = 0.$$

Then, the Nash-equilibrium strategies uniquely specified by

$$u^\star = -B_2^T\bar{P}_2(t)x(t) \qquad (11.24)$$

$$w^\star = -\frac{1}{\gamma^2}B_1^T\bar{P}_1 x(t) \qquad (11.25)$$

solve the infinite-horizon SFBMH2HINLCP for the system. Moreover, the optimal costs for the game associated with the system are given by

$$J_1(u^\star, w^\star) = \frac{1}{2}x^T(t_0)\bar{P}_1 x(t_0), \tag{11.26}$$

$$J_2(u^\star, w^\star) = \frac{1}{2}x^T(t_0)\bar{P}_2 x(t_0). \tag{11.27}$$

Proof: Take

$$Y(x,t) = \frac{1}{2}x^T \bar{P}_1 x, \quad \bar{P}_1 \le 0,$$

$$V(x,t) = \frac{1}{2}x^T \bar{P}_2 x, \quad \bar{P}_2 \ge 0.$$

and apply the result of the theorem. $\square$

Remark 11.1.2 *Sufficient conditions for the existence of the asymptotic solutions to the coupled algebraic-Riccati equations are discussed in Reference [222].*

The following proposition can also be proven.

Proposition 11.1.1 *Consider the nonlinear system (11.4) and the infinite-horizon SFBMH2HINLCP for this system. Suppose the following hold:*

(a1) $\{f + g_1 w^\star, h_1\}$ *is locally zero-state detectable;*

(a2) *there exists a pair of C^1 negative and positive-definite functions $\tilde{Y}, \tilde{V} : \tilde{N} \to \Re$ respectively, locally defined in a neighborhood $N \subset \mathcal{X}$ of the origin $x = 0$ satisfying the pair of coupled HJIEs (11.18), (11.19).*

Then $f - g_2 g_2^T \tilde{V}_x^T - \frac{1}{\gamma^2} g_1 g_1^T \tilde{Y}_x^T$ is locally asymptotically-stable.

Proof: Rewrite the HJIE (11.19) as

$$\tilde{V}_x(f(x) - g_2(x)g_2^T(x)\tilde{V}_x^T(x) - \frac{1}{2}g_1^T(x)g_1^T(x)\tilde{Y}_x(x)) = -\frac{1}{2}\|g_2^T(x)\tilde{V}_x^T(x)\|^2 - \frac{1}{2}\|h_1(x)\|^2$$

$$\Longleftrightarrow \dot{\tilde{V}} = -\frac{1}{2}\|g_2^T(x)\tilde{V}_x^T(x)\|^2 - \frac{1}{2}\|h_1(x)\|^2$$

$$\le 0.$$

Again the condition $\dot{\tilde{V}} \equiv 0 \, \forall t \ge t_s$, implies that $u^\star \equiv 0$, $h_1(x) \equiv 0 \, \forall t \ge t_s$. By Assumption (a1), this implies $\lim_{t\to\infty} x(t) = 0$, and by the LaSalle's invariance-principle, we conclude asymptotic-stability of the vector-field $f - g_2 g_2^T \tilde{V}_x^T - \frac{1}{\gamma^2} g_1 g_1^T \tilde{Y}_x^T$. $\square$

Remark 11.1.3 *The proof of the equivalent result for the linear system Σ^l can be pursued along the same lines. Moreover, many interesting corollaries could also be derived (see also Reference [179]).*

11.1.2 Extension to a General Class of Nonlinear Systems

In this subsection, we consider the *SFBMH2HINLCP* for a general class of nonlinear systems which are not necessarily affine, and extend the approach developed above to this class of systems. We consider the class of nonlinear systems described by

$$\Sigma : \begin{cases} \dot{x} &= F(x, w, u), \quad x(t_0) = x_0 \\ z &= Z(x, u) \\ y &= x \end{cases} \tag{11.28}$$

where all the variables have their previous meanings and dimensions, while $F : \mathcal{X} \times \mathcal{W} \times \mathcal{U} \to \Re^n$, $Z : \mathcal{X} \times \mathcal{U} \to \Re^s$ are smooth functions of their arguments. It is also assumed that $F(0,0,0) = 0$ and $Z(0,0) = 0$ and the system has a unique equilibrium-point at $x = 0$. The infinite-horizon $SFBMH2HINLCP$ for the above system can similarly be formulated as a dynamic two-person nonzero-sum differential game with the same cost functionals (11.7), (11.8). In addition, we also assume the following for simplicity.

Assumption 11.1.3 *The system matrices satisfy the following conditions:*

(i)

$$det \left[\left(\frac{\partial Z}{\partial u}(0,0) \right)^T \left(\frac{\partial Z}{\partial u}(0,0) \right) \right] \neq 0;$$

(ii)

$$Z(x,u) = 0 \Rightarrow u = 0$$

Define now the Hamiltonian functions $K_i : T^\star \mathcal{X} \times \mathcal{W} \times \mathcal{U} \to \Re$, $i = 1, 2$ corresponding to the cost functionals (11.7), (11.8) respectively:

$$K_1(x, w, u, \bar{Y}_x^T) = \bar{Y}_x(x)F(x, w, u) + \frac{1}{2}\gamma^2 \|w\|^2 - \|Z(x,u)\|^2$$

$$K_2(x, w, u, \bar{V}_x^T) = \bar{V}_x(x)F(x, w, u) + \frac{1}{2}\|Z(x,u)\|^2$$

for some smooth functions $\bar{Y}, \bar{V} : \mathcal{X} \to \Re$, and where $p_1 = \bar{Y}_x^T$, $p_2 = \bar{V}_x^T$ are the adjoint variables. Then, it can be shown that

$$\partial^2 K \big|_{w=0, u=0}(0) \triangleq \left[\begin{array}{cc} \frac{\partial^2 K_2}{\partial u^2} & \frac{\partial^2 K_2}{\partial w \partial u} \\ \frac{\partial^2 K_1}{\partial u \partial w} & \frac{\partial^2 K_1}{\partial w^2} \end{array} \right] \Bigg|_{w=0, u=0}(0)$$

$$= \left[\begin{array}{cc} \left(\frac{\partial Z}{\partial u}(0,0) \right)^T \left(\frac{\partial Z}{\partial u}(0,0) \right) & 0 \\ 0 & \gamma^2 I \end{array} \right]$$

is nonsingular by Assumption 11.1.3. Therefore, by the Implicit-function Theorem, there exists an open neighborhood $\bar{X}$ of $x = 0$ such that the equations

$$\frac{\partial K_1}{\partial w}(x, \bar{w}^\star(x), \bar{u}^\star(x)) = 0,$$

$$\frac{\partial K_2}{\partial u}(x, \bar{w}^\star(x), \bar{u}^\star(x)) = 0$$

have unique solutions $(\bar{u}^\star(x), \bar{w}^\star(x))$, with $\bar{u}^\star(0) = 0$, $\bar{w}^\star(0) = 0$. Moreover, the pair $(\bar{u}^\star, \bar{w}^\star)$ constitutes a Nash-equilibrium solution to the dynamic game (11.28), (11.7), (11.8). The following theorem summarizes the solution to the infinite-horizon problem for the general class of nonlinear systems (11.28).

Theorem 11.1.3 *Consider the nonlinear system (11.28) and the SFBMH2HINLCP for it. Suppose Assumption 11.1.3 holds, and also the following:*

(A1) *the pair $\{F(x,0,0), Z(x,0)\}$ is zero-state detectable;*

(A2) *there exists a pair of C^1 locally negative and positive-definite functions $\bar{Y}, \bar{V} : \bar{N} \to \Re$*

respectively, defined in a neighborhood $\bar{N}$ of $x = 0$, vanishing at $x = 0$ and satisfying the pair of coupled HJIEs:

$$\bar{Y}_x(x)F(x, \bar{w}^\star(x), \bar{u}^\star(x)) + \frac{1}{2}\gamma^2\|\bar{w}^\star(x)\|^2 - \frac{1}{2}\|Z(x, \bar{u}^\star(x))\|^2 = 0, \ \ \bar{Y}(0) = 0,$$

$$\bar{V}_x(x)F(x, \bar{w}^\star(x), \bar{u}^\star(x)) + \frac{1}{2}\|Z(x, \bar{u}^\star(x))\|^2 = 0, \ \ \bar{V}(0) = 0;$$

(A3) *the pair $\{F(x, \bar{w}^\star(x), 0), Z(x, 0)\}$ is zero-state detectable.*

Then, the state-feedback controls $(\bar{u}^\star(x), \bar{w}^\star(x))$ solve the dynamic game problem and the SFBMH2HINLCP for the system (11.28). Moreover, the optimal costs of the policies are given by

$$\begin{aligned} \bar{J}_1^\star(\bar{u}^\star, \bar{w}^\star) &= \bar{Y}(x_0), \\ \bar{J}_2^\star(\bar{u}^\star, \bar{w}^\star) &= \bar{V}(x_0). \end{aligned}$$

Proof: The proof can be pursued along the same lines as the previous results. $\square$

11.2 Discrete-Time Mixed $\mathcal{H}_2/\mathcal{H}_\infty$ Nonlinear Control

In this section, we discuss the state-feedback mixed $\mathcal{H}_2/\mathcal{H}_\infty$-control problem for discrete-time nonlinear systems. We begin with an affine discrete-time state-space model defined on $\mathcal{X} \subset \Re^n$ in coordinates $(x_1, \ldots, x_n)$:

$$\Sigma^{da} : \begin{cases} \dot{x}_{k+1} &= f(x_k) + g_1(x_k)w_k + g_2(x_k)u_k; \ \ x(k_0) = x^0 \\ z_k &= h_1(x_k) + k_{12}(x_k)u_k \\ y_k &= x_k, \end{cases} \tag{11.29}$$

where all the variables and system matrices have their previous meanings and dimensions respectively. We assume similarly that the system has a unique equilibrium at $x = 0$, and is such that $f(0) = 0$ and $h_1 0) = 0$. For simplicity, we similarly also assume the following hold for the system matrices.

Assumption 11.2.1 *The system matrices are such that*

$$\left. \begin{aligned} h_1^T(x)k_{12}(x) &= 0, \\ k_{12}^T(x)k_{12}(x) &= I. \end{aligned} \right\} \tag{11.30}$$

Again, as in the continuous-time case, the standard problem is to design a static state-feedback controller, $\mathbf{K}^{da}$, such that the $\mathcal{H}_2$-norm of the closed-loop system which is defined as

$$\|\mathbf{K}^{da} \circ \Sigma^{da}\|_{\ell_2} \overset{\Delta}{=} \sup_{0 \neq w_0 \in \mathcal{S}} \frac{\|z\|_{\mathcal{P}'}}{\|w_0\|_{\mathcal{S}'}},$$

and the $\mathcal{H}_\infty$-norm of the system defined by

$$\|\mathbf{K}^{da} \circ \Sigma^{da}\|_{\ell_\infty} \overset{\Delta}{=} \sup_{0 \neq w_1 \in \mathcal{P}'} \frac{\|z\|_2}{\|w_1\|_2},$$

are minimized over a time horizon $[k_0, K] \subset \mathbf{Z}$, where

$$\mathcal{P}' \triangleq \{w : w \in \ell_\infty, R_{ww}(k), S_{ww}(j\omega) \text{ exist for all } k \text{ and all } \omega \text{ resp.,}$$
$$\|w\|_{\mathcal{P}'} < \infty\}$$

$$\mathcal{S}' \triangleq \{w : w \in \ell_\infty, R_{ww}(k), S_{ww}(j\omega) \text{ exist for all } k \text{ and all } \omega \text{ resp.,}$$
$$\|S_{ww}(j\omega)\|_\infty < \infty\}$$

$$\|z\|_{\mathcal{P}'}^2 \triangleq \lim_{K \to \infty} \frac{1}{2K} \sum_{k=-K}^{K} \|z_k\|^2$$

$$\|w_0\|_{\mathcal{S}'}^2 = \|S_{w_0 w_0}(j\omega)\|_\infty$$

and $R_{ww}, S_{ww}(j\omega))$ are the autocorrelation and power spectral density matrices of w [152]. Notice also that $\|(.)\|_{\mathcal{P}'}$ and $\|(.)\|_{\mathcal{S}'}$ are seminorms. The spaces $\mathcal{P}'$ and $\mathcal{S}'$ are the discrete-time spaces of bounded-power and bounded-spectral signals respectively.

However, we do not solve the above standard problem in this section. Instead, we solve an associated suboptimal problem in which $w \in \ell_2[k_0, \infty)$ is a single disturbance, and the objective is to minimize the output energy $\|z\|_{\ell_2}$ subject to the constraint that $\|\Sigma^{da}\|_{\ell_\infty} \leq \gamma^*$ for some number $\gamma^* > 0$. The problem is more formally defined as follows.

Definition 11.2.1 *(Discrete-Time State-Feedback Mixed $\mathcal{H}_2/\mathcal{H}_\infty$ Nonlinear Control Problem (DSFBMH2HINLCP)).*

(A) *Finite-Horizon Problem $(K < \infty)$: Find (if possible!) a time-varying static state-feedback control law of the form:*

$$u = \tilde{\alpha}_{2d}(k, x_k), \quad \tilde{\alpha}_{2d}(k, 0) = 0, \quad k \in \mathbf{Z}$$

such that:

(a) *the closed-loop system*

$$\Sigma^{clda} : \begin{cases} x_{k+1} &= f(x_k) + g_1(x_k)w_k + g_2(x_k)\tilde{\alpha}_{2d}(x_k) \\ z_k &= h_1(x_k) + k_{21}(x_k)\tilde{\alpha}_{2d}(x_k) \end{cases} \tag{11.31}$$

is stable with $w = 0$ and has locally finite ℓ_2-gain from w to z less or equal to γ^*, starting from $x^0 = 0$, for all $k \in [k_0, K]$ and for a given number $\gamma^* > 0$.

(b) *the output energy $\|z\|_{\ell_2}$ of the system is minimized for all disturbances $w \in \mathcal{W} \subset \ell_2[k_0, \infty)$.*

(B) *Infinite-Horizon Problem $(K \to \infty)$: In addition to the items (a) and (b) above, it is also required that*

(c) *the closed-loop system Σ^{clda} defined above with $w \equiv 0$ is locally asymptotically-stable about the equilibrium-point $x = 0$.*

The problem is similarly formulated as a two-player nonzero-sum differential game with two cost functionals:

$$\min_{u \in \mathcal{U}, w \in \mathcal{W}} J_{1d}(u, w) = \frac{1}{2} \sum_{k=k_0}^{K} (\gamma^2 \|w_k\|^2 - \|z_k\|^2) \tag{11.32}$$

$$\min_{u \in \mathcal{U}, w \in \mathcal{W}} J_{2d}(u, w) = \frac{1}{2} \sum_{k=k_0}^{K} \|z_k\|^2. \tag{11.33}$$

Again, sufficient conditions for the solvability of the above dynamic game (11.32), (11.33), (11.29), and the existence of Nash-equilibrium strategies are given by the following pair of discrete-time Hamilton-Jacobi-Isaac's difference equations (DHJIEs):

$$W(x,k) = \inf_w \left\{ W(f(x)+g_1(x)w+g_2(x)u^\star, k+1) + \frac{1}{2}(\gamma^2\|w\|^2 - \|z_k^\star\|^2) \right\},$$
$$W(x, K+1) = 0, \tag{11.34}$$

$$U(x,k) = \inf_u \left\{ U(f(x)+g_1(x)w^\star+g_2(x)u, k+1) + \frac{1}{2}\|z_k^\star\|^2 \right\},$$
$$U(x, K+1) = 0, \tag{11.35}$$

for some smooth negative and positive-definite functions $W, U : \mathcal{X} \times \mathbf{Z} \to \Re$ respectively, and where $z_k^\star = h_1(x) + k_{12}(x)u^\star(x)$.

To solve the problem, we define the Hamiltonian functions $H_i : \mathcal{X} \times \mathcal{W} \times \mathcal{U} \times \Re \to \Re$, $i = 1, 2$ corresponding to the cost functionals (11.32), (11.32) respectively and the system equations (11.29):

$$H_1(x,w,u,W) = W(f(x)+g_1(x)w+g_2(x)u, k+1) - W(x) +$$
$$\frac{1}{2}(\gamma^2\|w\|^2 - \|z_k\|^2), \tag{11.36}$$
$$H_2(x,w,u,U) = U(f(x)+g_1(x)w+g_2(x)u, k+1) - U(x) +$$
$$\frac{1}{2}\|z_k\|^2. \tag{11.37}$$

Similarly, as in Chapter 8, let

$$\partial^2 H(x) \triangleq \begin{bmatrix} \frac{\partial^2 H_2}{\partial u^2} & \frac{\partial^2 H_2}{\partial w \partial u} \\ \frac{\partial^2 H_1}{\partial u \partial w} & \frac{\partial^2 H_1}{\partial w^2} \end{bmatrix}(x) \triangleq \begin{bmatrix} r_{11}(x) & r_{12}(x) \\ r_{21}(x) & r_{22}(x) \end{bmatrix}$$

$$F^\star(x) \triangleq f(x) + g_1(x)w^\star(x) + g_2(x)u^\star(x), \tag{11.38}$$

and therefore

$$\left. \begin{aligned} r_{11}(x) &= g_2^T(x)\frac{\partial^2 U}{\partial\lambda^2}\Big|_{\lambda=F^\star(x)} g_2(x) + I \\ r_{12}(x) &= g_2^T(x)\frac{\partial^2 U}{\partial\lambda^2}\Big|_{\lambda=F^\star(x)} g_1(x) \\ r_{21}(x) &= g_1^T(x)\frac{\partial^2 W}{\partial\lambda^2}\Big|_{\lambda=F^\star(x)} g_2(x) \\ r_{22}(x) &= \gamma^2 I + g_1^T(x)\frac{\partial^2 W}{\partial\lambda^2}\Big|_{\lambda=F^\star(x)} g_1(x). \end{aligned} \right\} \tag{11.39}$$

The following theorem then presents' sufficient conditions for the solvability of the finite-horizon problem.

Theorem 11.2.1 *Consider the discrete-time nonlinear system (11.29), and the finite-horizon DSFBMH2HINLCP with cost functionals (11.32), (11.33). Suppose there exists a pair of negative and positive-definite C^2 (with respect to the first argument)-functions $W, U : M \times \mathbf{Z} \to \Re$ locally defined in a neighborhood M of the origin $x = 0$, such that $W(0,k) = 0$ and $U(0,k) = 0$, and satisfying the coupled DHJIEs:*

$$W(x,k) = W(f(x)+g_1(x)w^\star+g_2(x)u^\star, k+1) + \frac{1}{2}(\gamma^2\|w\|^2 - \|z_k^\star\|^2),$$
$$W(x, K+1) = 0, \tag{11.40}$$

$$U(x,k) = U(f(x)+g_1(x)w^\star+g_2(x)u^\star, k+1) + \frac{1}{2}\|z_k^\star\|^2,$$
$$U(x, K+1) = 0, \tag{11.41}$$

together with the conditions

$$r_{22}(0) > 0, \quad det[r_{11}(0) - r_{22}^{-1}(0)r_{21}(0)] \neq 0. \tag{11.42}$$

Then the state-feedback controls defined implicitly by

$$w^\star = -g_1^T(x)\frac{1}{\gamma^2}\frac{\partial W}{\partial \lambda}\bigg|_{\lambda = f(x)+g_1(x)w^\star+g_2(x)u^\star} \tag{11.43}$$

$$u^\star = -g_2^T(x)\frac{\partial U}{\partial \lambda}\bigg|_{\lambda = f(x)+g_1(x)w^\star+g_2(x)u^\star} \tag{11.44}$$

solve the finite-horizon DSFBMH2HINLCP for the system. Moreover, the optimal costs are given by

$$J_{1d}^\star(u^\star, w^\star) = W(k_0, x_0), \tag{11.45}$$

$$J_{2d}^\star(u^\star, w^\star) = U(k_0, x_0). \tag{11.46}$$

Proof: We prove item (a) in Definition 11.2.1 first. Assume there exist solutions $W < 0$, $U > 0$ of the DHJIEs (11.40), (11.41), and consider the Hamiltonian functions $H_1(.,.,.,.)$, $H_2(.,.,.,.)$. Applying the necessary conditions for optimality

$$\frac{\partial H_1}{\partial w}(x, u, w) = 0, \quad \frac{\partial H_2}{\partial u}(x, u, w) = 0,$$

and solving these for $w^\star$, $u^\star$ respectively, we get the Nash-equilibrium strategies (11.43), (11.44). Moreover, if the conditions (11.42) are satisfied, then the matrix

$$\partial^2 H\big|_{w=0, u=0}(0) = \begin{bmatrix} r_{11}(0) & r_{12}(0) \\ r_{21}(0) & r_{22}(0) \end{bmatrix} =$$

$$\begin{bmatrix} I & r_{12}(0)r_{22}^{-1}(0) \\ 0 & I \end{bmatrix} \begin{bmatrix} r_{11}(0) - r_{12}(0)r_{22}^{-1}(0)r_{21}(0) & 0 \\ 0 & r_{22}(0) \end{bmatrix} \begin{bmatrix} I & 0 \\ r_{22}^{-1}(0)r_{21}(0) & I \end{bmatrix}$$

is nonsingular. Therefore, by the Implicit-function Theorem, there exist open neighborhoods X_1 of $x = 0$, W_1 of $w = 0$ and U_1 of $u = 0$, such that the equations (11.43), (11.44) have unique solutions.

Now suppose, $(u^\star, w^\star)$ have been obtained from (11.43), (11.44), then subsituting in the DHJIEs (11.34), (11.35) yield the DHJIEs (11.40), (11.41). Moreover, by Taylor-series expansion, we can write

$$H_1(x, w, u^\star(x), W) = H_1(x, w^\star(x).u^\star(x), W) + \frac{1}{2}(w - w^\star(x))^T[r_{22}(x) +$$

$$O(\|w - w^\star(x)\|)](w - w^\star(x)).$$

In addition, since $r_{22}(0) > 0$ implies $r_{22}(x) > 0$ for all x in a neighborhood X_2 of $x = 0$ by the Inverse-function Theorem [234], it then follows from above that there exists also a neighborhood W_2 of $w = 0$ such that

$$H_1(x, w, u^\star(x), W) \geq H_1(x, w^\star(x), u^\star(x), W) = 0 \quad \forall x \in X_2, \forall w \in W_2,$$

$$\Leftrightarrow W(f(x) + g_1(x)w + g_2(x)u^\star(x), k+1) - W(x, k) + \frac{1}{2}(\gamma^2\|w\|^2 - \|u^\star(x)\|^2 - \|h_1\|^2) \geq 0$$

$$\forall x \in X_2, \forall w \in W_2.$$

Setting now $w = 0$, we have

$$\tilde{W}(f(x) + g_2(x)u^\star(x), k+1) - \tilde{W}(x, k) \leq -\frac{1}{2}\|u^\star(x)\|^2 - \frac{1}{2}\|h_1(x)\|^2 \leq 0$$

for some function $\tilde{W} = -W > 0$. Hence, the closed-loop system is Lyapunov-stable. To prove item (b), consider the Hamiltonian function $H_2(., w^\star, ., .)$ and expand it in Taylor's-series:

$$H_2(x, w^\star, u, U) = H_2(x, w^\star, u^\star, U) + \frac{1}{2}(u - u^\star)^T[r_{11}(x) + O(\|u - u^\star\|)](u - u^\star).$$

Since

$$r_{11}(0) = I + g_2^T(0)\frac{\partial^2 W}{\partial \lambda^2}(0)g_2(0) \geq I,$$

again there exists a neighborhood $\tilde{X}_2$ of $x = 0$ such that $r_{11}(x) > 0$ by the Inverse-function Theorem. Therefore,

$$H_2(x, w^\star, u, U) \geq H_2(x, w^\star, u^\star, U) = 0 \quad \forall u \in \mathcal{U}$$

and the $\mathcal{H}_2$-cost is minimized.

Finally, we determine the optimal costs of the strategies. For this, consider the cost functional $J_{1d}(u^\star, w^\star)$ and write it as

$$
\begin{aligned}
J_{1d}(u, w) + W(x_{K+1}, K + 1) - W(x_{k_0}, k_0) &= \sum_{k=k_0}^{K} \left\{ \frac{1}{2}(\gamma^2\|w_k^\star\|^2 - \|z_k^\star\|^2) + \right. \\
&\qquad \left. W(x_{k+1}, k + 1) - W(x_k, k) \right\} \\
&= \sum_{k=k_0}^{K} H_1(x, w^\star, u^\star, W) = 0.
\end{aligned}
$$

Since $W(x_{K+1}, K + 1) = 0$, we have the result. Similarly, for $J_2(w^\star, u^\star)$, we have

$$
\begin{aligned}
J_{2d}(u, w) + U(x_{K+1}, K + 1) - U(x_{k_0}, k_0) &= \sum_{k=k_0}^{K} \left\{ \frac{1}{2}\|z_k^\star\|^2 + U(x_{k+1}, k + 1) - U(x_k, k) \right\} \\
&= \sum_{k=k_0}^{K} H_2(x, w^\star, u^\star, U) = 0,
\end{aligned}
$$

and since $U(x_{K+1}, K + 1) = 0$, the result also follows. $\square$

The above result can be specialized to the linear discrete-time system

$$\Sigma^{dl} : \begin{cases} \dot{x}_{k+1} &= Ax_k + B_1 w_k + B_2 u_k \\ z_k &= C_1 x_k + D_{12} w_k \\ y_k &= x_k \end{cases} \tag{11.47}$$

where all the variables and matrices have their previous meanings and dimensions. Then we have the following corollary to Theorem 11.2.1.

Corollary 11.2.1 *Consider the linear system Σ^{dl} under the Assumption 11.1.2. Suppose there exist $P_{1,k} < 0$ and $P_{2,k} > 0$ symmetric solutions of the cross-coupled discrete-Riccati* *difference-equations (DRDEs):*

$$
\begin{aligned}
P_{1,k} =\ & A^T\Big\{P_{1,k} - 2P_{1,k}B_1B_{\gamma,k}^{-1}\Gamma_{1,k} - 2P_{1,k}B_2\Lambda_k^{-1}B_2^T P_{2,k} + \\
& 2\Gamma_{1,k}^T B_{\gamma,k}^{-T}B_1 P_1 B_2\Lambda_k^{-1}B_2 P_{2,k}\Gamma_{2,k} + \Gamma_{1,k}^T B_{\gamma,k}^{-T}B_1 P_{1,k}B_1 B_{\gamma,k}^{-1}\Gamma_{1,k} + \\
& \Gamma_{2,k}^T P_{2,k}B_2\Lambda_k^{-T}B_2 P_{1,k}B_2\Lambda_k^{-1}B_2^T P_{2,k}\Gamma_{2,k} + \gamma^2\Gamma_{1,k}^T B_{\gamma,k}^{-T}B_{\gamma,k}^{-1}\Gamma_{1,k} - \\
& \Gamma_{2,k}^T P_{2,k}B_2\Lambda_k^{-T}\Lambda_k^{-1}B_2^T P_2\Gamma_{2,k}\Big\}A - C_1^T C_1, \quad P_{1,K} = 0 \quad\quad (11.48)
\end{aligned}
$$

$$
\begin{aligned}
P_{2,k} =\ & A^T\Big\{P_{2,k} - 2P_{2,k}B_1B_{\gamma,k}^{-1}\Gamma_{1,k} - 2P_{2,k}B_2\Lambda_k^{-1}B_2^T P_{2,k}\Gamma_{2,k} + \\
& 2\Gamma_{1,k}^T B_{\gamma,k}^{-T}B_1 P_{2,k}B_2\Lambda_k^{-1}B_{2,k}^T P_{2,k}\Gamma_{2,k} + \\
& \Gamma_{2,k}^T P_{2,k}B_2\Lambda_k^{-T}\Lambda_k^{-1}B_2^T P_{2,k}\Gamma_{2,k}\Big\}A + C_1^T C_1, \quad P_{2,K} = 0 \quad\quad (11.49)
\end{aligned}
$$

$$
B_{\gamma,k} := [\gamma^2 I - B_2\Lambda_k^{-1}B_2^T P_{2,k}B_1 + B_1^T P_{1,k}B_1] > 0 \quad\quad (11.50)
$$

for all k in $[k_0, K]$. Then, the Nash-equilibrium strategies uniquely specified by

$$
w_{l,k}^\star = -B_{\gamma,k}^{-1}\Gamma_{1,k}Ax_k, \quad k \in [k_0, K] \quad\quad (11.51)
$$
$$
u_{l,k}^\star = -\Lambda_k^{-1}B_2^T P_{2,k}\Gamma_{2,k}Ax_k, \quad k \in [k_0, K] \quad\quad (11.52)
$$

where

$$
\begin{aligned}
\Lambda_k &:= (I + B_2^T P_{2,k}B_2), \quad \forall k, \\
\Gamma_{1,k} &:= [B_1^T P_{1,k} - B_2\Lambda_k^{-1}B_2^T P_{2,k}] \quad \forall k, \\
\Gamma_{2,k} &:= [I - B_1B_{\gamma,k}^{-1}(B_1^T P_{1,k} - B_2\Lambda_k^{-1}B_2^T P_{2,k})] \quad \forall k,
\end{aligned}
$$

solve the finite-horizon DSFBMH2HINLCP for the system. Moreover, the optimal costs for the game are given by

$$
J_{1,l}(u^\star, w^\star) = \frac{1}{2}x_{k_0}^T P_{1,k_0}x_{k_0}, \quad\quad (11.53)
$$

$$
J_{2,l}(u^\star, w^\star) = \frac{1}{2}x_{k_0}^T P_{2,k_0}x_{k_0}. \qu\quad (11.54)
$$

Proof: Assume the solutions to the coupled HJIEs are of the form,

$$
W(x_k, k) = \frac{1}{2}x_k^T P_{1,k}x_k, \quad P_{1,k} < 0, \quad k = 1, \ldots, K, \quad\quad (11.55)
$$

$$
U(x_k, k) = \frac{1}{2}x_k^T P_{2,k}x_k, \quad P_{2,k} > 0, \quad k = 1, \ldots, K. \quad\quad (11.56)
$$

Then, the Hamiltonians $H_1(.,.,.,.)$, $H_2(.,.,.,.)$ are given by

$$
\begin{aligned}
H_{1,l}(x, w, u, W) =\ & \frac{1}{2}(Ax + B_1w + B_2u)^T P_{1,k}(Ax + B_1w + B_2u) - \\
& \frac{1}{2}x^T P_{1,k}x + \frac{1}{2}\gamma^2\|w\|^2 - \frac{1}{2}\|z\|^2,
\end{aligned}
$$

$$
\begin{aligned}
H_{2,l}(x, w, u, U) =\ & \frac{1}{2}(Ax + B_1w + B_2u)^T P_{2,k}(Ax + B_1w + B_2u) - \\
& \frac{1}{2}x^T P_{2,k}x + \frac{1}{2}\|z\|^2.
\end{aligned}
$$

Applying the necessary conditions for optimality, we get

$$\frac{\partial H_{1,l}}{\partial w} = B_1^T P_{1,k}(Ax + B_1w + B_2u) + \gamma^2 w = 0, \tag{11.57}$$

$$\frac{\partial H_{2,l}}{\partial u} = B_2^T P_{2,k}(Ax + B_1w + B_2u) + u = 0. \tag{11.58}$$

Solving the last equation for u we have

$$u = -(I + B_2^T P_{2,k} B_2)^{-1} B_2^T P_{2,k}(Ax + B_1w),$$

which upon substitution in the first equation gives

$$B_1^T P_{1,k}\left\{ Ax + B_1w - B_2(I + B_2^T P_{2,k} B_2)^{-1} B_2^T P_{2,k}(Ax + B_1w) \right\} + \gamma^2 w = 0$$
$$\Longleftrightarrow w_{l,k}^\star = -B_{\gamma,k}^{-1}[B_1^T P_{1,k} - B_2\Lambda_k^{-1} B_2^T P_{2,k}]Ax_k$$
$$= -B_{\gamma,k}^{-1}\Gamma_{1,k} Ax_k \quad k \in [k_0, K], \tag{11.59}$$

if and only if

$$B_{\gamma,k} := [\gamma^2 I - B_2\Lambda_k^{-1} B_2^T P_{2,k} B_1 + B_1^T P_{1,k} B_1] > 0 \ \forall k,$$

where

$$\Lambda_k := (I + B_2^T P_{2,k} B_2), \quad \forall k,$$
$$\Gamma_{1,k} := [B_1^T P_{1,k} - B_2\Lambda_k^{-1} B_2^T P_{2,k}] \ \forall k.$$

Notice that Λ_k is nonsingular for all k since $P_{2,k}$ is positive-definite. Now, substitute $w^\star$ in the expression for u to get

$$u_{l,k}^\star = -\Lambda_k^{-1} B_2^T P_{2,k}[I - B_1 B_{\gamma,k}^{-1}(B_1^T P_{1,k} - B_2\Lambda_k^{-1} B_2^T P_{2,k})]Ax_k, \quad k \in [k_0, K]$$
$$= -\Lambda_k^{-1} B_2^T P_{2,k}\Gamma_{2,k} Ax_k \tag{11.60}$$

where

$$\Gamma_{2,k} := [I - B_1 B_{\gamma,k}^{-1}(B_1^T P_{1,k} - B_2\Lambda_k^{-1} B_2^T P_{2,k})].$$

Finally, substituting $(u_{l,k}^\star, w_{l,k}^\star)$ in the DHJIEs (11.40), (11.41), we get the DRDEs (11.48), (11.49). The optimal costs are also obtained by substitution in (11.45), (11.46). $\square$

Remark 11.2.1 *Note, in the above Corollary 11.2.1 for the solution of the linear discrete-time problem, it is better to consider strictly positive-definite solutions of the DRDEs (11.48), (11.49) because the condition $B_{\gamma,k} > 0$ must be respected for all k.*

11.2.1 The Infinite-Horizon Problem

In this subsection, we consider similarly the infinite-horizon $DSFBMH2HINLCP$ for the affine discrete-time nonlinear system Σ^{da}. We let $K \to \infty$, and seek time-invariant functions and feedback gains that solve the $DSFBMH2HINLCP$. Again we require that the closed-loop system be locally asymptotically-stable, and for this, we need the following definition of detectability for the discrete-time system Σ^{da}.

Definition 11.2.2 *The pair $\{f, h\}$ is said to be locally zero-state detectable if there exists a neighborhood $\tilde{\mathcal{O}}$ of $x = 0$ such that, if x_k is a trajectory of $x_{k+1} = f(x_k)$ satisfying $x(k_0) \in \tilde{\mathcal{O}}$, then $h(x_k)$ is defined for all $k \geq k_0$, and $h(x_k) = 0$ for all $k \geq k_s$, implies $\lim_{k\to\infty} x_k = 0$. Moreover $\{f, h\}$ is said to be zero-state detectable if $\tilde{\mathcal{O}} = \mathcal{X}$.*

Theorem 11.2.2 *Consider the nonlinear system Σ^{da} defined by (11.29) and the infinite-horizon $DSFBMH2HINLCP$ with cost functionals (11.32), (11.8). Suppose*

(H1) *the pair $\{f, h_1\}$ is zero-state detectable;*

(H2) *there exists a pair of negative and positive-definite C^2-functions $\tilde{W}, \tilde{U} : \tilde{M} \times \mathbf{Z} \to \Re$ locally defined in a neighborhood $\tilde{M}$ of the origin $x = 0$, such that $\tilde{W}(0) = 0$ and $\tilde{U}(0) = 0$, and satisfying the coupled DHJIEs:*

$$\tilde{W}(f(x) + g_1(x)w^\star + g_2(x)u^\star) - \tilde{W}(x) + \frac{1}{2}(\gamma^2\|w\|^2 - \|z_k^\star\|^2) = 0, \quad W(0) = 0, \; (11.61)$$

$$\tilde{U}(f(x) + g_1(x)w^\star + g_2(x)u^\star) - \tilde{U}(x) + \frac{1}{2}\|z_k^\star\|^2 = 0, \quad U(0) = 0. \; (11.62)$$

together with the conditions

$$r_{22}(0) > 0, \quad det[r_{11}(0) - r_{22}^{-1}(0)r_{21}(0)] \neq 0. \tag{11.63}$$

Then, the state-feedback controls defined implicitly by

$$w^\star = -g_1^T(x)\frac{1}{\gamma^2}\frac{\partial\tilde{W}}{\partial\lambda}\bigg|_{\lambda=f(x)+g_1(x)w^\star+g_2(x)u^\star} \tag{11.64}$$

$$u^\star = -g_2^T(x)\frac{\partial\tilde{U}}{\partial\lambda}\bigg|_{\lambda=f(x)+g_1(x)w^\star+g_2(x)u^\star} \tag{11.65}$$

solve the infinite-horizon $DSFBMH2HINLCP$ for the system. Moreover, the optimal costs are given by

$$J_{1d}^\star(u^\star, w^\star) = \tilde{W}(x_0), \tag{11.66}$$

$$J_{2d}^\star(u^\star, w^\star) = \tilde{U}(x_0). \tag{11.67}$$

Proof: We only prove item (c) in the definition, since the proofs of items (a) and (b) are exactly similar to the finite-horizon problem. Accordingly, using similar manipulations as in the proof of item (a) of Theorem 11.2.1, it can be shown that with $w \equiv 0$,

$$\tilde{W}(f(x) + g_2(x)u^\star) - \tilde{W}(x) = -\frac{1}{2}\|z\|^2.$$

Therefore, the closed-loop system is Lyapunov-stable. Further, the condition $\tilde{W}(f(x) + g_2(x)u^\star(x)) \equiv \tilde{W}(x) \forall k \geq k_c$, for some $k_c \geq k_0$, implies that $u^\star \equiv 0$, $h_1(x) \equiv 0 \forall k \geq k_c$. By hypothesis $(H1)$, this implies $\lim_{t\to\infty} x_k = 0$, and by LaSalle's invariance-principle, we conclude asymptotic-stability. $\square$

    The above theorem can again be specialized to the linear system Σ^{dl} in the following corollary.

Corollary 11.2.2 *Consider the discrete linear system Σ^{dl} under the Assumption 11.1.2. Suppose there exist $\bar{P}_1 < 0$ and $\bar{P}_2 > 0$ symmetric solutions of the cross-coupled discrete-*

algebraic Riccati equations (DAREs):

$$
\begin{aligned}
\bar{P}_1 &= A^T \Big\{ \bar{P}_1 - 2\bar{P}_1 B_1 B_\gamma^{-1} \Gamma_1 - 2\bar{P}_1 B_2 \Lambda^{-1} B_2^T \bar{P}_2 + 2\Gamma_1^T B_\gamma^{-T} B_1 \bar{P}_1 B_2 \Lambda^{-1} B_2 \bar{P}_2 \Gamma_2 + \\
&\quad \Gamma_1^T B_\gamma^{-T} B_1 \bar{P}_1 B_1 B_\gamma^{-1} \Gamma_1 + \Gamma_2^T \bar{P}_2 B_2 \Lambda^{-T} B_2 \bar{P}_1 B_2 \Lambda^{-1} B_2^T \bar{P}_2 \Gamma_2 + \gamma^2 \Gamma_1^T B_\gamma^{-T} B_\gamma^{-1} \Gamma_1 - \\
&\quad \Gamma_2^T \bar{P}_2 B_2 \Lambda^{-T} \Lambda^{-1} B_2^T \bar{P}_2 \Gamma_2 \Big\} A - C_1^T C_1,
\end{aligned}
\tag{11.68}
$$

$$
\begin{aligned}
\bar{P}_2 &= A^T \Big\{ \bar{P}_2 - 2\bar{P}_2 B_1 B_\gamma^{-1} \Gamma_1 - 2\bar{P}_2 B_2 \Lambda^{-1} B_2^T \bar{P}_2 \Gamma_2 + \\
&\quad 2\Gamma_1^T B_\gamma^{-T} B_1 \bar{P}_2 B_2 \Lambda_k^{-1} B_2^T \bar{P}_2 \Gamma_2 + \Gamma_2^T \bar{P}_2 B_2 \Lambda^{-T} \Lambda^{-1} B_2^T \bar{P}_2 \Gamma_2 \Big\} A + C_1^T C_1.
\end{aligned}
\tag{11.69}
$$

$$
B_\gamma := [\gamma^2 I - B_2 \Lambda^{-1} B_2^T P_2 B_1 + B_1^T P_1 B_1] > 0
\tag{11.70}
$$

Then the Nash-equilibrium strategies uniquely specified by

$$
w_{l,k}^\star = -B_\gamma^{-1} \Gamma_1 A x_k,
\tag{11.71}
$$

$$
u_{l,k}^\star = \Lambda^{-1} B_2^T \bar{P}_2 \Gamma_2 A x_k,
\tag{11.72}
$$

where

$$
\begin{aligned}
\Lambda &:= (I + B_2^T \bar{P}_2 B_2), \\
\Gamma_1 &:= [B_1^T \bar{P}_1 - B_2 \Lambda^{-1} B_2^T \bar{P}_2], \\
\Gamma_2 &:= [I - B_1 B_\gamma^{-1} (B_1^T P_1 - B_2 \Lambda^{-1} B_2^T \bar{P}_2)],
\end{aligned}
$$

solve the infinite-horizon $DSFBMH2HINLCP$ for the system. Moreover, the optimal costs for the game are given by

$$
J_{1,l}(u^\star, w^\star) = \frac{1}{2} x_{k_0}^T \bar{P}_1 x_{k_0},
\tag{11.73}
$$

$$
J_{2,l}(u^\star, w^\star) = \frac{1}{2} x_{k_0}^T \bar{P}_2 x_{k_0}.
\tag{11.74}
$$

Proof: Take

$$
Y(x_k) = \frac{1}{2} x_k^T \bar{P}_1 x_k, \quad \bar{P}_1 < 0
$$

$$
V(x) = \frac{1}{2} x_k^T \bar{P}_2 x_k, \quad \bar{P}_2 > 0
$$

and apply the results of the theorem. $\square$

11.3 Extension to a General Class of Discrete-Time Nonlinear Systems

In this subsection, we similarly extend the results of the previous subsection to a more general class of nonlinear discrete-time systems which is not necessarily affine. We consider the following state-space model defined on $\mathcal{X} \subset \Re^n$ in local coordinates $(x_1, \ldots, x_n)$

$$
\Sigma : \begin{cases}
\dot{x}_{k+1} &= \tilde{F}(x_k, w_k, u_k), \quad x(t_0) = x_0 \\
z_k &= \tilde{Z}(x_k, u_k) \\
y_k &= x_k,
\end{cases}
\tag{11.75}
$$

where all the variables have their previous meanings, while $\tilde{F} : \mathcal{X} \times \mathcal{W} \times \mathcal{U} \to \mathcal{X}$, $\tilde{Z} : \mathcal{X} \times \mathcal{U} \to \Re^s$ are smooth functions of their arguments. In addition, we assume that $\tilde{F}(0,0,0) = 0$ and $\tilde{Z}(0,0) = 0$. Furthermore, define similarly the Hamiltonian functions corresponding to the cost functionals (11.32), (11.33), $\tilde{K}_i : \mathcal{X} \times \mathcal{W} \times \mathcal{U} \times \Re \to \Re, i = 1, 2$ respectively:

$$\widetilde{K}_1(x, w, u, \widetilde{W}) = \widetilde{W}(\tilde{F}(x, w, u)) - \widetilde{W}(x) + \frac{1}{2}\gamma^2\|w\|^2 - \|\tilde{Z}(x, u)\|^2,$$

$$\widetilde{K}_2(x, w, u, \bar{U}) = \tilde{U}(\tilde{F}(x, w, u)) - \tilde{U}(x) + \frac{1}{2}\|\tilde{Z}(x, u)\|^2,$$

for some smooth functions $\widetilde{W}, \tilde{U} : \mathcal{X} \to \Re$. In addition, define also

$$\partial^2 \widetilde{K}(x) \triangleq \left[\begin{array}{cc} \frac{\partial^2 \widetilde{K}_2}{\partial u^2} & \frac{\partial^2 \widetilde{K}_2}{\partial w \partial u} \\ \frac{\partial^2 \widetilde{K}_1}{\partial u \partial w} & \frac{\partial^2 \widetilde{K}_1}{\partial w^2} \end{array} \right] (x) = \left[\begin{array}{cc} s_{11}(x) & s_{12}(x) \\ s_{21}(x) & s_{22}(x) \end{array} \right],$$

where

$$s_{11}(0) = \left[\left(\frac{\partial \tilde{F}}{\partial u}\right)^T \frac{\partial^2 \tilde{U}}{\partial \lambda^2}(0)\frac{\partial \tilde{F}}{\partial u} + \left(\frac{\partial \tilde{Z}}{\partial u}\right)^T \frac{\partial \tilde{Z}}{\partial u} \right]_{x=0, w=0, u=0},$$

$$s_{12}(0) = \left[\left(\frac{\partial \tilde{F}}{\partial u}\right)^T \frac{\partial^2 \tilde{U}}{\partial \lambda^2}(0)\frac{\partial \tilde{F}}{\partial w} \right]_{x=0, w=0, u=0},$$

$$s_{21}(0) = \left[\left(\frac{\partial \tilde{F}}{\partial w}\right)^T \frac{\partial^2 \widetilde{W}}{\partial \lambda^2}(0)\frac{\partial \tilde{F}}{\partial u} \right]_{x=0, w=0, u=0},$$

$$s_{22}(0) = \left[\left(\frac{\partial \tilde{F}}{\partial w}\right)^T \frac{\partial^2 \widetilde{W}}{\partial \lambda^2}(0)\frac{\partial \tilde{F}}{\partial w} + \gamma^2 I \right]_{x=0, w=0, u=0}.$$

We then make the following assumption.

Assumption 11.3.1 *For the Hamiltonian functions, $\widetilde{K}_1$, $\widetilde{K}_2$, we assume*

$$s_{22}(0) > 0, \quad det[s_{11}(0) - s_{12}(0)s_{22}^{-1}s_{21}(0)] \neq 0.$$

Under the above assumption, the Hessian matrix $\partial^2 \widetilde{K}(0)$ is nonsingular, and therefore by the Implicit-function Theorem, there exists an open neighborhood M_0 of $x = 0$ such that the equations

$$\frac{\partial \widetilde{K}_1}{\partial w}(x, \tilde{w}^\star(x), \tilde{u}^\star(x)) = 0,$$

$$\frac{\partial \widetilde{K}_2}{\partial u}(x, \tilde{w}^\star(x), \tilde{u}^\star(x)) = 0$$

have unique solutions $\tilde{u}^\star(x)$, $\tilde{w}^\star(x)$, with $\tilde{u}^\star(0) = 0$, $\tilde{w}^\star(0) = 0$. Moreover, the pair $(\tilde{u}^\star, \tilde{w}^\star)$ constitutes a Nash-equilibrium solution to the dynamic game (11.32), (11.33), (11.75). The following theorem then summarizes the solution to the infinite-horizon problem for the general class of discrete-time nonlinear systems (11.75).

Theorem 11.3.1 *Consider the discrete-time nonlinear system (11.75) and the DSF-BMH2HINLCP for this system. Suppose Assumption 11.3.1 holds, and also the following:*

(Ad1) *the pair $\{\tilde{F}(x,0,0), \tilde{Z}(x,0)\}$ is zero-state detectable;*

(Ad2) *there exists a pair of C^2 locally negative and positive-definite functions $\widetilde{W}, \widetilde{U} : \tilde{M} \to \Re$ respectively, defined in a neighborhood $\tilde{M}$ of $x = 0$, vanishing at $x = 0$ and satisfying the pair of coupled DHJIEs:*

$$\widetilde{W}(\tilde{F}(x, \tilde{w}^\star(x), \tilde{u}^\star(x))) - \widetilde{W}(x) + \frac{1}{2}\gamma^2\|\tilde{w}^\star(x)\|^2 - \frac{1}{2}\|\tilde{Z}(x, \tilde{u}^\star(x))\|^2 = 0, \quad \widetilde{W}(0) = 0,$$

$$\widetilde{U}(\tilde{F}(x, \tilde{w}^\star(x), \tilde{u}^\star(x))) - \widetilde{U}(x) + \frac{1}{2}\|\tilde{Z}(x, \tilde{u}^\star(x))\|^2 = 0, \quad \widetilde{U}(0) = 0;$$

(A3) *the pair $\{\tilde{F}(x, \tilde{w}^\star(x), 0), \tilde{Z}(x, 0)\}$ is locally zero-state detectable.*

Then the state-feedback controls $(\tilde{u}^\star(x), \tilde{w}^\star(x))$ solve the dynamic game problem and the DSFBMH2HINLCP for the system (11.75). Moreover, the optimal costs of the policies are given by

$$\tilde{J}_{1d}^\star(\tilde{w}^\star, \tilde{u}^\star) = \widetilde{W}(x_0),$$
$$\tilde{J}_{2d}^\star(\tilde{w}^\star, \tilde{u}^\star) = \widetilde{U}(x_0).$$

Proof: The proof can be pursued along the same lines as the previous results. $\square$

11.4 Notes and Bibliography

This chapter is mainly based on the paper by Lin [180]. The approach adopted throughout the chapter was originally inspired by the paper by Limebeer et al. [179] for linear systems. The chapter mainly extended the results of the paper to the nonlinear case. But in addition, the discrete-time problem has also been developed. Finally, application of the results to tracking control for Robot manipulators can be found in [80].

12

Mixed $\mathcal{H}_2/\mathcal{H}_\infty$ Nonlinear Filtering

The $\mathcal{H}_\infty$ nonlinear filter has been discussed in chapter 8, and its advantages over the Kalman-filter have been mentioned. In this chapter, we discuss the mixed $\mathcal{H}_2/\mathcal{H}_\infty$-criterion approach for estimating the states of an affine nonlinear system in the spirit of Reference [179]. Many authors have considered mixed $\mathcal{H}_2/\mathcal{H}_\infty$-filtering techniques for linear systems [162, 257], [269]-[282], which enjoy the advantages of both Kalman-filtering and $\mathcal{H}_\infty$-filtering. In particular, the paper [257] considers a differential game approach to the problem, which is attractive and transparent. In this chapter, we present counterpart results for nonlinear systems using a combination of the differential game approach with a dissipative system's perspective. We discuss both the continuous-time and discrete-time problems.

12.1 Continuous-Time Mixed $\mathcal{H}_2/\mathcal{H}_\infty$ Nonlinear Filtering

The general set-up for the mixed $\mathcal{H}_2/\mathcal{H}_\infty$-filtering problem is shown Figure 12.1, where the plant is represented by an affine nonlinear system Σ^a, while $\mathbf{F}$ is the filter. The filter processes the measurement output y from the plant which is also corrupted by the noise signal w, and generates an estimate $\hat{z}$ of the desired variable z. The noise signal w entering the plant is comprised of two components, $w = \begin{pmatrix} w_0 \\ w_1 \end{pmatrix}$, with $w_0 \in \mathcal{S}$ a bounded-spectral signal (e.g., a white Gaussian noise signal); and $w_1 \subset \mathcal{P}$ is a bounded-power signal or $\mathcal{L}_2$-signal. Thus, the induced-norm (or gain) from the input w_0 to $\check{z}$ (the output error) is the $\mathcal{L}_2$-norm of the interconnected system $\mathbf{F} \circ \Sigma^a$, i.e.,

$$\|\mathbf{F} \circ \Sigma^a\|_{\mathcal{L}_2} \triangleq \sup_{0 \neq w_0 \in \mathcal{S}} \frac{\|\check{z}\|_{\mathcal{P}}}{\|w_0\|_{\mathcal{S}}}, \tag{12.1}$$

while the induced norm from w_1 to $\check{z}$ is the $\mathcal{L}_\infty$-norm of $\mathbf{P}^a$, i.e.,

$$\|\mathbf{F} \circ \Sigma^a\|_{\mathcal{L}_\infty} \triangleq \sup_{0 \neq w_1 \in \mathcal{P}} \frac{\|\check{z}\|_2}{\|w_1\|_2}. \tag{12.2}$$

The objective is to synthesize a filter, $\mathcal{F}$, for estimating the state $x(t)$ or a function of it, $z = h_1(x)$, from observations of $y(\tau)$ up to time t over a time horizon $[t_0, T]$, i.e.,

$$\mathsf{Y}_t \triangleq \{y(\tau) : \tau \leq t\}, \quad t \in [t_0, T],$$

such that the above pair of norms (12.1), (12.2) are minimized, while at the same time achieving asymptotic zero estimation error with $w \equiv 0$. In this context, the above norms will be interpreted as the $\mathcal{H}_2$ and $\mathcal{H}_\infty$-norms of the interconnected system.

However, in this chapter, we do not solve the above problem, instead we solve an associated $\mathcal{H}_2/\mathcal{H}_\infty$-filtering problem in which there is a single exogenous input $w \in \mathcal{L}_2[t_0, T]$

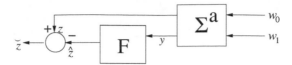

FIGURE 12.1

Set-Up for Mixed $\mathcal{H}_2/\mathcal{H}_\infty$ Nonlinear Filtering

and an associated $\mathcal{H}_2$-cost which represents the output energy of the system in $\check{z}$. More specifically, we seek to synthesize a filter, $\mathcal{F}$, for estimating the state $x(t)$ or a function of it, $z = h_1(x)$, from the observations Y_t such that the $\mathcal{L}_2$-gain from the input w to the penalty function $\check{z}$ (to be defined later) as well as the output energy defined by $\|\check{z}\|_{\mathcal{H}_2}$ are minimized, and at the same time achieving asymptotic zero estimation error with $w \equiv 0$.

Accordingly, the plant is represented by an affine nonlinear causal state-space system defined on a manifold $\mathcal{X} \subset \Re^n$ with zero control input:

$$\Sigma^a \; : \; \begin{cases} \dot{x} & = & f(x) + g_1(x)w; \;\; x(t_0) = x_0 \\ z & = & h_1(x) \\ y & = & h_2(x) + k_{21}(x)w, \end{cases} \tag{12.3}$$

where $x \in \mathcal{X}$ is the state vector, $w \in \mathcal{W} \subseteq \mathcal{L}_2([t_0, \infty), \Re^r)$ is an unknown disturbance (or noise) signal, which belongs to $\mathcal{W}$, the set of admissible noise/disturbance signals, $y \in \mathcal{Y} \subset \Re^m$ is the measured output (or observation) of the system, and belongs to $\mathcal{Y}$, the set of admissible outputs, $z \in \Re^s$ is the output of the system that is to be estimated.

The functions $f : \mathcal{X} \to V^\infty(\mathcal{X})$, $g_1 : \mathcal{X} \to \mathcal{M}^{n \times r}(\mathcal{X})$, $h_1 : \mathcal{X} \to \Re^s$, $h_2 : \mathcal{X} \to \Re^m$ and $k_{21} : \mathcal{X} \to \mathcal{M}^{m \times r}(\mathcal{X})$ are real C^∞-functions of x. Furthermore, we assume without any loss of generality that the system (12.3) has a unique equilibrium-point at $x = 0$ such that $f(0) = 0$, $h_1(0) = 0$, $h_2(0) = 0$. We also assume that there exists a unique solution $x(t, t_0, x_0, w), \forall t \in \Re$ for the system for all initial conditions x_0 and all $w \in \mathcal{W}$.

Again, since it is difficult to minimize exactly the $\mathcal{H}_\infty$-norm, in practice we settle for a suboptimal problem, which is to minimize $\|\check{z}\|_{\mathcal{H}_2}$ while rendering $\|\mathcal{F} \circ \Sigma^a\|_{\mathcal{H}_\infty} \leq \gamma^\star$. For nonlinear systems, this $\mathcal{H}_\infty$-constraint is interpreted as the $\mathcal{L}_2$-gain constraint and is defined as

$$\int_{t_0}^T \|\check{z}(\tau)\|^2 d\tau \leq \gamma^2 \int_{t_0}^T \|w(\tau)\|^2 d\tau, \;\; T > 0. \tag{12.4}$$

More formally, we define the local nonlinear (suboptimal) mixed $\mathcal{H}_2/\mathcal{H}_\infty$-filtering or state estimation problem as follows.

Definition 12.1.1 *(Mixed $\mathcal{H}_2/\mathcal{H}_\infty$ (Suboptimal) Nonlinear Filtering Problem (MH2HINL-FP)). Given the plant Σ^a and a number $\gamma^\star > 0$, find a filter $\mathcal{F} : \mathcal{Y} \to \mathcal{X}$ such that*

$$\hat{x}(t) = \mathcal{F}(Y_t)$$

and the output energy $\|\check{z}\|_{\mathcal{L}_2}$ is minimized while the constraint (12.4) is satisfied for all $\gamma \geq \gamma^\star$, for all $w \in \mathcal{W}$ and for all initial conditions $x(t_0) \in O.$ In addition with $w \equiv 0$, $\lim_{t \to \infty} \check{z}(t) = 0$.

Moreover, if the above conditions are satisfied for all $x(t_0) \in \mathcal{X}$, we say that $\mathcal{F}$ solves the $MH2HINLFP$ globally.

Remark 12.1.1 *The problem defined above is the finite-horizon filtering problem. We have the infinite-horizon problem if we let $T \to \infty$.*

12.1.1 Solution to the Finite-Horizon Mixed $\mathcal{H}_2/\mathcal{H}_\infty$ Nonlinear Filtering Problem

To solve the $MH2HINLFP$, we similarly consider the following Kalman-Luenberger filter structure

$$\Sigma^{af} : \begin{cases} \dot{\hat{x}} &= f(\hat{x}) + L(\hat{x},t)[y - h_2(\hat{x})], \quad \hat{x}(t_0) = \hat{x}_0 \\ \hat{z} &= h_1(\hat{x}) \end{cases} \tag{12.5}$$

where $\hat{x} \in \mathcal{X}$ is the estimated state, $L(.,.) \in \mathcal{M}^{n \times m}(\mathcal{X} \times \Re)$ is the error-gain matrix which is smooth and has to be determined, and $\hat{z} \in \Re^s$ is the estimated output of the filter. We can now define the estimation error or penalty variable, $\check{z}$, which has to be controlled as:

$$\check{z} = h_1(x) - h_1(\hat{x}).$$

Then we combine the plant (12.3) and estimator (12.5) dynamics to obtain the following augmented system:

$$\begin{aligned} \dot{\check{x}} &= \check{f}(\check{x}) + \check{g}(\check{x})w, \quad \check{x}(t_0) = (x_0^T \ \hat{x}_0^T)^T \\ \check{z} &= \check{h}(\check{x}) \end{aligned} \Bigg\}, \tag{12.6}$$

where

$$\check{x} = \begin{pmatrix} x \\ \hat{x} \end{pmatrix}, \ \check{f}(\check{x}) = \begin{pmatrix} f(x) \\ f(\hat{x}) + L(\hat{x},t)(h_2(x) - h_2(\hat{x})) \end{pmatrix},$$

$$\check{g}(\check{x}) = \begin{pmatrix} g_1(x) \\ L(\hat{x},t)k_{21}(x) \end{pmatrix}, \ \check{h}(\check{x}) = h_1(x) - h_1(\hat{x}).$$

The problem can then be formulated as a two-player nonzero-sum differential game (from Chapter 3, see also [59]) with two cost functionals:

$$J_1(L,w) = \frac{1}{2} \int_{t_0}^{T} [\gamma^2 \|w(\tau)\|^2 - \|\check{z}(\tau)\|^2] d\tau, \tag{12.7}$$

$$J_2(L,w) = \frac{1}{2} \int_{t_0}^{T} \|\check{z}(\tau)\|^2 d\tau. \tag{12.8}$$

Here, the first functional is associated with the $\mathcal{H}_\infty$-constraint criterion, while the second functional is related to the output energy of the system or $\mathcal{H}_2$-criterion. It can easily be seen that, by making $J_1 \geq 0$ then the $\mathcal{H}_\infty$-constraint $\|\mathcal{F} \circ \mathbf{P}^a\|_{\mathcal{H}_\infty} \leq \gamma$ is satisfied. A Nash-equilibrium solution [59] to the above game is said to exist if we can find a pair of strategies $(L^\star, w^\star)$ such that

$$J_1(L^\star, w^\star) \leq J_1(L^\star, w) \ \forall w \in \mathcal{W}, \tag{12.9}$$
$$J_2(L^\star, w^\star) \leq J_2(L, w^\star) \ \forall L \in \mathcal{M}^{n \times m}. \tag{12.10}$$

To arrive at a solution to this problem, we form the Hamiltonian function $H_i : T^\star(\mathcal{X} \times \mathcal{X}) \times \mathcal{W} \times \mathcal{M}^{n \times m} \to \Re$, $i = 1, 2$, associated with the two cost functionals:

$$H_1(\check{x}, w, L, Y_{\check{x}}^T) = Y_{\check{x}}(\check{x},t)(\check{f}(\check{x}) + \check{g}(\check{x})w) + \frac{1}{2}(\gamma^2 \|w\|^2 - \|\check{z}\|^2), \tag{12.11}$$

$$H_2(\check{x}, w, L, V_{\check{x}}^T) = V_{\check{x}}(\check{x},t)(\check{f}(\check{x}) + \check{g}(\check{x})w) + \frac{1}{2}\|\check{z}\|^2 \tag{12.12}$$

with the adjoint variables $p_1 = Y_{\check{x}}^T$, $p_2 = V_{\check{x}}^T$ respectively, for some smooth functions

Y, $V : \mathcal{X} \times \mathcal{X} \times \Re \to \Re$ and where $Y_{\breve{x}}$, $V_{\breve{x}}$ are the row-vectors of first partial-derivatives of the functions with respect to (wrt) $\breve{x}$ respectively. The following theorem from Chapter 3 then gives sufficient conditions for the existence of a Nash-equilibrium solution to the above game.

Theorem 12.1.1 *Consider the two-player nonzero-sum dynamic game (12.7)-(12.8),(12.6) of fixed duration $[t_0, T]$ under closed-loop memoryless perfect state information pattern. A pair of strategies $(w^\star, L^\star)$ provides a feedback Nash-equilibrium solution to the game, if there exists a pair of C^1-functions (in both arguments) Y, $V : \mathcal{X} \times \mathcal{X} \times \Re \to \Re$, satisfying the pair of HJIEs:*

$$Y_t(\breve{x}, t) = - \inf_{w \in \mathcal{W}} H_1(\breve{x}, w, L^\star, Y_{\breve{x}}), \quad Y(\breve{x}, T) = 0, \qquad (12.13)$$

$$V_t(\breve{x}, t) = - \min_{L \in \mathcal{M}^{n \times m}} H_2(\breve{x}, w^\star, L, V_{\breve{x}}), \quad V(\breve{x}, T) = 0. \qquad (12.14)$$

Based on the above theorem, it is now easy to find the above Nash-equilibrium pair for our game. The following theorem gives sufficient conditions for the solvability of the MH2HINLFP. For simplicity we make the following assumption on the plant.

Assumption 12.1.1 *The system matrices are such that*

$$k_{21}(x)g_1^T(x) = 0,$$
$$k_{21}(x)k_{21}^T(x) = I.$$

Remark 12.1.2 *The first of the above assumptions means that the measurement noise and the system noise are independent.*

Theorem 12.1.2 *Consider the nonlinear system (12.3) and the MH2HINLFP for it. Suppose the function h_1 is one-to-one (or injective) and the plant Σ^a is locally asymptotically-stable about the equilibrium point $x = 0$. Further, suppose there exists a pair of C^1 (with respect to both arguments) negative and positive-definite functions Y, $V : N \times N \times \Re \to \Re$ respectively, locally defined in a neighborhood $N \times N \subset \mathcal{X} \times \mathcal{X}$ of the origin $\breve{x} = 0$, and a matrix function $L : N \times \Re \to \mathcal{M}^{n \times m}$ satisfying the following pair of coupled HJIEs:*

$$Y_t(\breve{x}, t) + Y_x(\breve{x}, t)f(x) + Y_{\hat{x}}(\breve{x}, t)f(\hat{x}) - \frac{1}{2\gamma^2}Y_x(\breve{x})g_1(x)g_1^T(x)Y_x^T(\breve{x}) -$$
$$\frac{1}{2\gamma^2}Y_{\hat{x}}(\breve{x})L(\hat{x}, t)L^T(\hat{x}, t)Y_{\hat{x}}^T(\breve{x}) - \frac{1}{\gamma^2}Y_{\hat{x}}(\breve{x})L(\hat{x}, t)L^T(\hat{x}, t)V_{\hat{x}}^T(\breve{x}) -$$
$$\frac{1}{2}(h_1(x) - h_1(\hat{x}))^T(h_1(x) - h_1(\hat{x})) = 0, \quad Y(\breve{x}) = 0 \qquad (12.15)$$

$$V_t(\breve{x}, t) + V_x(\breve{x}, t)f(x) + V_{\hat{x}}(\breve{x}, t)f(\hat{x}) - \frac{1}{\gamma^2}V_x(\breve{x})g_1(x)g_1^T(x)Y_x^T(\breve{x}) -$$
$$\frac{1}{\gamma^2}V_{\hat{x}}(\breve{x})L(\hat{x}, t)L^T(\hat{x}, t)Y_{\hat{x}}^T - \gamma^2(h_2(x) - h_2(\hat{x}))^T(h_2(x) - h_2(\hat{x})) +$$
$$\frac{1}{2}(h_1(x) - h_1(\hat{x}))^T(h_1(x) - h_1(\hat{x})) = 0, \quad V(\breve{x}, T) = 0, \qquad (12.16)$$

together with the coupling condition

$$V_{\hat{x}}(\breve{x}, t)L(\hat{x}, t) = -\gamma^2(h_2(x) - h_2(\hat{x}))^T, \quad x, \hat{x} \in N. \qquad (12.17)$$

Then:

(i) *there exists locally a Nash-equilibrium solution $(w^\star, L^\star)$ for the game (12.7), (12.8), (12.6);*

(ii) *the augmented system (12.6) is dissipative with respect to the supply-rate $s(w, \breve{z}) = \frac{1}{2}(\gamma^2 \|w\|^2 - \|\breve{z}\|^2)$ and hence has finite $\mathcal{L}_2$-gain from w to $\breve{z}$ less or equal to γ;*

(iii) *the optimal costs or performance objectives of the game are $J_1^\star(L^\star, w^\star) = Y(\breve{x}_0, t_0)$ and $J_2^\star(L^\star, w^\star) = V(\breve{x}_0, t_0)$;*

(iv) *the filter Σ^{af} with the gain matrix $L(\hat{x}, t)$ satisfying (12.17) solves the finite-horizon $MH2HINLFP$ for the system locally in N.*

Proof: Assume there exist definite solutions Y, V to the HJIEs (12.15)-(12.16), and (i) consider the Hamiltonian function $H_1(., ., ., .)$ first:

$$H_1(\breve{x}, w, L, Y_{\breve{x}}^T) = Y_x f(x) + Y_{\hat{x}} f(\hat{x}) + Y_{\hat{x}} L(\hat{x}, t)(h_2(x) - h_2(\hat{x})) + Y_x g_1(x)w +$$
$$Y_{\hat{x}} L(\hat{x}, t)k_{21}(x)w - \frac{1}{2}\|\breve{z}\|^2 + \frac{1}{2}\gamma^2 \|w\|^2$$

where some of the arguments have been suppressed for brevity. Since it is quadratic and convex in w, we can apply the necessary condition for optimality, i.e.,

$$\left.\frac{\partial H_1}{\partial w}\right|_{w=w^\star} = 0,$$

to get

$$w^\star := -\frac{1}{\gamma^2}(g_1^T(x)Y_x^T(\breve{x}, t) + k_{21}^T(x)L^T(\hat{x}, t)Y_{\hat{x}}^T(\breve{x}, t)). \tag{12.18}$$

Substituting now $w^\star$ in the expression for $H_2(., ., ., .)$ (12.12), we get

$$H_2(\breve{x}, w^\star, L, V_{\breve{x}}^T) = V_x f(x) + V_{\hat{x}} f(\hat{x}) + V_{\hat{x}} L(\hat{x}, t)(h_2(x) - h_2(\hat{x})) - \frac{1}{\gamma^2} V_x g_1(x)g_1^T(x)Y_x^T -$$
$$\frac{1}{\gamma^2} V_{\hat{x}} L(\hat{x}, t)L^T(\hat{x}, t)Y_{\hat{x}}^T + \frac{1}{2}(h_1(x) - h_1(\hat{x}))^T(h_1(x) - h_1(\hat{x})).$$

Then completing the squares for L in the above expression for $H_2(., ., ., .)$, we get

$$H_2(\breve{x}, w^\star, L, V_{\breve{x}}^T) = V_x f(x) + V_{\hat{x}} f(\hat{x}) - \frac{1}{\gamma^2} V_x g_1(x)g_1^T(x)Y_x^T +$$
$$\frac{1}{2}(h_1(x) - h_1(\hat{x}))^T(h_1(x) - h_1(\hat{x})) +$$
$$\frac{1}{2\gamma^2}\left\|L^T(\hat{x}, t)V_{\hat{x}}^T + \gamma^2(h_2(x) - h_2(\hat{x}))\right\|^2 - \frac{\gamma^2}{2}\|h_2(x) - h_2(\hat{x})\|^2 -$$
$$\frac{1}{2\gamma^2}\left\|L^T(\hat{x}, t)V_{\hat{x}}^T + L^T(\hat{x}, t)Y_{\hat{x}}^T\right\|^2 + \frac{1}{2\gamma^2}\|L^T(\hat{x}, t)Y_{\hat{x}}^T\|^2.$$

Thus, choosing $L^\star$ according to (12.17) minimizes $H_2(., ., ., .)$ and renders the Nash-equilibrium condition

$$H_2(w^\star, L^\star) \leq H_2(w^\star, L) \quad \forall L \in \mathcal{M}^{n \times m}$$

satisfied. Moreover, substituting $(w^\star, L^\star)$ in (12.14) gives the HJIE (12.16).

Next, substitute $L^\star$ as given by (12.17) in the expression for $H_1(., ., ., .)$ and complete the squares to obtain:

$$H_1(\breve{x}, w, L^\star, Y_{\breve{x}}^T) = Y_x f(x) + Y_{\hat{x}} f(\hat{x}) - \frac{1}{\gamma^2} Y_{\hat{x}} L(\hat{x}, t)L^T(\hat{x}, t)V_{\hat{x}}^T -$$
$$\frac{1}{2\gamma^2} Y_x g_1(x)g_1^T(x)Y_x^T - \frac{1}{2\gamma^2} Y_{\hat{x}} L(\hat{x})L^T(\hat{x})Y_{\hat{x}}^T +$$
$$\frac{\gamma^2}{2}\left\|w + \frac{1}{\gamma^2}g_1^T(x)Y_x^T + \frac{1}{\gamma^2}k_{21}^T(x)L^T(\hat{x}, t)Y_{\hat{x}}^T\right\|^2 - \frac{1}{2}\|z\|^2.$$

Substituting now $w = w^\star$ as given by (12.18), we see that the second Nash-equilibrium condition

$$H_1(w^\star, L^\star) \le H_1(w, L^\star), \quad \forall w \in \mathcal{W}$$

is also satisfied. Therefore, the pair $(w^\star, L^\star)$ constitute a Nash-equilibrium solution to the two-player nonzero-sum dynamic game. Moreover, substituting $(w^\star, L^\star)$ in (12.13) gives the HJIE (12.15).

(ii) Consider the HJIE (12.15) and rewrite as:

$$\left\{ Y_t(\check{x}, t) + Y_x f(x) + Y_{\hat{x}} f(\hat{x}) + Y_{\hat{x}} L(\hat{x}, t)(h_2(x) - h_2(\hat{x}) + Y_x g_1(x) w + \right.$$

$$Y_{\hat{x}} L(\hat{x}, t) k_{21}(x) w - \frac{1}{2}\|\check{z}\|^2 + \frac{1}{2}\gamma^2 \|w\|^2 \Big\} - \frac{1}{2}\gamma^2 \|w\|^2 - Y_x g_1(x) w -$$

$$Y_{\hat{x}} L(\hat{x}, t) k_{21}(x) w - \frac{1}{2\gamma^2} Y_x g_1(x) g_1^T(x) Y_x^T - \frac{1}{2\gamma^2} Y_{\hat{x}} L(\hat{x}, t) L^T(\hat{x}, t) Y_{\hat{x}}^T = 0$$

$$\Longleftrightarrow \left\{ Y_t(\check{x}, t) + Y_{\check{x}}[\check{f}(\check{x}) + \check{g}(\check{x}) w] + \frac{1}{2}\gamma^2 \|w\|^2 - \frac{1}{2}\|\check{z}\|^2 \right\} - \frac{1}{2}\gamma^2 \|w - w^\star\|^2 = 0$$

$$\Longrightarrow \left\{ \check{Y}_t(\check{x}, t) + \check{Y}_{\check{x}}[\check{f}(\check{x}) + \check{g}(\check{x}) w] \right\} \le \frac{1}{2}\gamma^2 \|w\|^2 - \frac{1}{2}\|\check{z}\|^2 \qquad (12.19)$$

for some function $\check{Y} = -Y > 0$. Integrating now the last expression above from $t = t_0$ and $\check{x}(t_0)$ to $t = T$ and $\check{x}(T)$, we get the dissipation-inequality:

$$\check{Y}(\check{x}(T), T) - \check{Y}(\check{x}(t_0), t_0) \le \frac{1}{2} \int_{t_0}^{T} \{\gamma^2 \|w(t)\|^2 - \|\check{z}(t)\|^2\} dt, \qquad (12.20)$$

and hence the result.

(iii) Consider the cost functional $J_1(L, w)$ first, and rewrite it as

$$J_1(L, w) + Y(\check{x}(T), T) - Y(\check{x}(t_0), t_0)$$

$$= \int_{t_0}^{T} \left\{ \frac{1}{2}\gamma^2 \|w(t)\|^2 - \frac{1}{2}\|\check{z}(t)\|^2 + \dot{Y}(\check{x}, t) \right\} dt$$

$$= \int_{t_0}^{T} H_1(\check{x}, w, L, Y_{\check{x}}^T) dt$$

$$= \int_{t_0}^{T} \left\{ \frac{\gamma^2}{2}\|w - w^\star\|^2 + Y_{\hat{x}} L(\hat{x}, t)(h_2(x) - h_2(\hat{x})) + \frac{1}{\gamma^2} Y_{\hat{x}} L(\hat{x}, t) L^T(\hat{x}, t) V_{\hat{x}}^T \right\} dt,$$

where use has been made of the HJIE (12.15) in the above manipulations. Substitute now $(L^\star, w^\star)$ as given by (12.17), (12.18) respectively to get the result.

Similarly, consider the cost functional $J_2(L, w)$ and rewrite it as

$$J_2(L, w) - V(\check{x}(t_0), t_0) = \int_{t_0}^{T} \left\{ \frac{1}{2}\|\check{z}(t)\|^2 + \dot{V}(\check{x}, t) \right\} dt$$

$$= \int_{t_0}^{T} H_2(\check{x}, w, L, Y_{\check{x}}^T) dt.$$

Then substituting $(L^\star, w^\star)$ as given by (12.17), (12.18) respectively, and using the HJIE (12.16) the result also follows.

Finally, (iv) follows from (i)-(iii). $\square$

Remark 12.1.3 *Notice that by virtue of (12.17), the HJIE (12.16) can also be represented in the following form:*

$$V_t(\breve{x}, t) + V_x(\breve{x}, t)f(x) + V_{\hat{x}}(\breve{x}, t)f(\hat{x}) - \frac{1}{\gamma^2}V_x(\breve{x}, t)g_1(x)g_1^T(x)Y_x^T(\breve{x}, t) -$$
$$\frac{1}{\gamma^2}V_{\hat{x}}(\breve{x}, t)L(\hat{x}, t)L^T(\hat{x}, t)Y_{\hat{x}}^T(\breve{x}, t) - \frac{1}{\gamma^2}V_{\hat{x}}(\breve{x}, t)L(\hat{x}, t)L^T(\hat{x}, t)V_{\hat{x}}^T(\breve{x}, t) +$$
$$\frac{1}{2}(h_1(x) - h_1(\hat{x}))^T(h_1(x) - h_1(\hat{x})) = 0, \quad V(\breve{x}, T) = 0. \tag{12.21}$$

The above result (Theorem 12.1.2) can be specialized to the linear-time-invariant (LTI) system:

$$\Sigma^l : \begin{cases} \dot{x} &= Ax + G_1 w, \quad x(t_0) = x_0 \\ \breve{z} &= C_1(x - \hat{x}) \\ y &= C_2 x + D_{21} w, \end{cases} \tag{12.22}$$

where all the variables have their previous meanings, and $A \in \Re^{n \times n}$, $G_1 \in \Re^{n \times r}$, $C_1 \in \Re^{s \times n}$, $C_2 \in \Re^{m \times n}$ and $D_{21} \in \Re^{m \times r}$ are real constant matrices. We have the following corollary to Theorem 12.1.2.

Corollary 12.1.1 *Consider the LTI system Σ^l defined by (12.22) and the MH2HINLFP for it. Suppose C_1 is full-column rank and A is Hurwitz. Suppose further, there exist a negative and a positive-definite real-symmetric solutions $P_1(t)$, $P_2(t)$, $t \in [t_0, T]$ (respectively) to the coupled Riccati ordinary differential-equations (ODEs):*

$$-\dot{P}_1(t) = A^T P_1(t) + P_1(t)A -$$
$$\frac{1}{\gamma^2}[P_1(t) \ P_2(t)]\begin{bmatrix} L(t)L^T(t) + G_1 G_1^T & L(t)L^T(t) \\ L(t)L^T(t) & 0 \end{bmatrix}\begin{bmatrix} P_1(t) \\ P_2(t) \end{bmatrix} -$$
$$C_1^T C_1, \quad P_1(T) = 0, \tag{12.23}$$
$$-\dot{P}_2(t) = A^T P_2(t) + P_2(t)A -$$
$$\frac{1}{\gamma^2}[P_1(t) \ P_2(t)]\begin{bmatrix} 0 & L(t)L^T(t) + G_1 G_1^T \\ L(t)L^T(t) + G_1 G_1^T & 2L(t)L^T(t) \end{bmatrix}\begin{bmatrix} P_1(t) \\ P_2(t) \end{bmatrix} +$$
$$C_1^T C_1, \quad P_2(T) = 0, \tag{12.24}$$

together with the coupling condition

$$P_2(t)L(t) = -\gamma^2 C_2^T, \tag{12.25}$$

for some $n \times m$ matrix function $L(t)$ defined for all $t \in [t_0, T]$. Then:

(i) *there exists a Nash-equilibrium solution $(w^\star, L^\star)$ for the game given by*

$$w^\star := -\frac{1}{\gamma^2}(G_1^T + D_{21}^T L(t))P_1(t)(x - \hat{x}),$$
$$(x - \hat{x})^T P_2(t)L^\star = -\gamma^2(x - \hat{x})^T C_2^T;$$

(ii) *the augmented system*

$$\Sigma^{fl} : \begin{cases} \dot{\breve{x}} &= \begin{bmatrix} A & 0 \\ L^\star(t)C_2 & A - L^\star(t)C_2 \end{bmatrix}\breve{x} + \begin{bmatrix} G_1 \\ L^\star(t)D_{21} \end{bmatrix}w, \\ \breve{x}(t_0) &= \begin{bmatrix} x_0 \\ \hat{x}_0 \end{bmatrix} \\ \breve{z} &= [C_1 \ -C_1]\breve{x} := \breve{C}\breve{x} \end{cases}$$

has $\mathcal{H}_\infty$-norm from w to $\check{z}$ less than equal to γ;

(iii) *the optimal costs or performance objectives of the game are $J_1^\star(L^\star, w^\star) = \frac{1}{2}(x_0 - \hat{x}_0)^T P_1(t_0)(x_0 - \hat{x}_0)$ and $J_2^\star(L^\star, w^\star) = \frac{1}{2}(x_0 - \hat{x}_0)^T P_2(t_0)(x_0 - \hat{x}_0)$;*

(iv) *the filter Σ^{fl} with the gain matrix $L(t) = L^\star(t)$ satisfying (12.25) solves the finite-horizon MH2HINLFP for the system.*

Proof: Take

$$Y(\check{x}, t) = \frac{1}{2}(x - \hat{x})^T P_1(t)(x - \hat{x}), \quad P_1 = P_1^T < 0,$$

$$V(\check{x}, t) = \frac{1}{2}(x - \hat{x})^T P_2(t)(x - \hat{x}), \quad P_2 = P_2^T > 0,$$

and apply the result of the theorem. $\square$

12.1.2 Solution to the Infinite-Horizon Mixed $\mathcal{H}_2/\mathcal{H}_\infty$ Nonlinear Filtering

In this section, we discuss the infinite-horizon filtering problem, in which case we let $T \to \infty$. In this case, we seek a time-independent gain matrix $\hat{L}(\hat{x})$ and functions $Y, V : \widehat{N} \times \widehat{N} \subset \mathcal{X} \times \mathcal{X} \to \Re$ such that the HJIEs:

$$Y_x(\check{x})f(x) + Y_{\hat{x}}f(\hat{x}) - \frac{1}{2\gamma^2}Y_x(\check{x})g_1(x)g_1^T(x)Y_x^T(\check{x}) - \frac{1}{2\gamma^2}Y_{\hat{x}}(\check{x})\hat{L}(\hat{x})\hat{L}^T(\hat{x})Y_{\hat{x}}^T(\check{x}) -$$

$$\frac{1}{\gamma^2}Y_{\hat{x}}(\check{x})\hat{L}(\hat{x})\hat{L}^T(\hat{x})V_{\hat{x}}^T(\check{x}) - \frac{1}{2}(h_1(x) - h_1(\hat{x}))^T(h_1(x) - h_1(\hat{x})) = 0,$$

$$Y(0) = 0 \tag{12.26}$$

$$V_x(\check{x})f(x) + V_{\hat{x}}(\check{x})f(\hat{x}) - \frac{1}{\gamma^2}V_x(\check{x})g_1(x)g_1^T(x)Y_x^T(\check{x}) - \frac{1}{\gamma^2}V_{\hat{x}}(\check{x})\hat{L}(\hat{x})\hat{L}^T(\hat{x})Y_{\hat{x}}^T(\check{x}) -$$

$$\gamma^2(h_2(x) - h_2(\hat{x}))^T(h_2(x) - h_2(\hat{x})) + \frac{1}{2}(h_1(x) - h_1(\hat{x}))^T(h_1(x) - h_1(\hat{x})) = 0,$$

$$V(0) = 0 \tag{12.27}$$

are satisfied together with the coupling condition:

$$V_{\hat{x}}(\check{x})\hat{L}(\hat{x}) = -\gamma^2(h_2(x) - h_2(\hat{x}))^T, \quad x, \hat{x} \in \widehat{N}, \tag{12.28}$$

where $\hat{L}$ is the asymptotic value of L. It is also required in this case that the augmented system (12.6) is stable. Moreover, in this case, we can relax the requirement of asymptotic-stability for the original system (12.3) with a milder requirement of detectability which we define next.

Definition 12.1.2 *The pair $\{f, h\}$ is said to be locally zero-state detectable if there exists a neighborhood $\mathcal{O}$ of $x = 0$ such that, if $x(t)$ is a trajectory of $\dot{x}(t) = f(x)$ satisfying $x(t_0) \in \mathcal{O}$, then $h(x(t))$ is defined for all $t \geq t_0$ and $h(x(t)) \equiv 0$ for all $t \geq t_s$, for some $t_s \geq t_0$, implies $\lim_{t \to \infty} x(t) = 0$. Moreover, the system is zero-state detectable if $\mathcal{O} = \mathcal{X}$.*

Remark 12.1.4 *From the above definition, it can be inferred that, if h_1 is injective, then*

$$\{f, h_1\} \quad \text{zero-state detectable} \Rightarrow \{\check{f}, \check{h}\} \quad \text{zero-state detectable}$$

and conversely.

It is also desirable for a filter to be stable or admissible. The *"admissibility"* of a filter is defined as follows.

Definition 12.1.3 *A filter $\mathcal{F}$ is admissible if it is asymptotically (or internally) stable for any given initial condition $x(t_0)$ of the plant Σ^a, and with $w \equiv 0$*

$$\lim_{t \to \infty} \check{z}(t) = 0.$$

The following proposition can now be proven along the same lines as Theorem 12.1.2.

Proposition 12.1.1 *Consider the nonlinear system (12.3) and the infinite-horizon $MH2HINLFP$ for it. Suppose the function h_1 is one-to-one (or injective) and the plant Σ^a is zero-state detectable. Further, suppose there exists a pair of C^1 negative and positive-definite functions $Y, V : \hat{N} \times \hat{N} \to \Re$ respectively, locally defined in a neighborhood $\hat{N} \times \hat{N} \subset \mathcal{X} \times \mathcal{X}$ of the origin $\check{x} = 0$, and a matrix function $\hat{L} : \hat{N} \to \mathcal{M}^{n \times m}$ satisfying the pair of coupled HJIEs (12.26), (12.27) together with (12.28). Then:*

(i) *there exists locally a Nash-equilibrium solution $(\hat{w}^\star, \hat{L}^\star)$ for the game;*

(ii) *the augmented system (12.6) is dissipative with respect to the supply-rate $s(w, z) = \frac{1}{2}(\gamma^2 \|w\|^2 - \|\check{z}\|^2)$ and hence has $\mathcal{L}_2$-gain from w to $\check{z}$ less or equal to γ;*

(iii) *the optimal costs or performance objectives of the game are $J_1^\star(\hat{L}^\star, \hat{w}^\star) = Y(\check{x}_0)$ and $J_2^\star(\hat{L}^\star, \hat{w}^\star) = V(\check{x}_0)$;*

(iv) *the filter Σ^{af} with the gain matrix $\hat{L}(\hat{x}) = \hat{L}^\star(\hat{x})$ satisfying (12.28) is admissible and solves the infinite-horizon $MH2HINLFP$ locally in $\hat{N}$.*

Proof: Since the proof of items (i)-(iii) is the same as in Theorem 12.1.2, we only prove (iv) here. Using similar manipulations as in the proof of Theorem 12.1.2, we get an inequality similar to (12.19). This inequality implies that with $w = 0$,

$$\dot{Y} \le -\frac{1}{2}\|\check{z}\|^2. \tag{12.29}$$

Therefore, by Lyapunov's theorem, the augmented system is locally stable. Furthermore, for any trajectory of the system $\check{x}(t)$ such that $\dot{Y}(\check{x}) \equiv 0$ for all $t \ge t_c$, for some $t_c \ge t_0$, it implies that $\check{z} \equiv 0 \, \forall \ge t_c$. This in turn implies $h_1(x) = h_1(\hat{x})$, and $x(t) = \hat{x}(t) \, \forall t \ge t_c$ by the injectivity of h_1. This further implies that $h_2(x) = h_2(\hat{x}) \, \forall t \ge t_c$ and it is a trajectory of the free system:

$$\dot{\check{x}} = \left(\begin{array}{c} f(x) \\ f(\hat{x}) \end{array} \right).$$

By the zero-state detectability of $\{f, h_1\}$, we have $\lim_{t \to \infty} \check{x}(t) = 0$, and asymptotic-stability follows by LaSalle's invariance-principle. On the other hand, if we have strict inequality, $\dot{Y} < -\frac{1}{2}\|\check{z}\|^2$, asymptotic-stability follows immediately from Lyapunov's theorem, and $\lim_{t \to \infty} z(t) = 0$. Therefore, Σ^{af} is admissible. Combining now with items (i)-(iii), the conclusion follows. $\square$

Similarly, for the linear system Σ^l (12.22), we have the following corollary.

Corollary 12.1.2 *Consider the LTI system Σ^l defined by (12.22) and the $MH2HINLFP$ for it. Suppose C_1 is full column rank and (A, C_1) is detectable. Suppose further, there exist a* *negative and a positive-definite real-symmetric solutions P_1, P_2, (respectively) to the coupled algebraic-Riccati equations (AREs):*

$$A^T P_1 + P_1 A - \frac{1}{\gamma^2}[P_1 \ P_2]\begin{bmatrix} \hat{L}\hat{L}^T + G_1 G_1^T & \hat{L}\hat{L}^T \\ \hat{L}\hat{L}^T & 0 \end{bmatrix}\begin{bmatrix} P_1 \\ P_2 \end{bmatrix} - C_1^T C_1 = 0 \qquad (12.30)$$

$$A^T P_2 + P_2 A - \frac{1}{\gamma^2}[P_1 \ P_2]\begin{bmatrix} 0 & \hat{L}\hat{L}^T + G_1 G_1^T \\ \hat{L}\hat{L}^T + G_1 G_1^T & 2\hat{L}\hat{L}^T \end{bmatrix}\begin{bmatrix} P_1 \\ P_2 \end{bmatrix} + C_1^T C_1 = 0, \quad (12.31)$$

together with the coupling-condition

$$P_2 \hat{L} = -\gamma^2 C_2^T. \qquad (12.32)$$

Then:

(i) *there exists a Nash-equilibrium solution $(w^\star, L^\star)$ for the game;*

(ii) *the augmented system Σ^{lf} has $\mathcal{H}_\infty$-norm from w to $\check{z}$ less than equal to γ;*

(iii) *the optimal costs or performance objectives of the game are $J_1^\star(\hat{L}^\star, \hat{w}^\star) = \frac{1}{2}(x_0 - \hat{x}_0)^T P_1(x_0 - \hat{x}_0)$ and $J_2^\star(\hat{L}^\star, \hat{w}^\star) = \frac{1}{2}(x_0 - \hat{x}_0)^T P_2(x_0 - \hat{x}_0)$;*

(iv) *the filter Σ_{lf} with gain-matrix $\hat{L} = \hat{L}^\star \in \Re^{n \times m}$ satisfying (12.32) is admissible, and solves the infinite-horizon $MH2HINLFP$ for the linear system.*

We consider a simple example.

Example 12.1.1 *We consider a simple example of the following scalar system:*

$$\begin{aligned} \dot{x} &= -x^3, \quad x(0) = x_0, \\ z &= x \\ y &= x + w. \end{aligned}$$

We consider the infinite-horizon problem and the associated HJIEs. It can be seen that the system satisfies all the assumptions of Theorem 12.1.2 and Proposition 12.1.1. Then, substituting in the HJIEs (12.21), (12.26), and coupling condition (12.28), we get

$$-x^3 y_x - \hat{x}^3 y_{\hat{x}} - \frac{1}{2\gamma^2}l^2 y_{\hat{x}}^2 - \frac{1}{\gamma^2}l^2 v_{\hat{x}} y_{\hat{x}} - \frac{1}{2}(x - \hat{x})^2 = 0,$$

$$-x^3 v_x - \hat{x}^3 v_{\hat{x}} - \frac{1}{\gamma^2}l^2 v_{\hat{x}} y_{\hat{x}} - \gamma^2(x - \hat{x})^2 + \frac{1}{2}(x - \hat{x})^2 = 0,$$

$$v_{\hat{x}} l + \gamma^2(x - \hat{x}) = 0.$$

Looking at the above system of PDEs, we see that there are 5 variables: v_x, $v_{\hat{x}}$, y_x, $v_{\hat{x}}$, l and only 3 equations. Therefore, we make the following simplifying assumption. Let

$$v_x = -v_{\hat{x}}.$$

Then, the above equations reduce to

$$x^3 y_x + \hat{x}^3 y_{\hat{x}} + \frac{1}{2\gamma^2}l^2 y_{\hat{x}}^2 - \frac{1}{\gamma^2}l^2 v_x y_{\hat{x}} + \frac{1}{2}(x - \hat{x})^2 = 0, \qquad (12.33)$$

$$x^3 v_x - \hat{x}^3 v_x - \frac{1}{\gamma^2}l^2 v_x y_{\hat{x}} + \gamma^2(x - \hat{x})^2 - \frac{1}{2}(x - \hat{x})^2 = 0, \qquad (12.34)$$

$$v_x l - \gamma^2(x - \hat{x}) = 0. \qquad (12.35)$$

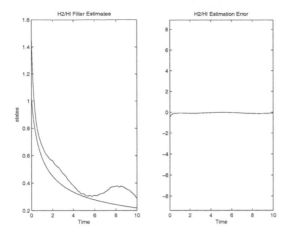

FIGURE 12.2
Nonlinear $\mathcal{H}_2/\mathcal{H}_\infty$-Filter Performance with Unknown Initial Condition; Reprinted from Int. J. of Robust and Nonlinear Control, vol. 19, no. 4, pp. 394-417, © 2009, "Mixed $\mathcal{H}_2/\mathcal{H}_\infty$ nonlinear filtering," by M. D.S. Aliyu and E. K. Boukas, with permission from Wiley Blackwell.

Subtract now equations (12.33) and (12.34) to get

$$(x^3 y_x + \hat{x}^3 y_{\hat{x}}) + (\hat{x}^3 - x^3)v_x + \frac{1}{2\gamma^2}l^2 y_{\hat{x}}^2 + (1 - \gamma^2)(x - \hat{x})^2 = 0, \qquad (12.36)$$

$$v_x l - \gamma^2(x - \hat{x}) = 0. \qquad (12.37)$$

Now let

$$v_x = (x - \hat{x}) \Rightarrow v_{\hat{x}} = -(x - \hat{x}), \;\; v(x, \hat{x}) = \frac{1}{2}(x - \hat{x})^2, \;\; and \;\; \Rightarrow l_{2/\infty}^* = \gamma^2.$$

Then, if we let $\gamma = 1$, the equation governing y_x (12.36) becomes

$$y_{\hat{x}}^2 + 2(x^3 y_x + \hat{x}^3 y_{\hat{x}}) + 2(\hat{x}^3 - x^3)(x - \hat{x}) = 0$$

and

$$y_x = -(x + \hat{x}), \;\; y_{\hat{x}} = -(x + \hat{x}),$$

approximately solves the above PDE (locally!). This corresponds to the solution

$$y(x, \hat{x}) = -\frac{1}{2}(x + \hat{x})^2.$$

Figure 12.2 shows the filter performance with unknown initial condition and with the measurement noise $w(t) = 0.5w_0 + 0.01\sin(t)$, where w_0 is zero-mean Gaussian-white with unit variance.

12.1.3 Certainty-Equivalent Filters (CEFs)

It should be observed from the previous two sections, 12.1.1, 12.1.2, that the filter gains (12.17), (12.28) may depend on the original state of the system which is to be estimated, except for the linear case where the gains are constants. As discussed in Chapter 8, this will make the filters practically impossible to construct. Therefore, the filter equation and gains must be modified to overcome this difficulty.

Furthermore, we observe that the number of variables in the HJIEs (12.15)-(12.27) is twice the dimension of the plant. This makes them more difficult to solve considering their notoriety, and makes the scheme less attractive. Therefore, based on these observations, we consider again in this section *certainty-equivalent* filters which are practically implementable, and in which the governing equations are of lower-order. We begin with the following definition.

Definition 12.1.4 *For the nonlinear system (12.3), we say that it is locally zero-input observable if for all states x_1, $x_2 \in U \subset \mathcal{X}$ and input $w(.) \equiv 0$*

$$y(., x_1, w) \equiv y(., x_2, w) \Longrightarrow x_1 = x_2$$

where $y(., x_i, w), i = 1, 2$ is the output of the system with the initial condition $x(t_0) = x_i$. Moreover, the system is said to be zero-input observable, if it is locally observable at each $x_0 \in \mathcal{X}$ or $U = \mathcal{X}$.

Consider now as in Chapter 8, the following class of filters:

$$\tilde{\Sigma}^{af} : \begin{cases} \dot{\hat{x}} &= f(\hat{x}) + g_1(\hat{x})w^\star + \tilde{L}(\hat{x}, y)[y - h_2(\hat{x}) - k_{21}(\hat{x})\tilde{w}^\star], \ \ \hat{x}(t_0) = \hat{x}_0 \\ \hat{z} &= h_2(\hat{x}) \\ \tilde{z} &= y - h_2(\hat{x}), \end{cases} \tag{12.38}$$

where $\tilde{L}(.,.) \in \mathcal{M}^{n \times m}$ is the filter gain matrix, $\tilde{w}^\star$ is the estimated worst-case system noise (hence the name certainty-equivalent filter) and $\tilde{z}$ is the new penalty variable. Then, consider the infinite-horizon mixed $\mathcal{H}_2/\mathcal{H}_\infty$ dynamic game problem with the cost functionals (12.7), (12.8), and with the above filter. We can define the corresponding Hamiltonians $\tilde{H}_i : T^\star \mathcal{X} \times \mathcal{W} \times T^\star \mathcal{Y} \times \mathcal{M}^{n \times m} \to \Re, i = 1, 2$ as

$$\tilde{H}_1(\hat{x}, w, y, \tilde{L}, \tilde{Y}_{\hat{x}}^T, \tilde{Y}_y^T) = \tilde{Y}_{\hat{x}}(\hat{x}, y)[f(\hat{x}) + g_1(\hat{x})w + \tilde{L}(\hat{x}, y)(y - h_2(\hat{x}) - k_{21}(\hat{x})w)] +$$
$$\tilde{Y}_y(\hat{x}, y)\dot{y} + \frac{1}{2}\gamma^2 \|w\|^2 - \frac{1}{2}\|\tilde{z}\|^2,$$

$$\tilde{H}_2(\hat{x}, w, y, \tilde{L}, \tilde{V}_{\hat{x}}^T, \tilde{V}_y^T) = \tilde{V}_{\hat{x}}(\hat{x}, y)[f(\hat{x}) + g_1(\hat{x})w + \tilde{L}(\hat{x}, y)(y - h_2(\hat{x}) - k_{21}(\hat{x})w))] +$$
$$\tilde{V}_y(\hat{x}, y)\dot{y} + \frac{1}{2}\|\tilde{z}\|^2,$$

for some smooth functions $\tilde{V}, \tilde{Y} : \mathcal{X} \times \mathcal{Y} \to \Re$, and where the adjoint variables are set as $\tilde{p}_1 = \tilde{Y}_{\hat{x}}^T$, $\tilde{p}_2 = \tilde{V}_{\hat{x}}^T$. Further, we have

$$\left. \frac{\partial \tilde{H}_1}{\partial w} \right|_{w=\tilde{w}^\star} = 0 \Longrightarrow \tilde{w}^\star = -\frac{1}{\gamma^2}[g_1(\hat{x}) - \tilde{L}(\hat{x}, y)k_{21}(\hat{x})]^T \tilde{Y}_{\hat{x}}^T(\hat{x}, y),$$

and repeating similar derivation as in the previous section, we can arrive at the following result.

Theorem 12.1.3 *Consider the nonlinear system (12.3) and the MH2HINLFP for it. Suppose the plant Σ^a is locally asymptotically-stable about the equilibrium point $x = 0$ and zero-input observable. Further, suppose there exists a pair of C^1 (with respect to both arguments) negative and positive-definite functions $\tilde{Y}, \tilde{V} : \tilde{N} \times \Upsilon \to \Re$ respectively, locally* *defined in a neighborhood $\widetilde{N} \times \Upsilon \subset \mathcal{X} \times \mathcal{Y}$ of the origin $(\hat{x}, y) = (0, 0)$, and a matrix function $\tilde{L} : \widetilde{N} \times \Upsilon \to \mathcal{M}^{n \times m}$ satisfying the following pair of coupled HJIEs:*

$$\tilde{Y}_{\hat{x}}(\hat{x}, y) f(\hat{x}) + \tilde{Y}_y(\hat{x}, y)\dot{y} - \frac{1}{2\gamma^2}\tilde{Y}_{\hat{x}}(\hat{x}, y)g_1(\hat{x})g_1^T(\hat{x})\tilde{Y}_{\hat{x}}^T(\hat{x}, y) -$$

$$\frac{1}{2\gamma^2}\tilde{Y}_{\hat{x}}(\hat{x}, y)\tilde{L}(\hat{x}, y)\tilde{L}^T(\hat{x}, y)\tilde{Y}_{\hat{x}}^T(\hat{x}, y) - \frac{1}{\gamma^2}\tilde{Y}_{\hat{x}}(\hat{x}, y)\tilde{L}(\hat{x}, y)\tilde{L}^T(\hat{x}, y)\tilde{V}_{\hat{x}}^T(\hat{x}, y) -$$

$$\frac{1}{2}(y - h_2(\hat{x}))(y - h_2(\hat{x})) = 0, \quad \tilde{Y}(0, 0) = 0, \quad \hat{x} \in \widetilde{N}, \; y \in \Upsilon \qquad (12.39)$$

$$\tilde{V}_{\hat{x}}(\hat{x}, y) f(\hat{x}) + \tilde{V}_y(\hat{x}, y)\dot{y} - \frac{1}{\gamma^2}\tilde{V}_{\hat{x}}(\hat{x}, y)g_1(\hat{x})g_1^T(\hat{x})\tilde{Y}_{\hat{x}}^T(\hat{x}, y) -$$

$$\frac{1}{\gamma^2}\tilde{V}_{\hat{x}}(\hat{x}, y)\tilde{L}(\hat{x}, y)\tilde{L}^T(\hat{x}, y)\tilde{Y}_{\hat{x}}^T(\hat{x}, y) - \frac{1}{\gamma^2}\tilde{V}_{\hat{x}}(\hat{x}, y)\tilde{L}(\hat{x}, y)\tilde{L}^T(\hat{x}, y)\tilde{V}_{\hat{x}}^T(\hat{x}, y) +$$

$$\frac{1}{2}(y - h_2(\hat{x}))^T(y - h_2(\hat{x})) = 0, \quad \tilde{V}(0, 0) = 0, \quad \hat{x} \in \widetilde{N}, \; y \in \Upsilon \qquad (12.40)$$

together with the coupling condition

$$\tilde{V}_{\hat{x}}(\hat{x}, y)\tilde{L}(\hat{x}, y) = -\gamma^2(y - h_2(\hat{x}))^T, \; \hat{x} \in \widetilde{N}, \; y \in \Upsilon. \qquad (12.41)$$

Then, the filter $\tilde{\Sigma}^{af}$ with the gain matrix $\tilde{L}(\hat{x}, y)$ satisfying (12.41) solves the infinite-horizon MH2HINLFP for the system locally in $\widetilde{N}$.

Proof: (Sketch) Using similar manipulations as in Theorem 12.1.2 and Prop 12.1.1, it can be shown that the existence of a solution to the coupled HJIEs (12.39), (12.40) implies the existence of a solution to the dissipation-inequality

$$\tilde{Y}(\hat{x}, y) - \tilde{Y}(\hat{x}_0, y_0) \le \frac{1}{2}\int_{t_0}^{\infty}(\gamma^2\|w\|^2 - \|\tilde{z}\|^2)dt$$

for some smooth function $\tilde{Y} = -\tilde{Y} > 0$. Again using similar arguments as in Theorem 12.1.2 and Prop 12.1.1, the result follows. In particular, with $w \equiv 0$ and $\tilde{Y}(\hat{x}, y) = 0 \Rightarrow \tilde{z} \equiv 0$, which in turn implies $x \equiv \hat{x}$ by the zero-input observability of the system. $\square$

Remark 12.1.5 *Comparing the HJIEs (12.39)-(12.40) with (12.26)-(12.27) we see that the dimensionality of the former is half. This is indeed significant. Moreover, the filter gain corresponding to the new HJIEs (12.41) does not depend on x.*

12.2 Discrete-Time Mixed $\mathcal{H}_2/\mathcal{H}_\infty$ Nonlinear Filtering

The set-up for this case is the same as the continuous-time case shown in Figure 12.1 with the difference that the variables and measurements are discrete, and is shown on Figure 12.3. Therefore, the plant is similarly described by an affine causal discrete-time nonlinear state-space system with zero input defined on an n-dimensional space $\mathcal{X} \subseteq \Re^n$ in coordinates $x = (x_1, \ldots, x_n)$:

$$\Sigma^{da} : \begin{cases} x_{k+1} &= f(x_k) + g_1(x_k)w_k; \; x(k_0) = x_0 \\ z_k &= h_1(x_k) \\ y_k &= h_2(x_k) + k_{21}(x_k)w_k \end{cases} \qquad (12.42)$$

where $x \in \mathcal{X}$ is the state vector, $w \in \mathcal{W}$ is the disturbance or noise signal, which belongs to the set $\mathcal{W} \subset \ell_2([k_0, \infty)\Re^r)$ of admissible disturbances or noise signals, the output $y \in \Re^m$ is the measured output of the system, while $z \in \Re^s$ is the output to be estimated. The functions

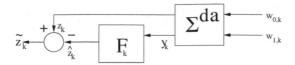

FIGURE 12.3
Set-Up for Discrete-Time Mixed $\mathcal{H}_2/\mathcal{H}_\infty$ Nonlinear Filtering

$f : \mathcal{X} \to \mathcal{X}$, $g_1 : \mathcal{X} \to \mathcal{M}^{n \times r}(\mathcal{X})$, $h_1 : \mathcal{X} \to \Re^s$, $h_2 : \mathcal{X} \to \Re^m$, and $k_{12} : \mathcal{X} \to \mathcal{M}^{s \times p}(\mathcal{X})$, $k_{21} : \mathcal{X} \to \mathcal{M}^{m \times r}(\mathcal{X})$ are real C^∞ functions of x. Furthermore, we assume without any loss of generality that the system (12.42) has a unique equilibrium-point at $x = 0$ such that $f(0) = 0$, $h_1(0) = h_2(0) = 0$.

The objective is again to synthesize a filter, $\mathcal{F}_k$, for estimating the state x_k (or more generally a function of it $z_k = h_1(x_k)$) from observations of y_i up to time k and over a time horizon $[k_0, K]$, i.e., from

$$Y_k \triangleq \{y_i : i \le k\}, \quad k \in [k_0, K]$$

such that the gains (or induced norms) from the inputs $w_0 := \{w_{0,k}\}$, $w_1 := \{w_{1,k}\}$, to the error or penalty variable $\check{z}$ defined respectively as the ℓ_2-norm and ℓ_∞-norm of the interconnected system $\mathcal{F}_k \circ \Sigma^{da}$ respectively, i.e.,

$$\|\mathcal{F}_k \circ \Sigma^{da}\|_{\ell_2} \triangleq \sup_{0 \ne w_0 \in \mathcal{S}'} \frac{\|\check{z}\|_{\mathcal{P}'}}{\|w_0\|_{\mathcal{S}'}}, \tag{12.43}$$

and

$$\|\mathcal{F}_k \circ \Sigma^{da}\|_{\ell_\infty} \triangleq \sup_{0 \ne w_1 \in \mathcal{P}'} \frac{\|\check{z}\|_2}{\|w_1\|_2} \tag{12.44}$$

are minimized, where

$$\mathcal{P}' \triangleq \{w : w \in \ell_\infty, \; R_{ww}(k), \; S_{ww}(j\omega) \text{ exist for all } k \text{ and all } \omega \text{ resp.,}$$
$$\|w\|_{\mathcal{P}'} < \infty\}$$

$$\mathcal{S}' \triangleq \{w : w \in \ell_\infty, \; R_{ww}(k), \; S_{ww}(j\omega) \text{ exist for all } k \text{ and all } \omega \text{ resp.,}$$
$$\|S_{ww}(j\omega)\|_\infty < \infty\}$$

$$\|z\|_{\mathcal{P}'}^2 \triangleq \lim_{K \to \infty} \frac{1}{2K} \sum_{k=-K}^{K} \|z\|^2$$

$$\|w_0\|_{\mathcal{S}'}^2 = \|S_{w_0 w_0}(j\omega)\|_\infty, \quad \|w_0\|_{\mathcal{S}'}^2 = \|S_{w_0 w_0}(j\omega)\|_\infty,$$

and R_{ww}, $S_{ww}(j\omega)$ are the autocorrelation and power spectral-density matrices of w [152]. In addition, if the plant is stable, we replace the induced ℓ-norms above by their equivalent $\mathcal{H}$-subspace norms.

The above problem is the standard discrete-time mixed $\mathcal{H}_2/\mathcal{H}_\infty$-filtering problem. However, as pointed out in the previous section, due to the difficulty of solving the above problem, we do not solve it either in this section. Instead, we solve the associated problem involving a single noise/disturbance signal $w \in \mathcal{W} \subset \ell_2[k_0, \infty)$, and minimize the output energy of the system, $\|\check{z}\|_{l_2}$, while rendering $\|\mathcal{F}_k \circ \Sigma^{da}\|_{\ell_\infty} \le \gamma^\star$ for a given number $\gamma^\star > 0$, for all $w \in \mathcal{W}$ and for all initial conditions $x^0 \in O \subset \mathcal{X}$. In addition, we also require that with $w_k \equiv 0$, the estimation error converges to zero, i.e., $\lim_{k \to \infty} \check{z}_k = 0$.

Again, for the discrete-time nonlinear system (12.42), the above $\mathcal{H}_\infty$ constraint is interpreted as the ℓ_2-gain constraint and is represented as

$$\sum_{k=k_0}^{K} \|\breve{z}_k\|^2 \leq \gamma^2 \sum_{k=k_0}^{K} \|w_k\|^2, \quad K > k_0 \in \mathbf{Z} \tag{12.45}$$

for all $w \in \mathcal{W}$, for all initial states $x^0 \in O \subset \mathcal{X}$.

The discrete-time mixed-$\mathcal{H}_2/\mathcal{H}_\infty$ filtering or estimation problem can then be defined formally as follows.

Definition 12.2.1 *(Discrete-Time Mixed $\mathcal{H}_2/\mathcal{H}_\infty$ (Suboptimal) Nonlinear Filtering Problem (DMH2HINLFP)). Given the plant Σ^{da} and a number $\gamma^* > 0$, find a filter $\mathcal{F}_k : \mathcal{Y} \to \mathcal{X}$ such that*

$$\hat{x}_{k+1} = \mathcal{F}_k(\mathsf{Y}_k)$$

and the output energy $\|\breve{z}\|_{\ell_2}$ is minimized, while the constraint (12.45) is satisfied for all $\gamma \geq \gamma^*$, for all $w \in \mathcal{W}$ and all $x^0 \in O$. In addition, with $w_k \equiv 0$, we have $\lim_{k\to\infty} \breve{z}_k = 0$.

Moreover, if the above conditions are satisfied for all $x^0 \in \mathcal{X}$, we say that $\mathcal{F}_k$ solves the DMH2HINLFP globally.

Remark 12.2.1 *The problem defined above is the finite-horizon filtering problem. We have the infinite-horizon problem if we let $K \to \infty$.*

12.2.1 Solution to the Finite-Horizon Discrete-Time Mixed $\mathcal{H}_2/\mathcal{H}_\infty$ Nonlinear Filtering Problem

We similarly consider the following class of estimators:

$$\Sigma^{daf} : \begin{cases} \hat{x}_{k+1} &= f(x_k) + L(\hat{x}_k, k)[y_k - h_2(x_k)], \quad \hat{x}(k_0) = \hat{x}^0 \\ \hat{z}_k &= h_1(\hat{x}_k) \end{cases} \tag{12.46}$$

where $\breve{x}_k \in \mathcal{X}$ is the estimated state, $L(.,.) \in \mathcal{M}^{n\times m}(\mathcal{X} \times \mathbf{Z})$ is the error-gain matrix which is smooth and has to be determined, and $\hat{z} \in \Re^s$ is the estimated output of the filter. We can now define the estimation error or penalty variable, $\breve{z}$, which has to be controlled as:

$$\breve{z}_k := z_k - \hat{z}_k = h_1(x_k) - h_1(\hat{x}_k).$$

Then, we combine the plant (12.42) and estimator (12.46) dynamics to obtain the following augmented system:

$$\left. \begin{array}{rcl} \breve{x}_{k+1} &=& \breve{f}(\breve{x}_k) + \breve{g}(\breve{x}_k)w_k, \quad \breve{x}(k_0) = (x^{0^T} \; \hat{x}^{0^T})^T \\ \breve{z}_k &=& \breve{h}(\breve{x}_k) \end{array} \right\}, \tag{12.47}$$

where

$$\breve{x}_k = \begin{pmatrix} x_k \\ \hat{x}_k \end{pmatrix}, \quad \breve{f}(\breve{x}) = \begin{pmatrix} f(x_k) \\ f(\hat{x}_k) + L(\hat{x}_k, k)(h_2(x_k) - h_2(\hat{x}_k)) \end{pmatrix},$$

$$\breve{g}(\breve{x}) = \begin{pmatrix} g_1(x_k) \\ L(\hat{x}_k, k)k_{21}(x_k) \end{pmatrix}, \quad \breve{h}(\breve{x}_k) = h_1(x_k) - h_1(\hat{x}_k).$$

The problem is then similarly formulated as a two-player nonzero-sum differential game with the following cost functionals:

$$J_1(L, w) = \frac{1}{2} \sum_{k=k_0}^{K} \{ \gamma^2 \|w_k\|^2 - \|\check{z}_k\|^2 \}, \tag{12.48}$$

$$J_2(L, w) = \frac{1}{2} \sum_{k_0}^{K} \|\check{z}_k\|^2, \tag{12.49}$$

where $w := \{w_k\}$. The first functional is associated with the $\mathcal{H}_\infty$-constraint criterion, while the second functional is related to the output energy of the system or $\mathcal{H}_2$-criterion. It is seen that, by making $J_1 \geq 0$, the $\mathcal{H}_\infty$ constraint $\|\mathcal{F}_k \circ \Sigma^{da}\|_{\mathcal{H}_\infty} \leq \gamma$ is satisfied. Then, similarly, a Nash-equilibrium solution to the above game is said to exist if we can find a pair $(L^\star, w^\star)$ such that

$$J_1(L^\star, w^\star) \leq J_1(L^\star, w) \ \ \forall w \in \mathcal{W}, \tag{12.50}$$

$$J_2(L^\star, w^\star) \leq J_2(L_k, w^\star) \ \ \forall L \in \mathcal{M}^{n \times m}. \tag{12.51}$$

Sufficient conditions for the solvability of the above game are well known (Chapter 3, also [59]), and are given in the following theorem.

Theorem 12.2.1 *For the two-person discrete-time nonzero-sum game (12.48)-(12.49), (12.47), under memoryless perfect information structure, there exists a feedback Nash-equilibrium solution if, and only if, there exist $2(K - k_0)$ functions $Y, V : N \subset \mathcal{X} \times \mathbf{Z} \to \Re$, $N \subset \mathcal{X}$ such that the following coupled recursive equations (discrete-time Hamilton-Jacobi-Isaacs equations (DHJIE)) are satisfied:*

$$Y(\check{x}, k) = \inf_{w \in \mathcal{W}} \left\{ \frac{1}{2} [\gamma^2 [\|w_k\|^2 - \|\check{z}_k(\check{x})\|^2] + Y(\check{x}_{k+1}, k + 1) \right\},$$
$$Y(\check{x}, K + 1) = 0, \quad k = k_0, \ldots, K, \forall \check{x} \in N \times N \tag{12.52}$$

$$V(\check{x}, k) = \min_{L \in \mathcal{M}^{n \times m}} \left\{ \frac{1}{2} \|\check{z}_k(\check{x})\|^2 + V(\check{x}_{k+1}, k + 1) \right\},$$
$$V(\check{x}, K + 1) = 0, \quad k = k_0, \ldots, K, \ \ \forall \check{x} \in N \times N \tag{12.53}$$

where $\check{x} = \check{x}_k$, $L = L(x_k, k)$, $w := \{w_k\}$.

Thus, we can apply the above theorem to derive sufficient conditions for the solvability of the $DMH2HINLFP$. To do that, we define the Hamiltonian functions $H_i : (\mathcal{X} \times \mathcal{X}) \times \mathcal{W} \times \mathcal{M}^{n \times m} \times \Re \to \Re$, $i = 1, 2$ associated with the cost functionals (12.48), (12.49) respectively:

$$H_1(\check{x}, w_k, L, Y) = Y(\check{f}(\check{x}) + \check{g}(\check{x})w_k, k + 1) - Y(\check{x}, k) + \frac{1}{2}\gamma^2 \|w_k\|^2 - \frac{1}{2}\|\check{z}_k\|^2, \tag{12.54}$$

$$H_2(\check{x}, w_k, L, V) = V(\check{f}(\check{x}) + \check{g}(\check{x})w_k, k + 1) - V(\check{x}, k) + \frac{1}{2}\|\check{z}_k\|^2 \tag{12.55}$$

for some smooth functions $Y, V : \mathcal{X} \to \Re$, $Y < 0$, $V > 0$ where the adjoint variables corresponding to the cost functionals (12.48), (12.49) are set as $p_1 = Y$, $p_2 = V$ respectively.

The following theorem then presents sufficient conditions for the solvability of the $DMH2HINLFP$ on a finite-horizon.

Theorem 12.2.2 *Consider the nonlinear system (12.42) and the DMH2HINLFP for it. Suppose the function h_1 is one-to-one (or injective) and the plant Σ^{da} is locally asymptotically-stable about the equilibrium-point $x = 0$. Further, suppose there exists a* *pair of C^2 negative and positive-definite functions $Y, V : N \times N \times \mathbf{Z} \to \Re$ respectively, locally defined in a neighborhood $N \times N \subset \mathcal{X} \times \mathcal{X}$ of the origin $\breve{x} = 0$, and a matrix function $L : N \times \mathbf{Z} \to \mathcal{M}^{n \times m}$ satisfying the following pair of coupled HJIEs:*

$$Y(\breve{x}, k) = Y(\breve{f}^\star(\breve{x}) + \breve{g}^\star(\breve{x})w_k^\star(\breve{x}), k+1) + \frac{1}{2}\gamma^2 \|w_k^\star(\breve{x})\|^2 - \frac{1}{2}\|\breve{z}_k(\breve{x})\|^2, \; Y(\breve{x}, K+1) = 0,$$
(12.56)

$$V(\breve{x}, k) = V(\breve{f}^\star(\breve{x}) + \breve{g}^\star(\breve{x})w_k^\star(\breve{x}), k+1) + \frac{1}{2}\|\breve{z}_k(\breve{x})\|^2, \;\; V(\breve{x}, K+1) = 0, \tag{12.57}$$

$k = k_0, \ldots, K$, *together with the side-conditions*

$$w_k^\star = -\frac{1}{\gamma^2}\breve{g}^T(\breve{x}) \frac{\partial^T Y(\lambda, k+1)}{\partial \lambda}\bigg|_{\lambda = \breve{f}(\breve{x}) + \breve{g}(\breve{x})w_k^\star}, \tag{12.58}$$

$$L^\star = \arg\min_L \{H_2(\breve{x}, w_k^\star, L, V)\}, \tag{12.59}$$

$$\frac{\partial^2 H_1}{\partial w^2}(\breve{x}, w_k, L^\star, Y)\bigg|_{\breve{x}=0} > 0, \tag{12.60}$$

$$\frac{\partial^2 H_2}{\partial L^2}(\breve{x}, w_k^\star, L, V)\bigg|_{\breve{x}=0} > 0, \tag{12.61}$$

where

$$\breve{f}^\star(\breve{x}) = \breve{f}(\breve{x})\big|_{L=L^\star}, \;\; \breve{g}^\star(\breve{x}) = \breve{g}(\breve{x})|_{L=L^\star}.$$

Then:

(i) *there exists locally a Nash-equilibrium solution $(w^\star, L^\star)$ for the game (12.48), (12.49), (12.42) locally in N;*

(ii) *the augmented system (12.47) is locally dissipative with respect to the supply-rate $s(w_k, \breve{z}_k) = \frac{1}{2}(\gamma^2\|w_k\|^2 - \|\breve{z}_k\|^2)$ and hence has ℓ_2-gain from w to $\breve{z}$ less or equal to γ;*

(iii) *the optimal costs or performance objectives of the game are $J_1^\star(L^\star, w^\star) = Y(\breve{x}^0, k_0)$ and $J_2^\star(L^\star, w^\star) = V(\breve{x}^0, k_0)$;*

(iv) *the filter Σ^{daf} with the gain matrix $L(\hat{x}_k, k)$ satisfying (12.59) solves the finite-horizon $DMH2HINLFP$ for the system locally in N.*

Proof: Assume there exist definite solutions Y, V to the DHJIEs (12.56)-(12.57), and (i) consider the Hamiltonian function $H_1(., ., ., .)$. Then applying the necessary condition for the worst-case noise, we have

$$\frac{\partial^T H_1}{\partial w}\bigg|_{w=w_k^\star} = \breve{g}^T(\breve{x}) \frac{\partial^T Y(\lambda, k+1)}{\partial \lambda}\bigg|_{\lambda = \breve{f}(\breve{x}) + \breve{g}(\breve{x})w_k^\star} + \gamma^2 w_k^\star = 0,$$

to get

$$w_k^\star := -\frac{1}{\gamma^2}\breve{g}^T(\breve{x}) \frac{\partial^T Y(\lambda, k+1)}{\partial \lambda}\bigg|_{\lambda = \breve{f}(\breve{x}) + \breve{g}(\breve{x})w^\star} := \alpha_0(\breve{x}, w_k^\star). \tag{12.62}$$

Thus, $w^\star$ is expressed implicitly. Moreover, since

$$\frac{\partial^2 H_1}{\partial w^2} = \breve{g}^T(\breve{x}) \frac{\partial^2 Y(\lambda, k+1)}{\partial \lambda^2}\bigg|_{\lambda = \breve{f}(\breve{x}) + \breve{g}(\breve{x})w_k^\star} \breve{g}(\breve{x}) + \gamma^2 I$$

is nonsingular about $(\breve{x}, w) = (0, 0)$, equation (12.62) has a unique solution $\alpha_1(\breve{x})$, $\alpha_1(0) = 0$ in the neighborhood $N_0 \times W_0$ of $(x, w) = (0, 0)$ by the Implicit-function Theorem [234].

Now, substitute $w^\star$ in the expression for $H_2(.,.,.,.)$ (12.55), to get

$$H_2(\breve{x}, w_k^\star, L, V) = V(\breve{f}(\breve{x}) + \breve{g}(\breve{x})w_k^\star(\breve{x}), k+1) - V(\breve{x}, k) + \frac{1}{2}\|\breve{z}_k(\breve{x})\|^2,$$

and let

$$L^\star = \arg\min_L \{H_2(\breve{x}, w_k^\star, L, V)\}.$$

Then by Taylor's theorem, we can expand $H_2(., w^\star, ., .)$ about $L^\star$ [267] as

$$
\begin{aligned}
H_2(\breve{x}, w^\star, L, Y) = {} & H_2(\breve{x}, w^\star, L^\star, Y_{\breve{x}}^T) + \\
& \frac{1}{2}Tr\left\{[I_n \otimes (L - L^\star)^T]\frac{\partial^2 H_2}{\partial L^2}(w^\star, L)[I_m \otimes (L - L^\star)^T]\right\} + \\
& O(\|L - L^\star\|^3).
\end{aligned}
$$

Therefore, taking $L^\star$ as in (12.59) and if the condition (12.61) holds, then $H_2(.,.,w^\star,.)$ is minimized, and the Nash-equilibrium condition

$$H_2(w^\star, L^\star) \leq H_2(w^\star, L) \quad \forall L \in \mathcal{M}^{n\times m}, k = k_0, \ldots, K$$

is satisfied. Moreover, substituting $(w^\star, L^\star)$ in (12.53) gives the DHJIE (12.57).

Now substitute $L^\star$ as given by (12.59) in the expression for $H_1(.,.,.,.)$ and expand it in Taylor's-series about $w^\star$ to obtain

$$
\begin{aligned}
H_1(\breve{x}, w_k, L^\star, Y) = {} & Y(\breve{f}^\star(\breve{x}) + \breve{g}^\star(\breve{x})w, k+1) - Y(x, k) + \frac{1}{2}\gamma^2\|w_k\|^2 - \frac{1}{2}\|\breve{z}_k\|^2 \\
= {} & H_1(\breve{x}, w_k^\star, L^\star, Y) + \frac{1}{2}(w_k - w_k^\star)^T\frac{\partial^2 H_2}{\partial w_k^2}(w_k, L^\star)(w_k - w_k^\star) + \\
& O(\|w_k - w_k^\star\|^3).
\end{aligned}
$$

Further, substituting $w = w^\star$ as given by (12.62) in the above, and if the condition (12.60) is satisfied, we see that the second Nash-equilibrium condition

$$H_1(w^\star, L^\star) \leq H_1(w, L^\star), \quad \forall w \in \mathcal{W}$$

is also satisfied. Therefore, the pair $(w^\star, L^\star)$ constitute a Nash-equilibrium solution to the two-player nonzero-sum dynamic game. Moreover, substituting $(w^\star, L^\star)$ in (12.52) gives the DHJIE (12.56).

(ii) The Nash-equilibrium condition

$$H_1(\breve{x}, w, L^\star, Y) \geq H_1(\breve{x}, w^\star, L^\star, Y) = 0 \quad \forall \breve{x} \in U, \forall w \in \mathcal{W}$$

implies

$$Y(\breve{x}, k) - Y(\breve{x}_{k+1}, k+1) \leq \frac{1}{2}\gamma^2\|w_k\|^2 - \frac{1}{2}\|\breve{z}_k\|^2, \quad \forall \breve{x} \in U, \forall w \in \mathcal{W}$$

$$\iff \breve{Y}(\breve{x}_{k+1}, k+1) - \breve{Y}(\breve{x}_k, k) \leq \frac{1}{2}\gamma^2\|w_k\|^2 - \frac{1}{2}\|\breve{z}_k\|^2, \quad \forall \breve{x} \in U, \forall w \in \mathcal{W} \quad (12.63)$$

for some positive-definite function $\breve{Y} = -Y > 0$. Summing now the above inequality from $k = k_0$ to $k = K$ we get the dissipation-inequality

$$\breve{Y}(\breve{x}_{K+1}, K+1) - \breve{Y}(x_{k_0}, k_0) \leq \sum_{k=k_0}^{K}\frac{1}{2}\gamma^2\|w_k\|^2 - \frac{1}{2}\|\breve{z}_k\|^2. \quad (12.64)$$

Thus, from Chapter 3, the system has ℓ_2-gain from w to $\breve{z}$ less or equal to γ.

(iii) Consider the cost functional $J_1(L, w)$ first, and rewrite it as

$$
\begin{aligned}
J_1(L, w) + Y(\breve{x}_{K+1}, K+1) - Y(\breve{x}(k_0), k_0) &= \sum_{k=k_0}^{K} \left\{ \frac{1}{2}\gamma^2 \|w_k\|^2 - \frac{1}{2}\|\breve{z}_k\|^2 + \right. \\
&\quad \left. Y(\breve{x}_{k+1}, k+1) - Y(\breve{x}_k, k) \right\} \\
&= \sum_{k=k_0}^{K} H_1(\breve{x}, w_k, L_k, Y).
\end{aligned}
$$

Substituting $(L^\star, w^\star)$ in the above equation and using the DHJIE (12.56) gives $H_1(\breve{x}, w^\star, L^\star, Y) = 0$ and the result follows. Similarly, consider the cost functional $J_2(L, w)$ and rewrite it as

$$
\begin{aligned}
J_2(L, w) + V(\breve{x}_{K+1}, K+1) - V(\breve{x}_{k_0}, k_0) &= \sum_{k=k_0}^{K} \left\{ \frac{1}{2}\|\breve{z}_k\|^2 + V(\breve{x}_{k+1}, k+1) - V(\breve{x}_k, k) \right\} \\
&= \sum_{k=k_0}^{K} H_2(\breve{x}, w_k, L_k, V).
\end{aligned}
$$

Since $V(\breve{x}, K+1) = 0$, substituting $(L^\star, w^\star)$ in the above and using the DHJIE (12.57) the result similarly follows.

(iv) Notice that the inequality (12.63) implies that with $w_k \equiv 0$,

$$
\breve{Y}(\breve{x}_{k+1}, k+1) - \breve{Y}(\breve{x}_k, k) \leq -\frac{1}{2}\|\breve{z}_k\|^2, \quad \forall \breve{x} \in \Upsilon, \tag{12.65}
$$

and since $\breve{Y}$ is positive-definite, by Lyapunov's theorem, the augmented system is locally stable. Finally, combining (i)-(iii), (iv) follows. $\square$

12.2.2 Solution to the Infinite-Horizon Discrete-Time Mixed $\mathcal{H}_2/\mathcal{H}_\infty$ Nonlinear Filtering Problem

In this subsection, we discuss the infinite-horizon filtering problem, in which case we let $K \to \infty$. Moreover, in this case, we seek a time-invariant gain $\hat{L}(\hat{x})$ for the filter, and consequently time-independent functions $Y, V : \tilde{N} \times \tilde{N} \to \Re$ locally defined in a neighborhood $\tilde{N} \times \tilde{N} \subset \mathcal{X} \times \mathcal{X}$ of $(x, \hat{x}) = (0, 0)$, such that the following steady-state DHJIEs:

$$
Y(\breve{f}^\star(\breve{x}) + \breve{g}^\star(\breve{x})\tilde{w}^\star(\breve{x})) - Y(\breve{x}) + \frac{1}{2}\gamma^2\|\tilde{w}^\star(\breve{x})\|^2 - \frac{1}{2}\|\breve{z}(\breve{x})\|^2 = 0, \; Y(0) = 0, \tag{12.66}
$$

$$
V(\breve{f}^\star(\breve{x}) + \breve{g}^\star(\breve{x})\tilde{w}^\star) - V(\breve{x}) + \frac{1}{2}\|\breve{z}(\breve{x})\|^2 = 0, \quad V(0) = 0, \tag{12.67}
$$

are satisfied together with the side-conditions:

$$
\tilde{w}^\star = \left. -\frac{1}{\gamma^2}\breve{g}^T(\breve{x})\frac{\partial^T Y(\lambda)}{\partial \lambda} \right|_{\lambda = \breve{f}(\breve{x}) + \breve{g}(\breve{x})\tilde{w}^\star} := \alpha_2(\breve{x}, \tilde{w}^\star), \tag{12.68}
$$

$$
\tilde{L}^\star(\hat{x}) = \arg \min_{\tilde{L}} \left\{ \tilde{H}_2(\breve{x}, w^\star, \tilde{L}, V) \right\}, \tag{12.69}
$$

$$
\left. \frac{\partial^2 \tilde{H}_1}{\partial w^2}(\breve{x}, w, \tilde{L}^\star, Y) \right|_{\breve{x}=0} > 0, \tag{12.70}
$$

$$\left.\frac{\partial^2 \widetilde{H}_2}{\partial \tilde{L}^2}(\breve{x}, \tilde{w}^\star, \tilde{L}, V)\right|_{\breve{x}=0} > 0, \qquad (12.71)$$

where $\tilde{w}^\star$, $\tilde{L}^\star$ are the asymptotic values of $w^\star$, $L^\star$ respectively,

$$\breve{f}^\star(\breve{x}) = \breve{f}(\breve{x})\Big|_{\tilde{L}=\tilde{L}^\star}, \quad \breve{g}^\star(\breve{x}) = \breve{g}(\breve{x})|_{\tilde{L}=\tilde{L}^\star},$$

$$\widetilde{H}_1(\breve{x}, w_k, \tilde{L}, Y) = Y(\breve{f}(\breve{x}) + \breve{g}(\breve{x})w_k) - Y(\breve{x}) + \frac{1}{2}\gamma^2\|w_k\|^2 - \frac{1}{2}\|\breve{z}_k\|^2,$$

$$\widetilde{H}_2(\breve{x}, w_k, \tilde{L}, V) = V(\breve{f}(\breve{x}) + \breve{g}(\breve{x})w_k) - V(\breve{x}) + \frac{1}{2}\|\breve{z}_k\|^2.$$

Again here, since the estimation is carried over an infinite-horizon, it is necessary to ensure that the augmented system (12.47) is stable with $w = 0$. However, in this case, we can relax the requirement of asymptotic-stability for the original system (12.42) with a milder requirement of detectability which we define next.

Definition 12.2.2 *The pair $\{f, h\}$ is said to be locally zero-state detectable if there exists a neighborhood $\mathcal{O}$ of $x = 0$ such that, if x_k is a trajectory of $x_{k+1} = f(x_k)$ satisfying $x(k_0) \in \mathcal{O}$, then $h(x_k)$ is defined for all $k \geq k_0$ and $h(x_k) = 0$, for all $k \geq k_s$, implies $\lim_{k\to\infty} x_k = 0$. Moreover $\{f, h\}$ is zero-state detectable if $\mathcal{O} = \mathcal{X}$.*

The "*admissibility*" of the discrete-time filter is similarly defined as follows.

Definition 12.2.3 *A filter $\mathcal{F}$ is admissible if it is asymptotically (or internally) stable for any given initial condition $x(k_0)$ of the plant Σ^{da}, and with $w \equiv 0$*

$$\lim_{k\to\infty} \breve{z}_k = 0.$$

The following proposition can now be proven along the same lines as Theorem 12.2.2.

Proposition 12.2.1 *Consider the nonlinear system (12.42) and the infinite-horizon $DMH2HINLFP$ for it. Suppose the function h_1 is one-to-one (or injective) and the plant Σ^{da} is zero-state detectable. Suppose further, there exists a pair of C^2 negative and positive-definite functions $Y, V : \widetilde{N} \times \widetilde{N} \to \Re$ respectively, locally defined in a neighborhood $\widetilde{N} \times \widetilde{N} \subset \mathcal{X} \times \mathcal{X}$ of the origin $\breve{x} = 0$, and a matrix function $\tilde{L} : \widetilde{N} \to \mathcal{M}^{n\times m}$ satisfying the pair of coupled DHJIEs (12.66), (12.67) together with (12.68)-(12.71). Then:*

(i) *there exists locally a Nash-equilibrium solution $(\tilde{w}^\star, \tilde{L}^\star)$ for the game;*

(ii) *the augmented system (12.47) is dissipative with respect to the supply-rate $s(w, \breve{z}) = \frac{1}{2}(\gamma^2\|w\|^2 - \|\breve{z}\|^2)$ and hence has ℓ_2-gain from w to $\breve{z}$ less or equal to γ;*

(iii) *the optimal costs or performance objectives of the game are $J_1^\star(\tilde{L}^\star, \tilde{w}^\star) = Y(\breve{x}^0)$ and $J_2^\star(\tilde{L}^\star, \tilde{w}^\star) = V(\breve{x}^0)$;*

(iv) *the filter Σ^{daf} with the gain matrix $L(\hat{x}) = \tilde{L}^\star(\hat{x})$ satisfying (12.69) solves the infinite-horizon $DMH2HINLFP$ locally in $\widetilde{N}$.*

Proof: Since the proof of items (i)-(iii) is similar to that of Theorem 12.2.2, we prove only item (iv).

(iv) Using similar manipulations as in the proof of Theorem 12.2.2, it can be shown that a similar inequality as (12.63) also holds. This implies that with $w_k \equiv 0$,

$$\breve{Y}(\breve{x}_{k+1}) - \breve{Y}(\breve{x}_k) \le -\frac{1}{2}\|\breve{z}_k\|^2, \quad \forall \breve{x} \in \widetilde{N} \tag{12.72}$$

and since $\breve{Y}$ is positive-definite, by Lyapunov's theorem, the augmented system is locally stable. Furthermore, for any trajectory of the system $\breve{x}_k$ such that $\breve{Y}(\breve{x}_{k+1}) - \breve{Y}(\breve{x}_k) = 0$ for all $k \ge k_c > k_0$, it implies that $z_k \equiv 0$. This in turn implies $h_1(x_k) = h_1(\hat{x}_k)$, and $x_k = \hat{x}_k \, \forall k \ge k_c$ since h_1 is injective. This further implies that $h_2(x_k) = h_2(\hat{x}_k) \, \forall k \ge k_c$ and it is a trajectory of the free system:

$$\breve{x}_{k+1} = \left(\begin{array}{c} f(x_k) \\ f(\hat{x}_k) \end{array} \right).$$

By zero-state of the detectability of $\{f, h_1\}$, we have $\lim_{k\to\infty} x_k = 0$, and we have internal stability of the augmented system with $\lim_{k\to\infty} z_k = 0$. Hence, Σ^{daf} is admissible. Finally, combining (i)-(iii), (iv) follows. $\square$

12.2.3 Approximate and Explicit Solution to the Infinite-Horizon Discrete-Time Mixed $\mathcal{H}_2/\mathcal{H}_\infty$ Nonlinear Filtering Problem

In this subsection, we discuss how the $DMH2HINLFP$ can be solved approximately to obtain explicit solutions [126]. We consider the infinite-horizon problem for this purpose, but the approach can also be used for the finite-horizon problem. For simplicity, we make the following assumption on the system matrices.

Assumption 12.2.1 *The system matrices are such that*

$$\begin{aligned} k_{21}(x)g_1^T(x) &= 0, \\ k_{21}(x)k_{21}^T(x) &= I. \end{aligned}$$

Consider now the infinite-horizon Hamiltonian functions

$$\begin{aligned} H_1(\breve{x}, w, \hat{L}, \tilde{Y}) &= \tilde{Y}(\breve{f}(\breve{x}) + \breve{g}(\breve{x})w) - \tilde{Y}(\breve{x}) + \frac{1}{2}\gamma^2\|w\|^2 - \frac{1}{2}\|\breve{z}\|^2, \\ H_2(\breve{x}, w, \hat{L}, \tilde{V}) &= \tilde{V}(\breve{f}(\breve{x}) + \breve{g}(\breve{x})w) - \tilde{V}(\breve{x}) + \frac{1}{2}\|\breve{z}\|^2, \end{aligned}$$

for some negative and positive-definite functions $\tilde{Y}, \tilde{V} : \widehat{N} \times \widehat{N} \to \Re$, $\widehat{N} \subset \mathcal{X}$ a neighborhood of $x = 0$, and where $\breve{x} = \breve{x}_k$, $w = w_k$, $z = z_k$. Expanding them in Taylor series[1] about $\hat{f}(\breve{x})$ up to first-order:

$$\begin{aligned} \widehat{H}_1(\breve{x}, w, \hat{L}, \tilde{Y}) = \; & \left\{ \tilde{Y}(\hat{f}(\breve{x})) + \tilde{Y}_x(\hat{f}(\breve{x}))g_1(x)w + \tilde{Y}_{\hat{x}}(\hat{f}(\breve{x}))[\hat{L}(\hat{x})(h_2(x) - h_2(\hat{x}) + k_{21}(x)w)] \right. \\ & \left. + O(\|\tilde{v}\|^2) \right\} - \tilde{Y}(\breve{x}) + \frac{1}{2}\gamma^2\|w\|^2 - \frac{1}{2}\|\breve{z}\|^2, \quad \forall \breve{x} \in \widehat{N} \times \widehat{N}, \; w \in \mathcal{W} \tag{12.73} \end{aligned}$$

$$\begin{aligned} \widehat{H}_2(\breve{x}, w, \hat{L}, \tilde{V}) = \; & \left\{ \tilde{V}(\hat{f}(\breve{x})) + \tilde{V}_x(\hat{f}(\breve{x}))g_1(x)w + \tilde{V}_{\hat{x}}(\hat{f}(\breve{x}))[\hat{L}(\hat{x})(h_2(x) - h_2(\hat{x}) + k_{21}(x)w)] \right. \\ & \left. + O(\|\tilde{v}\|^2) \right\} - \tilde{V}(\breve{x}) + \frac{1}{2}\|\breve{z}\|^2, \quad \forall \breve{x} \in \widehat{N} \times \widehat{N}, \; w \in \mathcal{W} \tag{12.74} \end{aligned}$$

---

[1] A second-order Taylor series approximation would be more accurate, but the first-order method gives a solution that is very close to the continuous-time case.

where $\tilde{Y}_x$, $\tilde{V}_x$ are the row-vectors of the partial-derivatives of $\tilde{Y}$ and $\tilde{V}$ respectively,

$$\tilde{v} = \left(\begin{array}{c} g_1(x)w \\ \hat{L}(\hat{x})[h_2(x) - h_2(\hat{x}) + k_{21}(x)w] \end{array} \right)$$

and

$$\lim_{\tilde{v} \to 0} \frac{O(\|\tilde{v}\|^2)}{\|\tilde{v}\|^2} = 0.$$

Then, applying the necessary conditions for the worst-case noise, we get

$$\frac{\partial H_1}{\partial w} = 0 \implies \hat{w}^\star := -\frac{1}{\gamma^2}[g_1^T(x)\tilde{Y}_x^T(\hat{f}(\breve{x})) + k_{21}^T(x)\hat{L}^T(\hat{x})\tilde{Y}_{\hat{x}}^T(\hat{f}(\breve{x}))]. \tag{12.75}$$

Now substitute $\hat{w}^\star$ in (12.74) to obtain

$$\begin{aligned} \widehat{H}_2(\breve{x}, \hat{w}^\star, \hat{L}, V) &\approx V(\hat{f}(\breve{x})) - \tilde{V}(\breve{x}) - \frac{1}{\gamma^2}\tilde{V}_x(\hat{f}(\breve{x}))g_1(x)g_1^T(x)\tilde{Y}_x^T(\hat{f}(\breve{x})) + \\ &\quad \tilde{V}_{\hat{x}}(\hat{f}(\breve{x}))\hat{L}(\hat{x})[h_2(x) - h_2(\hat{x})] - \frac{1}{\gamma^2}\tilde{V}_{\hat{x}}(\hat{f}(\breve{x}))\hat{L}(\hat{x})\hat{L}^T(\hat{x})\tilde{Y}_x^T(\hat{f}(\breve{x})) + \\ &\quad \frac{1}{2}\|z\|^2. \end{aligned}$$

Then, completing the squares for $\hat{L}$ in the above expression for $\widehat{H}_2(.,.,.,.)$, we have

$$\begin{aligned} \widehat{H}_2(\breve{x}, \hat{w}^\star, \hat{L}, \tilde{V}) &\approx \tilde{V}(\hat{f}(\breve{x})) - \tilde{V}(\breve{x}) - \frac{1}{\gamma^2}\tilde{V}_x(\hat{f}(\breve{x}))g_1(x)g_1^T(x)\tilde{Y}_x^T(\hat{f}(\breve{x})) + \frac{1}{2}\|z\|^2 + \\ &\quad \frac{1}{2\gamma^2}\left\|\hat{L}^T(\hat{x})\tilde{V}_{\hat{x}}^T(\hat{f}(\breve{x})) + \gamma^2(h_2(x) - h_2(\hat{x}))\right\|^2 - \frac{\gamma^2}{2}\|h_2(x) - h_2(\hat{x})\|^2 + \\ &\quad \frac{1}{2\gamma^2}\|\hat{L}^T(\hat{x})\tilde{Y}_{\hat{x}}^T(\hat{f}(\breve{x}))\|^2 - \frac{1}{2\gamma^2}\left\|\hat{L}^T(\hat{x})\tilde{V}_{\hat{x}}^T(\hat{f}(\breve{x})) + \hat{L}^T(\hat{x})\tilde{Y}_{\hat{x}}^T(\hat{f}(\breve{x}))\right\|^2. \end{aligned}$$

Therefore, taking $\hat{L}^\star$ as

$$\tilde{V}_{\hat{x}}(\hat{f}(\breve{x}))\hat{L}^\star(\hat{x}) = -\gamma^2(h_2(x) - h_2(\hat{x}))^T, \quad x, \hat{x} \in \widehat{N} \tag{12.76}$$

minimizes $\widehat{H}_2(.,.,.,.)$ and renders the Nash-equilibrium condition

$$\widehat{H}_2(\hat{w}^\star, \hat{L}^\star) \le \widehat{H}_2(w^\star, \hat{L}) \quad \forall \hat{L} \in \mathcal{M}^{n \times m}$$

satisfied.

Substitute now $\hat{L}^\star$ as given by (12.76) in the expression for $H_1(.,.,.,.)$ and complete the squares in w to obtain:

$$\begin{aligned} \widehat{H}_1(\breve{x}, w, \hat{L}^\star, \tilde{Y}) &= \tilde{Y}(\hat{f}(\breve{x})) - \tilde{Y}(\breve{x}) - \frac{1}{\gamma^2}\tilde{Y}_{\hat{x}}(\hat{f}(\breve{x}))\hat{L}(\hat{x})\hat{L}^T(\hat{x})\tilde{V}_{\hat{x}}^T(\hat{f}(\breve{x})) - \\ &\quad \frac{1}{2\gamma^2}\tilde{Y}_x(\hat{f}(\breve{x}))g_1(x)g_1^T(x)\tilde{Y}_x^T(\hat{f}(\breve{x})) - \frac{1}{2\gamma^2}\tilde{Y}_{\hat{x}}(\hat{f}(\breve{x}))\hat{L}(\hat{x})\hat{L}^T(\hat{x})\tilde{Y}_{\hat{x}}^T(\hat{f}(\breve{x})) \\ &\quad -\frac{1}{2}\|z\|^2 + \frac{\gamma^2}{2}\left\|w + \frac{1}{\gamma^2}g_1^T(x)\tilde{Y}_x^T(\hat{f}(\breve{x})) + \frac{1}{\gamma^2}k_{21}^T(x)\hat{L}^T(\hat{x})\tilde{Y}_{\hat{x}}^T(\hat{f}(\breve{x}))\right\|^2. \end{aligned}$$

Similarly, substituting $w = \hat{w}^\star$ as given by (12.75), we see that, the second Nash-equilibrium condition

$$\widehat{H}_1(\hat{w}^\star, \hat{L}^\star) \le H_1(w, \hat{L}^\star), \quad \forall w \in \mathcal{W}$$

is also satisfied. Thus, the pair $(\hat{w}^\star, \hat{L}^\star)$ constitutes a Nash-equilibrium solution to the two-player nonzero-sum dynamic game corresponding to the Hamiltonians $\widehat{H}_1(.,.,.,.)$ and $\widehat{H}_2(.,.,.,)$. With this analysis, we have the following important theorem.

Theorem 12.2.3 *Consider the nonlinear system (12.42) and the infinite-horizon DMH2HINLFP for it. Suppose the function h_1 is one-to-one (or injective) and the plant Σ^{da} is zero-state detectable. Suppose further, there exists a pair of C^1 negative and positive-definite functions $\tilde{Y}, \tilde{V} : \widehat{N} \times \widehat{N} \to \Re$ respectively, locally defined in a neighborhood $\widehat{N} \times \widehat{N} \subset \mathcal{X} \times \mathcal{X}$ of the origin $\breve{x} = 0$, and a matrix function $\hat{L} : \widehat{N} \to \mathcal{M}^{n \times m}$ satisfying the pair of coupled DHJIEs:*

$$\tilde{Y}(\widehat{f}(\breve{x})) - \tilde{Y}(\breve{x}) - \frac{1}{2\gamma^2}\tilde{Y}_x(\widehat{f}(\breve{x}))g_1(x)g_1^T(x)\tilde{Y}_x^T(\widehat{f}(\breve{x})) - \frac{1}{2\gamma^2}\tilde{Y}_{\hat{x}}(\widehat{f}(\breve{x}))\hat{L}(\hat{x})\hat{L}^T(\hat{x})\tilde{Y}_{\hat{x}}^T(\widehat{f}(\breve{x})) -$$
$$\frac{1}{\gamma^2}\tilde{Y}_{\hat{x}}(\widehat{f}(\breve{x}))\hat{L}(\hat{x})\hat{L}^T(\hat{x})\tilde{V}_{\hat{x}}^T(\widehat{f}(\breve{x})) -$$
$$\frac{1}{2}(h_1(x) - h_1(\hat{x}))^T(h_1(x) - h_1(\hat{x})) = 0, \quad \tilde{Y}(0) = 0, \qquad (12.77)$$
$$\tilde{V}(\widehat{f}(\breve{x})) - \tilde{V}(\breve{x}) - \frac{1}{\gamma^2}\tilde{V}_x(\widehat{f}(\breve{x}))g_1(x)g_1^T(x)\tilde{Y}_x^T(\widehat{f}(\breve{x})) - \frac{1}{\gamma^2}\tilde{V}_{\hat{x}}(\widehat{f}(\breve{x}))\hat{L}(\hat{x})\hat{L}^T(\hat{x})\tilde{Y}_{\hat{x}}^T(\widehat{f}(\breve{x})) -$$
$$\gamma^2(h_2(x) - h_2(\hat{x}))^T(h_2(x) - h_2(\hat{x})) +$$
$$\frac{1}{2}(h_1(x) - h_1(\hat{x}))^T(h_1(x) - h_1(\hat{x})) = 0, \quad \tilde{V}(0) = 0, \qquad (12.78)$$

together with the coupling condition (12.76). Then:

(i) *there exists locally in $\widehat{N}$ a Nash-equilibrium solution $(\hat{w}^\star, \hat{L}^\star)$ for the dynamic game corresponding to (12.48), (12.49), (12.47);*

(ii) *the augmented system (12.47) is locally dissipative with respect to the supply rate $s(w, \breve{z}) = \frac{1}{2}(\gamma^2\|w\|^2 - \|\breve{z}\|^2)$ in $\widehat{N}$, and hence has ℓ_2-gain from w to $\breve{z}$ less or equal to γ;*

(iii) *the optimal costs or performance objectives of the game are approximately $J_1^\star(\hat{L}^\star, \hat{w}^\star) = \tilde{Y}(\breve{x}_0)$ and $J_2^\star(\hat{L}^\star, \hat{w}^\star) = \tilde{V}(\breve{x}_0)$;*

(iv) *the filter Σ^{daf} with the gain-matrix $\hat{L}(\hat{x}) = \hat{L}^\star(\hat{x})$ satisfying (12.76) solves the infinite-horizon DMH2HINLFP for the system locally in $\widehat{N}$.*

Proof: Part (i) has already been shown above. To complete it, we substitute $(\hat{L}^\star, \hat{w}^\star)$ in the DHJIEs (12.66), (12.67) with $\widehat{H}_1(.,.,.,.,.)$, $\widehat{H}_2(.,.,.,.)$ replacing $H_1(.,.,.,.,.)$, $H_2(.,.,.,.,.)$ respectively, to get the DHJIEs (12.77), (12.78) respectively.
(ii) Consider the time-variation of $\tilde{Y}$ along a trajectory of the system (12.47) with $\hat{L} = \hat{L}^\star$:

$$\begin{aligned}
\tilde{Y}(\breve{x}_{k+1}) &= \tilde{Y}(\breve{f}^\star(x) + \breve{g}^\star(x)w) \quad \forall \breve{x} \in \widehat{N}, \forall w \in \mathcal{W} \\
&\approx \tilde{Y}(\widehat{f}(\breve{x})) + \tilde{Y}_x(\widehat{f}(\breve{x}))g_1(x)w + \tilde{Y}_{\hat{x}}(\widehat{f}(\breve{x}))[\hat{L}^\star(\hat{x})(h_2(x) - h_2(\hat{x}) + k_{21}(x)w)] \\
&= \tilde{Y}(\widehat{f}(\breve{x})) - \frac{1}{2\gamma^2}\tilde{Y}_x(\widehat{f}(\breve{x}))g_1(x)g_1^T(x)\tilde{Y}_x^T(\widehat{f}(\breve{x})) - \\
&\quad \frac{1}{\gamma^2}\tilde{Y}_{\hat{x}}(\widehat{f}(\breve{x}))\hat{L}^\star(\hat{x})\hat{L}^{\star T}(\hat{x})\tilde{V}_{\hat{x}}^T(\widehat{f}(\breve{x})) - \frac{1}{2\gamma^2}\tilde{Y}_{\hat{x}}(\widehat{f}(\breve{x}))\hat{L}^\star(\hat{x})\hat{L}^{\star T}(\hat{x})\tilde{Y}_{\hat{x}}^T(\widehat{f}(\breve{x})) + \\
&\quad \frac{\gamma^2}{2}\left\|w + \frac{1}{\gamma^2}g_1^T(x)\tilde{Y}_x^T(\widehat{f}(\breve{x})) + \frac{1}{\gamma^2}k_{21}^T(x)\hat{L}^{\star T}(\hat{x})\tilde{Y}_{\hat{x}}^T(\widehat{f}(\breve{x}))\right\|^2 - \frac{\gamma^2}{2}\|w\|^2
\end{aligned}$$

$$
\begin{aligned}
= \quad & \tilde{Y}(\breve{x}) + \frac{1}{2}\|\breve{z}\|^2 - \frac{\gamma^2}{2}\|w\|^2 + \frac{\gamma^2}{2}\left\|w + \frac{1}{\gamma^2}g_1^T(x)\tilde{Y}_x^T(\widehat{f}(\breve{x})) + \right. \\
& \left. \frac{1}{\gamma^2}k_{21}^T(x)\hat{L}^{\star T}(\hat{x})\tilde{Y}_{\hat{x}}^T(\widehat{f}(\breve{x}))\right\|^2 \\
\geq \quad & \tilde{Y}(\breve{x}) + \frac{1}{2}\|\breve{z}\|^2 - \frac{\gamma^2}{2}\|w\|^2 \quad \forall \breve{x} \in \widehat{N}, \; \forall w \in \mathcal{W}
\end{aligned}
$$

where use has been made of the first-order Taylor-approximation, equation (12.76), and the DHJIE (12.77) in the above manipulations. The last inequality further implies that

$$
\widetilde{Y}(\breve{x}_{k+1}) - \widetilde{Y}(\breve{x}) \leq \frac{\gamma^2}{2}\|w\|^2 - \frac{1}{2}\|\breve{z}\|^2 \quad \forall \breve{x} \in \widehat{N}, \; \forall w \in \mathcal{W}
$$

for some $\widetilde{Y} = -\tilde{Y} > 0$, which is the infinitesimal dissipation-inequality [180]. Therefore, the system has ℓ_2-gain $\leq \gamma$. The proof of asymptotic-stability can now be pursued along the same lines as in Proposition 12.2.1.

The proofs of items (iii)-(iv) are similar to those in Theorem 12.2.2. $\square$

Remark 12.2.2 *The benefits of the Theorem 12.2.3 can be summarized as follows. First and foremost is the benefit of the explicit solutions for computational purposes. Secondly, the approximation is reasonably accurate, as it captures a great deal of the dynamics of the system. Thirdly, it greatly simplifies the solution as it does away with extra sufficient conditions (see e.g., the conditions (12.60), (12.61) in Theorem 12.2.1). Fourthly, it opens the way also to develop an iterative procedure for solving the coupled DHJIEs.*

Remark 12.2.3 *In view of the coupling condition (12.76), the DHJIE can be represented as*

$$
\tilde{V}(\widehat{f}(\breve{x})) - \tilde{V}(\breve{x}) - \frac{1}{\gamma^2}\tilde{V}_x(\widehat{f}(\breve{x}))g_1(x)g_1^T(x)\tilde{Y}_x^T(\widehat{f}(\breve{x})) - \frac{1}{\gamma^2}\tilde{V}_{\hat{x}}(\widehat{f}(\breve{x}))\hat{L}(\hat{x})\hat{L}^T(\hat{x})\tilde{V}_{\hat{x}}^T(\widehat{f}(\breve{x})) -
$$
$$
\frac{1}{\gamma^2}\tilde{V}_{\hat{x}}(\widehat{f}(\breve{x}))\hat{L}(\hat{x})\hat{L}^T(\hat{x})\tilde{Y}_{\hat{x}}^T(\widehat{f}(\breve{x})) +
$$
$$
\frac{1}{2}(h_1(x) - h_1(\hat{x}))^T(h_1(x) - h_1(\hat{x})) = 0, \quad \tilde{V}(0) = 0. \tag{12.79}
$$

The result of the theorem can similarly be specialized to the linear-time-invariant (LTI) system:

$$
\Sigma^{dl} \; : \; \begin{cases} \dot{x}_{k+1} & = \; Ax_k + G_1 w_k, \quad x(k_0) = x^0 \\ \breve{z}_k & = \; C_1(x_k - \hat{x}_k) \\ y_k & = \; C_2 x_k + D_{21}w_k, \end{cases} \tag{12.80}
$$

where all the variables have their previous meanings, and $F \in \Re^{n \times n}$, $G_1 \in \Re^{n \times r}$, $C_1 \in \Re^{s \times n}$, $C_2 \in \Re^{m \times n}$ and $D_{21} \in \Re^{m \times r}$ are constant real matrices. We have the following corollary to Theorem 12.2.3.

Corollary 12.2.1 *Consider the LTI system Σ^{dl} defined by (12.80) and the DMH2HINLFP for it. Suppose C_1 is full column rank and A is Hurwitz. Suppose further, there exist a negative and a positive-definite real-symmetric solutions $\widehat{P}_1$, $\widehat{P}_2$ (respectively) to the coupled discrete-algebraic-Riccati equations (DAREs):*

$$
A^T \widehat{P}_1 A - \widehat{P}_1 - \frac{1}{2\gamma^2}A^T \widehat{P}_1 G_1 G_1^T \widehat{P}_1 A - \frac{1}{2\gamma^2}A^T \widehat{P}_1 \hat{L}\hat{L}^T \widehat{P}_1 A - \frac{1}{\gamma^2}A^T \widehat{P}_1 \hat{L}\hat{L}^T \widehat{P}_2 A - C_1^T C_1 = 0 \tag{12.81}
$$

$$A^T \widehat{P}_2 A - \widehat{P}_2 - \frac{1}{\gamma^2} A^T \widehat{P}_2 G_1 G_1^T \widehat{P}_2 A - \frac{1}{\gamma^2} A^T \widehat{P}_2 \hat{L} \hat{L}^T \widehat{P}_1 A - \frac{1}{\gamma^2} A^T \widehat{P}_2 \hat{L} \hat{L}^T \widehat{P}_2 A + C_1^T C_1 = 0$$

$$(12.82)$$

together with the coupling condition:

$$A^T \widehat{P}_2 \hat{L} = -\gamma^2 C_2^T. \tag{12.83}$$

Then:

(i) *there exists a Nash-equilibrium solution $(\hat{w}_l^\star, \hat{L}^\star)$ for the game given by*

$$\hat{w}^\star = -\frac{1}{\gamma^2}(G_1^T + D_{21}^T \hat{L}^\star)\widehat{P}_1 A(x - \hat{x}),$$

$$(x - \hat{x})^T A^T \widehat{P}_2 \hat{L}^\star = -\gamma^2 (x - \hat{x})^T C_2^T;$$

(ii) *the augmented system*

$$\Sigma^{dlf} : \begin{cases} \breve{x}_{k+1} = \begin{bmatrix} A & 0 \\ \hat{L}^\star C_2 & A - \hat{L}^\star C_2 \end{bmatrix} \breve{x}_k + \begin{bmatrix} G_1 \\ \hat{L}^\star D_{21} \end{bmatrix} w, \quad \breve{x}(k_0) = \begin{bmatrix} x^0 \\ \hat{x}^0 \end{bmatrix} \\ \breve{z}_k = [C_1 \quad -C_1]\breve{x}_k := \breve{C}\breve{x}_k \end{cases}$$

has $\mathcal{H}_\infty$-norm from w to $\breve{z}$ less than or equal to γ;

(iv) *the optimal costs or performance objectives of the game are approximately $J_1^\star(\hat{L}^\star, \hat{w}_k^\star) = \frac{1}{2}(x^0 - \hat{x}^0)^T \widehat{P}_1 (x^0 - \hat{x}^0)$ and $J_2^\star(\hat{L}^\star, \hat{w}^\star) - \frac{1}{2}(x^0 - \hat{x}^0)^T \widehat{P}_2 (x^0 - \hat{x}^0)$;*

(iv) *the filter $\mathcal{F}$ defined by*

$$\Sigma_{ldf} : \quad \hat{x}_{k+1} = A\hat{x}_k + \hat{L}(y - C_2\hat{x}_k), \quad \hat{x}(k_0) = \hat{x}^0$$

with the gain matrix $\hat{L} = \hat{L}^\star$ satisfying (12.83) solves the infinite-horizon $DMH2HINLFP$ for the discrete-time linear system.

Proof: Take:

$$\tilde{Y}(\breve{x}) = \frac{1}{2}(x - \hat{x})^T \widehat{P}_1 (x - \hat{x}), \quad \widehat{P}_1 = \widehat{P}_1^T < 0,$$

$$\tilde{V}(\breve{x}) = \frac{1}{2}(x - \hat{x})^T \widehat{P}_2 (x - \hat{x}), \quad \widehat{P}_2 = \widehat{P}_2^T > 0,$$

and apply the result of the theorem. $\square$

12.2.4 Discrete-Time Certainty-Equivalent Filters (CEFs)

Again, it should be observed as in the continuous-time case Sections 12.2.1, 12.2.2 and 12.2.3, the filter gains (12.59), (12.69), (12.76) may also depend on the original state, x, of the system which is to be estimated. Therefore in this section, we develop the discrete-time counterparts of the results of Section 12.1.3.

Definition 12.2.4 *For the nonlinear system (12.42), we say that it is locally zero-input observable, if for all states x_k, $x_{k'} \in U \subset \mathcal{X}$ and input $w(.) \equiv 0$,*

$$y(\bar{k}, x_k, w) = y(\bar{k}, x_{k'}, w) \Longrightarrow x_k = x_{k'}$$

where $y(., x, w)$ is the output of the system with the initial condition $x(k_0) = x$. Moreover, *the system is said to be zero-input observable if it is locally observable at each $x_k \in \mathcal{X}$ or $U = \mathcal{X}$.*

We similarly consider the following class of certainty-equivalent filters:

$$
\widetilde{\Sigma}^{af} : \begin{cases}
\hat{x}_{k+1} &= f(\hat{x}_k) + g_1(\hat{x}_k)w_k^\star + \widetilde{L}(\hat{x}_k, y_k)[y - h_2(\hat{x}_k) - k_{21}(\hat{x}_k)\tilde{w}_k^\star]; \\
&\quad \hat{x}(k_0) = \hat{x}^0 \\
\hat{z}_k &= h_2(\hat{x}_k) \\
\tilde{z}_k &= y_k - h_2(\hat{x}_k),
\end{cases}
\tag{12.84}
$$

where $\widetilde{L}(.,.) \in \mathcal{M}^{n \times m}$ is the gain of the filter, $\tilde{w}^\star$ is the estimated worst-case system noise (hence the name certainty-equivalent filter) and $\tilde{z}$ is the new penalty variable. Then, if we consider the infinite-horizon mixed $\mathcal{H}_2/\mathcal{H}_\infty$ dynamic game problem with the cost functionals (12.48), (12.49) and the above filter, we can similarly define the associated corresponding approximate Hamiltonians (as in Section 12.2.3) $\widetilde{H}_i : \mathcal{X} \times \mathcal{W} \times \mathcal{Y} \times \mathcal{M}^{n \times m} \times \Re \to \Re$, $i = 1, 2$ as

$$
\begin{aligned}
\widehat{K}_1(\hat{x}, w, y, \widetilde{L}, \widetilde{Y}) &= \widetilde{Y}(\widehat{f}(\hat{x}), y) - \widetilde{Y}(\hat{x}, y_{k-1}) + \widetilde{Y}_{\hat{x}}(\hat{x}, y)[f(\hat{x}) + g_1(\hat{x})w + \\
&\quad \widetilde{L}(\hat{x}, y)(y - h_2(\hat{x}) - k_{21}(\hat{x})w)] + \frac{1}{2}\gamma^2\|w\|^2 - \frac{1}{2}\|\tilde{z}\|^2 \\
\widehat{K}_2(\hat{x}, w, y, \widetilde{L}, \widetilde{V}) &= \widetilde{V}(\widehat{f}(\hat{x}), y) - \widetilde{V}(\hat{x}, y_{k-1}) + \widetilde{V}_{\hat{x}}(\hat{x}, y)[f(\hat{x}) + g_1(\hat{x})w + \\
&\quad \widetilde{L}(\hat{x}, y)(y - h_2(\hat{x}) - k_{21}(\hat{x})w)] + \frac{1}{2}\|\tilde{z}\|^2
\end{aligned}
$$

for some smooth functions $\widetilde{V}, \widetilde{Y} : \mathcal{X} \times \mathcal{Y} \to \Re$, where $\hat{x} = \hat{x}_k$, $w = w_k$, $y = y_k$, $\tilde{z} = \tilde{z}_k$, and the adjoint variables are set as $\tilde{p}_1 = \widetilde{Y}$, $\tilde{p}_2 = \widetilde{V}$. Then

$$
\left. \frac{\partial \widehat{K}_1}{\partial w} \right|_{w = \tilde{w}^\star} = [g_1(\hat{x}) - \widetilde{L}(\hat{x}, y)k_{21}(\hat{x})]^T \widetilde{Y}_{\hat{x}}^T(\widehat{f}(\hat{x}), y) + \gamma^2 w = 0
$$

$$
\implies \tilde{w}^\star = -\frac{1}{\gamma^2}[g_1(\hat{x}) - \widetilde{L}(\hat{x}, y)k_{21}(\hat{x})]^T \widetilde{Y}_{\hat{x}}^T(\widehat{f}(\hat{x}), y).
$$

Consequently, repeating the steps as in Section 12.2.3 and Theorem 12.2.3, we arrive at the following result.

Theorem 12.2.4 *Consider the nonlinear system (12.42) and the DMH2HINLFP for it. Suppose the plant Σ^{da} is locally asymptotically-stable about the equilibrium point $x = 0$ and zero-input observable. Suppose further, there exists a pair of C^1 (with respect to the first argument) negative and positive-definite functions $\widetilde{Y}, \widetilde{V} : \widetilde{N} \times \Upsilon \to \Re$ respectively, locally defined in a neighborhood $\widetilde{N} \times \Upsilon \subset \mathcal{X} \times \mathcal{Y}$ of the origin $(\hat{x}, y) = (0, 0)$, and a matrix function $\widetilde{L} : \widetilde{N} \times \Upsilon \to \mathcal{M}^{n \times m}$ satisfying the following pair of coupled DHJIEs:*

$$
\begin{aligned}
&\widetilde{Y}(\widehat{f}(\hat{x}), y) - \widetilde{Y}(\hat{x}, y_{k-1}) - \tfrac{1}{2\gamma^2}\widetilde{Y}_{\hat{x}}(\widehat{f}(\hat{x}), y)g_1(\hat{x})g_1^T(\hat{x})\widetilde{Y}_{\hat{x}}^T(\widehat{f}(\hat{x}), y) - \\
&\tfrac{1}{2\gamma^2}\widetilde{Y}_{\hat{x}}(\hat{x}, y)\widetilde{L}(\hat{x}, y)\widetilde{L}^T(\hat{x}, y)\widetilde{Y}_{\hat{x}}^T(\hat{x}, y) - \tfrac{1}{\gamma^2}\widetilde{Y}_{\hat{x}}(\widehat{f}(\hat{x}), y)\widetilde{L}(\hat{x}, y)\widetilde{L}^T(\hat{x}, y)\widetilde{V}_{\hat{x}}^T(\widehat{f}(\hat{x}), y) - \\
&\quad \tfrac{1}{2}(y - h_2(\hat{x}))^T(y - h_2(\hat{x})) = 0, \quad \widetilde{Y}(0, 0) = 0, \tag{12.85} \\
&\widetilde{V}_{\hat{x}}(\widehat{f}(\hat{x}), y) - \widetilde{V}(\hat{x}, y_{k-1}) - \tfrac{1}{\gamma^2}\widetilde{V}_{\hat{x}}(\widehat{f}(\hat{x}), y)g_1(\hat{x})g_1^T(\hat{x})\widetilde{Y}_{\hat{x}}^T(\hat{x}, y) - \\
&\tfrac{1}{\gamma^2}\widetilde{V}_{\hat{x}}(\widehat{f}(\hat{x}), y)\widetilde{L}(\hat{x}, y)\widetilde{L}^T(\hat{x}, y)\widetilde{Y}_{\hat{x}}^T(\widehat{f}(\hat{x}), y) - \tfrac{1}{\gamma^2}\widetilde{V}_{\hat{x}}(\widehat{f}(\hat{x}), y)\widetilde{L}(\hat{x}, y)\widetilde{L}^T(\hat{x}, y)\widetilde{V}_{\hat{x}}^T(\widehat{f}(\hat{x}), y) + \\
&\quad \tfrac{1}{2}(y - h_2(\hat{x})^T(y - h_2(\hat{x})) = 0, \quad \widetilde{V}(0, 0) = 0 \tag{12.86}
\end{aligned}
$$

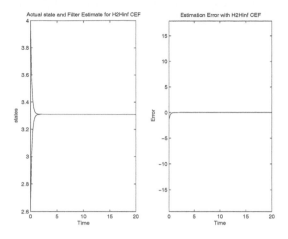

FIGURE 12.4
Discrete-Time $\mathcal{H}_2/\mathcal{H}_\infty$-Filter Performance with Unknown Initial Condition and ℓ_2-Bounded Disturbance; Reprinted from Int. J. of Robust and Nonlinear Control, RNC1643 (published online August) © 2010, "Discrete-time Mixed $\mathcal{H}_2/\mathcal{H}_\infty$ Nonlinear Filtering," by M. D. S. Aliyu and E. K. Boukas, with permission from Wiley Blackwell.

$\hat{x} \in \tilde{N}$, $y \in \Upsilon$, *together with the coupling condition*

$$\tilde{V}_{\hat{x}}(\hat{f}(\hat{x}), y)\tilde{L}(\hat{x}, y) = -\gamma^2 (y - h_2(\hat{x}))^T, \quad \hat{x} \in \tilde{N} \ y \in \Upsilon. \tag{12.87}$$

Then the filter $\tilde{\Sigma}^{af}$ with the gain matrix $\tilde{L}(\hat{x}, y)$ satisfying (12.87) solves the infinite-horizon DMH2HINLFP for the system locally in $\tilde{N}$.

Proof: Follows the same lines as Theorem 12.2.3. □

Remark 12.2.4 *Comparing the HJIEs (12.85)-(12.86) with (12.77)-(12.78) reveals that they are similar, but we have gained by reducing the dimensionality of the PDEs, and more importantly, the filter gain does not depend on x any longer.*

12.3 Example

We consider a simple example to illustrate the result of the previous section.

Example 12.3.1 *We consider the following scalar system*

$$\begin{aligned} x_{k+1} &= x_k^{\frac{1}{5}} + x_k^{\frac{1}{3}} \\ y_k &= x_k + w_k \end{aligned}$$

where $w_k = e^{-0.3k} \sin(0.25\pi k)$ is an ℓ_2-bounded disturbance.
Approximate solutions of the coupled DHJIEs (12.85) and (12.86) can be calculated using an iterative approach. With $\gamma = 1$, and $g_1(x) = 0$, we can rewrite the coupled DHJIEs as

$$\tilde{Y}^{j+1}(\hat{x}, y) \triangleq \tilde{Y}^j(\hat{x}, y_{k-1}) + \frac{1}{2}(y-x)^2 = \tilde{Y}^j(\hat{f}, y) - \frac{1}{2}\tilde{Y}_{\hat{x}}^{j^2}(\hat{f}, y)l^{j^2} - \tilde{Y}_{\hat{x}}^j(\hat{f}, y)\tilde{V}_{\hat{x}}^j(\hat{f}, y)l^{j^2}, \tag{12.88}$$

$$\tilde{V}^{j+1}(\hat{x}, y) \triangleq \tilde{V}^j(\hat{x}, y_{k-1}) - \frac{1}{2}(y-x)^2 = \tilde{V}^j(\hat{f}, y) - \tilde{V}_{\hat{x}}^{j^2}(\hat{f}, y)l^{j^2} - \tilde{Y}_{\hat{x}}^j(\hat{f}, y)\tilde{V}_{\hat{x}}^j(\hat{f}, y)l^{j^2}, \tag{12.89}$$

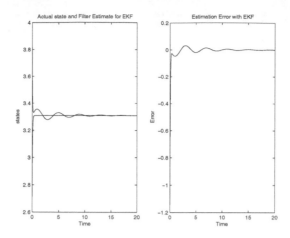

FIGURE 12.5

Extended-Kalman-Filter Performance with Unknown Initial Condition and ℓ_2-Bounded Disturbance; Reprinted from Int. J. of Robust and Nonlinear Control, RNC1643 (published online August) © 2010, "Discrete-time mixed $\mathcal{H}_2/\mathcal{H}_\infty$ nonlinear filtering," by M. D. S. Aliyu and E. K. Boukas, with permission from Wiley Blackwell.

$j = 0, 1, \ldots$. Then, with the initial filter gain $l^0 = 1$, initial guess for solutions as $\widetilde{Y}^0(\hat{x}, y) = -\frac{1}{2}(\hat{x}^2 + y^2)$, $\widetilde{V}^0(\hat{x}, y) = \frac{1}{2}(\hat{x}^2 + y^2)$ respectively, we perform one iteration of the above recursive equations (12.88), (12.89) to get

$$\widetilde{Y}^1 = -\frac{1}{2}[(\hat{x}^{1/5} + \hat{x}^{1/3})^2 + y^2] + \frac{1}{2}(\hat{x}^{1/5} + \hat{x}^{1/3})^2,$$

$$\widetilde{V}^1 = \frac{1}{2}[(\hat{x}^{1/5} + \hat{x}^{1/3})^2 + y^2].$$

Therefore,

$$\widetilde{Y}^1(\hat{x}, y_{k-1}) = -\frac{1}{2}(y - x)^2 - \frac{1}{2}y^2,$$

$$\widetilde{V}^1(\hat{x}, y_{k-1}) = \frac{1}{2}(y - x)^2 + \frac{1}{2}[(\hat{x}^{1/5} + \hat{x}^{1/3})^2 + y^2].$$

We can use the above approximate solution $\widetilde{V}^1(\hat{x}, y_{k-1})$ to the DHJIE to estimate the states of the system, since the gain of the filter depends only on $\widetilde{V}^1_{\hat{x}}(\hat{x}, y)$. Consequently, we can compute the filter gain as

$$l(x_k, y_k) \approx -\frac{(y_k - x_k)}{y_k - x_k^{\frac{1}{2}} - x_k^{\frac{1}{3}}}. \tag{12.90}$$

The result of simulation with this filter is then shown in Figure 12.4. We also compare this result with that of an extended-Kalman filter for the same system shown in Figure 12.5. It can clearly be seen that, the mixed $\mathcal{H}_2/\mathcal{H}_\infty$ filter performance is superior to that of the EKF.

12.4 Notes and Bibliography

This chapter is entirely based on the authors' contributions [18, 19, 21]. The reader is referred to these references for more details.

13

Solving the Hamilton-Jacobi Equation

In this chapter, we discuss some approaches for solving the Hamilton-Jacobi-equations (HJE) associated with the optimal control problems for affine nonlinear systems discussed in this book. This has been the biggest bottle-neck in the practical application of nonlinear $\mathcal{H}_\infty$-control theory. There is no systematic numerical approach for solving the HJIEs. Various attempts have however been made in this direction in the past three decades. Starting with the work of Lukes [193], Glad [112], who proposed a polynomial approximation approach, Van der Schaft [264] applied the approach to the HJIEs and developed a recursive approach which was also refined by Isidori [145]. Since then, many other authors have proposed similar approaches for solving the HJEs [63, 80, 125, 260, 286]. However, all the contributions so far in the literature are local.

The main draw-backs with the above approaches for solving the HJIE are that, (i) they are not closed-form, and convergence of the sequence of solutions to a closed-form solution cannot be guaranteed; (ii) there are no efficient methods for checking the positive-definiteness of the solution; (iii) they are sensitive to uncertainties and pertubations in the system; and (iv) also sensitive to the initial condition. Thus, the global asymptotic stability of the closed-loop system cannot be guaranteed.

Therefore, more refined solutions that will guarantee global asymptotic-stability are required if the theory of nonlinear $\mathcal{H}_\infty$-control is to yield any fruits. Thus, with this in mind, more recently, Isidori and Lin [145] have shown that starting from a solution of an algebraic-Riccati-equation (ARE) related to the linear $\mathcal{H}_\infty$ problem, if one is free to choose a state-dependent weight of control input, it is possible to construct a global solution to the HJIE for a class of nonlinear systems in strict-feedback form. A parallel approach using a backstepping procedure and inverse-optimality has also been proposed in [96], and in [167]-[169] for a class of strict-feedback systems.

In this chapter, we shall review some major approaches for solving the HJE, and present one algorithm that may yield global solutions. The chapter is organized as follows. In Section 13.1, we review some popular polynomial and Taylor-series approximation methods for solving the HJEs. Then in Section 13.2, we discuss a factorization approach which may yield exact and global solutions. The extension of this approach to Hamiltonian mechanical systems is then discussed in Section 13.3. Examples are given throughout to illustrate the usefulness of the various methods.

13.1 Review of Some Approaches for Solving the HJBE/HJIE

In this section, we review some major approaches for solving the HJBE and HJIE that have been proposed in the literature. In [193], a recursive procedure for the HJBE for a class of nonlinear systems is proposed. This was further refined and generalized for affine nonlinear systems by Glad [112]. To summarize the method briefly, we consider the following affine nonlinear system

$$\Sigma_a : \quad \dot{x}(t) = a(x) + b(x)u \tag{13.1}$$

under the quadratic cost functional:

$$\min \int_0^\infty \left[l(x) + \frac{1}{2} u^T R u \right] dt, \tag{13.2}$$

where $a : \Re^n \to \Re^n$, $b : \Re^n \to \Re^{n \times k}$, $l : \Re^n \to \Re$, $l \geq 0$, a, b, $l \in C^\infty$, $0 < R \in \Re^{k \times k}$. Then, the HJBE corresponding to the above optimal control problem is given by

$$v_x(x)a(x) - \frac{1}{2} v_x(x)b(x)R^{-1}b^T(x)v_x^T(x) + l(x) = 0, \quad v(0) = 0, \tag{13.3}$$

for some smooth $C^2(\Re^n)$ positive-semidefinite function $v : \Re^n \to \Re$, and the optimal control is given by

$$u^\star = k(x) = -R^{-1}b^T(x)v_x^T(x).$$

Further, suppose there exists a positive-semidefinite solution to the algebraic-Riccati-equation (ARE) corresponding to the linearization of the system (13.1), then it can be shown that there exists a real analytic solution v to the HJBE (13.3) in a neighborhood $\mathcal{O}$ of the origin [193]. Therefore, if we write the linearization of the system and the cost function as

$$
\begin{aligned}
l(x) &= \frac{1}{2} x^T Q x + l_h(x) \\
a(x) &= A x + a_h(x) \\
b(x) &= B + b_h(x) \\
v(x) &= \frac{1}{2} x^T P x + v_h(x)
\end{aligned}
$$

where $Q = \frac{\partial l}{\partial x}(0)$, $A = \frac{\partial a}{\partial x}(0)$, $B = b(0)$, $P = \frac{\partial v}{\partial x}(0)$ and l_h, a_h, b_h, v_h contain higher-order terms. Substituting the above expressions in the HJBE (13.3), it splits into two parts:

$$A^T P + P A - P B R^{-1} B^T P + Q = 0 \tag{13.4}$$

$$v_{hx}(x)A_c x + v_x(x)a_h(x) - \frac{1}{2} v_{hx}(x)BR^{-1}B^T v_{hx}^T(x) - \frac{1}{2} v_x(x)\beta_h(x)v_x^T(x) + l_h(x) = 0 \tag{13.5}$$

where

$$A_c = A - BR^{-1}B^T P, \quad \beta_h(x) = b(x)R^{-1}b^T(x) - BR^{-1}B^T$$

and β_h contains terms of degree 1 and higher. Equation (13.4) is the ARE of linear-quadratic control for the linearized system. Thus, if the system is stabilizable and detectable, there exists a unique positive-semidefinite solution to this equation. Hence, it represents the *first-order approximation* to the solution of the HJBE (13.3). Now letting the superscript (m) denote the *m*th-order terms, (13.5) can be written as

$$-v_{hx}^{m+1}(x)A_c x = [v_x(x)a_h(x) - \frac{1}{2} v_{hx}(x)BR^{-1}B^T v_{hx}^T(x) - \frac{1}{2} v_x(x)\beta_h(x)v_x^T(x)]^m + l_h^m(x). \tag{13.6}$$

The RHS contains only $m-th$, $(m-1)-th, \ldots$-order terms of v. Therefore, equation (13.5) defines a linear system of equations for the $(m+1)th$ order coefficients with the RHS containing previously computed terms. Thus, v^{m+1} can be computed recursively from (13.6). It can also be shown that the system is nonsingular as soon as A_c is a stable matrix. This is satisfied vacuously if P is a stabilizing solution to the ARE (13.4).

We consider an example.

Example 13.1.1 *Consider the second-order system*

$$\dot{x}_1 = -x_1^2 + x_2 \tag{13.7}$$
$$\dot{x}_2 = -x_2 + u \tag{13.8}$$

with the cost functional

$$J = \int_0^\infty \frac{1}{2}(x_1^2 + x_2^2 + u^2)dt.$$

The solution of the ARE (13.4) gives the quadratic (or second-order) approximation to the solution of the HJBE:

$$V^{[2]}(x) = x_1^2 + x_1x_2 + \frac{1}{2}x_2^2.$$

Further, the higher-order terms are computed recursively up to fourth-order to obtain

$$V \approx V^{[2]} + V^{[3]} + V^{[4]} = x_1^2 + x_1x_2 + \frac{1}{2}x_2^2 - 1.593x_1^3 - 2x_1^2x_2 - 0.889x_1x_2^2 -$$
$$0.148x_2^3 + 1.998x_1^4 + 2.778x_1^3x_2 + 1.441x_1^2x_2^2 +$$
$$0.329x_1x_2^3 + 0.0288x_2^4.$$

The above algorithm has been refined by Van der Schaft [264] for the HJIE (6.16) corresponding to the state-feedback $\mathcal{H}_\infty$-control problem:

$$V_x(x)f(x) + V_x(x)[\frac{1}{\gamma^2}g_1(x)g_1^T(x) - g_2(x)g_2^T(x)]V_x^T(x) + \frac{1}{2}h_1^T(x)h_1(x) = 0, \quad V(0) = 0, \tag{13.9}$$

for some smooth C^1-function $V : M \to \Re$, $M \subset \Re^n$. Suppose there exists a solution $P \geq 0$ to the ARE

$$F^TP + PF + P[\frac{1}{\gamma^2}G_1G_1^T - G_2G_2^T]P + H_1^TH_1 = 0 \tag{13.10}$$

corresponding to the linearization of the HJIE (13.9), where $F = \frac{\partial f}{\partial x}(0)$, $G_1 = g_1(0)$, $G_2 = g_2(0)$, $H_1 = \frac{\partial h_1}{\partial x}(0)$. Let V be a solution to the HJIE, and if we write

$$V(x) = \frac{1}{2}x^TPx + V_h(x)$$
$$f(x) = Fx + f_h(x)$$
$$\frac{1}{2}[\frac{1}{\gamma^2}g_1(x)g_1^T(x) - g_2(x)g_2^T(x)] = \frac{1}{2}[\frac{1}{\gamma^2}G_1G_1^T - G_2G_2^T] + R_h(x)$$
$$\frac{1}{2}h_1^T(x)h_1(x) = \frac{1}{2}x^TH_1^TH_1x + \theta_h(x)$$

where V_h, f_h, R_h and θ_h contain higher-order terms. Then, similarly the HJIE (13.9) splits into two parts, (i) the ARE (13.10); and (ii) the higher-order equation

$$-\frac{\partial V_h(x)}{\partial x}F_{cl}x = \frac{\partial V(x)}{\partial x}f_h(x) + \frac{1}{2}\frac{\partial V_h(x)}{\partial x}[\frac{1}{\gamma^2}G_1G_1^T - G_2G_2^T]\frac{\partial^T V_h(x)}{\partial x}(x) +$$
$$\frac{1}{2}\frac{\partial V(x)}{\partial x}R_h(x)\frac{\partial^T V(x)}{\partial x} + \theta_h(x) \tag{13.11}$$

where

$$F_{cl} \triangleq F - G_2G_2^TP + \frac{1}{\gamma^2}G_1G_1^TP.$$

We can rewrite (13.11) as

$$-\frac{\partial V_h^m(x)}{\partial x} F_{cl} x = H_m(x),\tag{13.12}$$

where $H_m(x)$ denotes the m-th order terms on the RHS, and thus if $P \geq 0$ is a stabilizing solution of the ARE (13.10), then the above Lyapunov-equation (13.12) can be integrated for V_h^m to get,

$$V_h^m(x) = \int_0^\infty H_m(e^{F_{cl}t} x)dt.$$

Therefore, V_h^m is determined by H_m. Consequently, since H_m depends only on $V^{(m-1)}$, $V^{(m-2)}, \ldots, V^2 = \frac{1}{2} x^T P x$, V_h^m can be computed recursively using (13.12) and starting from V^2, to obtain V similarly as

$$V \approx V^{[2]} + V^{[3]} + V^{[4]} + \ldots$$

The above approximate approach for solving the HJBE or HJIE has a short-coming; namely, there is no guarantee that the sequence of solutions will converge to a smooth positive-semidefinite solution. Moreover, in general, the resulting solution obtained cannot be guaranteed to achieve closed-loop asymptotic-stability or global $\mathcal{L}_2$-gain $\leq \gamma$ since the functions f_h, R_h, θ_h are not exactly known, rather they are finite approximations from a Taylor-series expansion. The procedure can also be computationally intensive especially for the HJIE as various values of γ have to be tried. The above procedure has been refined in [145] for the measurement-feedback case, and variants of the algorithm have also been proposed in other references [63, 154, 155, 286, 260].

13.1.1 Solving the HJIE/HJBE Using Polynomial Expansion and Basis Functions

Two variants of the above algorithms that use basis function approximation are given in [63] using the Galerkin approximation and [260] using a series solution. In [260] the following series expansion for V is used

$$V(x) = V^{[2]}(x) + V^{[3]}(x) + \ldots\tag{13.13}$$

where $V^{[k]}$ is a *homogeneous-function* of order k in n-scalar variables $x_1, x_2, \ldots, x_n$, i.e., it is a linear combination of

$$N_k^n \triangleq \left(\begin{array}{c} n+k-1 \\ k \end{array} \right)$$

terms of the form $x_1^{i_1} x_2^{i_2} \ldots x_n^{i_n}$, where i_j is a nonnegative integer for $j = 1, \ldots, n$ and $i_1 + i_2 + \ldots + i_n = k$. The vector whose components consist of these terms is denoted by $x^{[k]}$; for example,

$$x^{[1]} = \left[\begin{array}{c} x_1 \\ x_2 \end{array} \right], \quad x^{[2]} = \left[\begin{array}{c} x_1^2 \\ x_1 x_2 \\ x_2^2 \end{array} \right], \ldots.$$

To summarize the approach, rewrite HJIE (13.9) as

$$V_x(x)f(x) + \frac{1}{2} V_x(x)G(x)R^{-1}G^T(x)V_x^T(x) + Q(x) = 0\tag{13.14}$$

where

$$G(x) = [g_1(x) \ g_2(x)], \quad R = \left[\begin{array}{cc} \gamma^2 & 0 \\ 0 & -1 \end{array} \right], \quad Q(x) = \frac{1}{2} x^T C^T C x$$

and let $\nu = [w^T \quad u^T]^T$, $\nu_\star = [w_\star^T \quad u_\star^T]^T = [l_\star^T \quad k_\star^T]$. Then we can expand f, G, ν as homogeneous functions:

$$
\begin{aligned}
f(x) &= f^{[1]}(x) + f^{[2]}(x) + \dots \\
G(x) &= G^{[0]}(x) + G^{[1]}(x) + \dots \\
\nu_\star(x) &= \nu_\star^{[1]}(x) + \nu_\star^{[2]}(x) + \dots
\end{aligned}
$$

where

$$
\nu_\star^{[k]} = -R^{-1} \sum_{j=0}^{k-1} (G^{[j]})^T (V_x^{[k+1-j]})^T, \tag{13.15}
$$

$f^{[1]}(x) = Fx$, $G^{[0]} = [G_1 \; G_2]$, $G_1 = g_1(0)$, $G_2 = g_2(0)$. Again rewrite HJIE (13.14) as

$$
V_x(x)f(x) + \frac{1}{2}\nu_\star(x)R\nu_\star(x) + Q(x) = 0, \tag{13.16}
$$

$$
\nu_\star(x) + R^{-1}G^T(x)V_x^T(x) = 0. \tag{13.17}
$$

Substituting now the above expansions in the HJIE (13.16), (13.17) and equating terms of order $m \geq 2$ to zero, we get

$$
\sum_{k=0}^{m-2} V_x^{[m-k]} f^{[k+1]} + \frac{1}{2}\sum_{k=1}^{m-1} \nu_\star^{[m-k]} R\nu_\star^{[k]} + Q^{[m]} = 0. \tag{13.18}
$$

It can be checked that, for $m = 2$, the above equation simplifies to

$$
V_x^{[2]} f^{[1]} + \frac{1}{2}\nu_\star^{[1]} R\nu_\star^{[1]} + Q^{[2]} = 0,
$$

where

$$
f^{[1]}(x) = Ax, \quad \nu_\star^{[1]}(x) = -R^{-1}B^T V_x^{[2]T}(x), \quad Q^{[2]}(x) = \frac{1}{2}x^T C^T Cx, \quad B = [G_1 \; G_2].
$$

Substituting in the above equation, we obtain

$$
V_x^{[2]}(x)Ax + \frac{1}{2}V_x^{[2]}(x)BR^{-1}B^T V_x^{[2]}(x) + \frac{1}{2}x^T C^T Cx = 0.
$$

But this is the ARE corresponding to the linearization of the system, hence

$$
V^{[2]}(x) = x^T Px,
$$

where $P = P^T > 0$ solves the ARE (13.10) with $A_\star = A - BR^{-1}B^T P$ Hurwitz. Moreover,

$$
\nu_\star^{[1]}(x) = -B^T Px, \quad k_\star^{[1]}(x) = -G_1^T Px.
$$

Consider now the case $m \geq 3$, and rewrite (13.18) as

$$
\sum_{k=0}^{m-2} V_x^{[m-k]} f^{[k+1]} + \frac{1}{2}\nu_\star^{[m-1]T} R\nu_\star^{[1]} + \frac{1}{2}\sum_{k=2}^{m-2} \nu_\star^{[m-k]T} R\nu_\star^{[k]} = 0. \tag{13.19}
$$

Then, using

$$
\nu_\star^{[m-1]T} = -\sum_{k=0}^{m-2} V_x^{[m-k]} G^{[k]} R^{-1},
$$

equation (13.19) can be written as

$$V_x^{[m]} \bar{f}^{[1]} = \frac{1}{2} \sum_{k=1}^{m-2} V_x^{[m-k]} \bar{f}^{[k+1]} - \frac{1}{2} \sum_{k=2}^{m-2} (\nu_\star^{[m-k]})^T R \nu_\star^{[k]} \qquad (13.20)$$

where

$$\bar{f}(x) \triangleq f(x) + G(x)\nu_\star^{[1]}(x).$$

Equation (13.20) can now be solved for any $V^{[m]}$, $m \geq 3$. If we assume every $V^{[m]}$ is of the form $V^{[m]} = C_m x^{[m]}$, where $C_m \in \Re^{1 \times N_m^n}$ is a row-vector of unknown coefficients, then substituting this in (13.20), we get a system of N_m^n linear equations in the unknown entries of C_m. It can be shown that if the eigenvalues of $A_\star = A - BB^T P$ are nonresonant,[1] then the system of linear equations has a unique solution for all $m \geq 3$. Moreover, since $A_\star$ is stable, the approximation V is analytic and $V^{[m]}$ converges finitely.

To summarize the procedure, if we start with $V^{[2]}(x) = \frac{1}{2}x^T P x$ and $\nu_\star^{[2]}(x) = -R^{-1}B^T P x$, equations (13.20), (13.15) could be used to compute recursively the sequence of terms

$$V^{[3]}(x), \; \nu_\star^{[2]}(x), \; V^{[4]}(x), \; \nu_\star^{[3]}(x), \ldots$$

which converges point-wise to V and $\nu_\star$. We consider an example of the nonlinear benchmark problem [71] to illustrate the approach.

Example 13.1.2 *[260]. The system involves a cart of mass M which is constrained to translate along a straight horizontal line. The cart is connected to an inertially fixed point via a linear spring as shown in Figure 13.1. A pendulum of mass* **m** *and inertia* **I** *which rotates about a vertical line passing through the cart mass-center is also mounted. The dynamic equations for the system after suitable linearization are described by*

$$\left. \begin{aligned} \ddot{\xi} + \xi &= \varepsilon(\dot{\theta}^2 \sin\theta - \ddot{\theta}\cos\theta) + w \\ \ddot{\theta} &= -\varepsilon\ddot{\xi}\cos\theta + u \end{aligned} \right\} \qquad (13.21)$$

where ξ is the displacement of the cart, u is the normalized input torque, w is the normalized spring force which serves as a disturbance and θ is the angular position of the proof body. The coupling between the translational and the rotational motions is governed by the parameter ε which is defined by

$$\varepsilon = \frac{me}{\sqrt{(I + me^2)(M + m)}}, \quad 0 \leq \varepsilon \leq 1,$$

where e is the eccentricity of the pendulum, and $\varepsilon = 0$ if and only if $e = 0$. In this event, the dynamics reduces to

$$\left. \begin{aligned} \ddot{\xi} + \xi &= w \\ \ddot{\theta} &= u \end{aligned} \right\} \qquad (13.22)$$

and is not stabilizable, since it is completely decoupled. However, if we let $x := [x_1 \; x_2 \; x_3 \; x_4]^T = [\xi \; \dot{\xi} \; \theta \; \dot{\theta}]^T$, then the former can be represented in state-space as

$$\dot{x} = \begin{bmatrix} x_2 \\ \frac{-x_1 + \varepsilon x_4^2 \sin x_3}{1 - \varepsilon^2 \cos^2 x_3} \\ x_4 \\ \frac{\varepsilon \cos x_3 (x_1 - \varepsilon x_4^2 \sin x_3)}{1 - \varepsilon^2 \cos^2 x_3} \end{bmatrix} + \begin{bmatrix} 0 \\ \frac{1}{1 - \varepsilon^2 \cos^2 x_3} \\ 0 \\ \frac{-\varepsilon \cos x_3}{1 - \varepsilon^2 \cos^2 x_3} \end{bmatrix} w + \begin{bmatrix} 0 \\ \frac{-\varepsilon \cos x_3}{1 - \varepsilon^2 \cos^2 x_3} \\ 0 \\ \frac{1}{1 - \varepsilon^2 \cos^2 x_3} \end{bmatrix} u, \qquad (13.23)$$

where $1 - \varepsilon^2 \cos^2 x_3 \neq 0$ for all x_3 and $\varepsilon < 1$.

---

[1] A set of eigenvalues $\{\lambda_1, \ldots, \lambda_n\}$ is said to be resonant if $\sum_{j=1}^n i_j \lambda_j = 0$ for some nonnegative integers $i_1, i_2, \ldots, i_n$ such that $\sum_{j=1}^n i_j > 0$. Otherwise, it is called nonresonant.

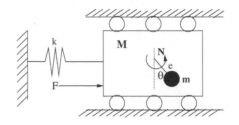

FIGURE 13.1

Nonlinear Benchmark Control Problem; Originally adopted from Int. J. of Robust and Nonlinear Control, vol. 8, pp. 307-310 © 1998, "A benchmark problem for nonlinear control design," by R. T. Bupp et al., with permission from Wiley Blackwell.

We first consider the design of a simple linear controller that locally asymptotically stabilizes the system with $w = 0$. Consider the linear control law

$$u = -k_1\theta - k_2\dot{\theta}, \quad k_1 > 0, k_2 > 0 \tag{13.24}$$

and the Lyapunov-function candidate

$$V(x) = \frac{1}{2}x^T P(\theta)x = \frac{1}{2}\dot{\xi}^2 + \frac{1}{2}\dot{\theta}^2 + \varepsilon\dot{\xi}\dot{\theta}\cos\theta + \frac{1}{2}\xi^2 + \frac{1}{2}k_1\theta^2$$

where

$$P(\theta) = \begin{bmatrix} 1 & 0 & 0 & 0 \\ 0 & 1 & 0 & \varepsilon\cos\theta \\ 0 & 0 & k_1 & 0 \\ 0 & \varepsilon\cos\theta & 0 & 1 \end{bmatrix}.$$

The eigenvalues of $P(\theta)$ are $\{1, k_1, 1 \pm \varepsilon\cos\theta\}$, and since $0 \le \varepsilon < 1$, $P(\theta)$ is positive-definite. Moreover, along trajectories of the the closed-loop system (13.23), (13.24),

$$\dot{V} = -k_2\dot{\theta}^2 \le 0.$$

Hence, the closed-loop system is stable. In addition, using LaSalle's invariance-principle, local asymptotic-stability can be concluded.

Design of Nonlinear Controller

Next, we apply the procedure developed above to design a nonlinear $\mathcal{H}_\infty$-controller by solving the HJIE. For this, we note the system specifications and constraints:

$$|\xi| \le 1.282, \quad \varepsilon = 0.2, \quad |u| \le 1.411,$$

and we choose

$$C = diag\{1, \sqrt{0.1}, \sqrt{0.1}, \sqrt{0.1}\}.$$

We first expand $f(x)$ and $G(x)$ in the HJIE (13.14) using the basis functions $x^{[1]}, x^{[2]}, x^{[3]}, \ldots$

as

$$f(x) \approx \begin{bmatrix} x_2 \\ -\frac{25}{24}x_1 + \frac{5}{24}x_4^2 x_3 + \frac{25}{576}x_1 x_3^2 + \dots \\ x_4 \\ \frac{5}{24}x_1 - \frac{1}{24}x_4^2 x_3 - \frac{65}{576}x_1 x_3^2 + \dots \end{bmatrix}$$

$$G(x) \approx \begin{bmatrix} 0 & 0 \\ \frac{25}{24} - \frac{25}{576}x_3^2 + \dots & -\frac{5}{24} + \frac{65}{576}x_3^2 + \dots \\ 0 & 0 \\ -\frac{5}{24} + \frac{65}{576}x_3^2 + \dots & \frac{25}{24} - \frac{25}{576}x_1 x_3^2 + \dots \end{bmatrix}.$$

Then the linearized plant about $x = 0$ is given by the following matrices

$$A = \frac{\partial f}{\partial x}(0) = \begin{bmatrix} 0 & 1 & 0 & 0 \\ -\frac{25}{24} & 0 & 0 & 0 \\ 0 & 0 & 0 & 1 \\ \frac{5}{24} & 0 & 0 & 0 \end{bmatrix}, \quad B_1 = G_1(0) = \begin{bmatrix} 0 \\ \frac{25}{24} \\ 0 \\ -\frac{5}{24} \end{bmatrix},$$

$$B_2 = G_2(0) = \begin{bmatrix} 0 \\ -\frac{5}{24} \\ 0 \\ \frac{25}{24} \end{bmatrix}.$$

We can now solve the ARE (13.10) with $H_1 = C$. It can be shown that for all $\gamma > \gamma^\star = 5.5$, the Riccati equation has a positive-definite solution. Choosing $\gamma = 6$, we get the solution

$$P = \begin{bmatrix} 19.6283 & -2.8308 & -0.7380 & -2.8876 \\ -2.8308 & 15.5492 & 0.8439 & 1.9915 \\ -0.7380 & 0.8439 & 0.3330 & 0.4967 \\ -2.8876 & 1.9915 & 0.4967 & 1.4415 \end{bmatrix} > 0$$

yielding

$$A_\star = \begin{bmatrix} 0 & 1.0000 & 0 & 0 \\ -1.6134 & 0.2140 & 0.0936 & 0.2777 \\ 0 & 0 & 0 & 1.000 \\ 2.7408 & 1.1222 & -0.3603 & -1.1422 \end{bmatrix}$$

with eigenvalues $\{-0.0415 \pm i1.0156, -0.4227 \pm i0.3684\}$. The linear feedbacks resulting from the above P are also given by

$$\nu_\star^{[1]}(x) = [l_\star^{[1]} \; k_\star^{[1]}]^T = \begin{bmatrix} -0.0652x_1 + 0.4384x_2 + 0.0215x_3 + 0.0493x_4 \\ 2.4182x_1 + 1.1650x_2 - 0.3416x_3 - 1.0867x_4 \end{bmatrix}.$$

Higher-Order Terms

Next, we compute the higher-order terms in the expansion. Since $A_\star$ is Hurwitz, its eigenvalues are nonresonant, and so the system of linear equations for the coefficient matrix in (13.20) has a unique solution and the series converges. Moreover, for this example,

$$f^{[2k]}(x) = 0, \quad G^{[2k-1]}(x) = 0, \quad k = 1, 2, \dots$$

Thus, $V^{[3]}(x) = 0$, and $\nu_\star^{[2]}(x) = 0$. The first non-zero higher-order terms are

$$\nu_\star^{[3]}(x) = -R^{-1}[B^T V_x^{[4]T}(x) + G^{[2]T}(x)],$$
$$V_x^{[4]}(x)A_\star x = -V_x^{[2]}(x)[f^{[3]}(x) + G^{[2]}(x)\nu_\star^{[1]}(x)].$$

The above system of equations can now be solved to yield

$$
\begin{aligned}
V^{[4]}(x) \;=\; & 162.117x_1^4 + 91.1375x_2^4 - 0.9243x_4x_3^2x_1 - 0.4143x_4x_3^2x_2 - \\
& 0.1911x_4x_3^3 + 59.6096x_4^2x_1^2 - 66.0818x_4^2x_2x_1 + 42.1915x_4^2x_2^2 - \\
& 8.7947x_4^2x_3x_1 + 0.6489x_4^4 - 151.8854x_4x_1^3 + 235.0386x_4x_2x_1^2 - \\
& 193.9184x_4x_2^2x_1 + 96.2534x_4x_2^3 + 43.523x_4x_3x_1^2 - 45.9440x_4x_3x_2x_1 - \\
& 258.0269x_2x_1^3 + 330.6741x_2^2x_1^2 - 186.0128x_2^3x_1 - 49.3835x_3x_1^3 + \\
& 92.8910x_3x_2x_1^2 - 70.7044x_3x_2^2x_1 - 0.2068x_3^3x_2 + 37.1118x_4x_3x_2^2 + \\
& 9.1538x_4^2x_3x_2 - 0.2863x_4^2x_3^2 - 9.6367x_4^3x_1 + 8.0508x_4^3x_2 + \\
& 0.7288x_4^3x_3 + 46.1405x_3x_2^3 + 9.4792x_3^2x_1^2 - 6.2154x_3^2x_2x_1 + \\
& 8.4205x_3^2x_2^2 - 0.4965x_3^3x_1 - 0.0156x_3^4
\end{aligned}
$$

and

$$
\nu_\star^{[3]}(x) \triangleq [l_\star^{[3]} \;\; k_\star^{[3]}]^T =
$$

$$
\left[
\begin{array}{l}
0.1261x_4^2x_3 - 0.0022x_4x_3^2 - 0.8724x_4^2x_1 + 1.1509x_4^2x_2 + 0.1089x_4^3 + \\
1.0208x_4x_3x_2 + 1.8952x_3x_2^2 + 0.2323x_3^2x_2 + 3.0554x_4x_1^2 - 5.2286x_4x_2x_1 + \\
3.9335x_4x_2^2 - 0.6138x_4x_3x_1 - 1.9128x_3x_2x_1 - 0.0018x_3^3 + \\
8.8880x_2x_1^2 - 7.5123x_2^2x_1 + 1.2179x_3x_1^2 - 3.2935x_1^3 + 4.9956x_2^3 - 0.0928x_1x_3^2 \\
\\
-0.1853x_4^2x_3 + 0.0929x_4x_3^2 + 8.1739x_4^2x_1 - 3.7896x_4^2x_2 - 0.5132x_4^3 - \\
1.8036x_4x_3x_2 - 4.9101x_3x_2^2 + 0.3018x_3^2x_2 - 37.6101x_4x_1^2 + 28.4355x_4x_2x_1 - \\
13.8702x_4x_2^2 + 4.3753x_4x_3x_1 + 9.1991x_3x_2x_1 + 0.0043x_3^3 - 53.5255x_2x_1^2 + \\
42.8702x_2^2x_1 - 12.9923x_3x_1^2 + 52.2292x_1^3 - 12.1580x_2^3 + 0.02812x_1x_3^2
\end{array}
\right] .
$$

Remark 13.1.1 *The above computational results give an approximate solution to the HJIE and the control law up to order three. Some difficulties that may be encountered include how to check the positive-definiteness of the solutions in general. Locally around $x = 0$ however, the positive-definiteness of P is reassuring, but does not guarantee global positive-definiteness and a subsequent global asymptotic-stabilizing controller. The computational burden of the algorithm also limits its attractiveness.*

In the next section we discuss exact methods for solving the HJIE which may yield global solutions.

13.2   A Factorization Approach for Solving the HJIE

In this section, we discuss a factorization approach that may yield exact global solutions of the HJIE for the class of affine nonlinear systems. We begin with a discussion of sufficiency conditions for the existence of exact solutions to the HJIE (13.9) which are provided by the Implicit-function Theorem [157]. In this regard, let us write HJIE (13.9) in the form:

$$
HJI(x, V_x) = 0, \quad x \in M \subset \mathcal{X}, \tag{13.25}
$$

where $HJI : T^\star M \to \Re$. Then we have the following theorem.

Theorem 13.2.1 *Assume that $V \in C^2(M)$, and the functions $f(.)$, $g_1(.)$, $g_2(.)$, $h(.)$ are smooth $C^2(M)$ functions. Then $HJI(.,.)$ is continuously-differentiable in an open neighborhood $N \times \Psi \subset T^\star M$ of the origin. Furthermore, let $(\bar{x}, \bar{V}_x)$ be a point in $N \times \Psi$ such* *that $HJI(\bar{x}, \bar{V}_x) = 0$ and the $\mathcal{F}$-derivative of $HJI(.,.)$ with respect to V_x is nonzero, i.e., $\frac{\partial}{\partial V_x} HJI(\bar{x}, \bar{V}_x) \neq 0$, then there exists a continuously-differentiable solution:*

$$V_x(x) = \overline{HJI}(x) \tag{13.26}$$

for some function $\overline{HJI} : N \to \Re$, of HJIE (13.9) in $N \times \Psi$.

Proof: The proof of the above theorem follows from standard results of the Implicit-function Theorem [291]. This can also be shown by linearization of HJI around $(\bar{x}, \bar{V}_x)$; the existence of such a point is guaranteed from the linear $\mathcal{H}_\infty$-control results [292]. Accordingly,

$$HJI(x, V_x) = HJI(\bar{x}, \bar{V}_x) + (V_x - \bar{V}_x)\frac{\partial HJI}{\partial V_x}(\bar{x}, \bar{V}_x) + (x - \bar{x})^T \frac{\partial HJI}{\partial x}(\bar{x}, \bar{V}_x) + HOT = 0 \tag{13.27}$$

where HOT denote higher-order terms. Since $HJI(\bar{x}, \bar{V}_x) = 0$, it follows from (13.27) that there exists a ball $B(\bar{x}, \bar{V}_x; r) \in N \times \Psi$ of radius $r > 0$ centered at $(\bar{x}, \bar{V}_x)$ such that in the limit as $r \to 0$, $HOT \to 0$ and V_x can be expressed in terms of x. $\square$

Remark 13.2.1 *Theorem 13.2.1 is only an existence result, and hence is not satisfactory, in the sense that it does not guarantee the uniqueness of V_x and it is only a local solution.*

The objective of the approach then is to find an expression for V_x from the HJIE so that V can be recovered from it by carrying out the *line-integral* $\int_0^x V_x(\sigma)d\sigma$. The integration is taken over any path joining the origin to x. For convenience, this is usually done along the axes as:

$$V(x) = \int_0^{x_1} V_{x_1}(y_1, 0, \ldots, 0)dy_1 + \int_0^{x_2} V_{x_2}(x_1, y_2, \ldots, 0)dy_2 + \ldots + \int_0^{x_n} V_{x_n}(x_1, x_2, \ldots, y_n)dy_n.$$

In addition, to ensure that V_x is the gradient of the scalar function V, it is necessary and sufficient that the Hessian matrix V_{xx} is symmetric for all $x \in N$. This will be referred to as the *"curl-condition"*:

$$\frac{\partial V_{x_i}(x)}{\partial x_j} = \frac{\partial V_{x_j}(x)}{\partial x_i}, i, j = 1, \ldots, n. \tag{13.28}$$

Further, to account for the HOT in the Taylor-series expansion above, we introduce a parameter $\zeta = \zeta(x)$ into the solution (13.26) as

$$V_x(x) = \overline{HJI}(x, \zeta).$$

The next step is to search for $\xi \in TM$ such that the HJIE (13.9) is satisfied together with the curl conditions (13.28).

To proceed, let

$$\mathcal{Q}_\gamma(x) = [\frac{1}{\gamma^2}g_1(x)g_1^T(x) - g_2(x)g_2^T(x)], \ x \in M,$$

then we have the following result.

Theorem 13.2.2 *Consider the HJIE (13.9) and suppose there exists a vector field $\zeta : N \to TN$ such that*

$$\zeta^T(x)\mathcal{Q}_\gamma^+(x)\zeta(x) - f^T(x)\mathcal{Q}_\gamma^+(x)f(x) + h^T(x)h(x) \leq 0; \ \forall x \in N, \tag{13.29}$$

where the matrix Q_γ^+ is the generalized-inverse of Q_γ, then

$$V_x(x) = -(f(x) \pm \zeta(x))^T Q_\gamma^+(x), \ x \in N \qquad (13.30)$$

satisfies the HJIE (13.9).

Proof: By direct substitution and using the properties of generalized inverses [232]. □

Remark 13.2.2 ζ *is referred to as the "discriminant-factor" of the system or the HJIE and (13.29) as the "discriminant-equation." Moreover, since $V_x(0) = 0$, we require that $\zeta(0) = 0$. This also holds for any equilibrium-point x_e of the system Σ.*

Consider now the Hessian matrix of V from the above expression (13.30) which is given by:

$$V_{xx}(x) = -\left(\frac{\partial f(x)}{\partial x} \pm \frac{\partial \zeta(x)}{\partial x}\right)^T Q_\gamma^+(x) - \left(I_n \otimes (f(x) \pm \zeta(x))^T\right)\frac{\partial Q_\gamma^+(x)}{\partial x}, \qquad (13.31)$$

where

$$\frac{\partial Q_\gamma^+(x)}{\partial x} = \left[\frac{\partial Q_\gamma^+(x)}{\partial x_1}, \dots, \frac{\partial Q_\gamma^+}{\partial x_n}\right]^T, \quad \frac{\partial f}{\partial x} = f_x(x) = [\frac{\partial f_1}{\partial x}, \dots, \frac{\partial f_n}{\partial x}],$$

$$\frac{\partial \zeta}{\partial x} = \zeta_x(x) = [\frac{\partial \zeta_1}{\partial x}, \dots, \frac{\partial \zeta_n}{\partial x}], \quad h_x(x) = \frac{\partial h}{\partial x}(x) - [\frac{\partial h_1}{\partial x}, \dots, \frac{\partial h_m}{\partial x}].$$

Then, the Curl-conditions $V_{xx}(x) = V_{xx}^T(x)$ will imply

$$\left(\frac{\partial f(x)}{\partial x} \pm \frac{\partial \zeta(x)}{\partial x}\right)^T Q_\gamma^+(x) + \left(I_n \otimes (f(x) \pm \zeta(x))^T\right)\frac{\partial Q_\gamma^+(x)}{\partial x} =$$

$$Q_\gamma^+(x)\left(\frac{\partial f(x)}{\partial x} \pm \frac{\partial \zeta(x)}{\partial x}\right) + \frac{\partial Q_\gamma^+(x)}{\partial x}\left(I_n \otimes (f(x) \pm \zeta(x))\right) \qquad (13.32)$$

which reduces to

$$\left(\frac{\partial f(x)}{\partial x} \pm \frac{\partial \zeta(x)}{\partial x}\right)^T Q_\gamma^+ = Q_\gamma^+\left(\frac{\partial f(x)}{\partial x} \pm \frac{\partial \zeta(x)}{\partial x}\right), \quad x \in N \qquad (13.33)$$

if $Q_\gamma^+(x)$ is a constant matrix.

Equations (13.32), (13.33) are a system of $\frac{n(n-1)}{2}$ first-order PDEs in n unknowns, ζ, which could be solved for ζ up to an arbitrary vector $\lambda \in TN$ (cf. these conditions with the variable-gradient method for finding a Lyapunov-function [157]).

Next, the second requirement $V_{xx} \geq 0$ will imply that

$$\left(\frac{\partial f(x)}{\partial x} \pm \frac{\partial \zeta(x)}{\partial x}\right)^T Q_\gamma^+(x) + \left(I_n \otimes (f(x) \pm \zeta(x))^T\right)\frac{\partial Q_\gamma^+(x)}{\partial x} \leq 0, \ \forall x \in N, \qquad (13.34)$$

and can only be imposed after a solution to (13.32) has been obtained.

Let us now specialize the above results to the a linear system

$$\Sigma_l: \quad \dot{x} = Fx + G_1 w + G_2 u$$

$$z(t) = \begin{bmatrix} H_1 x \\ u \end{bmatrix},$$

where $F \in \Re^{n \times n}$, $G_1 \in \Re^{n \times r}$, and $G_2 \in \Re^{n \times k}$, $H_1 \in \Re^{m \times n}$ are constant matrices. Then, in this case, $\zeta(x) = \Gamma x$, $\Gamma \in \Re^{n \times n}$, $Q_\gamma = (\frac{1}{\gamma^2} G_1 G_1^T - G_2 G_2^T)$. Thus, V_x is given by

$$V_x(x) = -(Fx \pm \Gamma x)^T Q_\gamma^+ = -x^T (F \pm \Gamma)^T Q_\gamma^+ \qquad (13.35)$$

where Γ satisfies:

$$\Gamma^T Q_\gamma^+ \Gamma - F^T Q_\gamma^+ F + H^T H = 0.$$

Now define

$$\Delta = F^T Q_\gamma^+ F - H_1^T H_1,$$

then Γ is given by

$$\Gamma^T Q_\gamma^+ \Gamma = \Delta. \qquad (13.36)$$

This suggests that Γ is a coordinate-transformation matrix, which in this case is called a *congruence transformation* between Q_γ^+ and Δ. Moreover, from (13.35), V_{xx} is given by

$$V_{xx} = -(F \pm \Gamma)^T Q_\gamma^+ \qquad (13.37)$$

(cf. with the solution of the linear $\mathcal{H}_\infty$-Riccati equation $P = X_2 X_1^{-1}$, where the columns of $\begin{bmatrix} X_1 \\ X_2 \end{bmatrix}$ span $\Lambda_{\bar{H}_\gamma}^-$, the stable-eigenspace of the Hamiltonian-matrix associated with the linear $\mathcal{H}_\infty$ problem, with X_1 invertible [68, 292]). A direct connection of the above results with the Hamiltonian matrices approach for solving the ARE can be drawn from the following fact (see also Theorem 13.2, page 329 in [292]): If $X \in \mathbf{C}^{n \times n}$ is a solution of the ARE

$$A^T X + XA + XRX + Q = 0, \qquad (13.38)$$

then there exist matrices X_1, $X_2 \in \mathcal{C}^{n \times n}$, with X_1 invertible, such that $X = X_2 X_1^{-1}$ and the columns of $\begin{bmatrix} X_1 \\ X_2 \end{bmatrix}$ form a basis for an n-dimensional invariant-subspace of H defined by:

$$H \triangleq \begin{bmatrix} A & R \\ -Q & -A^T \end{bmatrix}.$$

In view of the above fact, we now prove that the columns of $\begin{bmatrix} I \\ P \end{bmatrix} \triangleq \begin{bmatrix} I \\ -(F \pm \Gamma)^T Q_\gamma^+ \end{bmatrix}$ span an n-dimensional invariant-subspace of the Hamiltonian matix

$$H_\gamma^l = \begin{bmatrix} F & (\frac{1}{\gamma^2} G_1 G_1^T - G_2 G_2^T) \\ -H^T H & -F^T \end{bmatrix} \triangleq \begin{bmatrix} F & Q_\gamma \\ -H^T H & -F^T \end{bmatrix} \qquad (13.39)$$

corresponding to the ARE (13.38).

Theorem 13.2.3 *Suppose there exists a Γ that satisfies (13.36) and such that $P = -(F \pm \Gamma)^T Q_\gamma^+$ is symmetric, then P is a solution of ARE (13.10). Moreover, if Q_γ is nonsingular, then the columns of $\begin{bmatrix} I \\ P \end{bmatrix}$ span an n-dimensional invariant subspace of H_γ^l. Otherwise, of $\begin{bmatrix} Q_\gamma^+ & 0 \\ 0 & I \end{bmatrix} H_\gamma^l$.*

Proof: The first part of the theorem has already been shown. For the second part, using the symmetry of P we have

$$
\begin{bmatrix} Q_\gamma^+ & 0 \\ 0 & I \end{bmatrix} \begin{bmatrix} F & Q_\gamma \\ -H^T H & -F^T \end{bmatrix} \begin{bmatrix} I \\ -(F \pm \Gamma)^T Q_\gamma^+ \end{bmatrix}
$$

$$
= \begin{bmatrix} Q_\gamma^+ & 0 \\ 0 & I \end{bmatrix} \begin{bmatrix} F & Q_\gamma \\ -H^T H & -F^T \end{bmatrix} \begin{bmatrix} I \\ -Q_\gamma^+(F \pm \Gamma) \end{bmatrix}
$$

$$
= \begin{bmatrix} Q_\gamma^+ & 0 \\ 0 & I \end{bmatrix} \begin{bmatrix} I \\ -(F \pm \Gamma)^T Q_\gamma^+ \end{bmatrix} (\mp \Gamma). \tag{13.40}
$$

Hence, P defined as above indeed spans an n-dimensional invariant-subspace of H_γ^l if Q_γ is nonsingular. $\square$

13.2.1 Worked Examples

In this subsection, we consider worked examples solved using the approach outlined in the previous section.

Example 13.2.1 *Consider the following system:*

$$
\begin{aligned}
\dot{x}_1 &= -x_1^3(t) - x_2(t) \\
\dot{x}_2 &= x_2 + u + w \\
z &= \begin{bmatrix} x_2 \\ u \end{bmatrix}.
\end{aligned}
$$

Let $\gamma = 2$, then

$$
f(x) = \begin{bmatrix} -x_1^3 + x_2 \\ x_2 \end{bmatrix}; \ G_1 = \begin{bmatrix} 0 \\ 1 \end{bmatrix}; \ G_2 = \begin{bmatrix} 0 \\ 1 \end{bmatrix}; \ h(x) = \begin{bmatrix} 0 \\ x_2 \end{bmatrix},
$$

$$
Q_\gamma = \begin{bmatrix} 0 & 0 \\ 0 & -\frac{3}{4} \end{bmatrix}, \ Q_\gamma^+ = \begin{bmatrix} 0 & 0 \\ 0 & -\frac{4}{3} \end{bmatrix}.
$$

Substituting the above functions in (13.29), (13.33), (13.34), we get

$$
-4\zeta_2^2 + 4x_2^2 + 3x_2^2 \le 0; \quad \forall x \in N_e \tag{13.41}
$$

$$
\zeta_{1,x_2}(x) = -1 \ \forall x \in N_e \tag{13.42}
$$

$$
\begin{bmatrix} 0 & \frac{4}{3} - \frac{4}{3}\zeta_{1,x_2} \\ 0 & -\frac{4}{3} - \frac{4}{3}\zeta_{2,x_2} \end{bmatrix} \le 0; \quad \forall x \in N_e. \tag{13.43}
$$

Solving the above system we get

$$
\zeta_2(x) = \pm \frac{\sqrt{7}}{2} x_2, \ \ \zeta_1(x) = -x_2 + \phi(x_1)
$$

where $\phi(x_1)$ is some arbitrary function which we can take as $\Phi(x) = 0$ without any loss of generality. Thus,

$$
V_x(x) = -(f(x) \pm \zeta(x))^T Q^+ = (0 \quad \frac{8 \pm 4\sqrt{7}}{6} x_2).
$$

Finally, integrating the positive term in the expression for V_x above from 0 to x, we get

$$
V(x) = \frac{2 + \sqrt{7}}{3} x_2^2
$$

which is positive-semidefinite.

The next example will illustrate a general transformation approach for handling the discriminant equation/inequality and symmetry condition.

Example 13.2.2 *Consider the following example with the disturbance affecting the first state equation and a weighting on the states:*

$$
\begin{aligned}
\dot{x}_1 &= -x_1^3 - x_2 + w_1 \\
\dot{x}_2 &= x_1 + x_2 + u + w_2 \\
z &= \begin{bmatrix} Q^{1/2} x \\ R^{1/2} u \end{bmatrix}
\end{aligned}
$$

where $Q = diag\{q_1, q_2\} \geq 0$, $R = r > 0$ are weighting matrices introduced to make the HJIE solvable. The state-feedback HJIE associated with the above system is thus given by

$$
V_x(x) f(x) + \frac{1}{2} V_x(x) [\frac{1}{\gamma^2} g_1(x) g_1^T(x) - g_2(x) R^{-1} g_2^T(x)] V_x^T(x) + \frac{1}{2} x^T Q x \leq 0, \qquad (13.44)
$$

with $V(0) = 0$, $\mathcal{Q}_\gamma(x) = [\frac{1}{\gamma^2} g_1(x) g_1^T(x) - g_2(x) R^{-1} g_2^T(x)]$. Then

$$
f(x) = \begin{bmatrix} -x_1^3 - x_2 \\ x_1 + x_2 \end{bmatrix}, \; G_2 = \begin{bmatrix} 0 \\ 1 \end{bmatrix}, \; G_1 = I_2, \; \mathcal{Q}_\gamma = \begin{bmatrix} \frac{1}{\gamma^2} & 0 \\ 0 & \frac{r - \gamma^2}{r \gamma^2} \end{bmatrix}, \; \mathcal{Q}_\gamma^+ = \begin{bmatrix} \gamma^2 & 0 \\ 0 & \frac{r \gamma^2}{r - \gamma^2} \end{bmatrix}.
$$

Substituting the above functions in (13.29), (13.33), (13.34), we get

$$
\gamma^2 \zeta_1^2 + \frac{r \gamma^2}{r - \gamma^2} \zeta_2^2 - \gamma^2 (x_1^3 + x_2)^2 - \frac{r \gamma^2}{r - \gamma^2} (x_1 + x_2)^2 + q_1 x_1^2 + q_2 x_2^2 = 0 \quad (13.45)
$$

$$
\frac{r \gamma^2}{r - \gamma^2} \zeta_{2,x_1} - \zeta_{1,x_2} = \frac{\gamma^2}{r - \gamma^2} \qquad (13.46)
$$

$$
\begin{bmatrix} \gamma^2 (\zeta_{1,x_1} - 3x_1^2) & \frac{r \gamma^2}{r - \gamma^2} (\zeta_{2x_1} + 1) \\ \gamma^2 (\zeta_{1x_2} - 1) & \frac{r \gamma^2}{r - \gamma^2} (\zeta_{2,x_2} + 1) \end{bmatrix} \leq 0. \qquad (13.47)
$$

One way to handle the above algebraic-differential system is to parameterize ζ_1 and ζ_2 as:

$$
\zeta_1(x) = ax_1 + bx_2, \;\; \zeta_2(x) = cx_1 + dx_2 + ex_1^3,
$$

where $a, b, c, d, e \in \Re$ are constants, and to try to solve for these constants. This approach may not however work for most systems. We therefore illustrate next a more general procedure for handling the above system.

Suppose we choose r, γ such that $\frac{r \gamma^2}{r - \gamma^2} > 0$ *(usually we take $r >> 1$ and $0 < \gamma < 1$).* We now apply the following transformation to separate the variables:

$$
\zeta_1(x) = \frac{1}{\gamma} \rho(x) \cos \theta(x), \;\; \zeta_2(x) = \sqrt{\frac{(r - \gamma^2)}{r \gamma^2}} \rho(x) \sin \theta(x).
$$

where ρ, $\theta : N \to \Re$ are C^2 functions. Substituting in the equation (13.45), we get

$$
\rho(x) = \pm \sqrt{\frac{r \gamma^2}{r - \gamma^2} (x_1 + x_2)^2 - \gamma^2 (x_1^3 + x_2)^2 - q_1 x_1^2 - q_2 x_2^2}.
$$

Thus, for the HJIE (13.44) to be solvable, it is necessary that γ, r, q_1, q_2 are chosen such that the function under the square-root in the above equation is positive for all $x \in N$ so that ρ is real. As a matter of fact, the above expression defines N, i.e., $N = \{x \mid \rho \in \Re\}$.

If however we choose r, γ such that $\frac{r\gamma^2}{r-\gamma^2} < 0$, then we must paramerize ζ_1, ζ_2 as $\zeta_1(x) = \frac{1}{\gamma}\rho(x)\cosh\theta(x)$, $\zeta_2(x) = \sqrt{\frac{(r-\gamma^2)}{r\gamma^2}}\rho(x)\sinh\theta(x)$. A difficulty also arises when $Q_\gamma^+(x)$ is not diagonal, e.g., if $Q_\gamma^+ = \begin{bmatrix} a & b \\ 0 & c \end{bmatrix}$, $a, b, c \in \Re$. Then $\zeta^T(x)Q_\gamma^+\zeta(x) = i\zeta_1^2(x) + j\zeta_1(x)\zeta_2(x) + k\zeta_2^2(x)$, for some i, j, $k \in \Re$. The difficulty here is created by the cross-term $j\zeta_1\zeta_2$ as the above parameterization cannot lead to a simplification of the problem. However, we can use a completion of squares to get

$$\zeta^T(x)Q_\gamma^+\zeta(x) = (\sqrt{i}\zeta_1(x) + \frac{j}{2\sqrt{i}}\zeta_2(x))^2 + (k - \frac{j^2}{4i})\zeta_2^2(x)$$

(assuming $i > 0$, otherwise, pull out the negative sign outside the bracket). Now we can define $\left(\sqrt{i}\zeta_1(x) + \frac{j}{2\sqrt{i}}\zeta_2(x)\right) = \rho(x)\cos\theta(x)$, and $\zeta_2(x) = (k - \frac{j^2}{4i})^{-1/2}\rho(x)\sin\theta(x)$. Thus, in reality,

$$\zeta_1(x) = \rho(x)\left(\frac{1}{\sqrt{i}}\cos\theta(x) - \frac{j}{2i}(k - \frac{j^2}{4i})^{1/2}\sin\theta(x)\right).$$

Next, we determine $\theta(.)$ from (13.46). Differentiating ζ_1, ζ_2 with respect to x_2 and x_1 respectively and substituting in (13.46), we get

$$\beta(\rho_{x_1}(x)\sin\theta(x) + \rho(x)\theta_{x_1}(x)\cos\theta(x)) -$$
$$\kappa(\rho_{x_2}(x)\cos\theta(x) - \rho(x)\theta_{x_2}(x)\sin\theta(x)) = \eta \qquad (13.48)$$

where $\beta = \frac{1}{\gamma}\sqrt{\frac{r}{(r-\gamma^2)}}$, $\kappa = \frac{1}{\gamma}$, $\eta = \frac{\gamma^2}{r-\gamma^2}$. This first-order PDE in θ can be solved using the method of "characteristics" discussed in Chapter 4 (see also the reference [95]). However, the geometry of the problem calls for a simpler approach. Moreover, since θ is a free parameter that we have to assign to guarantee that V_{xx} is symmetric and positive-(semi) definite, there are many solutions to the above PDE. One solution can be obtained as follows. Rearranging the above equation, we get

$$(\beta\rho_{x_1}(x) + \kappa\rho(x)\theta_{x_2}(x))\sin\theta(x) + (-\kappa\rho_{x_2}(x) + \beta\rho(x)\theta_{x_1}(x))\cos\theta(x) = \eta. \qquad (13.49)$$

If we now assign

$$(\beta\rho_{x_1}(x) + \kappa\rho(x)\theta_{x_2}(x)) = \frac{\eta}{2}\sin\theta(x),$$
$$(-\kappa\rho_{x_2}(x) + \beta\rho(x)\theta_{x_1}(x)) = \frac{\eta}{2}\cos\theta(x),$$

then we see that (13.49) is satisfied. Further, squaring both sides of the above equations and adding, we get

$$\frac{4}{\eta^2}(\beta\rho_{x_1}(x) + \kappa\rho(x)\theta_{x_2}(x))^2 + \frac{4}{\eta^2}(-\kappa\rho_{x_2}(x) + \beta\rho(x)\theta_{x_1}(x))^2 = 1 \qquad (13.50)$$

which is the equation of an ellipse in the coordinates θ_{x_1}, θ_{x_2}, centered at $\frac{\kappa\rho_{x_2}(x)}{\beta\rho(x)}$, $-\frac{\beta\rho_{x_1}(x)}{\kappa\rho(x)}$ and radii $\frac{\eta}{2\beta\rho(x)}$, $\frac{\eta}{2\kappa\rho(x)}$ respectively. Thus, any point on this ellipse will give the required gradient for θ. One point on this ellipse corresponds to the following gradients in θ:

$$\theta_{x_1}(x) = \frac{\kappa\rho_{x_2}(x)}{\beta\rho(x)} + \frac{1}{\sqrt{2}}\left(\frac{\eta}{2\beta\rho(x)}\right)$$
$$\theta_{x_2}(x) = -\frac{\beta\rho_{x_1}(x)}{\kappa\rho(x)} + \frac{1}{\sqrt{2}}\left(\frac{\eta}{2\kappa\rho(x)}\right)).$$

Hence, we can finally obtain θ as

$$\theta(x) = \int_{0+}^{x_1} \left(\frac{\kappa \rho_{x_2}(x)}{\beta \rho(x)} + \frac{1}{\sqrt{2}}\left(\frac{\eta}{2\beta\rho(x)}\right) \right)\Bigg|_{x_2=0} dx_1 + \int_{0+}^{x_2} \left(-\frac{\beta \rho_{x_1}(x)}{\kappa\rho(x)} + \frac{1}{\sqrt{2}}\left(\frac{\eta}{2\kappa\rho(x)}\right) \right) dx_2$$

The above integral can be evaluated using Mathematica or Maple. The result is very complicated and lengthy, so we choose not to report it here.

Remark 13.2.3 *Note, any available method can also be used to solve the symmetry PDE in θ (13.48), as the above approach may be too restricted and might not yield the desired solution. Indeed, a general solution would be more desirable. Moreover, any solution should be checked against the positive-(semi)definite condition (13.47) to see that it is satisfied. Otherwise, some of the design parameters, r, γ, q_i, should be adjusted to see that this condition is at least satisfied.*

Finally, we can compute V as

$$V(x) = -\int_{0+}^{x} \left(f(x) + \rho(x) \begin{bmatrix} \cos\theta(x) \\ \sin\theta(x) \end{bmatrix} \right)^T Q_\gamma^+ dx. \tag{13.51}$$

Remark 13.2.4 *It may not be necessary to compute V explicitly, since the optimal control $u^\star = \alpha(x)$ is only a function of V_x. What is more important is to check that the positive-(semi)definiteness condition (13.47) is locally satisfied around the origin $\{0\}$. Then the V function corresponding to this V_x will be a candidate solution for the HJIE. However, we still cannot conclude at this point that it is a stabilizing solution. In the case of the above example, it can be seen that by setting $\zeta_1(x) = \rho(x)\cos\theta(x)$, $\zeta_2(x) = \rho(x)\sin\theta(x)$, and their derivatives equal to 0, the inequality (13.47) is locally satisfied at the origin $\{0\}$.*

Example 13.2.3 *In this example, we consider the model of a satellite considered in [46, 87, 154, 169] (and the references there in). The equations of motion of the spinning satellite are governed by two subsystems; namely, a kinematic model and a dynamic model. The configuration space of the satellite is a six dimensional manifold the tangent bundle of $SO(3)$, or $TSO(3)$, (where $SO(3)$ is the special orthogonal linear matrix group). The equations of motion are given by:*

$$\dot{R} = RS(\omega) \tag{13.52}$$

$$J\dot{\omega} = S(\omega)J\omega + u + Pd, \tag{13.53}$$

where $\omega \in \Re^3$ is the angular velocity vector about a fixed inertial reference frame with three principal axes and having the origin at the center of gravity of the satellite, $R \in SO(3)$, is the orientation matrix of the satellite, $u \in \Re^3$ is the control torque input vector, and d is the vector of external disturbances on the spacecraft. $P = diag\{P_1, P_2, P_3\}$, $P_i \in \Re$, $i = 1, 2, 3$, is a constant gain matrix, J is the inertia matrix of the system, and $S(\omega)$ is the skew-symmetric matrix

$$S(\omega) = \begin{bmatrix} 0 & -\omega_3 & \omega_2 \\ \omega_3 & 0 & -\omega_1 \\ -\omega_2 & \omega_1 & 0 \end{bmatrix}.$$

We consider the control of the angular velocities governed by the dynamic subsystem (13.53). By letting $J = diag\{I_1, I_2, I_3\}$, where $I_1 > 0$, $I_2 > 0$, $I_3 > 0$ and without any loss of *generality assuming $I_1 \neq I_2 \neq I_3$ are the principal moments of inertia, the subsystem (13.53) can be represented as:*

$$
\begin{aligned}
I_1\dot{\omega}_1(t) &= (I_2 - I_3)\omega_2(t)\omega_3(t) + u_1 + P_1 d_1(t) \\
I_2\dot{\omega}_2(t) &= (I_3 - I_1)\omega_3(t)\omega_1(t) + u_2 + P_2 d_2(t) \\
I_3\dot{\omega}_3(t) &= (I_1 - I_2)\omega_1(t)\omega_2(t) + u_3 + P_3 d_3(t).
\end{aligned}
$$

Now define

$$
A_1 = \frac{(I_2 - I_3)}{I_1}, \quad A_2 = \frac{(I_3 - I_1)}{I_2}, \quad A_3 = \frac{(I_1 - I_2)}{I_3},
$$

then the above equations can be represented as:

$$
\begin{aligned}
\dot{\omega}(t) \quad &:= \quad f(\omega) + B_1 d(t) + B_2 v(t) \\
&= \quad \begin{bmatrix} A_1\omega_2\omega_3 \\ A_2\omega_1\omega_3 \\ A_3\omega_1\omega_2 \end{bmatrix} + \begin{bmatrix} b_1 & 0 & 0 \\ 0 & b_2 & 0 \\ 0 & 0 & b_3 \end{bmatrix} \begin{bmatrix} d_1 \\ d_2 \\ d_3 \end{bmatrix} + \\
&\qquad \begin{bmatrix} 1 & 0 & 0 \\ 0 & 1 & 0 \\ 0 & 0 & 1 \end{bmatrix} \begin{bmatrix} u_1 \\ u_2 \\ u_3 \end{bmatrix},
\end{aligned} \tag{13.54}
$$

where

$$
b_1 = \frac{P_1}{I_1}, \quad b_2 = \frac{P_2}{I_2}, \quad b_3 = \frac{P_3}{I_3}.
$$

In this regard, consider the output function:

$$
z = h(\omega) = \begin{bmatrix} c_1\omega_1 \\ c_2\omega_2 \\ c_3\omega_3 \end{bmatrix} \tag{13.55}
$$

where c_1, c_2, c_3 are design parameters. Applying the results of Section 13.2, we have

$$
Q_\gamma = (\frac{1}{\gamma^2}B_1 B_1^T - B_2 B_2^T) = \begin{bmatrix} \frac{b_1^2}{\gamma^2} - 1 & 0 & 0 \\ 0 & \frac{b_2^2}{\gamma^2} - 1 & 0 \\ 0 & 0 & \frac{b_3^2}{\gamma^2} - 1 \end{bmatrix},
$$

$$
V_\omega(\omega) = -(f(\omega) \pm \zeta(\omega))^T Q_\gamma^+ \tag{13.56}
$$

and (13.29) implies

$$
\zeta^T(\omega)Q_\gamma^+\zeta(\omega) - f^T(\omega)Q_\gamma^+ f(\omega) + h^T(\omega)h(\omega) \leq 0.
$$

Upon substitution, we get

$$
(\frac{\gamma^2}{b_1^2 - \gamma^2})\zeta_1^2(\omega) + (\frac{\gamma^2}{b_2^2 - \gamma^2})\zeta_2^2(\omega) + (\frac{\gamma^2}{b_3^2 - \gamma^2})\zeta_3^2(\omega) - (\frac{\gamma^2}{b_1^2 - \gamma^2})A_1^2\omega_2^2\omega_3^2 -
$$

$$
(\frac{\gamma^2}{b_2^2 - \gamma^2})A_2^2\omega_1^2\omega_3^2 - (\frac{\gamma^2}{b_3^2 - \gamma^2})A_3^2\omega_1^2\omega_2^2 + c_1^2\omega_1^2 + c_2^2\omega_2^2 + c_3^2\omega_3^2 \leq 0.
$$

Further, substituting in (13.33), we have the following additional constraints on ζ:

$$
A_1\omega_3 + \zeta_{1,\omega_2} = \zeta_{2,\omega_1} + A_2\omega_3, \quad A_1\omega_2 + \zeta_{1,\omega_3} = A_3\omega_2 + \zeta_{3,\omega_1}, \quad A_2\omega_1 + \zeta_{3,\omega_1} = A_3\omega_1 + \zeta_{1,\omega_3}.
$$

$$\tag{13.57}$$

Thus, for any $\gamma > b = \max_i\{b_i\}, i = 1, 2, 3$, and $c_i = \sqrt{\left|\left(\frac{\gamma^2}{b_i^2 - \gamma^2}\right)\right|}, i = 1, 2, 3$, and under the assumption $A_1 + A_2 + A_3 = 0$ (see also [46]), we have the following solution

$$
\begin{aligned}
\zeta_1 &= -A_1\omega_2\omega_3 + \omega_1 \\
\zeta_2 &= -A_2\omega_1\omega_3 + \omega_2 \\
\zeta_3 &= -A_3\omega_1\omega_2 + \omega_3.
\end{aligned}
$$

Thus,

$$
V_\omega(\omega) = -(f(\omega) + \zeta(\omega))^T Q_\gamma^+ = \left[\left(\frac{\gamma^2}{\gamma^2 - b_1^2}\right)\omega_1 \quad \left(\frac{\gamma^2}{\gamma^2 - b_2^2}\right)\omega_2 \quad \left(\frac{\gamma^2}{\gamma^2 - b_3^2}\right)\omega_3\right]
$$

and integrating from 0 to ω, we get

$$
V(\omega) = \frac{1}{2}\left(\left(\frac{\gamma^2}{\gamma^2 - b_1^2}\right)\omega_1^2 + \left(\frac{\gamma^2}{\gamma^2 - b_2^2}\right)\omega_2^2 + \left(\frac{\gamma^2}{\gamma^2 - b_3^2}\right)\omega_3^2\right)
$$

which is positive-definite for any $\gamma > b$.

Remark 13.2.5 *Notice that the solution of the discriminant inequality does not only give us a stabilizing feedback, but also the linearizing feedback control. However, the linearizing terms drop out in the final expression for V, and consequently in the expression for the optimal control $u^\star = -B_2^T V_\omega^T(\omega)$. This clearly shows that cancellation of the nonlinearities is not optimal.*

13.3   Solving the Hamilton-Jacobi Equation for Mechanical Systems and Application to the Toda Lattice

In this section, we extend the factorization approach discussed in the previous section to a class of Hamiltonian mechanical systems, and then apply the approach to solve the Toda lattice equations discussed in Chapter 4. Moreover, in Chapter 4 we have reviewed Hamilton's transformation approach for integrating the equations of motion by introducing a canonical transformation which can be generated by a generating function also known as Hamilton's principle function. This led to the Hamilton-Jacobi PDE which must be solved to obtain the required transformation generating function. However, as has been discussed in the previous sections, the HJE is very difficult to solve, except for the case when the Hamiltonian function is such that the equation is separable. It is therefore our objective in this section to present a method for solving the HJE for a class of Hamiltonian systems that may not admit a separation of variables.

13.3.1   Solving the Hamilton-Jacobi Equation

In this subsection, we propose an approach for solving the Hamilton-Jacobi equation for a fairly large class of Hamiltonian systems, and then apply the appproach to the $\mathcal{A}_2$ Toda-lattice as a special case. To present the approach, let the configuration space of the class of Hamiltonian systems be a smooth n-dimensional manifold M with local coordinates $q = (q_1, \ldots, q_n)$, i.e., if (φ, U) is a coordinate chart, we write $\varphi = q$ and $\dot{q}_i = \frac{\partial}{\partial q_i}$ in the tangent bundle $TM|_U = TU$. Further, let the class of systems under consideration be represented by Hamiltonian functions $H : T^*M \to \Re$ of the form:

$$H(q,p) = \frac{1}{2}\sum_{i=1}^{n} p_i^2 + V(q), \tag{13.58}$$

where $(p_1(q),\ldots,p_n(q)) \in T_q^*M$, and together with $(q_1,\ldots,q_n)$ form the $2n$ symplectic-coordinates for the phase-space T^*M of any system in the class, while $V : M \to \Re_+$ is the potential function which we assume to be nonseparable in the variables $q_i, i = 1,\ldots,n$. The time-independent HJE corresponding to the above Hamiltonian function is given by

$$\frac{1}{2}\sum_{i=1}^{n}\left(\frac{\partial W}{\partial q_i}\right)^2 + V(q) = h, \tag{13.59}$$

where $W : M \to \Re$ is the Hamilton's characteristic-function for the system, and h is the energy constant. We then have the following main theorem for solving the HJE.

Theorem 13.3.1 *Let M be an open subset of $\Re^n$ which is simply-connected[2] and let $q = (q_1,\ldots,q_n)$ be the coordinates on M. Suppose $\rho, \theta_i : M \to \Re$ for $i = 1,\ldots,\lfloor\frac{n+1}{2}\rfloor$; $\theta = (\theta_1,\cdots,\theta_{\lfloor\frac{n+1}{2}\rfloor})$; and $\zeta_i : \Re \times \Re^{\lfloor\frac{n+1}{2}\rfloor} \to \Re$ are C^2 functions such that*

$$\frac{\partial\zeta_i}{\partial q_j}(\rho(q),\theta(q)) = \frac{\partial\zeta_j}{\partial q_i}(\rho(q),\theta(q)), \quad \forall i,j - 1,\ldots,n, \tag{13.60}$$

and

$$\frac{1}{2}\sum_{i=1}^{n}\zeta_i^2(\rho(q),\theta(q)) + V(q) = h \tag{13.61}$$

is solvable for the functions ρ, θ. Let

$$\omega^1 = \sum_{i=1}^{n}\zeta_i(\rho(q),\theta(q))dq_i,$$

and suppose C is a path in M from an initial point q_0 to an arbitrary point $q \in M$. Then

(i) *ω^1 is closed;*

(ii) *ω^1 is exact;*

(iii) *if $W(q) = \int_C \omega^1$, then W satisfies the HJE (13.59).*

Proof: (i)

$$d\omega^1 = \sum_{j=1}^{n}\sum_{i=1}^{n}\frac{\partial}{\partial q_j}\zeta_i(\rho(q),\theta(q))dq_j \wedge dq_i,$$

which by (13.60) implies $d\omega^1 = 0$. Hence, ω^1 is closed.[3] (ii) Since by (i) ω^1 is closed, then by the simple-connectedness of M, ω^1 is also exact. (iii) By (ii) ω^1 is exact. Therefore, the integral $W(q) = \int_C \omega^1$ is independent of the path C, and W corresponds to a scalar function. Furthermore, $dW = \omega^1$ and $\frac{\partial W}{\partial q_i} = \zeta_i(\rho(q),\theta(q))$. Thus, substituting this in the HJE (13.59) and if (13.61) holds, then W satisfies it. $\square$

---

[2]A subset of a set is simply connected if a loop inside it can be continuously shrinked to a point.
[3]A 1-form $\sigma : TM \to \Re$ is closed if $d\sigma = 0$. It is exact if $\sigma = dS$ for some smooth function $S : M \to \Re$.

In the next corollary we shall construct explicitly the functions $\zeta_i, i = 1, \ldots, n$ in the above theorem.

Corollary 13.3.1 *Assume the dimension n of the system is 2, and M, ρ, θ are as in the hypotheses of Theorem 13.3.1, and that conditions (13.60), (13.61) are solvable for θ and ρ. Also, define the functions $\zeta_i, i = 1, 2$ postulated in the theorem by $\zeta_1(q) = \rho(q) \cos \theta(q)$, $\zeta_2(q) = \rho(q) \sin \theta(q)$. Then, if*

$$\omega^1 = \sum_{i=1}^{2} \zeta_i(\rho(q), \theta(q)) dq_i,$$

$W = \int_C \omega^1$, and $q : [0,1] \to M$ is a parametrization of C such that $q(0) = q_0$, $q(1) = q$, then

(i) *W is given by*

$$W(q, h) = \gamma \int_0^1 \sqrt{(h - V(q(s)))} \left[\cos \theta(q(s)) q_1'(s) + \sin \theta(q(s)) q_2'(s) \right] ds \qquad (13.62)$$

where $\gamma = \pm\sqrt{2}$ and $q_i' = \frac{dq_i(s)}{ds}$.

(ii) *W satisfies the HJE (13.59).*

Proof: (i) If (13.60) is solvable for the function θ, then substituting the functions $\zeta_i(\rho(q), \theta(q)), i = 1, 2$ as defined above in (13.61), we get immediately

$$\rho(q) = \pm\sqrt{2(h - V(q))}.$$

Further, by Theorem 13.3.1, ω^1 given above is exact, and $W = \int_C \omega^1 dq$ is independent of the path C. Therefore, if we parametrize the path C by s, then the above line integral can be performed coordinate-wise with W given by (13.62) and $\gamma = \pm\sqrt{2}$. (ii) follows from Theorem 13.3.1. $\square$

Remark 13.3.1 *The above corollary gives one explicit parametrization that may be used. However, because the number of parameters available in the parametrization are limited, the above parametrization is only suitable for systems with $n = 2$. Other types of parametrizations that are suitable could also be employed.*

If however the dimension n of the system is 3, then the following corollary gives another parametrization for solving the HJE.

Corollary 13.3.2 *Assume the dimension n of the system is 3, and M, ρ, are as in the hypotheses of Theorem 13.3.1. Let $\zeta_i : \Re \times \Re \times \Re \to \Re$, $i = 1, 2, 3$ be defined by $\zeta_1(q) = \rho(q) \sin \theta(q) \cos \varphi(q)$, $\zeta_2(q) = \rho(q) \sin \theta(q) \sin \varphi(q)$, $\zeta_3(q) = \rho(q) \cos \theta(q)$, and assume (13.60) are solvable for θ and φ, while (13.61) is solvable for ρ. Then, if*

$$\omega^1 = \sum_{i=1}^{3} \zeta_i(\rho(q), \theta, \varphi) dq_i,$$

$W = \int_C \omega^1$, and $q : [0,1] \to M$ is a parametrization of C such that $q(0) = q_0$, $q(1) = q$, then

(i) *W is given by*

$$W(q,h) = \gamma \int_0^1 \sqrt{(h - V(q(s)))} \Big\{ \sin\theta(q(s))\cos\varphi(q(s))q_1'(s) +$$

$$\sin\theta(q(s))\sin\varphi(q(s))q_2'(s) + \cos\theta(q(s))q_3'(s) \Big\} ds, \qquad (13.63)$$

where $\gamma = \pm\sqrt{2}$.

(ii) *W satisfies the HJE (13.59).*

Proof: Proof follows along the same lines as Corollary 13.3.1. $\square$

Remark 13.3.2 *Notice that the parametrization employed in the above corollary is now of a spherical nature.*

If the HJE (13.59) is solvable, then the dynamics of the system evolves on the n-dimensional *Lagrangian-submanifold* $\tilde{N}$ which is an immersed-submanifold of maximal dimension, and can be locally parametrized as the graph of the function W, i.e.,

$$\tilde{N} = \{(q, \frac{\partial W}{\partial q}) \,|\, q \in N \subset M, \, W \text{ is a solution of HJE (13.59)}\}.$$

Moreover, for any other solution W' of the HJE, the volume enclosed by this surface is invariant. This is stated in the following proposition.

Proposition 13.3.1 *Let $N \subset M$ be the region in M where the solution W of the HJE (13.59) exists. Then, for any orientation of M, the volume-form of $\tilde{N}$*

$$\omega^n = \left(\sqrt{1 + \sum_{j=1}^n (\frac{\partial W}{\partial q_j})^2}\right) dq_1 \wedge dq_2 \ldots \wedge dq_n$$

is given by

$$\omega^n = \left(\sqrt{1 + 2(h - V(q))}\right) dq_1 \wedge dq_2 \ldots \wedge dq_n.$$

Proof: From the HJE (13.59), we have

$$\sqrt{1 + \sum_{j=1}^n (\frac{\partial W}{\partial q_j})^2} = \sqrt{1 + 2(h - V(q))}, \quad \forall q \in N$$

$$\Updownarrow$$

$$\omega^n = \left(\sqrt{1 + \sum_{j=1}^n (\frac{\partial W}{\partial q_j})^2}\right) dq_1 \wedge \ldots \wedge dq_n = \left(\sqrt{1 + 2(h - V(q))}\right) dq_1 \wedge \ldots \wedge dq_n. \square$$

We now apply the above ideas to solve the HJE for the two-particle nonperiodic $\mathcal{A}_2$ Toda-lattice described in Chapter 4.

13.3.2   Solving the Hamilton-Jacobi Equation for the $\mathcal{A}_2$-Toda System

Recall the Hamiltonian function and the canonical equations for the nonperiodic Toda lattice from Chapter 4:

$$H(q,p) = \frac{1}{2}\sum_{j=1}^{n} p_j^2 + \sum_{j=1}^{n-1} e^{2(q_j - q_{j+1})}. \tag{13.64}$$

Thus, the canonical equations for the system are given by

$$\left.\begin{array}{rcl}
\frac{dq_j}{dt} & = & p_j \quad j = 1, \ldots, n, \\
\frac{dp_1}{dt} & = & -2e^{2(q_1 - q_2)}, \\
\frac{dp_j}{dt} & = & -2e^{2(q_j - q_{j+1})} + 2e^{2(q_{j-1} - q_j)}, \quad j = 2, \ldots, n-1, \\
\frac{dp_n}{dt} & = & 2e^{2(q_{n-1} - q_n)}.
\end{array}\right\} \tag{13.65}$$

Consequently, the two-particle system (or $\mathcal{A}_2$ system) is given by the Hamiltonian (13.64):

$$H(q_1, q_2, p_1, p_2) = \frac{1}{2}(p_1^2 + p_2^2) + e^{2(q_1 - q_2)}, \tag{13.66}$$

and the HJE corresponding to this system is given by

$$\frac{1}{2}\left\{\left(\frac{\partial W}{\partial q_1}\right)^2 + \left(\frac{\partial W}{\partial q_2}\right)^2\right\} + e^{2(q_1 - q_2)} = h_2. \tag{13.67}$$

The following proposition then gives a solution of the above HJE corresponding to the $\mathcal{A}_2$ system.

Proposition 13.3.2 *Consider the HJE (13.67) corresponding to the $\mathcal{A}_2$ Toda lattice. Then, a solution for the HJE is given by*

$$\begin{aligned}
W(q_1', q_2', h_2) & = \cos\left(\frac{\pi}{4}\right)\int_1^{q_1'} \rho(q)dq_1 + m\sin\left(\frac{\pi}{4}\right)\int_1^{q_1'} \rho(q)dq_1 \\
& = (1+m)\left\{\frac{\sqrt{h_2 - e^{-2(b+m-1)}} - \sqrt{h_2}\tanh^{-1}\left[\frac{\sqrt{h_2 - e^{-2(b+m-1)}}}{\sqrt{h_2}}\right]}{m-1} - \right. \\
& \quad \left. \frac{\sqrt{h_2 - e^{-2b - 2(m-1)q_1'}} - \sqrt{h_2}\tanh^{-1}\left[\frac{\sqrt{h_2 - e^{-2b - 2(m-1)q_1'}}}{\sqrt{h_2}}\right]}{m-1}\right\}, \quad q_1 > q_2
\end{aligned}$$

and

$$\begin{aligned}
W(q_1', q_2', h_2) & = \cos\left(\frac{\pi}{4}\right)\int_1^{q_1'} \rho(q)dq_1 + m\sin\left(\frac{\pi}{4}\right)\int_1^{q_1'} \rho(q)dq_1 \\
& = (1+m)\left\{\frac{\sqrt{h_2 - e^{-2(b-m+1)}} - \sqrt{h_2}\tanh^{-1}\left[\frac{\sqrt{h_2 - e^{-2(b-m+1)}}}{\sqrt{h_2}}\right]}{m-1} - \right. \\
& \quad \left. \frac{\sqrt{h_2 - e^{-2b + 2(1-m)q_1'}} - \sqrt{h_2}\tanh^{-1}\left[\frac{\sqrt{h_2 - e^{-2b + 2(1-m)q_1'}}}{\sqrt{h_2}}\right]}{m-1}\right\}, \quad q_2 > q_1.
\end{aligned}$$

Furthermore, a solution for the system equations (13.65) for the $\mathcal{A}_2$ with the symmetric initial $q_1(0) = -q_2(0)$ and $\dot{q}_1(0) = \dot{q}_2(0) = 0$ is

$$q(t) = -\frac{1}{2}\log\sqrt{h_2} + \frac{1}{2}\log[\cosh 2\sqrt{h_2}(\beta - t)], \tag{13.68}$$

where h_2 is the energy constant and

$$\beta = \frac{1}{2\sqrt{h_2}} \tanh^{-1}\left(\frac{2\dot{q}_1^2(0)}{\sqrt{2h_2}}\right).$$

Proof: Applying the results of Theorem 13.3.1, we have $\frac{\partial W}{\partial q_1} = \rho(q)\cos\theta(q)$, $\frac{\partial W}{\partial q_2} = \rho(q)\sin\theta(q)$. Substituting this in the HJE (13.67), we immediately get

$$\rho(q) = \pm\sqrt{2(h_2 - e^{2(q_1 - q_2)})}$$

and

$$\rho_{q_2}(q)\cos\theta(q) - \theta_{q_2}\rho(q)\sin\theta(q) = \rho_{q_1}(q)\sin\theta(q) + \theta_{q_1}\rho(q)\cos\theta(q). \tag{13.69}$$

The above equation (13.69) is a first-order PDE in θ and can be solved by the method of *characteristics* developed in Chapter 4. However, the geometry of the system allows for a simpler solution. We make the simplifying assumption that θ is a constant function. Consequently, equation (13.69) becomes

$$\rho_{q_2}(q)\cos\theta = \rho_{q_1}(q)\sin\theta \Rightarrow \tan\theta = \frac{\rho_{q_2}(q)}{\rho_{q_1}(q)} = -1 \Rightarrow \theta = -\frac{\pi}{4}.$$

Thus,

$$p_1 = \rho(q)\cos(\frac{\pi}{4}),$$
$$p_2 = -\rho(q)\sin(\frac{\pi}{4}),$$

and integrating dW along the straightline path from $(1,-1)$ on the line segment

$$L: \quad q_2 = \frac{q_2'+1}{q_1'-1}q_1 + (1 + \frac{q_2'+1}{q_1'-1}) \triangleq mq_1 + b$$

to some arbitrary point (q_1', q_2') we get

$$
\begin{aligned}
W(q_1', q_2', h_2) &= \cos(\frac{\pi}{4})\int_1^{q_1'}\rho(q)dq_1 + m\sin(\frac{\pi}{4})\int_1^{q_1'}\rho(q)dq_1 \\
&= (1+m)\Bigg\{\frac{\sqrt{h_2 - e^{-2(b+m-1)}} - \sqrt{h_2}\tanh^{-1}[\frac{\sqrt{h_2 - e^{-2(b+m-1)}}}{\sqrt{h_2}}]}{m-1} - \\
&\qquad \frac{\sqrt{h_2 - e^{-2b-2(m-1)q_1'}} - \sqrt{h_2}\tanh^{-1}[\frac{\sqrt{h_2 - e^{-2b-2(m-1)q_1'}}}{\sqrt{h_2}}]}{m-1}\Bigg\}.
\end{aligned}
$$

Similarly, if we integrate from point $(-1,1)$ to (q_1', q_2'), we get

$$
\begin{aligned}
W(q_1', q_2', h_2) &= \cos(\frac{\pi}{4})\int_{-1}^{q_1'}\rho(q)dq_1 + m\sin(\frac{\pi}{4})\int_{-1}^{q_1'}\rho(q)dq_1 \\
&= (1+m)\Bigg\{\frac{\sqrt{h_2 - e^{-2(b-m+1)}} - \sqrt{h_2}\tanh^{-1}[\frac{\sqrt{h_2 - e^{-2(b-m+1)}}}{\sqrt{h_2}}]}{m-1} - \\
&\qquad \frac{\sqrt{h_2 - e^{-2b+2(1-m)q_1'}} - \sqrt{h_2}\tanh^{-1}[\frac{\sqrt{h_2 - e^{-2b+2(1-m)q_1'}}}{\sqrt{h_2}}]}{m-1}\Bigg\}.
\end{aligned}
$$

Finally, from (4.11) and (13.66), we can write

$$\dot{q}_1 = p_1 = \rho(q)\cos\left(\frac{\pi}{4}\right),$$
$$\dot{q}_2 = p_2 = -\rho(q)\sin\left(\frac{\pi}{4}\right).$$

Then $\dot{q}_1 + \dot{q}_2 = 0$ which implies that $q_1 + q_2 = k$, a constant, and by our choice of initial conditions, $k = 0$ or $q_1 = -q_2 = -q$. Now integrating the above equations from $t = 0$ to t we get

$$\frac{1}{2\sqrt{h_2}}\tanh^{-1}\left(\frac{\rho(q)}{\sqrt{2h_2}}\right) = \frac{1}{2\sqrt{h_2}}\tanh^{-1}\left(\frac{\rho(q(0))}{\sqrt{2h_2}}\right) - t,$$
$$\frac{1}{2\sqrt{h_2}}\tanh^{-1}\left(\frac{\rho(q)}{\sqrt{2h_2}}\right) = -\frac{1}{2\sqrt{h_2}}\tanh^{-1}\left(\frac{\rho(q(0))}{\sqrt{2h_2}}\right) + t.$$

Then, if we let

$$\beta = \frac{1}{2\sqrt{h_2}}\tanh^{-1}\left(\frac{\rho(q(0))}{\sqrt{2h_2}}\right),$$

and upon simplification, we get

$$
\begin{aligned}
q_1 - q_2 &= \frac{1}{2}\log\left[h_2\left(1 - \tanh^2 2\sqrt{h_2}(\beta - t)\right)\right] \\
&= \frac{1}{2}\log\left[h_2\, sech^2\, 2\sqrt{h_2}(\beta - t)\right].
\end{aligned}
\tag{13.70}
$$

Therefore,

$$
\begin{aligned}
q(t) &= -\frac{1}{2}\log\sqrt{h_2} - \frac{1}{2}\log[sech\, 2\sqrt{h_2}(\beta - t)] \\
&= -\frac{1}{2}\log\sqrt{h_2} + \frac{1}{2}\log[cosh\, 2\sqrt{h_2}(\beta - t)].
\end{aligned}
$$

Now, from (13.67) and (13.70), (13.70),

$$\rho(q(0)) = \dot{q}_1^2(0) + \dot{q}_2^2(0),$$

and in particular, if $\dot{q}_1(0) = \dot{q}_2(0) = 0$, then $\beta = 0$. Consequently,

$$q(t) = -\frac{1}{2}\log\sqrt{h_2} + \frac{1}{2}\log(cosh\, 2\sqrt{h_2}t) \tag{13.71}$$

which is of the form (4.57) with $v = \sqrt{h}$. $\square$

13.4   Notes and Bibliography

The material of Section 13.1 is based on the References [193, 112, 264], and practical application of the approach to a flexible robot link can be found in the Reference [286]. While the material of Section 13.1.1 is based on Reference [71]. Similar approach using Galerkin's approximation can be found in [63], and an extension of the approach to the DHJIE can also be found in [125].

The factorization approach presented in Section 13.2 is based on the author's contribution [10], and extension of the approach to stochastic systems can be found also in [11]. This approach is promising and is still an active area of research.

Furthermore, stochastic HJBEs are extensively discussed in [287, 98, 91, 167], and numerical algorithms for solving them can be found in [173].

Various practical applications of nonlinear $\mathcal{H}_\infty$-control and methods for solving the HJIE can be found in the References [46, 80, 87, 154, 155, 286]. An alternative method of solving the HJIE using backstepping is discussed in [96], while other methods of optimally controlling affine nonlinear systems that do not use the HJE, in particular using inverse-optimality, can be found in [91, 167]-[169].

The material of Section 13.3 is also solely based on the author's contribution [16]. More general discussions of HJE applied to mechanical systems can be found in the well-edited books [1, 38, 127, 115, 122, 200]. While the Toda lattice is discussed also extensively in the References [3]-[5], [72, 123, 201].

Generalized and viscosity solutions of Hamilton-Jacobi equations which may not necessarily be smooth are discussed extensively in the References [56, 83, 95, 97, 98, 186].

Finally, we remark that solving HJEs is in general still an active area of research, and many books have been written on the subject. There is not still a single approach that could be claimed to be satisfactory, all approaches have their advantages and disadvantages.

A

Proof of Theorem 5.7.1

The proof uses a back-stepping procedure and an inductive argument. Thus, under the assumptions (i), (ii) and (iv), a global change of coordinates for the system exists such that the system is in the strict-feedback form [140, 153, 199, 212]. By augmenting the ρ linearly independent set

$$z_1 = h(x), \ z_2 = L_f h(x), \ldots, z_\rho = L_f^{\rho-1} h(x)$$

with an arbitrary $n-\rho$ linearly independent set $z_{\rho+1} = \psi_{\rho+1}(x), \ldots, z_n = \psi_n$ with $\psi_i(0) = 0$, $\langle d\psi_i, \mathcal{G}_{\rho-1} \rangle = 0$, $\rho + 1 \leq i \leq n$. Then the state feedback

$$u = \frac{1}{L_{g_2} L_f^{\rho-1} h(x)}(v - L_f^\rho h(x))$$

globally transforms the system into the form:

$$\left. \begin{array}{rcl} \dot{z}_i & = & z_{i+1} + \Psi_i^T(z_1, \ldots, z_i)w \quad 1 \leq i \leq \rho - 1, \\ \dot{z}_\rho & = & v + \Psi_\rho^T(z_1, \ldots, z_\rho)w \\ \dot{z}_\mu & = & \psi(z) + \Xi^T(z)w, \\ y & = & z_1, \end{array} \right\} \tag{A.1}$$

where $z_\mu = (z_{\rho+1}, \ldots, z_n)$. Moreover, in the z-coordinates we have

$$\mathcal{G}_j = span\{\frac{\partial}{\partial z_{\rho-j}}, \ldots, \frac{\partial}{\partial z_\rho}\}, \ 0 \leq j \leq \rho - 1,$$

so that by condition (iii) the system equations (A.1) can be represented as

$$\left. \begin{array}{rcl} \dot{z}_i & = & z_{i+1} + \Psi_i^T(z_1, \ldots, z_i, z_\mu)w \quad 1 \leq i \leq \rho - 1, \\ \dot{z}_\rho & = & v + \Psi_\rho^T(z_1, \ldots, z_\rho, z_\mu)w \\ \dot{z}_\mu & = & \psi(z_1, z_\mu) + \Xi^T(z_1, z_\mu)w. \end{array} \right\} \tag{A.2}$$

Now, define

$$z_2^\star = -z_1 - \frac{1}{4}\lambda z_1(1 + \Psi_1^T(z_1, z_\mu)\Psi_1(z_1, z_\mu))$$

and set $z_2 = z_2^\star(z_1, z_\mu, \lambda)$ in (A.2). Then consider the time derivative of the function:

$$V_1 = \frac{1}{2}z_1^2$$

along the trajectories of the closed-loop system:

$$
\begin{aligned}
\dot{V}_1 &= -z_1^2 - \frac{1}{4}\lambda z_1^2(1 + \Psi_1^T\Psi_1) + z_1\Psi_1^T w \\
&= -z_1^2 - \lambda\left[\frac{1}{4}z_1^2(1 + \Psi_1^T\Psi_1) + z_1\Psi_1^T\lambda - \frac{z_1}{\lambda}\Psi_1^T w + \frac{(\Psi_1^T w)^2}{\lambda^2(1 + \Psi_1^T\Psi_1)}\right] + \\
&\quad \frac{(\Psi_1^T w)^2}{\lambda(1 + \Psi_1^T\Psi_1)} \\
&= -z_1^2 - \lambda\left(\frac{1}{2}z_1\sqrt{1 + \Psi_1^T\Psi_1} - \frac{\Psi_1^T w}{\lambda\sqrt{1 + \Psi_1^T\Psi_1}}\right)^2 + \frac{(\Psi_1^T w)^2}{\lambda(1 + \Psi_1^T\Psi_1)} \\
&\leq -z_1^2 + \frac{(\Psi_1^T w)^2}{\lambda(1 + \Psi_1^T\Psi_1)} \leq -z_1^2 + \frac{1}{\lambda}\frac{\|\Psi_1^T\|^2}{1 + \|\Psi_1^T\|^2}\|w\|^2 \\
&\leq -z_1^2 + \frac{1}{\lambda}\|w\|^2.
\end{aligned}
\tag{A.3}
$$

Now if $\rho = 1$, then we set $v = z_2^\star(z_1, z_\mu, \lambda)$ and integrating (A.3) from t_0 with $z(t_0) = 0$ to some t, we have

$$
V_1(x(t)) - V_1(0) \leq -\int_{t_0}^t y^2(\tau)d\tau + \frac{1}{\lambda}\int_{t_0}^t \|w(t)\|^2 d\tau
$$

and since $V_1(0) = 0$, $V_1(x) \geq 0$, the above inequality implies that

$$
\int_{t_0}^t y^2(\tau)d\tau \leq \frac{1}{\lambda}\int_{t_0}^t \|w(\tau)\|^2 d\tau.
$$

For $\rho > 1$, we first prove the following lemma.

Lemma A.0.1 *Suppose for some index i, $1 \leq i \leq \rho$, for the system*

$$
\left.
\begin{aligned}
\dot{z}_1 &= z_2 + \Psi_1^T(z_1, z_\mu)w \\
&\;\;\vdots \\
\dot{z}_i &= z_{i+1} + \Psi_i^T(z_1, \ldots, z_i, z_\mu)w
\end{aligned}
\right\},
\tag{A.4}
$$

there exist i functions

$$
z_j^\star = z_j^\star(z_1, \ldots, z_{j-1}, \lambda), \quad z_j^\star = z^\star(0, \ldots, 0, \lambda) = 0, \quad 2 \leq j \leq i+1,
$$

such that the function

$$
V_i = \frac{1}{2}\sum_{j=1}^i \tilde{z}_j^2,
$$

where

$$
\tilde{z}_1 = z_1, \quad \tilde{z}_j = z_j - z_j^\star(z_1, \ldots, z_{j-1}, z_\mu, \lambda), \quad 2 \leq j \leq i,
$$

has time derivative

$$
\dot{V}_i = -\sum_{j=1}^i \tilde{z}_j^2 + \frac{c}{\lambda}\|w\|^2
$$

along the trajectories of (A.4) with $z_{i+1} = z_{i+1}^\star$ and for some real number $c > 0$. Then, for the system

$$
\left.
\begin{aligned}
\dot{z}_1 &= z_2 + \Psi_1^T(z_1, z_\mu)w \\
&\;\;\vdots \\
\dot{z}_i &= z_{i+2} + \Psi_{i+1}^T(z_1, \ldots, z_{i+1}, z_\mu)w
\end{aligned}
\right\},
\tag{A.5}
$$

there exists a function

$$z_{i+2}^{\star}(z_1, \ldots, z_{i+1}, z_r, \lambda), \quad z_{i+2}^{\star}(0, \ldots, \lambda) = 0$$

such that the function

$$V_{i+1} = \frac{1}{2} \sum_{j=1}^{i+1} \tilde{z}_j^2$$

where

$$\tilde{z}_j = z_j - z_j^{\star}(z_1, \ldots, z_i, z_\mu, \lambda), \quad 1 \le j \le i+1,$$

has time derivative

$$\dot{V}_i = -\sum_{j=1}^{i+1} \tilde{z}_j^2 + \frac{c+1}{\lambda} \|w\|^2$$

along the trajectories of (A.5) with $z_{i+2} = z_{i+2}^{\star}$.

Proof: Consider the function

$$V_{i+1} = \frac{1}{2} \sum_{j=1}^{i+1} \tilde{z}_j^2$$

with $z_{i+2} = z_{i+2}^{\star}(z_1, \ldots, z_{i+1}, z_\mu, \lambda)$ in (A.4), and using the assumption in the lemma, we have

$$
\begin{aligned}
\dot{V}_{i+1} \le \ & -\sum_{j=1}^{i} \tilde{z}_j^2 + \frac{c}{\lambda} \|w\|^2 + \tilde{z}_{i+1}\Big(\tilde{z}_i + \Psi_{i+1}^T w - \sum_{j=1}^{i} \frac{\partial z_{i+1}^{\star}}{\partial z_j}(z_{j+1} + \Psi_j^T w) - \\
& \frac{\partial z_{i+1}^{\star}}{\partial z_\mu}(\psi + \Xi^T w) + z_{i+2}^{\star} \Big).
\end{aligned}
\tag{A.6}
$$

Let

$$\alpha_1(z_1, \ldots, z_{i+1}, z_\mu) = \tilde{z}_i - \sum_{j=1}^{i} \frac{\partial z_{i+1}^{\star}}{\partial z_j} z_{j+1} - \frac{\partial z_{i+1}^{\star}}{\partial z_\mu} \psi,$$

$$\alpha_2(z_1, \ldots, z_{i+1}, z_\mu) = \Psi_{i+1} - \sum_{j=1}^{i} \frac{\partial z_{i+1}^{\star}}{\partial z_j} \Psi_j - \Xi\left(\frac{\partial z_{i+1}^{\star}}{\partial z_\mu} \psi \right)^T,$$

$$z_{i+2}^{\star}(z_1, \ldots, z_{i+1}, z_\mu) = -\alpha_1 - \tilde{z}_{i+1} - \frac{1}{4}\lambda \tilde{z}_{i+1}(1 + \alpha_2^T \alpha_2)$$

and subsituting in (A.6), we get

$$
\begin{aligned}
\dot{V}_{i+1} \;\leq\;& -\sum_{j}^{i+1} \tilde{z}_j^2 + \tilde{z}_{i+1}\alpha_2^T w - \frac{1}{4}\lambda \tilde{z}_{i+1}^2 (1+\alpha_2^T\alpha_2) + \frac{c}{\lambda}\|w\|^2 \\[2mm]
=\;& -\sum_{j=1}^{i+1} \tilde{z}_j^2 - \lambda \left(\frac{1}{4}\tilde{z}_{i+1}^2 (1+\alpha_2^T\alpha_2) - \frac{\tilde{z}_{i+1}}{\lambda}\alpha_2^T w + \frac{(\alpha_2^T w)^2}{\lambda^2(1+\alpha_2^T\alpha_2)} \right) + \\[2mm]
& \frac{(\alpha_2^T w)^2}{\lambda(1+\alpha_2^T\alpha_2)} + \frac{c}{\lambda}\|w\|^2 \\[2mm]
=\;& -\sum_{j=1}^{i+1} \tilde{z}_j^2 - \lambda \left(\frac{1}{2}\tilde{z}_{i+1}\sqrt{1+\alpha_2^T\alpha_2} - \frac{\alpha_2^T w}{\lambda\sqrt{1+\alpha_2^T\alpha_2}} \right)^2 + \frac{(\alpha_2^T w)^2}{\lambda(1+\alpha_2^T\alpha_2)} + \\[2mm]
& \frac{c}{\lambda}\|w\|^2 \\[2mm]
\leq\;& -\sum_{j=1}^{i+1} \tilde{z}_j^2 + \frac{c+1}{\lambda}\|w\|^2
\end{aligned}
$$

as claimed. $\square$

To conclude the proof of the theorem, we note that the result of the lemma holds for $\rho = 1$ with $c = 1$. We now apply the result of the lemma $(\rho - 1)$ times to arrive at the feedback contol

$$
v = z_{\rho+1}^*(z_1, \ldots, z_\rho, z_\mu, \lambda)
$$

and the function

$$
V_\rho = \frac{1}{2}\sum_{j=1}^{\rho} \tilde{z}_j^2
$$

has time derivative

$$
\dot{V}_\rho = -\sum_{j=1}^{\rho} \tilde{z}_j^2 + \frac{\rho}{\lambda}\|w\|^2 \tag{A.7}
$$

along the trajectories of (A.4) with $z_{i+1} = z_{i+1}^*(z_1, \ldots, z_\rho, z_\mu, \lambda)$, $1 \leq i \leq \rho$.

Finally, integrating (A.7) from t_0 and $z(t_0) = 0$ to some t, we get

$$
V(x(t)) - V(x(t_0)) \leq -\int_{t_0}^{t} y^2(\tau)d\tau - \sum_{j=2}^{\rho}\int_{t_0}^{t}\tilde{z}_j^2(\tau)d\tau + \frac{\rho}{\lambda}\int_{t_0}^{t}\|w(\tau)\|^2 d\tau,
$$

and since $V_\rho(0) = 0$, $V_\rho(x) \geq 0$, we have

$$
\int_{t_0}^{t} y^2(\tau)d\tau \leq \frac{\rho}{\lambda}\int_{t_0}^{t}\|w(\tau)\|^2 d\tau,
$$

and the $\mathcal{L}_2$-gain can be made arbitrarily small since λ is arbitrary. This concludes the proof of the theorem. $\square$

B

Proof of Theorem 8.2.2

We begin with a preliminary lemma.

Consider the nonlinear time-varying system:

$$\dot{x}(t) = \phi(x(t), w(t)), \quad x(0) = x_0 \tag{B.1}$$

where $x(t) \in \Re^n$ is the state of the system and $w(t) \in \Re^r$ is an input, together with the cost functional:

$$J_i(x(.), w(.)) = \int_0^\infty \mu_i(x(t), w(t)) dt, \quad i = 0, \dots, l. \tag{B.2}$$

Suppose the following assumptions hold:

(a1) $\phi(., .)$ is continuous with respect to both arguments;

(a2) For all $\{x(t), w(t)\} \in \mathcal{L}_2[0, \infty)$, the integrals (B.2) above are bounded:

(a3) For any given $\epsilon > 0$, $w \in \mathcal{W}$, there exists a $\delta > 0$ such that for all $\|x_0\| \le \delta$, it implies:

$$|J_i(x_1(t), w(t)) - J_i(x_2(t), w(t))| < \epsilon, \quad i = 0, \dots, l,$$

where $x_1(.)$, $x_2(.)$, are any two trajectories of the system corresponding to the initial conditions $x_0 \ne 0$.

We then have the following lemma [236].

Lemma B.0.2 *Consider the nonlinear system (B.1) together with the integral functionals (B.2) satisfying the Assumptions (a1)-(a3). Then $J_0(x(.), w(.)) \ge 0 \ \forall \{x(.), w(.)\} \in \mathcal{X} \times \mathcal{W}$ subject to $J_i(x(.), w(.)) \ge 0$, $i = 1, \dots, l$ if, and only if, there exist a set of numbers $\tau_i \ge 0$, $i = 0, \dots, l$, $\sum_{i=0}^l \tau_i > 0$ such that*

$$\tau_0 J_0(x(.), w(.)) \ge \sum_{i=1}^l \tau_i J_i(x(.), w(.))$$

for all $\{x(.), w(.)\} \in \mathcal{X} \times \mathcal{W}$, $x(.)$ a trajectory of (B.1).

We now present the proof of the Theorem.

Combine the system (8.25) and filter (8.26) dynamics into the augmented system:

$$\Sigma_{f,0}^{a,\Delta} : \begin{cases} \dot{x} &= f(x) + H_1(x)v + g_1(x)w; \quad x(0) = 0 \\ \dot{\xi} &= a(\xi) + b(\xi)h_2(x) + b(\xi)H_2(x)v + b(\xi)k_{21}(x)w; \quad \xi(0) = 0 \\ z &= h_1(x) \\ \hat{z} &= c(\xi) \\ y &= h_2(x) + H_2(x)v + k_{21}(x)w \end{cases} \tag{B.3}$$

363

where $v \in \mathcal{W}$ is an auxiliary disturbance signal that excites the uncertainties. The objective is to render the following functional:

$$J_0(x(.), \xi(.), v(.)) = \int_0^\infty (\|w\|^2 - \|\tilde{z}\|^2)dt \geq 0. \tag{B.4}$$

Let $J_1(x(.), \xi(.), v(.))$ represent also the integral quadratic constraint:

$$J_1(x(.), \xi(.), v(.)) = \int_0^\infty (\|E(x)\|^2 - \|v(t)\|^2)dt \geq 0.$$

Then, by Lemma B.0.2, the Assumption (a1) holds for all $x(.)$, $w(.)$ and $v(.)$ satisfying (a2) if, and only if, there exists numbers τ_0, τ_1, $\tau_0 + \tau_1 > 0$, such that

$$\tau_0 J_0(.) - \tau_1 J_1(.) \geq 0 \quad \forall w(.), v(.) \in \mathcal{W}$$

for all $x(.)$, $\xi(.)$ satisfying (a1). It can be shown that both τ_0 and τ_1 must be positive by considering the following two cases:

1. $\tau_1 = 0$. Then $\tau_0 > 0$ and $J_0 \geq 0$. Now set $w = 0$ and by (a3) we have that $\tilde{z} = 0 \; \forall t \geq 0$ and $v(.)$. However this contradicts Assumption (a2), and therefore $\tau_1 > 0$.

2. $\tau_0 = 0$. Then $J_1 < 0$. But this immediately violates the assertion of the lemma. Hence, $\tau_0 = 0$.

Therefore, there exists $\tau > 0$ such that

$$J_0(.) - \tau J_1(.) \geq 0 \quad \forall w(.), v(.) \in \mathcal{W} \tag{B.5}$$

$$\Updownarrow \tag{B.6}$$

$$\int_0^\infty \left(\left\| \begin{bmatrix} w(t) \\ \tau v(t) \end{bmatrix} \right\|^2 - \left\| \begin{bmatrix} h_1(x) \\ \tau E(x) \end{bmatrix} - \begin{bmatrix} c(\xi) \\ 0 \end{bmatrix} \right\|^2 \right) \geq 0 \tag{B.7}$$

We now prove the necessity of the Theorem.

(Necessity:) Suppose (8.27) holds for the augmented system (B.3), i.e.,

$$\int_0^T \|z(t) - \hat{z}(t)\|^2 dt \leq \gamma^2 \int_0^T \|w(t)\|^2 dt \quad \forall w \in \mathcal{W}. \tag{B.8}$$

Then we need to show that

$$\int_0^T \|z_s(t) - \hat{z}_s(t)\|^2 dt \leq \gamma^2 \int_0^T \|w(t)\|^2 dt \quad \forall w_s \in \mathcal{W}. \tag{B.9}$$

for the scaled system. Accordingly, let

$$w_s = 0 \quad \forall t > T$$

without any loss of generality. Then choosing

$$\begin{bmatrix} w(.) \\ \tau v(.) \end{bmatrix} = w_s$$

we seek to make the trajectory of (B.3) identical to that of (8.29) with the filter (8.30). The result now follows from (B.7).

(Sufficiency:) Conversely, suppose (B.9) holds for the augmented system (B.3). Then we need to show that (B.8) holds for the system (8.25). Indeed, for any $w, v \in \mathcal{W}$, we can assume $w(t) = 0 \, \forall t > T$ without any loss of generality. Choosing now

$$w_s = \left[\begin{array}{c} w(.) \\ \tau v(.) \end{array} \right] \quad \forall t \in [0, T]$$

and $w_s(t) = 0 \, \forall t > T$. Then, (B.9) implies (B.7) and hence (B.4). Since $w(.)$ is truncated, we get (B.8). $\square$

Bibliography

[1] R. Abraham, and J. Marsden. *Foundations of Mechanics*. Benjamin Cummings, 2nd Ed., Massachusetts, 1978.

[2] H. Abou-Kandil, G. Frieling, and G. Jank, "Solutions and asymptotic behavior of coupled Riccati equations in jump linear systems," IEEE Transactions on Automatic Control, vol. 39, no. 8, pp. 1631-1636, 1994.

[3] M. Adler and P. van Moerbeke, "Completely integrable systems, Euclidean lie algebras, and curves," Advances in Math., 38, 267-317, 1980.

[4] M. Adler and P. van Moerbeke, "Linearization of Hamiltonian systems, Jacobi varieties, and representation theory," Advances in Math., 38, 318-379, 1980.

[5] M. Adler and P. van Moerbeke, "The Toda lattice, Dynkin diagrams, singularities, and Abelian varieties," Inventiones in Math., 103, 223-278, 1991.

[6] M. D. S. Aliyu, "Robust $\mathcal{H}_\infty$-control of nonlinear systems under matching," in Proceedings, Conference on Decision and Control, Tampa, Florida, 1998.

[7] M. D. S. Aliyu, "Minimax guaranteed cost control of nonlinear systems," Int. J. of Control, vol. 73, no. 16, pp. 1491-1499, 2000.

[8] M. D. S. Aliyu, "Dissipativity and stability of nonlinear jump systems," in Proceedings, American Control Conference, June 2-4, San Diego, 1999.

[9] M. D. S. Aliyu, "Passivity and stability of nonlinear systems with Markovian jump parameters," in Proceeding, American Control Conference, 2-4 June, 1999.

[10] M. D. S. Aliyu, "An approach for solving the Hamilton-Jacobi-Isaac equations arising in nonlinear $\mathcal{H}_\infty$-control," Automatica, vol. 39, no. 5, pp. 877-884, 2003.

[11] M. D. S. Aliyu, "A transformation approach for solving the Hamilton-Jacobi-Bellman equations in deterministic and stochastic nonlinear control," Automatica, vol. 39, no. 5, 2003.

[12] M. D. S. Aliyu and E. K. Boukas, "$\mathcal{H}_\infty$-control for Markovian jump nonlinear systems," in Proceeding, 37th IEEE Conference on Decision and Control, Florida, USA, 1998.

[13] M. D. S. Aliyu and E. K. Boukas, "Finite and infinite-horizon $\mathcal{H}_\infty$-control for nonlinear stochastic systems," IMA J. of Mathematical Control and Information, vol. 17, no. 3, pp. 265-279, 2000.

[14] M. D. S. Aliyu and E. K. Boukas, "Robust $\mathcal{H}_\infty$-control of nonlinear systems with Markovian jumps," IMA J. of Mathematical Control and Information, vol. 17, no. 3, pp. 295-303, 2000.

[15] M. D. S. Aliyu and J. Luttamaguzi, "On the bounded-real and positive-real conditions for affine nonlinear state-delayed systems," Automatica, vol. 42, no. 2, pp. 357-362, 2006.

[16] M. D. S. Aliyu and L. Smolinsky, "A parametrization approach for solving the Hamilton-Jacobi-equation and application to the $\mathcal{A}_2$-Toda lattice," J. of Nonlinear Dynamics and Systems Theory, vol. 5, no. 4, pp. 323-344, 2005.

[17] M. D. S. Aliyu, "On discrete-time $\mathcal{H}_\infty$ nonlinear filtering and approximate solution," Personal Communication.

[18] M. D. S. Aliyu and E. K. Boukas, "Mixed $\mathcal{H}_2/\mathcal{H}_\infty$ nonlinear filtering," in Proceedings, European Control Conference, Kos, Greece, 2007.

[19] M. D. S. Aliyu and E. K. Boukas, "Mixed $\mathcal{H}_2/\mathcal{H}_\infty$ nonlinear filtering," Int. J. of Robust and Nonlinear Control, vol. 19, no. 4, pp. 394-417, 2009.

[20] M. D. S. Aliyu and E. K. Boukas, "Discrete-time mixed $\mathcal{H}_2/\mathcal{H}_\infty$ nonlinear filtering," in Proceedings, American Control Conference, Seattle WA, USA, 2008.

[21] M. D. S. Aliyu and E. K. Boukas, "Discrete-time mixed $\mathcal{H}_2/\mathcal{H}_\infty$ nonlinear filtering", To appear in Int. J. of Robust and Nonlinear Control, (published online, 26 AUG 2010, DOI: 10.1002/rnc.1643).

[22] M. D. S. Aliyu and J. Luttamaguzi, "3-DOF nonlinear $\mathcal{H}_\infty$-estimators," Int. J. of Control, vol. 83, no. 6, 1136-1144, 2010.

[23] M. D. S. Aliyu and E. K. Boukas, "2-DOF-PD discrete-time nonlinear $\mathcal{H}_\infty$ filtering," Int. J. of Robust and Nonlinear Control, vol. 20, no. 7, pp. 818-833, 2010.

[24] M. D. S. Aliyu, E. K. Boukas, and Y. Chinniah, "Discrete-time PI-nonlinear $\mathcal{H}_\infty$ estimators," IEEE Transactions on Automatic Control, vol. 55, no. 3, pp. 784-790, 2010.

[25] M. D. S. Aliyu and E. K. Boukas, "Kalman filtering for singularly-perturbed nonlinear systems," Submitted to IET Control Theory and Applications.

[26] M. D. S. Aliyu and E. K. Boukas, "$\mathcal{H}_\infty$ estimation for nonlinear singularly perturbed systems," To appear in Int. J. of Robust and Nonlinear Control, (published online, 5 MAY 2010, DOI: 10.1002/rnc.1593).

[27] M. D. S. Aliyu and J. Luttamaguzi, "A note on static output-feedback for affine nonlinear systems," Asian J. of Control, vol. 13, no. 2, pp. 1-7, 2010.

[28] M. D. S. Aliyu, "2-DOF nonlinear $\mathcal{H}_\infty$ certainty equivalent filters," Accepted in Optimal Control Applications and Methods.

[29] W. F. Ames. *Nonlinear Partial-Differential Equations in Engineering, Vols. I, II*, Mathematics in Engineering and Science Series, Academic Press, New York, 1972.

[30] B. D. O. Anderson and S. Vongtpanitlerd. *Network Analysis and Synthesis: A Modern Systems Theory Approach.* Prentice-Hall, Englewood Cliffs, New Jersey, 1973.

[31] B. D. O. Anderson, "A systems theory criterion for positive-real matrices," SIAM J. on Control, vol. 5, pp. 171-182, 1967.

[32] B. D. O. Anderson and J. B. Moore, "An algebraic structure of generalized positive real matrices," SIAM J. on Control, vol. 6, no. 4, pp. 615-621, 1968.

[33] B. D. O. Anderson and J. B. Moore. *Optimal Control: Linear Quadratic Methods.* 2nd Ed., Prentice Hall, Englewood Cliffs, New Jersey, 1986.

[34] B. D. O. Anderson, M. R. James and D. J. N. Limebeer, "Robust stabilization of nonlinear systems via normalized coprime factor representations," Automatica, vol. 34, no. 12, pp. 1593-1599, 1998.

[35] B. D. O. Anderson and J. B. Moore. *Optimal Filtering.* Prentice Hall, Englewood Cliffs, New Jersey, 1979.

[36] I. Arasaratnam, S. Haykin and R. J. Elliot, "Discrete-time nonlinear filtering algorithms using Gauss-Hermite quadrature", in Proceeding of IEEE, vol. 95, no. 5, 2007.

[37] L. Arnold. *Stochastic Differential Equations: Theory and Applications.* John Wiley & Sons, USA 1973.

[38] V. I. Arnold. *Mathematical Methods of Classical Mechanics.* Springer Verlag, New York, 1985.

[39] W. Assawinchaichote, S. K. Nguang and P. Shi, "$\mathcal{H}_\infty$ Output-feedback control design for nonlinear singularly-perturbed systems: an LMI approach," Automatica. Vol. 40, no. 12. 2147-2152, 2004.

[40] W. Assawinchaichote and S. K. Nguang, "$\mathcal{H}_\infty$ filtering for fuzzy nonlinear singularly-perturbed systems with pole placement constraints: an LMI approach," IEEE Transactions on Signal Processing, vol. 52, pp. 1659-1667, 2004.

[41] W. Assawinchaichote and S. K. Nguang, "Robust $\mathcal{H}_\infty$ fuzzy filter design for uncertain nonlinear singularly-perturbed systems with Markovian jumps: an LMI approach," J. of Information Sciences, vol. 177, pp. 1699-1714, 2007.

[42] A. Astolfi, "Singular $\mathcal{H}_\infty$-control for nonlinear systems," Int. J. of Robust and Nonlinear Control, vol. 7, pp. 727-740, 1997.

[43] A. Astolfi, "On the relation between state feedback and full-information regulators in singular nonlinear $\mathcal{H}_\infty$-control," IEEE Transactions on Automatic Control, vol. 42, no. 7, pp. 984-988, 1997.

[44] A. Astolfi, "Parametrization of output-feedback controllers that satisfy an $\mathcal{H}_\infty$ bound," in Proceedings, 2nd European Control Conference, Groningen, The Netherlands, pp. 74-77, 1993.

[45] A. Astolfi and P. Colaneri, "A Hamilton-Jacobi setup for the static output-feedback stabilization of nonlinear systems," IEEE Transactions on Automatic Control, vol. 47, no. 12, pp. 2038-2041, 2002.

[46] A. Astolfi and A. Rapaport, "Robust stabilization of the angular momentum of a rigid body," Systems and Control Letters, vol. 34, pp. 257-264, 1998.

[47] M. Athans and P. L. Falb. *Optimal Control Theory: An Introduction.* McGraw-Hill, NY, 1966.

[48] A. Bagchi. *Optimal Control of Stochastic Systems.* Prentice Hall, UK, 1993.

[49] J. A. Ball, "Inner-outer factorization of nonlinear operators," J. of Functional Analysis, vol. 104, pp. 363-413, 1992.

[50] J. A. Ball, P. Kachroo, and A. J. Krener, "$\mathcal{H}_\infty$ tracking control for a class of nonlinear systems", IEEE Transactions on Automatic Control, vol. 44, no. 6, pp. 1202-1207, 1999.

[51] J. A. Ball and J. W. Helton, "Nonlinear $\mathcal{H}_\infty$ control theory for stable plants," Mathematics of Control, Signals, and Systems, vol. 5, pp. 233-261, 1992.

[52] J. A. Ball and J. W. Helton, "Factorization of nonlinear systems: towards a theory for nonlinear $\mathcal{H}_\infty$ control," Proceedings IEEE Conference on Decision and Control, Austin, TX, pp. 2376-2381, 1988.

[53] J. A. Ball, J. W. Helton and M. L. Walker, "$\mathcal{H}_\infty$-control for nonlinear systems via output feedback," IEEE Transactions on Automatic Control, vol. 38, pp. 546-559, 1993.

[54] J. A. Ball and A. J. van der Schaft, "J-Inner-outer factorization, j-spectral factorization and robust control for nonlinear systems," IEEE Transactions Automatic Control, vol. 41, no. 3, pp. 379-392, 1996.

[55] L. Baramov and H. Kimura, "Nonlinear local J-lossless conjugation and factorization," Int. J. of Robust and Nonlinear Control, vol. 6, pp. 869-893, 1996.

[56] M. Bardi and I. Capuzzo-Dolcetta. *Optimal Control and Viscosity Solutions of Hamilton-Jacobi-Bellman Equations.* Systems and Control Foundations, Birkhauser, Boston, 1997.

[57] T. Basar and P. Bernhard. *H_∞-Optimal Control and Related Minimax Design Problems.* Systems and Control Foundations and Applications, 2nd Ed., Birkhauser, Boston, 1995.

[58] T. Basar and P. Bernhard. *$\mathcal{H}_\infty$ Optimal Control and Related Minimax Design.* Birkhauser, New York, 1991.

[59] T. Basar and G. J. Olsder. *Dynamic Noncooperative Game Theory.* Mathematics in Science and Engineering, vol. 160, 1st Ed., Academic Press, UK, 1982.

[60] V. Barbu and G. Prata Da. *Hamilton-Jacobi Equations in Hilbert Space.* Pitman Advanced Publishing Program, London, 1983.

[61] S. Battilotti, "Robust stabilization of nonlinear systems with pointwise norm-bounded uncertainties: A control Lyapunov function approach," IEEE Transactions Automatic Control, vol. 44, no. 1, pp. 3-17, 1999.

[62] S. Battilotti, "Global output regulation and disturbance-attenuation with global stability via measurement feedback for a class of nonlinear systems," IEEE Transactions on Automatic control, vol. 41, no. 3, pp. 315-327, 1996.

[63] R. W. Beard, G. N. Saridas and J. T. Wen, "Galerkin approximations of the generalized HJB equation," Automatica, vol. 33, no. 12, pp. 2159-2177, 1997.

[64] R. Bellman. *Dynamic Programming.* Princeton University Press, Princeton, New Jersey, 1957.

[65] S. H. Benton. *The Hamilton-Jacobi Equation: A Global Approach.* Academic Press, New York, 1977.

[66] N. Berman and U. Shaked, "$\mathcal{H}_\infty$ nonlinear filtering", Int. J. of Robust and Nonlinear Control, vol. 6, pp. 281-296, 1996.

[67] D. S. Bernstein and W. H. Haddad, "LQG control with $\mathcal{H}_\infty$ performance bound," IEEE Transactions on Automatic control, vol. 34, no. 3, pp. 293-305, 1997.

[68] S. Bittanti, A. J. Laub and J. C. Willems. *The Riccati Equation.* Springer-Verlag, Germany, 1991.

[69] S. Boyd, L. El-Ghaoui, E. Feron and V. Balakrishnan. *Linear Matrix Inequalities in Systems and Control Theory.* SIAM, Philadelphia, 1994.

[70] R. S. Bucy, "Linear and nonlinear filtering," IEEE Proceedings, vol. 58, no. 6, pp. 854-864, 1970.

[71] R. T. Bupp, D. S. Bernstein and V. T. Coppola, "A benchmark problem for nonlinear control design," Int. J. of Robust and Nonlinear Control, vol. 8, pp. 307-310, 1998.

[72] O. I. Bogoyavlensky, "On pertubation of the periodic Toda lattice," Communications in Mathematical Physics, 51, pp. 201-209, 1976.

[73] A. E. Bryson and Y. C. Ho. *Applied Optimal Control.* Halsted Press, John Wiley & Sons, New York, 1975.

[74] C. I. Byrnes and W. Lin, "Losslessness, feedback equivalence, and global stabilization of discrete-time nonlinear systems," IEEE Transactions on Automatic Control, vol. 39, no. 1, pp. 83-97, 1994.

[75] C. I. Byrnes and W. Lin, "Discrete-time lossless systems, feedback equivalence, and passivity," Proceedings, IEEE Conference on Decision and Control, pp. 1775-1781, San Antonio, Texas, 1993.

[76] C. I. Byrnes and W. Lin, "On discrete-time nonlinear Control," Proceedings, IEEE Conf. on Dec. and Control, pp. 2990-2996, San Antonio, Texas, 1993.

[77] C. I. Byrnes, A. Isidori and J. C. Willems, "Passivity, feedback equivalence and global stabilization of minimum phase nonlinear systems," IEEE Transactions Automatic Control, vol. 36, no. 11, pp. 1228-1240, 1991.

[78] S. S. Chang S. and T. K. C. Peng, "Adaptive guaranteed cost control of systems with uncertain parameters," IEEE Transactions on Automatic Control, vol. 17, pp. 474-483, 1972.

[79] K. W. Chang, "Singular pertubation of a general boundary-value problem", SIAM J. Math Anal., vol. 3, pp. 520-526, 1972.

[80] B.-S. Chen and Y.-C. Chang, "Nonlinear mixed $\mathcal{H}_2/\mathcal{H}_\infty$ control for robust tracking of robotic systems," Int. J. Control, vol. 67, no. 6, pp. 837-857, 1998.

[81] X. Chen and K. Zhou, "Multi-objective $\mathcal{H}_2/\mathcal{H}_\infty$-control design", SIAM J. on Control and Optimization, vol. 40, no. 2, pp. 628-660, 2001.

[82] J. H. Chow and P. V. Kokotovic, "A Decomposition of near-optimal regulators for systems with slow and fast modes," IEEE Trans Autom Control, vol. 21, pp. 701-705, 1976.

[83] F. H. Clarke, Y. S. Ledyaev, R. J. Stern and P. R. Wolenski. *Nonsmooth Analysis and Control Theory*. Springer Verlag, New York, 1998.

[84] O. L. V. Costa, "Mean-square stabilizing solutions for the discrete-time coupled algebraic Riccati equations," IEEE Transaction on Automatic Control, vol. 41, no. 4, pp. 593-598, 1996.

[85] D. Cox, J. Little, and D. O'Shea. *Ideals, Varieties, and Algorithms*. 2nd edition, Springer-Verlag, New York, 1997.

[86] M. Cromme, J. M. Pedersen and M. P. Petersen, "Nonlinear $\mathcal{H}_\infty$ state feedback controllers: computation of valid region," Proceedings, American Control Conference, Albuquerque, New Mexico, pp. 391-395, 1997.

[87] M. Dalsmo and O. Egeland, "State feedback $\mathcal{H}_\infty$-suboptimal control of a rigid spacecraft," IEEE Transactions on Automatic Control, vol. 42, no. 8, 1997.

[88] M. Dalsmo and W. C. A. Maas, "Singular $\mathcal{H}_\infty$-suboptimal control for a class of nonlinear cascade systems," Automatica, vol. 34, no. 12, pp. 1531-1537, 1998.

[89] C. E. De-souza and L. Xie, "On the discrete-time bounded-real lemma with application in the characterization of static state-feedback $\mathcal{H}_\infty$-controllers," Systems and Control Letters, vol. 18, pp. 61-71, 1992.

[90] A. S. Debs and M. Athans, "On the optimal angular velocity control of asymmetric vehicles," IEEE Transactions on Automatic Control, vol. 14, pp. 80-83, 1969.

[91] H. Deng and M. Krstic, "Output-feedback stochastic nonlinear stabilization," IEEE Transactions on Automatic control, vol. 44, no. 2, pp. 328-333, 1999.

[92] J. C. Doyle, K. Glover, P. Khargonekhar and B. A. Francis, "State-space solutions to standard $\mathcal{H}_2$ and $\mathcal{H}_\infty$-control problems," IEEE Transactions on Automatic Control, vol. 34, pp. 831-847, 1989.

[93] J. C. Doyle, K. Zhou, K. Glover and B. Bodenheimer, "Mixed $\mathcal{H}_2/\mathcal{H}_\infty$ performance objectives II: optimal control," IEEE Transactions on Automatic Control, vol. 39, no. 8, pp 1575-1587, 1994.

[94] J. Dieudonne. *Elements d'Analyse Vol. 1, Foundements de l'Analyse Modern*. Gauthier-Villars, Paris, 1972.

[95] L. C. Evans. *Partial-Differential Equations*. Graduate Text in Maths., AMS, Providence, Rhode Island, 1998.

[96] K. Ezal, Z. Pan and P. V. Kokotovic, "Locally optimal and robust backstepping design", IEEE Transactions on Automatic Control, vol. 45, no. 2, pp. 260-271, 2000.

[97] I. J. Fialho and T. T. Georgiou, "Worst-case analysis of nonlinear systems", IEEE Transactions on Automatic Control, vol. 44, no. 6, pp. 1180-1196.

[98] W. H. Fleming and H. M. Soner. *Controlled Markov Processes and Viscosity Solutions*. Springer Verlag, New York, 1993.

[99] W. H. Fleming and R. W. Rishel. *Deterministic and Stochastic Optimal Control*. Springer Verlag, New York, 1975.

[100] P. Florchinger, "A passive systems approach to feedback stabilization of nonlinear stochastic systems", SIAM J. on Control and Optimiz., vol. 37, no. 6, pp. 1848-1864, 1999.

[101] B. C. Francis. *A Course in $\mathcal{H}_\infty$-Control.* Lecture Notes in Control and Information, Springer Verlag, New York, 1987.

[102] R. Freeman and P. V. Kokotovic, "Inverse optimality in robust stabilization," SIAM J. on Control and Optimization, vol. 34, no. 4, pp. 1365-1391, 1996.

[103] E. Fridman and U. Shaked, "On regional nonlinear $\mathcal{H}_\infty$-filtering," Systems and Control Letters, vol. 29, pp. 233-240, 1997.

[104] E. Fridman, "State-feedback $\mathcal{H}_\infty$-control of nonlinear singularly perturbed systems," Int. J. of Robust and Nonlinear Control, vol. 11, pp. 1115-1125, 2001.

[105] E. Fridman, "A descriptor system approach to nonlinear singularly perturbed optimal control problem," Automatica, vol. 37, pp. 543-549, 2001.

[106] E. Fridman, "Exact slow-fast decomposition of the nonlinear singularly perturbed optimal control problem," Systems and Control Letters, vol. 40, pp. 121-131, 2000.

[107] A. Friedman. *Differential Games.* Pure and Applied Mathematics Series, Wiley-Interscience, New York, 1971.

[108] G. Frieling, G. Jank and H. Abou-Kandil, "On the global existence of solutions to coupled matrix Riccati equations in closed-loop Nash games," IEEE Transactions on Automatic Control, vol. 41, no. 2, pp. 264-269, 1996.

[109] J. Fritz. *Partial Differential Equations.* Springer-Verlag Applied Mathematical Sciences Series, 3rd Ed., New York, 1978.

[110] P. Gahinet, A. Nemirovski, A. J. Laub, and M. Chilali. *Matlab LMI Control Toolbox.* MathWorks Inc., Natick, Massachusetts, 1995.

[111] Z. Gajic and M. T. Lim, "A new filtering method for linear singularly perturbed systems," IEEE Transactions Automatic Control, vol. 39, no. 9, pp. 1952-1955, 1994.

[112] S. T. Glad, "Robustness of nonlinear state-feedback: a survey," Automatica, vol. 23, pp. 425-435, 1987.

[113] J. C. Geromel, C. E. De-souza and R. E. Skelton, "Static output feedback controllers: Stability and convexity," IEEE Transactions on Automatic Control, vol. 43, no. 1, pp. 120-125, 1998.

[114] H. H. Goldstine. *A History of the Calculus of Variations from the 17th through to the 19th Century.* Springer Verlag, New York, 1980.

[115] H. Goldstein. *Classical Mechanics.* Addison-Wesley, Massachusetts, 1950.

[116] A. Graham. *Kronecker Products and Matrix Calculus with Applications.* Ellis Horwood, Chichester, UK, 1981.

[117] M. K. Green, "$\mathcal{H}_\infty$ controller synthesis by J-lossless coprime factorization," SIAM J. on Control and Optimization, vol. 30, pp. 522-547, 1992.

[118] M. K. Green, K. Glover, D. J. N. Limebeer and J. C. Doyle, "A J-spectral factorization approach to $\mathcal{H}_\infty$ control," SIAM J. on Control and Optimization, vol. 28, pp. 1350-1371, 1988.

[119] M. S. Grewal and A. P. Andrews. *Kalman Filtering: Theory and Practice.* Prentice Hall Information and Systems Science Series, Englewood Cliffs New Jersey, 1993.

[120] M. J. Grimble, D. Ho and A. El-Sayed, "Solution of the $\mathcal{H}_\infty$-filtering problem for discrete-time systems", IEEE Transactions Acoustic Speech and Signal Processing, vol. 38, pp. 1092-1104, 1990.

[121] G. Gu, "Stabilizability conditions of multivariable uncertain systems via output feedback," IEEE Transactions on Automatic Control, vol. 35, no. 7, pp. 925-927, Aug. 1990.

[122] S. W. Groesberg. *Advanced Mechanics.* John Wiley & Sons, New York, 1968.

[123] M. A. Guest. *Harmonic Maps, Loop Groups, and Integrable Systems.* London Mathematical Society Text 38, Cambridge University Press, UK, 1997.

[124] H. Guillard, "On nonlinear $\mathcal{H}_\infty$-control under sampled measurements", IEEE Transactions Automatic Control, vol. 42, no. 6, pp. 880-885, 1998.

[125] H. Guillard, S. Monaco and D. Normand-Cyrot, "Approximate solutions to nonlinear discrete-time $\mathcal{H}_\infty$-control," IEEE Transactions on Automatic Control, vol. 40, no. 12, pp. 2143-2148, 1995.

[126] H. Guillard, S. Monaco and D. Normand-Cyrot, "$\mathcal{H}_\infty$-control of discrete-time nonlinear systems," Int. J. of Robust and Nonlinear Control, vol. 6, pp. 633-643, 1996.

[127] P. Hagedorn. *Nonlinear Oscillations.* Oxford Engineering Science Series, New York, 1988.

[128] J. Hammer, "Fractional representations of nonlinear systems: a simplified approach," Int. J. Control, vol. 46, pp. 455-472, 1987.

[129] J. W. Helton and M. R. James. *Extending $\mathcal{H}_\infty$-Control to Nonlinear Systems: Control of Nonlinear Systems to Achieve Performance Objectives.* SIAM Frontiers in Applied Mathematics, Philadelphia, 1999.

[130] M. Hestenes, *Calculus of Variation and Optimal Control.* Springer Verlag, New York, 1980.

[131] D. Hill and P. Moylan, "The stability of nonlinear dissipative systems", IEEE Transactions on Automatic Control, vol. 21, pp. 708-711, 1976.

[132] D. Hill and P. Moylan, "Connection between finite-gain and asymptotic stability," IEEE Transactions on Automatic Control, vol. 25, no. 5, pp. 931-935, Oct. 1980.

[133] D. Hill and P. Moylan, "Stability results for nonlinear feedback systems," Automatica, vol. 13, pp. 377-382, 1977.

[134] D. Hill and P. Moylan, "Dissipative dynamical systems: Basic input-output and state properties," J. of Franklin Institute, vol. 309, pp. 327-357, 1980.

[135] J. I. Imura, T. Sugie, and T. Yoshikawa, "Characterization of the strict bounded-real condition for nonlinear systems," IEEE Transactions on Automatic Control, vol. 42, no. 10, pp. 1459-1464, 1997.

[136] R. Isaacs. *Differential Games*. SIAM Series in Applied Mathematics, John Wiley & Sons, New York, 1965.

[137] R. Isaacs, "Differential games: their scope, nature, and future", Journal of Optimization Theory and Applications, vol. 3, no. 5, pp. 283-292, 1969.

[138] A. Isidori, "Feedback control of nonlinear systems," Int. J. of Robust and Nonlinear Control, vol. 2, pp. 291-311, 1992.

[139] A. Isidori, "$\mathcal{H}_\infty$ control via measurement feedback for affine nonlinear systems," Int. J. of Robust and Nonlinear Control, vol. 4, pp. 553-574, 1994.

[140] A. Isidori. *Nonlinear Control Systems*. Springer Verlag, 3rd Ed., New York, 1995.

[141] A. Isidori and A. Astolfi, "Disturbance-attenuation and $\mathcal{H}_\infty$ control via measurement feedback for nonlinear systems," IEEE Transactions on Automatic Control, vol. 37, pp. 1283-1293, Sept. 1992.

[142] A. Isidori and A. Astolfi, "Nonlinear $\mathcal{H}_\infty$-control via measurement feedback," J. of Mathematical Systems, Estimation and Control, vol. 2, no. 1, pp. 31-44, 1992.

[143] A. Isidori and C. I. Byrnes, "Output regulation of nonlinear systems," IEEE Transactions Automatic Control, vol. 35, no. 2, pp. 131-140, 1990.

[144] A. Isidori and W. Lin, "Global $\mathcal{L}_2$-gain design for a class of nonlinear systems," Systems and Control Letters, vol. 34, no. 5, pp. 245-252, 1998.

[145] A. Isidori and W. Kang, "$\mathcal{H}_\infty$ control via measurement feedback for general class of nonlinear systems," IEEE Transactions on Automatic Control, vol. 40, no. 3, pp. 466-472, 1995.

[146] A. Isidori, A. J. Krener, G. Gori-Giorgi and S. Monaco, "Nonlinear decoupling via state-feedback: a differential geometric approach," IEEE Transactions on Automatic Control, vol. 26, pp. 331-345, 1981.

[147] A. Isidori and T. J. Tarn, "Robust regulation for nonlinear systems with gain-bounded uncertainties," IEEE Transactions on Automatic Control, vol. 40, no. 10, pp. 1744-1754, 1995.

[148] H. Ito, "Robust control for nonlinear systems with structured $\mathcal{L}_2$-gain bounded uncertainty," Systems and Control Letters, vol. 28, no. 3, 1996.

[149] M. R. James, "Partially observed differential games, infinite-dimensional HJI equations and nonlinear $\mathcal{H}_\infty$-control", SIAM J. on Control and Optimization, vol. 34, no. 4, pp. 1342-1364, 1996.

[150] M. R. James and J. S. Baras, "Robust $\mathcal{H}_\infty$ output-feedback control for nonlinear systems," IEEE Transaction Automatic Control, vol. 40, no. 6, pp. 1007-1017, 1995.

[151] M. R. James, J. S. Baras and R. J. Elliot, "Risk-sensitive control and dynamic games for partially observed discrete-time nonlinear systems," IEEE Transactions Automatic Control, vol. 39, no. 4, pp. 780-792, 1994.

[152] Jazwinski A. H., *Stochastic Processes and Filtering Theory*, Dover Publ., New York, 1998.

[153] I. Kanellakopoulos, P. V. Kokotovic and A. S. Morse, "Systematic design of adaptive controllers for feedback linearizable systems," IEEE Transactions Automatic Control, vol. 36, pp. 1241-1253, 1991.

[154] W. Kang, "Nonlinear $\mathcal{H}_\infty$-control and its application to rigid spacecraft," IEEE Transactions on Automatic Control, vol. 40, No. 7, pp. 1281-1285, 1995.

[155] W. Kang, P. K. De and A. Isidori, "Flight control in windshear via nonlinear $\mathcal{H}_\infty$ methods," Proceedings, 31st IEEE Conference on Decision and Control, Tucson, Arizona, pp. 1135-1142, 1992.

[156] R. P. Kanwal. *Generalized Functions: Theory and Techniques.* Math of Science and Engineering, vol. 171, Academic Press, New York, 1983.

[157] H. K. Khalil. *Nonlinear Systems.* Macmillan Academic Pub.,New York, 1992.

[158] H. K. Khalil. *Nonlinear Systems.* 2nd Ed., Macmillan, New York, 2000.

[159] P. P. Khargonekar, I. Petersen and K. Zhou, "Robust stabilization of uncertain linear systems: quadratic stabilizability and $\mathcal{H}_\infty$ control," IEEE Transactions on Automatic Control, vol. 35, pp. 356-361, 1990.

[160] P. P. Khargonekar, I. Petersen and M. A. Rotea, "$\mathcal{H}_\infty$ optimal control with state-feedback," IEEE Transactions on Automatic Control, vol. 33, no. 8, pp. 786-788, 1988.

[161] P. P. Khargonekar and M. A. Rotea, "Mixed $\mathcal{H}_2/\mathcal{H}_\infty$-control: a convex optimization approach," IEEE Transactions on Automatic Control, vol. 36, no. 7, pp. 824-837, 1991.

[162] P. P. Khargonekar, M. Rotea and E. Baeyens, "Mixed $\mathcal{H}_2/\mathcal{H}_\infty$-filtering," Int. J. of Robust and Nonlinear Control, vol. 6, no. 4, pp. 313-330, 1996.

[163] H. Kimura, Y. Lu and R. Kawatani, "On the structure of $\mathcal{H}_\infty$-control systems and related extensions," IEEE Transactions on Automatic Control, vol. 36, no. 6, pp. 653-667, 1991.

[164] D. Kirk. *Optimal Control Theory.* Prentice Hall, New York, 1972.

[165] P. V. Kokotovic, H. K. Khalil and J. O'Reilly. *Singular Pertubation Methods in Control: Analysis and Design.* Academic Press, Orlando, Florida, 1986.

[166] A. J. Krener, "Necessary and sufficient conditions for nonlinear worst-case ($\mathcal{H}_\infty$)-control and estimation," J. of Math. Systems Estimation and Control, vol. 4, no. 4, pp. 1-25, 1994.

[167] M. Krstic and H. Deng. *Stabilization of Nonlinear Uncertain Systems.* Springer Verlag, London, 1998.

[168] M. Krstic and Z. H. Li, "Inverse optimal design of input-to-state stable nonlinear controllers", IEEE Transactions on Automatic Control, vol. 43, no. 3, pp. 336-350, 1998.

[169] M. Krstic and P. Tsiotras, "Inverse optimal stabilization of a rigid spacecraft", IEEE Transactions Automatic Control, vol. 44, no. 5, pp. 1042-1049, 1999.

[170] M. Krstic, I. Kanellakopoulos, and P. V. Kokotovic. *Nonlinear and Adaptive Control Design*. John Wiley & Sons New York, 1985.

[171] V. Kucera and C. E. De-souza, "Necessary and sufficient conditions for output-feedback stabilizability," Automatica, vol. 31, pp. 1357-1359, 1995.

[172] H. J. Kushner, "Nonlinear filtering: the exact dynamical equation satisfied by the conditional mode," IEEE Transactions on Automatic Control, vol. 12, no. 3, pp. 1967.

[173] H. Kushner and P. Dupuis. *Numerical Methods for Stochastic Control Problems in Continuous-Time*. Springer Verlag, New York, 1993.

[174] H. Kwakernaak and R. Sivans. *Linear Optimal Control*. Wiley Interscience, New York, 1972.

[175] E. B. Lee and B. Markus. *Foundation of Optimal Control*. SIAM Series in Applied Math, John Wiley, New York, 1967.

[176] T. C. Lee, B. S. Chen and T. S. Lee, "Singular $\mathcal{H}_\infty$-control in nonlinear systems: positive-semidefinite storage functions and separation principle," Int. J. of Robust and Nonlinear Control, vol. 7, pp. 881-897, 1997.

[177] G. Leitman. *The Calculus of Variation and Optimal Control*. Plenum Press, New York 1981.

[178] M. T. Lim and Z. Gajic, "Reduced-order $\mathcal{H}_\infty$ optimal filtering for systems with slow and fast modes," IEEE Transactions on Circuits and Systems I: Fundamental Thry and Appl., vol. 47, no. 2, pp. 250-254, 2000.

[179] D. J. N. Limebeer, B. D. O. Anderson and B. Hendel, "A Nash game approach to the mixed $\mathcal{H}_2/\mathcal{H}_\infty$-control problem," IEEE Transactions on Automatic Control, vol. 39, no. 4, pp. 824-839, 1994.

[180] W. Lin, "Mixed $\mathcal{H}_2/\mathcal{H}_\infty$-control for nonlinear systems," Int. J. of Control, vol. 64, no. 5, pp. 899-922, 1996.

[181] W. Lin. *Synthesis of Discrete-Time Nonlinear Systems*. DSc Dissertation, Washington University, St. Louis, 1993.

[182] W. Lin and C. I. Byrnes, "Dissipativity, $\mathcal{L}_2$-gain and $\mathcal{H}_\infty$-control for discrete-time nonlinear systems," Proceedings, American Control Conf., Baltimore, Maryland, pp. 2257-2260, 1994.

[183] W. Lin and C. I. Byrnes, "$\mathcal{H}_\infty$-control of discrete-time nonlinear systems," IEEE Transactions on Automatic Control, vol. 41, no. 4, pp. 494-509, 1996.

[184] W. Lin and C. I. Byrnes, "Discrete-time nonlinear $\mathcal{H}_\infty$ control with measurement feedback," Automatica, vol. 31, no. 3, pp. 419-434, 1995.

[185] W. Lin and C. I. Byrnes, "KYP lemma, state-feedback and dynamic output-feedback in discrete-time bilinear systems," Systems and Control Letters, vol. 23, pp. 127-136, 1994.

[186] P. L. Lions. *Generalized Solutions of Hamilton-Jacobi Equations*. Pitman Research Notes in Maths, No. 69, Pitman, Boston, 1982.

[187] R. Lozano, B. Brogliato, B. Maschke and O. Egeland. *Dissipative Systems Analysis and Control: Theory and Applications*. Springer Verlag, London, 2000.

[188] W.-M. Lu, "A state-space approach to parameterization of stabilizing controllers for nonlinear systems," IEEE Transactions on Automatic Control, vol. 40, no. 9, pp. 1576-1588, 1995.

[189] W.-M. Lu, "$\mathcal{H}_\infty$-control of nonlinear time-varying systems with finite-time horizon," Int. J. Control, vol. 64, no. 2, pp. 241-262, 1997.

[190] W.-M. Lu and J. C. Doyle, "$\mathcal{H}_\infty$-control of nonlinear systems via output-feedback: controller parameterization," IEEE Transactions on Automatic Control, vol. 39, no. 9, pp. 2517-2521, 1994.

[191] W.-M. Lu and J. C. Doyle, "$\mathcal{H}_\infty$-control of nonlinear systems: A convex characterization," IEEE Transactions on Automatic Control, vol. 40, no. 9, pp. 1668-1675, 1995.

[192] W.-M. Lu and J. C. Doyle, "Robustness analysis and synthesis for nonlinear uncertain systems," IEEE Transactions on Automatic Control, vol. 42, no. 12, pp. 1654-1666, 1997.

[193] D. L. Lukes, "Optimal regulation of nonlinear dynamical systems," SIAM J. of Control, vol. 7, pp. 75-100, 1969.

[194] W. C. A. Maas and A. J. van der Schaft, "Nonlinear $\mathcal{H}_\infty$-control: the singular case," Int. J. of Robust and Nonlinear Control, vol. 6, pp. 669-689, 1996.

[195] J. M. Maciejowski. *Multivariable Feedback Design*. Addison-Wesley, Reading, Massachusetts, 1989.

[196] M. Hestenes. *Conjugate Direction Methods in Optimization*. Springer Verlag, New York, 1980.

[197] S. I. Marcus, "Optimal nonlinear estimation for a class of discrete-time stochastic systems", IEEE Transactions on Automatic Control, vol. 24, no. 2, pp. 297-302, 1979.

[198] R. Marino, W. Respondek and A. J. van der Schaft, "Almost-disturbance-decoupling for single-input single-output nonlinear systems," IEEE Transactions on Automatic Control, vol. 34, no. 9, pp. 1013-1016, 1989.

[199] R. Marino, W. Respondek, A. J. van der Schaft and P. Tomei, "Nonlinear $\mathcal{H}_\infty$ almost disturbance decoupling", Systems and Control Letters, vol. 23, pp. 159-168, 1994.

[200] J. Marsden and T. S. Ratiu. *Introduction to Mechanics and Symmetry*. Springer-Verlag, 1999.

[201] A. Mcdaniel and L. Smolinsky, "The flow of the $\mathcal{G}_2$ periodic Toda lattice", J. of Mathematical Physics, vol. 38, no. 2, pp. 926-945, 1997.

[202] P. Moylan, "On the implications of passivity in a class of nonlinear systems", IEEE Transactions Automatic Control, vol. 19, no. 4, pp. 327-357, 1980.

[203] R. E. Mortensen, "Maximum-likelihood recursive nonlinear filtering," J. of Optimization Theory and Application, vol. 2, no. 6, pp. 386-394, 1968.

[204] M. Munoz-Lecanda and F. J. Yaniz-Fernandez, "Dissipative control of mechanical systems: a geometric approach," SIAM J. on Control and Optimization, vol. 40, no. 5, pp. 1505-1516, 2002.

[205] D. Mustapha, "Relations between maximum entropy $\mathcal{H}_\infty$ control and combined $\mathcal{H}_\infty$/LQG control," Systems and Control Letters, vol. 12, pp. 193-203.

[206] D. Mustapha, K. Glover and D. J. N. Limebeer, "Solutions to the $\mathcal{H}_\infty$ general distance problem which minimize an entropy integral," Automatica, vol. 27, pp. 193-199, 1991.

[207] K. M. Nagpal and P. Khargonekar, "Filtering and smoothing in an $\mathcal{H}_\infty$ setting", IEEE Transaction on Automatic Control, vol. 36, no. 2, pp. 152-166, 1991.

[208] S. K. Nguang, "Robust nonlinear $\mathcal{H}_\infty$ output-feedback control," IEEE Transactions on Automatic Control, vol. 41, no. 7, pp. 1003-1007, 1996.

[209] S. K. Nguang, "Global robust nonlinear $\mathcal{H}_\infty$-control of a class of nonlinear systems," Int. J. of Robust and Nonlinear Control, vol. 7, pp. 75-84, 1997.

[210] S. K. Nguang and M. Fu, "$\mathcal{H}_\infty$-filtering for known and uncertain nonlinear systems," Proceedings, IFAC Symposium on Robust Control Design, Rio de Janeiro, Brazil, pp. 347-352, 1994.

[211] S. K. Nguang and M. Fu, "Robust nonlinear $\mathcal{H}_\infty$ filtering," Automatica, Vol. 32, no. 8, pp. 1195-1199, 1996.

[212] H. Nijmeijer and A. J. van der Schaft. *Nonlinear Dynamical Control Systems*. Springer Verlag, New York, 1990.

[213] Y. Orlov and L. Acho, "Nonlinear $\mathcal{H}_\infty$-control of time-varying systems: a unified distribution-based formalism for continuous and sampled-data measurement feedback design," IEEE Transactions on Automatic Control, vol. 46, no. 4, pp. 638-643, 2001.

[214] A. D. B. Paice and J. B. Moore, "The Youla-Kucera parametrization for nonlinear system," Systems and Control Letters, vol. 14, pp. 121-129, 1990.

[215] A. D. B. Paice and A. J. van der Schaft, "The class of stabilizing plant controller pairs," IEEE Transactions on Automatic Control, vol. 41, no. 5, pp. 634-645, 1996.

[216] Z. Pan and T. Basar, "$\mathcal{H}_\infty$ optimal control for singularly-perturbed systems-Part I: perfect state measurements," Automatica, vol. 29, no. 2, pp. 401-423, 1993.

[217] Z. Pan and T. Basar, "$\mathcal{H}_\infty$ optimal control for singularly-perturbed systems-Part II: imperfect state measurements," IEEE Transactions on Automatic Control, vol. 39, pp. 280-300, 1994.

[218] Z. Pan and T. Basar, "Time-scale separation and robust controller design for uncertain nonlinear singularly-perturbed systems under perfect state measurements," Int. J. of Robust and Nonlinear Control, vol. 6, pp. 585-608, 1996.

[219] Z. Pan and T. Basar, "Adaptive controller design for tracking and disturbance-attenuation in parametric strict feedback nonlinear systems," IEEE Transactions on Automatic Control, vol. 43, no. 8, pp. 1066-1083, 1998.

[220] G. P. Papavassilopoulos and G. J. Olsder, "On linear-quadratic, closed-loop, no-memory Nash game," J. of Optimization Theory and Appl., vol. 42, no. 4, pp. 551-560, 1984.

[221] G. P. Papavassilopoulos, J. V. Medanic and J. B. Cruz, "On the existence of Nash strategies and solutions to coupled Riccati equations in linear-quadratic Nash games," J. of Optimization Theory and Appl., vol. 28, pp. 49-76, 1979.

[222] G. P. Papavassilopoulos, J. V. Medanic and J. B. Cruz, "On the existence of Nash equilibrium solutions to coupled Riccati equations in linear quadratic games," Journal of Optimization Theory and Appl., vol. 28, no. 1, pp. 49-76, 1979.

[223] L. Pavel and F. W. Fairman, "Robust stabilization of nonlinear plants: an $\mathcal{L}_2$-approach," Int. J. Robust and Nonlinear Control, vol. 6, no. 7, pp. 691-726, 1996.

[224] L. Pavel and F. W. Fairman, "Nonlinear $\mathcal{H}_\infty$-control: a J-dissipative approach", IEEE Transactions on Automatic Control, vol. 42, no. 12, pp. 1636-1653, 1997.

[225] L. Pavel and F. W. Fairman, "Controller reduction for nonlinear plants: an $\mathcal{L}_2$ approach," Int. J. of Robust and Nonlinear Control, vol. 7, pp. 475-505, 1997.

[226] I. Petersen, "A stabilization algorithm for a class of uncertain linear systems," Systems and Control Letters, vol. 8, pp. 351-357, 1987.

[227] I. Petersen and D. Macfarlane, "Optimal guaranteed cost control and filtering for uncertain linear systems," IEEE Transactions on Automatic Control, vol. 39, pp. 1971-1977, 1994.

[228] H. K. Pillai and J. C. Willems, "Lossless and dissipative distributed systems," SIAM J. on Control and Optimization, vol. 40, no. 5, pp. 1406-1430, 2002.

[229] L. S. Pontryagin, V. Boltyanskii, R. Gamkrelidge and E. Mishchenko. *The Mathematical Theory of Optimal Processes*. Interscience Publishers, Inc., New York, 1962.

[230] W. Rudin. *Principles of Mathematical Analysis*. McGraw Hill, Singapore, 1976.

[231] H. Rund. *The Hamilton-Jacobi Theory in the Calculus of Variations*. Van Nostrand, London, 1966.

[232] T. L. Saaty. *Modern Nonlinear Equations*. McGraw-Hill, New York, 1967.

[233] M. G. Safonov, E. A. Joncheere, M. Verma and D. J. N. Limebeer, "Synthesis of positive-real multivariable feedback systems," Int. J. Control, vol. 45, no. 3, pp. 817-842, 1987.

[234] S. S. Sastry. *Nonlinear Systems: Analysis, Stability and Control*. Springer Verlag, New York, 1999.

[235] A. Savkin and I. Petersen, "Optimal guaranteed cost control of discrete-time uncertain nonlinear systems", IMA J. of Math Control and Info., vol. 14, pp. 319-332, 1997.

[236] A. V. Savkin and I. R. Petersen, "Nonlinear versus linear control in the absolute stabilizability of uncertain linear systems with structured uncertainty" Proceedings, IEEE Conference on Decision and Control, San Antonio, Texas, pp. 172-178, 1993.

[237] C. W. Scherer, "$\mathcal{H}_\infty$-optimizaton without assumptions on finite or infinite zeros," SIAM Journal of Control and Optimization, vol. 30, pp. 143-166, 1992.

[238] C. W. Scherer, "Multiobjective $\mathcal{H}_2/\mathcal{H}_\infty$-control," IEEE Transactions on Automatic Control, vol. 40, no. 6, pp. 1054-1062, 1995.

[239] C. W. Scherer, P. Gahinet and M. Chilali, "Multi-objective output-feedback control via LMI optimization," IEEE Transactions on Automatic Control, vol. 42, no. 7, pp. 896-911, 1997.

[240] J. M. A. Scherpen and A. J. van der Schaft, "Normalized coprime factorization and balancing for unstable nonlinear systems," Int. J. of Control, vol. 60, no. 6, pp. 1193-1222, 1994.

[241] S. P. Sethi and Q. Zhang. *Hierachical Decision Making in Stochastic Manufacturing Systems*. Springer Verlag, New York, 1997.

[242] U. Shaked, "$\mathcal{H}_\infty$ minimum error state estimation of linear stationary processes", IEEE Transactions on Automatic Control, vol. 35, no. 5, pp. 554-558, 1990.

[243] U. Shaked and Y. Theodor, "$\mathcal{H}_\infty$ optimal estimation: a tutorial", Proceedings, 31st Conference on Decision and Control, Tucson, Arizona, pp. 2278-2286, 1992.

[244] U. Shaked and N. Berman, "$\mathcal{H}_\infty$ nonlinear filtering of discrete-time processes ", IEEE Transactions Signal Processing, vol. 43, no. 9, pp. 2205-2209, 1995.

[245] T. Shen and K. Tamura, "Robust $\mathcal{H}_\infty$-control of uncertain nonlinear systems via state feedback", Transactions on Automatic Control, vol. 40, no. 4, pp. 766-769, 1995.

[246] X. Shen and L. Deng, "Decomposition solution of $\mathcal{H}_\infty$ filter gain in singularly-perturbed systems", Signal Processing, vol. 55, pp. 313-320, 1996.

[247] M. Shergei and U. Shaked, "Robust $\mathcal{H}_\infty$ nonlinear control via measurement-feedback," Int. J. of Robust and Nonlinear Control, vol. 7, pp. 975-987, 1997.

[248] E. Sontag, "Smooth stabilization implies coprime-factorization", IEEE Transactions on Automatic Control, vol. 34, No. 4, pp. 435-443, 1989.

[249] P. Soravia, "$\mathcal{H}_\infty$-control of nonlinear systems, differential games and viscosity solutions of HJEs," SIAM J. on Control and Optimization, vol. 34, no. 4, pp. 1071-1097, 1996.

[250] A. W. Starr and Y. C. Ho, "Nonzero-sum differential games," Journal of Optimization Theory and Application, vol. 3, no. 3, pp. 184-206, 1969.

[251] A. W. Starr and Y. C. Ho, "Further properties of nonzero-sum differential games," Journal of Optimization Theory and Application, vol. 3, no. 4, pp. 207-219, 1969.

[252] A. A. Stoorvogel. *The $\mathcal{H}_\infty$-Control Problem: A State-Space Approach*. PhD Dissertation, Technical University of Eindhoven, 1990, Prentice Hall, Englewood Cliffs, New Jersey, 1992.

[253] A. A. Stoorvogel and H. L. Trentelmann, "The quadratic matrix inequality in singular $\mathcal{H}_\infty$-control with state-feedback," SIAM J. on Control and Optimization, vol. 28, pp. 1190-1208, 1990.

[254] W. Sun, P. P. Khargonekar and D. Shim, "Solution to the the positive-real control problem for linear time-invariant systems," IEEE Transactions Automatic Control, vol. 39, no. 10, pp. 2034-2046, 1994.

[255] S. Suzuki, A. Isidori and T. J. Tarn, "$\mathcal{H}_\infty$-control of nonlinear systems with sampled measurement," J. of Mathematical Systems, Estimation and Control, vol. 5, no. 2, pp. 1-12, 1995.

[256] H. Tan and W. J. Rugh, "Nonlinear overtaking optimal control: sufficiency, stability and approximation," IEEE Transactions on Automatic Control, vol. 43, no. 12, pp. 1703-1718, 1998.

[257] Y. Theodor and U. Shaked, "A dynamic game approach to mixed $\mathcal{H}_\infty/\mathcal{H}_2$ estimation," Int. J. of Robust and Nonlinear Control, vol. 6, no. 4, pp. 331-345, 1996.

[258] M. Toda. *Theory of Nonlinear Lattices.* Springer-Verlag, Heidelberg, 1981.

[259] H. Trentelman and J. C. Willems, "Synthesis of dissipative systems using quadratic differential forms," IEEE Transactions on Automatic Control, vol. 47, no. 1, pp. 53-69, 2002.

[260] P. Tsiotras, M. Corless and M. Rotea, "An $\mathcal{L}_2$ disturbance-attenuation solution to the nonlinear benchmark problem," Int. J. of Robust and Nonlinear Control, vol. 8, pp. 311-330, 1998.

[261] H. D. Tuan and S. Hosoe, "On linearization technique in robust nonlinear $\mathcal{H}_\infty$ control," Systems and Control Letters, vol. 27, no. 1, 1996.

[262] H. D. Tuan, "On robust $\mathcal{H}_\infty$-control for nonlinear discrete and sampled data systems," IEEE Transactions on Automatic Control, vol. 43, no. 5, pp. 715-719, 1998.

[263] A. J. van der Schaft, "On a state-space approach to nonlinear $\mathcal{H}_\infty$-control," Systems and Control Letters, vol. 16, pp. 1-8, 1991.

[264] A. J. van der Schaft, "$\mathcal{L}_2$-gain analysis of nonlinear systems and nonlinear state feedback $\mathcal{H}_\infty$-control," IEEE Transactions on Automatic Control, vol. 37, no. 6, pp. 770-784, 1992.

[265] A. J. Van der Schaft, "Robust stabilization of nonlinear systems via stable kernel representation with $\mathcal{L}_2$-gain bounded uncertainty," Sytems and Control Letters, vol. 24, pp. 75-81, 1995.

[266] A. J. van der Schaft. *$\mathcal{L}_2$-Gain and Passivity Techniques in Nonlinear Control.* Lecture Notes in Control and Information Sciences, Springer Verlag, vol. 218, 1996.

[267] W. J. Vetter, "Matrix calculus operations and Taylor expansions," SIAM Review vol. 15, pp. 352-367, 1973.

[268] M. Vidyasagar. *Nonlinear Systems Analysis.* 2nd Ed., Prentice Hall, Englewood Cliffs, New Jersey, 1995.

[269] Z. Wang and H. Unbehauen, "Robust $\mathcal{H}_2/\mathcal{H}_\infty$ state estimation for systems with error variance constraints: the continuous-time case", IEEE Transactions on Automatic control, vol. 44, no. 5, pp. 1061-1065, May, 1999.

[270] Z. Wang and B. Huang, "Robust $\mathcal{H}_2/\mathcal{H}_\infty$-filtering for linear systems with error variance constraints," IEEE Transactions Signal Processing, vol. 48, no. 8, pp. 2463-2467, 2000.

[271] E. Whittaker. *A Treatise on the Analytical Dynamics of Particles and Rigid Bodies.* Dover, New York, 1944.

[272] J. C. Willems, "Least squares stationary optimal control and the algebraic Riccati equation," IEEE Transactions on Automatic Control, vol. 16, no. 6, pp. 621-634, 1971.

[273] J. C. Willems, "The construction of Lyapunov functions for input-output stable systems," SIAM J. on Control, vol. 9, pp. 105-134, 1971.

[274] J. C. Willems, "Dissipative dynamical systems: Part I: general theory, Part II: linear systems with quadratic supply rates," Archives for Rational Mechanics and Analysis, vol. 45, no. 5, pp. 321-393, 1972.

[275] J. C. Willems and H. Trentelman, "Synthesis of dissipative systems using quadratic differential forms," IEEE Transactions on Automatic Control, vol. 47, no. 1, pp. 53-69, 2002.

[276] J. C. Willems, "Almost invariant-subspaces: an approach to high-gain feedback - Part I: Almost controlled invariant subspaces, Part II: Almost conditionally invariant subspaces," IEEE Transactions on Automatic Control, vol. 26, no. 1, pp. 235-252, Oct. 1982, and vol. 27, no. 5, pp. 1071-1084, 1983.

[277] W. H. Wonham. *Linear Multivariable Control: A Geometric Approach.* Springer Verlag, New York, 1979.

[278] H.-N. Wu and H.-Y. Zhang, "Reliable $\mathcal{L}_2/\mathcal{H}_\infty$ fuzzy static output feedback control for nonlinear systems with sensor faults," Automatica, vol. 41, pp. 1925-1932, 2005.

[279] L. Xie, C. E. De souza and Y. Wang, "Robust filtering for a class of discrete-time nonlinear systems: An $\mathcal{H}_\infty$ approach," Int. J. of Robust and Nonlinear Control, vol. 6, pp. 297-312, 1996.

[280] L. Xie, C. E. De souza and Y. Wang, "Robust control of discrete-time uncertain systems," Automatica, vol. 29, pp. 1133-1137, 1993.

[281] I. Yaesh and U. Shaked, "Game theory approach to optimal linear estimation in the minimum $\mathcal{H}_\infty$-norm sense," Proceedings, 31st Conference on Decision and Control, Tampa, FLorida, pp. 421-425, 1989.

[282] F. Yang, Z. Wang, Y. S. Hung and H. Shu, "Mixed $\mathcal{H}_2/\mathcal{H}_\infty$-filtering for uncertain systems with regional pole assignment," IEEE Transactions on Aerospace and Electronic Systems, vol. 41, no. 2, pp. 438-448, 2005.

[283] G. H. Yang and J. Dong, "$\mathcal{H}_\infty$ filtering for fuzzy singularly-perturbed systems," IEEE Transactions on Systems, Man and Cybernetics-Part B, vol. 38, no. 5, 1371-1389, 2008.

[284] G.-H. Yang, J. Lam and J. Wang, "Robust control of uncertain nonlinear systems," Proceedings, 35th Conference on Decision and Control, Kobe, pp. 823-828, 1996.

[285] G.-H. Yang, J. Lam, and J. Wang, "Reliable $\mathcal{H}_\infty$ control of affine nonlinear systems," IEEE Transactions on Automatic Control, vol. 43, no. 8, pp. 1112-1116, 1998.

[286] M. J. Yazdanpanah, K. Khorasani and R. V. Patel, "Uncertainty compensation for a flexible-link manipulator using nonlinear $\mathcal{H}_\infty$ Control," Int. J. Control, vol. 69, no. 6, pp. 753-771, 1998.

[287] J. Yong and X. Y. Zhou. *Stochastic Controls, Hamiltonian Systems and Hamilton-Jacobi-Bellman Equations.* Springer Verlag, New York, 1999.

[288] C.-F. Yung, Y.-P. Lin, and F. B. Yeh, "A family of nonlinear output-feedback $\mathcal{H}_\infty$ controllers," IEEE Transactions on Automatic Control, vol. 41, no. 2, pp. 232-236, 1996.

[289] C.-F. Yung, J. L. Wu, and T. T. Lee, "$\mathcal{H}_\infty$-control for more general class of nonlinear systems," IEEE Transactions Automatic Control, vol. 43, no. 12, pp. 1724-1727, 1998.

[290] G. Zames, "Feedback and optimal sensitivity, model reference transformations, multiplicative semi-norms and approximate inverses," IEEE Transactions on Automatic Control, vol. 26, no. 2, pp. 301-320, 1981.

[291] E. Zeidler. *Nonlinear Functional Analysis and its Applications, Part I: Fixed-Point Theorems.* Springer Verlag, New York, 1986.

[292] K. Zhou, J. C. Doyle, and K. Glover. *Robust and optimal Control.* Prentice Hall, Englewood Cliffs, New Jersey, 1996.

[293] K. Zhou, K. Glover, B. Bodenheimer, and J. C. Doyle, "Mixed $\mathcal{H}_2/\mathcal{H}_\infty$ performance objectives I: robust performance analysis," IEEE Transactions on Automatic Control, vol. 39, no. 8, pp 1564-1574, 1994.

Index